THE WOMAN'S CARPENTRY BOOK

THE WOMAN'S CARPENTRY BOOK

Building Your Home from the Ground Up

JEANNE TETRAULT, editor

CAROL NEWHOUSE, photographer

BILLIE MIRACLE, illustrator

1980 Anchor Press/Doubleday, Garden City, New York

Jeanne Tetrault grew up on the East Coast. After attending college, she spent seven years homestead-farming with another woman, raising goats, sheep, and poultry. During that time, Jeanne Tetrault belonged to the feminist collective that began and nurtured *Country Women* magazine. These experiences resulted in her first book, *Country Women: A Handbook for the New Farmer.*

Jeanne is presently working as a professional carpenter in California and does remodeling, additions, foundation work, and roofing, among other things. She has built several small houses and a barn, combining learning to design with practical carpentry. She is also studying solar design and teaching women's building classes.

CONTRIBUTORS

Writers: Marion Andrews; Harriet Bye; Cheree Conrad; Priscilla Danzig; Alison Dykstra; Elana; Marsha Emerman; Susan Liebl; Sharon Marks; Julie Mendel; Anne Chivvis Moore; Kathy Mulholland; Nelly; Frances Rominski; Sky; Sally Smith; Jeanne Tetrault; Marga Waldech; Susun Weed; Jan Zaitlin. (All writing not otherwise credited was done by Jeanne Tetrault.)

Photographers: Kathryn Biglow; Cathy Cade; Janet Cole; Dian; Dorothy Dobbyn; Marsha Emerman; Lynda Koolish; Lisa Law; Rebecca Lawson; Sharon Marks; Billie Miracle; Ruth Mountaingrove; Carol Newhouse; La Rosa Ponderosa; Lyn Suzanne; Jeanne Tetrault.

by Jeanne Tetrault and Sherry Thomas
COUNTRY WOMEN

The Woman's Carpentry Book is published simultaneously in hard and paper covers.

Interior designed by Virginia M. Soulé

Hardcover ISBN 0-385-17183-8
Paperback ISBN 0-385-14269-2
Library of Congress Catalog Card Number 78-22801

Printed in the United States of America

This book is dedicated

— to our mothers, who have loved and raised strong daughters

— to our daughters: may they grow up believing in themselves and one another

— and to our sisters, whose courage and visions are building new worlds

Contents

INTRODUCTION xiii

part

I

WOMEN'S WORK
(Photography Section) 2

MY INCREDIBLE WOMAN-BUILT HOUSE—*Susun Weed* 13

THE KEY TO CARPENTRY: TOOLS 24

Hand Tools 24
Tools for Measuring and Marking 24
Tools for Fastening Things Together 37
Tools for Body Protection 47
Hand-held Power Tools 49
Woodworking Machinery—*Jan Zaitlin* 57

A ROUND HOUSE IS SOMETHING SPECIAL—*Marga Waldech* 67

A SECOND KEY: MATERIALS 78

Lumber and Lumber Products 78
Nails, Screws, and Other Fasteners 83
Flashing and Caulking 90
Adhesives 92
Insulation and Weather Stripping 93

LOG CABIN: BUILDING TOGETHER—*Marion Andrews* 97

WHAT DID WE MISS IN HIGH SCHOOL SHOP? 102

Putting Pieces Together 103
Locating Studs Behind the Wall 107

Determining a Bearing Wall 107
Adding a Partition Wall 108
Dealing with a Sticky Door 109
Replacing a Foundation Post 109
Replacing Sash Weights and Cords in Double-hung Windows 110
High School Shop Was Never Like This 111

SOME SIMPLE PROJECTS 112

Sawhorses 112
Building a Gate 114
Handmade Dutch Door—*Susun Weed* 118

IMAGINING
THE REAL HOUSE—*Sharon Marks* 121

CHANGING YOUR SPACE—IN A DAY OR TWO 126

Replacing Your Door 126
Hanging a Door: New Construction 131
Adding a Window 133

CASA ADOBE—*Elana* 137

HOUSE THOUGHTS 144

Designing an Energy-efficient House—*Alison Dykstra* 144
Let the Light Shine In—*Harriet Bye* 149

WHY START WITH
SOMETHING EASY?—*Susan Liebl and Julie Mendel* 157

FIRST STEPS FIRST 165

Simple Building Layout 165
Building with Piers 169
Continuous Concrete Foundation 180

STAR BUILDING—*Frances Rominski* 190

FROM YOUR FOUNDATION UP 200

Posts and Girders, Sills and Floor Joists 200
Subfloors 206
Rough-framing the Walls 208

A HAND-BUILT HOUSE IS A
SPECIAL KIND OF HOME—*Marsha Emerman* 217

ADDING THE ROOF 231

Framing Up a Gable Roof 231
Roof Sheathing 237

MOVING THE "BIG HOUSE"—*Nelly* 241

EXTERIORS 248

Sheathing and Siding 248

ATYPICAL A-FRAME—*Kathy Mulholland* 259

KEEPING OUT THE RAIN 268

Asphalt and Fiberglass Shingles—*Marion Andrews* 268
Wood Shingles and Shakes—*Harriet Bye* 271
Roll Roofing 273

MAKING MAGIC: YURT BUILDING—*Sky* 278

INTERIORS 286

Gypsum Wallboard/Sheetrock 286
Rock Wall—*Cheree Conrad* 292
Wood Paneling 294
Some Alternatives for Floors 297

HEXAGON BUILDING—*Sally Smith* 300

MOVING BEYOND: EARNING A LIVING 320

Union Work, Union Wages—*Anne Chivvis Moore* 320
Furniture Maker/Designer—*Jan Zaitlin* 330
Skylark Construction Co.: Breakthrough for Women—*Sky* 336

part

WOMEN'S WORK
(Photography Section) 341

THE LANGUAGE OF CARPENTRY:
TERMS AND DEFINITIONS 354

BUILDING CODE TABLES 361

How to Use These Tables 361
Tables 362

RESOURCES 374

Books, Tools, Miscellany 379

INDEX 381

Introduction

Over a span of several years, this book has grown like a well-loved house: first in fanciful sketches, next in concrete basic sections, later spurting additions and changing shapes. It began as a vague notion that a carpentry book by and for women should exist. Most women grow up with little or no access to tools and their workings. Carpentry is a skill reserved for the little boys who receive their "Junior Tool Boxes," for the adolescent boys who have an open door to the "shop" at school, and for the young men who are encouraged apprentices in the building trade unions. Women are conditioned and restrained into considering it a "natural" state of affairs that others will shape and control their environments: the houses we live in have been designed, built, repaired, and remodeled almost exclusively by men.

Most carpentry books continue this process of alienation and mystification. They present carpentry as a smooth system of accomplishments resting upon an assumed basis of familiarity with tools and materials. The illustrations show cleanly attired young men casually demonstrating the "correct" way to do something. The materials pictured are always top-grade; the workers look unruffled and confident. The textbook explanations are matter-of-fact and often so sketchy (loaded with assumptions that you know how to use a tool or do some simple process) that they are unusable. Little wonder that the "uninitiated" beginning carpenter who decides at age twenty or thirty or fifty to learn something about building on her own is intimidated! And so the idea of THE WOMAN'S CARPENTRY BOOK took root—a book for those of us who never learned how to handle a saw until we decided to build that gate . . . or put in a window . . . or build a house! The book began as a deliberate attempt to share, to make accessible the basics of carpentry and building.

Structuring a book is in some ways harder than building a house. As material collected, there were problems of form and voice: how to incorporate all that there was to share; how to start; how to use the many different voices, the writings of women who had built houses, who did woodworking and carpentry and contracting. The pieces that were written by women who had built their own houses were the most exciting. The majority of these women had started out with little or no experience at building—they were determined to do it anyway. The creation and construction of each house had a singular powerful effect: the fact of "I *can*" in the face of years of "you shouldn't." For some of these first-time builders, the house in itself was sufficient; for others, it was a jumping-off point for a new occupation. But the special common desire to share the process, mental and physical, of building one's own house began to shape the book. The center grew clearer: a book of women's carpentry centered in women-built houses. . . . As it grew, it turned out to be a celebration of women-built houses, a new definition of "women's work"; the book that had been imagined had a life and will of its own.

The book begins with Susun Weed's account of her house building. Hers was a months-long learning experience that was sometimes solitary and often shared with other women. Her writing doesn't include all of the details of pole foundations or post and beam construction, but she takes you through her process with a sense of determination and the joy of accomplishing what she'd imagined. Her house is a simple enough shelter—yet it is sturdy, unique, and somehow magical. From there the book weaves itself with a weft of practical sections, "how to" information, step-by-step procedures. The supporting warp consists of a series of pieces written by women who have built houses of all shapes, sizes, and styles all over the country. From the A-frame in New York State to the octagon in Hawaii to the adobe house in New Mexico, each house has its unique story, told by its builder. The book opens and closes with photograph sections that should change what you imagine to be "women's work"—most of the visual intent of the book focuses and refocuses here, in a celebration of women as builders, cabinet and furniture makers, teachers, creators.

The practical sections of the book start with the most basic discussions: tools and materials. Sharing a tool, the workings of a tool, the "secrets" of a trade, means sharing the potential power and productivity of that tool or trade. It is a gift of independence, autonomy, and self-sufficiency. All women will not want to become carpenters, build their own houses, or even learn how to rehang the front door, but breaking down the exclusiveness of men's control of these skills is a necessary step that women are taking and sharing with one another. There are also described simple projects that the "beginner" can take on with confidence: building a gate or a pair of sawhorses, hand-making a dutch door. Next come some basic changes you might like to make in your house: replacing a door, putting in a window. These projects take more discipline and more skill with tools, but are easily manageable if you have faith in yourself and have taken the time to grow familiar with your tools. "House Thoughts" is a reflective pause mid-book—some ideas on house design, energy use, light, and space. Some practical information sneaks in with Harriet Bye's "Let the Light Shine In"—details on fixing old windows, replacing glass, and so on.

The remainder of the book follows the logic of house building, beginning with layout and proceeding through foundations, framing, roofing, siding, and so on. All of this material may be adapted for remodeling and repair work. For instance, you might well have to reroof an existing house before you're ever faced with choosing a roofing material for your new house. Much of the material may be used for other projects, too. "Simple Building Layout" and "Building with Piers" will provide you with some basics for adding a porch or deck to your house; the "Interiors" section might be a source for information on Sheetrocking an already existing wall. Throughout, the intent has been to provide clear, understandable details for common carpentry projects you might want to do.

THE WOMAN'S CARPENTRY BOOK is by no means a complete guide—there are virtual *libraries* of books about building and no one book could even attempt to be really comprehensive. You may consider this book to be a series of possibilities—alternatives—information. A "Resources" section lists other books that will be of interest and help, and throughout the text there are references to these books whenever it seemed that more specific information was relevant—or when this book couldn't cover some of the ways of doing a job, or a particular aspect of carpentry that you might need. I have referred more than once to Dale McCormick's book *Against the Grain: A Carpentry Manual for Women.* Hers is a wonderful resource, a sister volume written by a woman carpenter for other women. I'd like to think that THE WOMAN'S

Carpentry Book and *Against the Grain* can stand by side on your bookshelf, each offering its special insights, information, and strengths.

As the structuring of this book gradually evolved, the problem of "voices" seemed to solve itself. The book is clearly a collective effort. The woman who writes about building a yurt is obviously not the same as the woman writing about shingling her roof—their styles of writing and expression are as different as the carpentry they talk about. The personal voices emerge more powerfully in some instances; in others, the details of work or materials seem to take precedence. Throughout, we've tried to share both the work and the feeling of the work, to make the essential connection between the two.

The final "Moving Beyond" section is intended for those of us who have found carpentry and working with wood to be interesting and challenging enough to pursue it as a trade, a life's work—and for other women starting out. Being a woman in a traditionally male field is not easy, much as you might love the work and be willing to struggle with the difficulties of setting up a business. We hope that more and more women will be encouraged to take up work that can be redefined as belonging to us all.

This is a book for women who build, who imagine building, who raise beams, who puzzle over plans and pick up hammers. It is a book for women who love the balanced solid tool, the subtle fine strength of wood. It is a book for women who think "maybe . . ." and dare to try. Above all, this is a book for women offered by women in pride and excitement: may we grow in our capabilities, our resourcefulness, our visions, and our skills. May we truly claim our houses as our homes.

part I

WOMEN'S WORK

(Photo by Jeanne Tetrault)

(Photo by Carol Newhouse)

(Photo by Carol Newhouse)

(Photo by Carol Newhouse)

(Photo by Jeanne Tetrault)

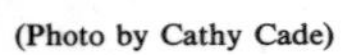

(Photo by Cathy Cade)

(Photo by Lynda Koolish)

(Photo by Carol Newhouse)

(Photo by Billie Miracle)

(Photo by Carol Newhouse)

(Photo by Carol Newhouse)

(Photo by Carol Newhouse)

(Photo by Sharon Marks)

(Photo by Carol Newhouse)

(Photo by Lynda Koolish)

(Photo by Carol Newhouse)

(Photo by Cathy Cade)

(Photo by Carol Newhouse)

(Photo by Carol Newhouse)

My Incredible Woman-built House

Of the many reasons—political, economic, spiritual, emotional, ecological—why I built a house, a major one was monetary. Seeking ways to live on less and less cash and thus have more time for my life at home, my daughter, my garden and animals, I listed my expenses and considered how to cut them down. At the top of the list was rent. I dreamed of buying a piece of land, but soon realized that investing my tiny savings in land was no exit from the money maze—I would simply have a whole new set of needs (developing water, building a house, making mortgage payments). I asked my landlady if she would consider selling me the cottage I was renting, but she needed the rent money to help cover taxes and wouldn't think of selling. Together, we came up with the idea of my building a house on her land. I would pay a small portion of the land taxes but would not have to pay rent; the house would be hers when/if I left. I would have room for my garden, my goats, and general use of the land. It seemed perfect . . . and preposterous:

Me?! Build a house?!

I'd never built anything before. I'd helped my men while they built things, but the screech of the power tools and the elaborate and mystifying plans kept me at a distance, feeling that I could never do all that. How could I build a house? Who would teach me? Who would help me? Could I do it for $1,000? Could I do it at all?

Then I became possessed: "I am going to build a house!" While everyone else was skeptical, I thought about nothing but building a house. I thought about the kind of house I wanted to build. It would have to be easy to build. It should keep me warm and dry and safe from the wind. It should be big enough to spread out in—no more tiny rooms. It would be filled with sunlight. A round or many-sided house intrigued me until I realized that every extra corner or curve would be more trouble to build and seal against the elements. Rectangles and squares seemed the easiest shapes to build. I never considered building a dome, having had enough second-hand experiences of leaking domes to be wary. And I didn't want to spend January nights at minus twenty degrees in a tepee!

My winter excursions to the library helped me to decide upon a simple rectangular house with an equally simple shed roof. I'd slant the roof down toward the north, so the front of my house would have southern sunlight; I'd make the roof overhang in the back of the house to protect my woodpile from rain and snow. There were some things I didn't want as part of my house: paint, plywood, concrete, electricity, bulldozers, fiberglass. And I hoped to build with the help of my daughter and women friends. The house I dreamed would be hand-built, woman-built, love-built. . . .

It was a very long and rainy spring. I spent a lot of my time at home standing

(Photo by Lyn Suzanne)

at the window looking at the rain, and willing it to stop. On sunny days I walked the land, looking for my building site. I decided to move out of the cottage into a tent, to save the rent money (and put it toward my house) and also to be able to live on various parts of the land until I found my perfect spot. I spent time visiting hand-built houses in my area and talking with their builders. One person offered me the use of a chain-saw mill to "flat" some timbers (this involves sawing a long, perfectly planed surface along a tree or timber so that milled lumber can be fastened to it). Flatted timbers supported the floor, lofts, and roof of this person's house and looked incredibly beautiful and solid. How great they would look in my house, too! Every house and person I visited gave me wonderful ideas for my house. In my mind, it began to acquire detail: wood stove in the middle, dutch door on the side, trees surrounding it. I decided to make a foundation of posts and to use post and beam construction methods. I planned to have exposed ceiling beams of flatted timbers and to build a loft. At this stage, I went to see an architect.

The architect was a friend of a friend, and I visited him in his hand-built house. Too small for me, I thought: mine will be bigger! Fat books full of charts and numbers helped us to determine live loads (thirteen friends at a party) and dead loads (snow on the roof), and strengths of different woods of various diameters, and how far a given thickness could span. We calculated just how long and how thick the trees should be that I would cut for the girders and posts and other supports of my house. I left with two pages of notes and designs and a clearer than ever mental picture of my house. As I drove home down the winding mountain road, I noticed some discarded telephone poles lying in the weeds by the edge of the road. The best posts to use for foundations, I recalled, are wood poles that have been pressure-treated with creosote . . . for example, telephone poles. I knew that I was ready to start working on the foundation.

The rain finally let up in June. The garden was tilled and planted. I had narrowed my site choices down to two: one seemed more practical, but the other was more beautiful and attracted me much more. I decided to follow some advice from the architect: "Build on the site you like and practical considerations will take care of themselves." Although the practical considerations don't exactly take care of them-

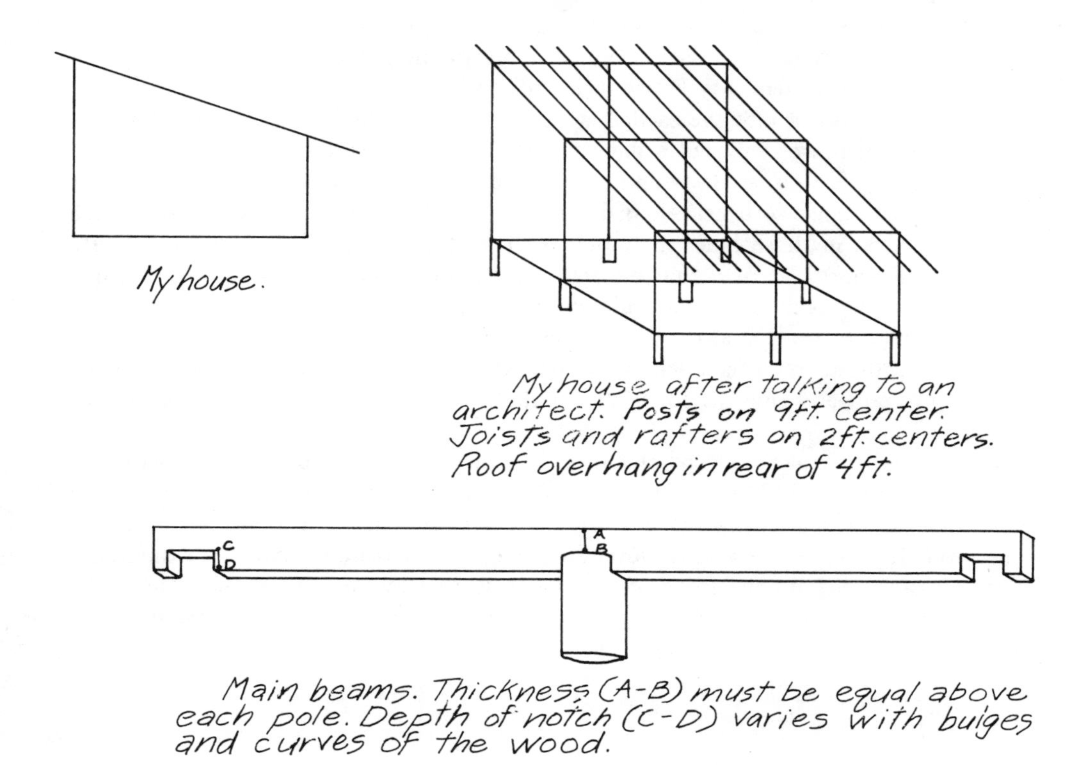

My house.

My house after talking to an architect. Posts on 9ft. center. Joists and rafters on 2ft. centers. Roof overhang in rear of 4ft.

Main beams. Thickness (A-B) must be equal above each pole. Depth of notch (C-D) varies with bulges and curves of the wood.

selves, I'm glad I followed that advice every time I look out of my windows!

Summer was upon me, with fall and winter not far behind. Every woman we knew had been invited to come and spend Summer Solstice on the land. I wanted to start digging the foundation on June 21 with a hundred hands to help! Nine holes needed to be dug, and they had to be 3½′ deep to stay below the frost line in these cold mountains in the winter. With enough of us taking turns, I reasoned, those deep holes could be quickly done. I gathered all of the shovels, post-hole diggers, pickaxes, and iron poles I could find in anticipation of frenzied hole digging on the first day of summer. But before anyone could dig, I realized, we had to know *where* to dig. I'd chosen my site, and now I had to lay out the design of the house with string and stakes. The front line for the house was easy—it paralleled the sweep of a mountain and faced slightly southwest. I fantasized watching the sun and moon move across my bedroom and kitchen windows. Then I tried to lay out the other three sides of the house. I knew from my reading that this initial laying out was critical. I wanted to do it exactly right, but after tangling myself up in string and moving my stakes several times, I was still far short of perfection. The next day I enlisted the help of a friend. How much easier it was with two! In a couple of hours we had the strings stretched out and nine stakes set into the ground at the spots where the holes were to be dug.

Summer Solstice swung around and women from all over arrived. We sang and ate and smoked and danced; we walked in the woods; we looked at the stars and hugged each other; and we even worked on digging those holes! The dirt was easy to dig for about 12″; then we hit hard, hard clay and boulders—the stubbornness of earth long settled in one spot. When everyone left, I had nine holes started, but none completed.

It took me until the end of July to finish digging those foundation holes. Some days I had to work in town, and how I resented being there when the sun was shining hotly. Some days it rained and I wished I were in town making money. When it was fair and I was home, I chose a hole and set myself to it. Sometimes a friend, Jody, would come and read her stories to me while I dug; sometimes she would sing. And I made up a song to sing to myself while I hacked at the hard-packed clay and rocks. A second friend, Sky, would sit quietly or dig with me some days. She always lent her strong arms and back when there was a rock to haul up out of a hole. In the midst of this digging, the three of us would go at dusk to the architect's road. With a handsaw we cut the telephone poles into manageable pieces and made off with them. My house was really begun!

Forty inches down into the earth I was, with forty blisters on my hands. Enough! I could dig no more. I placed a flat rock at the bottom of each hole and dropped the poles into their places. Jody and Sky helped me visit the roadworks depot, where we filled trash cans with gravel to fill in around the poles. I had seen how much water collected in those holes in the clay earth at each rain—one hole even had a spring in it!—and I learned that concrete draws and holds the water, thus causing even creosoted wood to rot too quickly. With the poles set in gravel, they would have a chance to drain and dry out in between winter snows, spring and fall rains, and summer thunderstorms. I was filled with excitement every morning when I stepped outside my tent and saw the foundation poles growing (so it seemed) up out of the earth!

Next I had to make the tops of the poles exactly even and exactly level with one another so that I could lay up the big timbers that would be the girders supporting my house. I spent a frustrating day with my tape measure and lots of string and a line level. It seemed like a fairly level site, but there was more than 3′ of difference between the back and the front of the site. I started by cutting off the pole at the highest point about 1′ above the ground level. This would leave me room to get under the house and would ensure that I complied with the building code in my area. Although I had not applied for a building permit (my landlady and I agreed that we could not afford the huge increase in taxes that another house on the land would incite), I did comply with the building code as much as possible, in case a building inspector came to the site. If my house wanted to collapse from old age I could accept that, but no one was going to tear her down because she didn't meet code standards! Barbara, the woman who had helped me lay the site stakes, came to my rescue again when I thought I'd never figure out how to cut all the poles so that their tops were level with one another. With the aid of her chain saw, some shims, and a carpenter's level "extended" with a long board (my line level and string were too inaccurate over these long spans), we finally got the pole tops reasonably level.

With blessings and songs, we began to fell the trees that would be the structural timbers of the house. I had read a lot about seasoning wood, but the people I'd talked to advised me not to season the timbers unless I was going to use seasoned wood throughout the construction. Use all green wood or all seasoned wood, they said; that way, the structure isn't stressed where a green piece shrinks and the seasoned wood attached to it doesn't. An all-green-wood house will shrink and season as a unit and be stronger through that process. We dragged the trees over to the site with the able assistance of my four-wheel-drive car and began stripping bark from them. The nine trees that I chose for the posts going up inside the house were left with their bark; I wanted the feeling of having real trees in my living space. Using the drawknife to debark the rest of the trees was a favorite task while it lasted. Early in the morning, before I left for work, I'd sit in the sun and sing and debark. This was the only chance I got for this pleasant work, for my daughter loved to peel the glistening, sappy bark away in long strips, and I loved having her do it. I was very happy that the muscles and skills of a nine-year-old could help build this house.

After debarking, we flatted the timbers. It took about two hours to set up and flat each timber with the chain-saw mill, and the work was difficult and extremely noisy. Midway, I relinquished the idea of sawing my own floor joists and ceiling rafters, and felt satisfied to have three big ash trees flatted for the floor girders and one smaller one to hold up the ceiling. I treated the debarked and flatted trees with a homemade mixture of equal parts penta (pentachlorophenol) and used motor oil to discourage rot and insect larvae. The woodpeckers keep the poles that I left barked clean and free of bugs and give us joyful mornings into the bargain.

It took eight women to lift the girders into place atop the foundation poles; flatted ash trees are heavy. We marked where the foundation posts hit the girders and chiseled out notches to make a tight fit. This sounds easy, but it involved picking up each girder about a dozen times before the three notches in it were cut level and to the right size and depth. When the girders were chiseled out just right and set in place for the last time, I sat astride each one in turn and used my brace and bit to drill holes through each girder into the foundation pole beneath it. Taking turns with a sledgehammer, we pounded gigantic spikes (slightly larger than the drilled holes and 2″ longer than the thickness of the beam) into the wood. I'd been told that gravity would contribute more to holding the house down than any amount of nails or spikes, but this step made me feel much more secure about the stability of my house.

The first load of rough-cut lumber was delivered July 31. (There was a mill nearby where I could buy any amount of this lumber at low cost.) A friend who had built herself a house the year before came over and helped me stack it so that it wouldn't warp or mold. The wood was so green and wet that it took both of us to pick up a 10′ 2×4! I didn't know it then, but I made the worst mistake of the whole project right there. I didn't know how to estimate how much wood I needed for the whole house, and I didn't allow myself any margin for error and for the inevitable poor pieces in any batch of rough-cut lumber. When a hefty helper broke a floor joist, I had to drive to the mill for another one. When I ran out of lumber in mid-October, I had to wait four weeks before I could get the mill to cut me more. It was a frustrating delay—having to wait while the last warm, sunny days slipped away . . . and those autumn nights were very chilly in my tent!

Working with a willing friend or by myself, I got the floor joists all laid across the girders and conquered the mystique of toenailing. Then I discovered I had done it wrong! I spent a restless night in my tent. What to do? When the sunlight returned, I found myself outside with the sledgehammer. I knocked loose the two joists that had been positioned wrongly, cut them to size, and nailed them back in the correct places. This was a turning point for me and my life. I now realized that I didn't have to live with my mistakes. I could correct them! I felt suffused with power: the power to change my mistakes instead of dragging along with them weighing me down.

In another week the floor was done. I put blocking between the floor joists first, to keep them from twisting. The next step was to lay down my floorboards. I used full 1″-thick, rough-cut pine planks in varying widths (as they came from the mill). These were placed at right angles to my floor joists and nailed in place with special cement-coated nails that would not pull out as the wood dried and shrank. The floorboards were so sappy that they squirted at us as my daughter and I and friends pounded the nails in. We didn't cut the boards to length as we went along; instead, we left them extending beyond the edges and trimmed off all of the excess with a chain saw when the entire floor was done. Some adventurous friends slept out on the floor of my house that night, but I went back to sleep in my tent. I wanted it to be a "real" house before I spent the night there!

Now I was ready to raise the posts (ash trees) into position. I cut the front ones 14′ long, the back ones 8′ long, and the middle ones 11′ long—all as accurately and evenly as I could with a chain saw. I called up all of the women who said they wanted to help and asked them to come over the next day—goddess granting sunshine. What a festive day it dawned! As women gathered, I laid down the sill (long 2×4's nailed around the perimeter of the floor, with some extra short pieces wherever the poles were to rest). What exhilaration and excitement as we pushed the first 14′ ash tree up into place! Some of us held it while others quickly nailed the braces (long pieces of 1×6 pine) into place. We had thoughtfully nailed the braces to the trees while they were lying down, two for each tree, so that the loose ends had only to be nailed against the sill to hold the posts in place. Some of the women stood out in the field directing us as we attempted to set the posts perpendicular

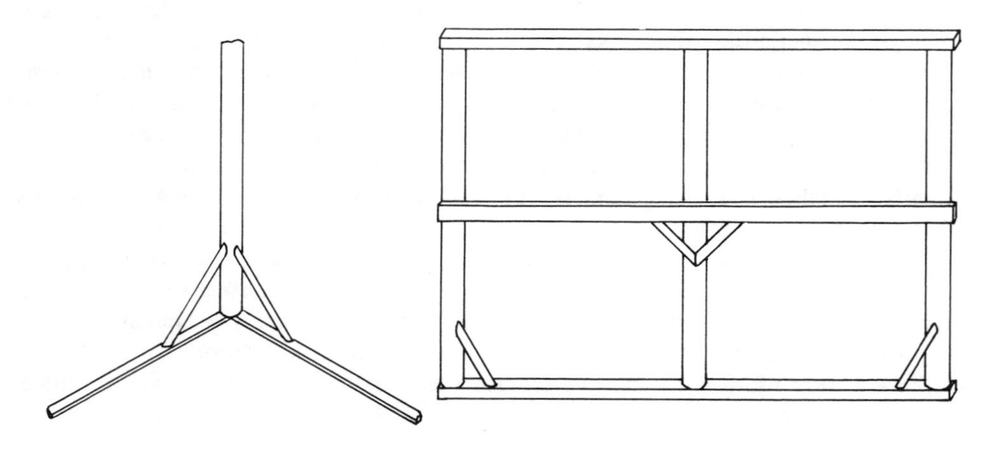

to the floor: "To the left! Too much—back to the right! A little more! There—hold it!" I had picked the "straightest" trees to use as posts, but no tree grows really straight. The curving lines of those trunks made it impossible to use a level to set them up perpendicularly; our keen eyes and our sense of balance were all we had to rely on to do it right. When in doubt, *I'd tilt the upright pole a bit in toward the middle of the house.* This helps to counteract all of the outward stress that the roof places on the posts. I had nightmares that the wind would topple the posts (onto my tent!), but the temporary bracing held fast through thunder and wind until I cut some small maples, which my daughter skinned, to use as permanent braces.

Putting up those permanent braces was tricky: cutting two 45° angles into a round piece of wood without twisting it isn't easy! My landlady worked with me as we sawed and rasped and axed and cursed. It took us four days to brace the four corners. The maple weathered a beautiful gray and checked, sometimes seeming to split all the way through, and held the corners up solidly and perfectly. As an extra precaution, we toenailed the posts onto their 2×4 pads with nails all around their bottom edges.

The next step was to put up the beams that would support the loft and hold the posts even more firmly in place. Again I called on my friends and prayed for some more sunshine. On the appointed (and sunny) day, we gathered with our hammers and ladders and set to work. It seemed like a simple task to nail a 20′-long 2×6 into the posts, but it isn't simple at all to hold a 20′ piece of wood 8′ in the air and parallel to the floor. And it isn't simple to nail anything into an ash tree! It's also incredibly hard to drive a spike while standing on a ladder. While three women tried their best to hold the beam right on the measured marks, I climbed my ladder and drilled with my brace and bit into that hard ash wood. We took turns holding the beam up and pounding spikes in with the sledgehammer. Across the middle poles we did the same thing; but since I didn't have another 20′ 2×6, we used two 10′ 2×6's and lap-jointed them together. After both beams were spiked into place on the poles, I added 2×4 bracing. I used weathered pieces to match the gray tones of the ash bark, and put two nails at each end so they wouldn't pivot.

The next workday I laid floor joists for the loft across those beams. I made the end joists a structural part of the house by wedging them as tightly as possible between the front and middle poles. By now I felt like an old hand at house building! I marked out the spacing for the joists, toenailed them in, and blocked between them. More 1″ pine in varying widths was laid down as a floor and scaffolding for the roof work. Every day, bit by bit, my daughter nailed down those boards; I helped her by cutting them to length as she nailed.

Even before the floor boards were nailed down on the loft, I began to work on the roof. Building a roof is just like building a floor—and I'd already done that twice, so I figured it would be easy. But I couldn't lift the plates and rafters into place by myself. And all my friends were getting a little bored with house building—that is, my house building; they had their own lives to build. I needed someone to work with me every day. Someone who could work with me after I got home from town; someone who could work for two sunny hours in the morning even if it rained all afternoon; someone I wouldn't have to call up and ask for help. I focused my energy on drawing that woman to me. I didn't know where she'd come from, but I knew she would come. If I was going to finish my house before winter, then she *had* to come.

I started on the roof: on a calm autumn afternoon six women came to help me. We hung our toes over the edge of the loft and put up the front plate. Actually, it was plates: two 20′-long rough-cut 2×6 plates, cut from the longest pine the mill had. Three women stood along the length of the plates and lifted and nailed them into place while three more women held their waists for safety. We were all scared of heights—what were we doing almost 20′ above the ground trying to nail this wood together squarely and accurately? Was I building a house or was it building

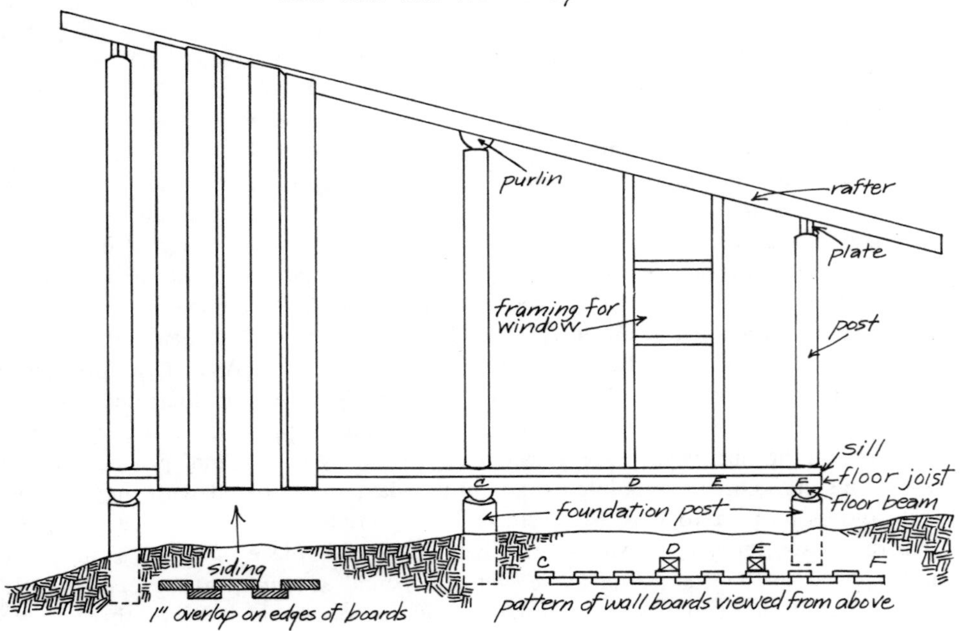

me? For the middle plate (the purlin) I used a flatted ash tree, notched the way I'd notched the floor girders to fit snugly over the top of the poles. It was heavier work than placing and nailing in the front plates, but easier because we only had to lift it waist-high. I drilled and spiked the purlin in just as I had done right below when building the floor. Then we tackled the back plate: two more 20′-long rough-cut 2×6's from the mill's longest pine. It was tricky working with those long and heavy pieces of wood while we balanced on ladders, but we weren't as fearful 9′ off the ground as we had been when we were 20′ high, so we managed to set it up easily and nail it down. Now I was ready for the rafters!

I studied rafter cuts by candlelight: the plumb cut and the bird's-mouth cut. I practiced with the framing square on scraps of wood. I couldn't afford to make a mistake on the rafters, because I hadn't ordered any extra rafter wood; I needed eleven 10′-long rafters and eleven 14′-long rafters, and that's all I ordered. I decided to make a rafter "model." Two things thwarted me, both connected with my choice of a tree as a purlin. First, because the purlin was not straight but nicely, though slightly, curved, I discovered that each rafter would have to be measured and cut separately to conform to that curve. A model that fit nicely at one end didn't fit at all in the middle or at the other end! This meant a lot of extra lifting of heavy rafters and extra measuring time to ensure that each cut was just right. Second, when I started to put up the rear rafters, I found there was a "dent" in the roof! After puzzling it out for an entire frustrating day, a friend and I at last saw that the purlin, because it was a whole tree (although therefore stronger than milled lumber) was only 4″ thick above the poles while the front and back plates were 6″ thick. This 2″ discrepancy between the thickness of the purlin and the thickness of the plates was throwing the whole roof out of line! The next morning, we corrected our mistake—with the sledgehammer! We could have shimmed up 2″ on the purlin had we discovered our mistake before all the front rafters had been laid up and nailed, but we hadn't. So we knocked the back plate off and sawed 6″ off the tops of the back poles—by hand! We felt uncomfortable using the chain saw while perched on a ladder, and we felt more than uncomfortable that night and the next day as our aching muscles tormented us, but this maneuver nicely leveled the roof line. I'm getting a little ahead of my story, however.

In mid-September, River had stopped by to drop off a friend. They had been in Tennessee all summer and River was on her way West, but she was curious to see, before she left, what had grown from those holes she'd helped dig three months earlier at the summer solstice. We walked through the field to the half-built structure, jumped up onto the floor, and climbed a ladder up to the loft. I had three rafters in place then. I put a piece of wood across the rafters: a pretend roof to shield us from the drizzling rain as we talked. Some enchantment drew her; the house held her fast; before morning it was settled: River would stay and help me finish the house. (I agreed to feed her and my landlady offered her an unused room in her house, rent-free for one month.)

We worked well together. We laid up the rafters in the warmth of Indian summer days. We found much pleasure in building and being together. It was hard, heavy work—and incredibly satisfying to see the pattern of the rafters grow against the cloud-moving blue sky. Every day—except when I had to work in town—one or two more rafters would join the pattern, until it was, at last, complete. We rested, lying on our backs in the loft and looking at the clouds roofing us over, until the urge to have a "real" roof grew too strong, and then we began to pass up and cut and nail onto the rafter 1″ pine boards. We started at the rear of the house and moved forward day by day until the "roof" was on. Then we waterproofed and insulated this bare wood roof. An overlapping double layer of black felt paper was stapled tightly to the boards. What beautiful views we had from the roof of the woods bursting into color!

Because I wanted to see my rafters from inside the house, I splurged and bought Techfoam insulation. This foil-covered urethane foam is only 1″ thick, but insulates nearly as well as 6′′ of fiberglass. The actual roofing material covers the Techfoam. Originally I had planned to use roll roofing, but a nearby roofing supplier was selling odd lots of shingles for less than the cost of roll roofing, and I didn't mind having a varicolored roof. This also neatly solved the problem of how to get heavy rolls of roofing up to the roof; shingles can be passed up in small bits. We laid the black paper over the whole roof at once, as it would be undamaged by rain (but not wind!—some blew off and had to be replaced), but we had to lay out the sheets of Techfoam and cover them immediately with shingles to avoid weather damage. Since the Techfoam sheets were 4′×8′, we worked in 4′ swaths going across the roof. Handling and cutting the Techfoam was hard if the wind blew at all, but nailing down shingles was easy and the work pleasant. My sister paid me a rare visit and joined us on the roof to nail shingles; my friend Elza visited and nailed in shingles, too! Of all the tasks involved in building my house, I think I enjoyed rhythmically shingling up on the roof in the sky, leaves falling down, best of all.

When it rained my roof kept me dry. But when it blew I longed for walls! All summer I had been on the lookout for windows and had collected six, with the help of friends. I walked around the "inside" of my house and decided where each window would go and which ones would open and close and which ones would be fixed. On the northeast side—the back of the house—I saw a small fixed window and a door. The southwestern, facing front, would fill my house with light with six windows upstairs and two big windows downstairs, all opening or removable. The northwest side would have a sunset-viewing window upstairs, and two windows downstairs, one fixed and one opening. The southeast side clearly took the brunt of the wicked weather, but it had a wonderful view. I limited myself to two small windows, both fixed, one up and one down, and foolishly decided to put my dutch door in that wall. It is lovely in the summer, but I never use it in the winter when the "back" door, which gives me access to the woodpile and doesn't have icy steps, is more practical.

River and I framed in all of these door and window openings with rough-cut 2×4's. We added an extra inch all around in the measurements of opening window and door frames to leave room for a finished frame, windowsills, and movement of the finished window or door. Instead of doing all the framing at once, we worked on one wall at a time: plumbing and leveling and nailing up the rough frames and then nailing on the wallboards.

My walls are a variation on board-and-batten that I call board-and-board. The 1″-thick pine boards of the walls are nailed, on the sides of the house, to the end rafter and the end joist, and, on the front and back, to the plate above and the sill below. They are also nailed to the rough framing whenever possible; and they are nailed together along their overlapping edges. I used galvanized "shake" nails for this edge nailing, as the design of the nails prevents them from working out of the wood as it shrinks and adapts to stresses. Since we were using boards of various widths, it became a mathematical game to figure out how many boards and what widths would fit across a given space. In the illustration, the given spaces are C to D, D to E, and E to F. The number of boards, times two, is subtracted from the total width of the boards to figure the actual distance the boards will cover. Or subtract 2″ from each board, then add the widths, then add 2″ to the total, and you will also have the actual width. We did the back wall first since it seemed the easiest: all the cuts were straight, little ladder hanging was needed, and there were only two openings in the wall. Here's how we did it: measure the distance from the post to the nearest 2×4; figure out the board puzzle as explained above; measure the height of the wall and cut the first board; plumb it and nail it up; put next to the first board a "spacer" 2″ smaller than the next board, an "outer" one; measure, cut, and nail up the next "inner" board; remove the spacer and measure, cut, and nail up the "outer" board; nail the edges of the outer board to the edges of the inner boards; and so on.

The walls went up steadily. The house, which had seemed so small with stars and sunlight as a roof and leaves as walls, grew and took on new dimensions as the walls and roof closed it in. When we started the second wall, we suddenly had a corner! You could stand in the corner and be totally out of the wind and the rain! Incredible! The boards for the side walls had to be cut on their top edges to the same slant as the roof line. We did this easily and accurately with the aid of a movable bevel, or "angle finder": a handy carpenter's tool with a movable arm that can be set to any angle and locked into place; then it's used just like a right angle to mark off the cutting line on the board. It was another mathematical puzzle to figure out just the right length for each board as they got longer and longer toward the front of the house. When we got to the loft section, we nailed the wall up in two parts: from the loft joist to the floor joist, and from the rafter to the loft joist (in front: from the plate to the loft beam, and from the beam to the sill of the floor).

We did as much work—figuring and cutting—as we could "inside" the house, for by mid-October it was cold; the days were getting shorter. On All Hallows' Eve, I relinquished to the cold my desire to keep fiberglass out of my house. Faithful friends helped me staple 4″-thick batts to the underside of my floor. (I wanted to use Techfoam, but it is too flammable to be used as floor insulation, and I couldn't afford it.) The floorboards had already shrunk so much that grass and earth could be seen through the cracks. The corners kept the wind out, but the floor didn't! It was a bitterly cold day; we shouldn't have been flat on our backs on the cold damp ground, breathing in fiberglass, and fighting to hold it against the wind. But there was little else to do—I'd run out of building wood.

While I impatiently waited for the mill to get around to cutting the tiny amount of wood I needed to finish the walls, I visited a nearby used-lumber yard. The old man who ran it had piles of treasures for builders: old windows, old doors, and all sorts of interesting pieces salvaged from barns and burned buildings. We bargained over the ten windows I wanted (eight matching ones, 5′ high by 2½′ wide, for the sunny loft, and a pair of tall thin ones to fit into an already framed space whose window had broken in a windstorm (it was just a piece of unframed glass). I drove a hard bargain and got them all for $60 and went jubilantly home to use my last pieces of 2×4 to frame out the window spaces in the loft.

Another week and still no lumber. I wrapped the space from the bottom of the house to the ground in plastic to cut the cold draft from the floor even more. I braced the front and back middle posts. I spent tearful hours struggling with stovepipe until I got my Ashley heating stove installed. I cut firewood. River and I used leftover pine boards to build the back door. I blocked the holes in the wall where the "outer" boards don't fit against the sill. And still no wood.

Thanksgiving Eve night it snowed: a wet heavy snow; and my tent collapsed. Luckily I was sleeping in the cottage with River and my daughter was visiting a friend. The next day it rained. On Friday, in the chill dusk, the lumber truck drove up to my "house" and dumped its load into the mud. *Blood, sweat, and tears build a house,* a friend told me when I started. There was very little blood but sweat and tears aplenty in the November of this house building.

I spent Winter Solstice with women friends old and new and slept in my house for the first time two days later. The windows were just rough frames covered with plastic; the dutch door was a huge hole covered with plastic; and the loft was sunny indeed with one wall—a small one—covered in plastic! But I slept peacefully and soundly that night, in my house that kept out the wind and the rain and the snow, in my truly incredible hand-built, woman-built, love-built house!

It is Candlemas now, more than a year later. My house is more incredible and much warmer than it was then. All of the windows (except one) are now "in" and the dutch door is built and hung. Billie, who showed me how to put up plastic over my windows last year so it wouldn't blow off, gave me her wonderful wood cookstove, and I can eat at home now. There's a gutter out back, and next year I'll collect rainwater for washing and cooking—gravity-fed plumbing! And I've relented even more in my fiberglass war—a friend gifted me with rolls and rolls of insulation that she found in an old building she was tearing down. I've let the nasty stuff into my house mostly in hopes of stanching my daughter's tears; she hates to wake up in the morning when it's below freezing in the house, and if it's been below zero outside all night, that's likely. I still don't have electricity and I plan to keep it that way. But we no longer have to expose ourselves to the frigid winter—a charming chamber pot has saved us!

I figure I'm past the break-even point financially now: if I had stayed in the cottage I would have paid $3,300 in rent for the past twenty-two months; instead, I have spent about $3,000 in building supplies, taxes, and rent to River, who now lives in the cottage, for use of her refrigerator, shower, plumbing, and care-free oil heat on nights when I'm too tired and cold to make a fire. I have no fuel bills and no electricity bills—although I spend about $75 a year on my chain saw and $10 a year on candles. I feel that I've achieved my goal of living for less money—and I've learned how to build a house, too! And the sun and the moon move across my bedroom windows just as I imagined they would!

The Key to Carpentry: Tools

HAND TOOLS

Having recently built a small house from the ground up using only hand tools, I have a renewed respect for their efficiency, versatility—and quietness! Hand tools are designed for virtually every aspect of carpentry work or woodworking, and learning to use a hand tool for each step of marking, cutting, fastening together, and finishing will give you a special sense of connection both with the materials you handle and with the process of building or creating. For the sake of organization, hand tools are divided into several broad categories: tools for measuring and marking; tools for cutting, scoring, shaping, and smoothing; tools for fastening things together; and demolition tools. I've included a few "miscellaneous" tools that don't really fit into any of the listed categories.

In buying hand tools, it's usually a good idea to choose a quality tool that may be a little more expensive initially but will make your work much easier and will last you much longer. Avoid the "handyman specials"—for reasons beyond the sexist semantics! These tools are usually of a poor quality, made with cheap materials, and they neither work as well nor last as long as the better versions of the same tool. You'll find yourself frustrated with their inefficiency or convinced that you're a poor carpenter when in many cases you could do the work that seems so hard if you had a good tool. Good tools don't make a good carpenter, but they certainly help! There is, of course, an exception to the general buy-good-tools rule: now and then you'll do well to pick up a cheap tool to use for rough work that might damage your better tool. For instance, you'll want to have a cheap chisel to use on wood that has embedded nails that might pit or chip the chisel blade. You might want a cheap saw for the same reason.

Throughout the tool section, I've stressed tool care—so much so that you may find yourself wearying of being reminded to put a delicate tool in the special compartment of your toolbox, or to avoid stepping on a tool (this seems rather obvious!). But my experience has been that tools are abused and ruined by a simple lack of tool consciousness, and that developing a habit of caring for your tools takes a certain consistent discipline. A tool that has been poorly cared for makes carpentry twice as hard as it needs to be. A rusty saw won't cut the way a clean, sharp saw will; a chipped chisel is hard to guide through a straight cut. By the time you pass from simple enchantment with your tools to a kind of tool obsession/tool fever, you'll probably be quite fanatical about keeping things in perfect condition. Meanwhile, some reminders might be of value.

TOOLS FOR MEASURING AND MARKING

Tape Measures

Measuring tapes for carpentry work have steel blades that are retracted into metal or plastic casings when not in use. The blade of a tape may vary from 10′ to 100′ long. For general work, a 16′ tape is handy and not too long for quick retraction and a compact housing. A 50′ or 100′ tape is useful for layout work, but must be wound back into its case (the smaller tapes have a spring mechanism for quick return). These longer tapes are not as clearly marked for precision work as are the shorter tapes. In choosing a tape, look for a good flexible blade (the wider the blade, the less flexible) with clear markings (usually in graduations of $\frac{1}{16}$″; special marks for $\frac{1}{2}$″, inch, etc.; special markings for 16″ centers are helpful, too) and a locking mechanism that allows you to fix the blade at any length outside the housing. My favorite tape is the 16′ Stanley Powerlock II; among other features, it is very durable, comfortable to handle, and well graduated. The tape itself doesn't wear out before its spring mechanism (I've used other tapes that have—the markings wore very quickly under daily use and the tape was soon useless). The only problem with this tape is a short "breaking-in" period for the very powerful return mechanism; you can get some nasty finger lashing if you don't let the tape in bit by bit when you're first using it. This extra power in the return mechanism will prove itself a bonus once you've broken the tape in. It will outlast most of the other brands that might be less snappy to begin with!

In using any tape measure, you should be aware of the play in the end of the blade; the metal hook that allows you to fix the blade by "catching" over the edge of a piece of wood is attached to the blade with tiny metal rivets that are not quite rigid in position. This allows a play of about 1/32″. If the end of the blade is hooked over a piece of wood and pulled, the reading you take at any point along the blade will include this extra 1/32″ or so (the exact amount varies from tape to tape). If you then take a reading or make a mark with the blade end butted into something or not extended, this reading or mark will be 1/32″ short. In any precision work, this tiny variation can make a lot of difference! It may mean that a corner won't quite meet, or that two pieces will vary slightly in length, or that a cut won't work as it should. So whenever you use your tape, use it consistently. Either make sure that you take each and every measurement with the tape fully extended, or that you make allowances for that extra 1/32″ if the tape is not extended.

Another matter to consider in using your tape measure is the accuracy of the mark you "extend" from the tape itself to the surface you are measuring and marking. If you simply make a pencil mark corresponding to the exact point on the tape, the thickness of that mark will give you a slight inaccuracy. The usual and more precise way to make the mark is to draw a little V. The point of the V should hit the tape at exactly the point you want to mark. This is another instance where a seemingly insignificant difference can have a noticeable effect on your finished work. A joint that won't fit tightly or a miter that remains slightly open can be the direct result of a careless mark made early in your work. It's good discipline to make a habit of marking all measurements with a V as though they were critically important. Your work will be more accurate and smoother—and when you need precision, it will automatically be there.

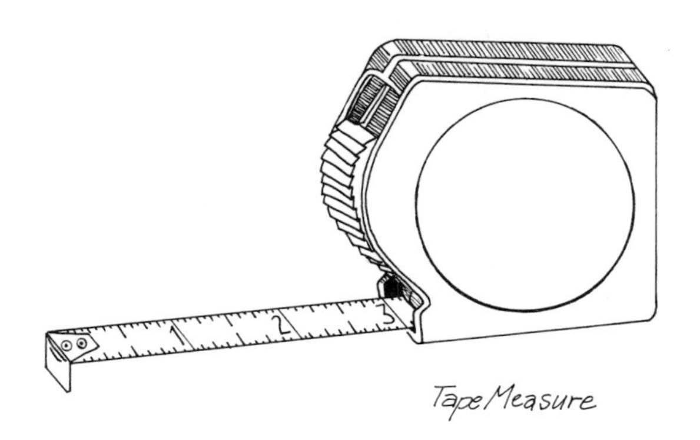

Tape Measure

Because of its protective housing, your tape measure can take more abuse than some of your other tools. It can be carried in the general toolbox and can survive short falls and even being stepped on! It *can't* take being bent around extreme corners, or being twisted into contortions that will crack the protective coating on the blade or permanently bend the blade itself. Pulling it out too far is another form of abuse that will shorten the life of any tape. It may or may not work immediately after this treatment, but the spring return will be weakened if not right out broken. It's a good idea to pull the tape *very gently* if you really need those last few inches. Letting it feed too quickly back into its casing can also damage the spring; a good habit is to use the lock mechanism or your thumb on the tape itself to control the reentry.

Keeping the tape dry is another prerequisite to a long work life for this tool. The protective coating (usually Mylar) will prevent rusting, but if you retract a wet or muddy tape without cleaning/drying it, you can damage the inner workings vital to its functioning. Take that extra moment to dry your tape if it's gotten wet.

Despite the best of care, your tape measure will wear out much faster than some other "basic" tools. First to go is either the tape itself or the spring return mechanism. If the tape wears out, or gets bent or broken somehow, you can replace it with a "refill" tape. This is a questionable option. Usually by the time the tape has gotten worn the spring return is also worn, so that a new tape might easily outlast the old spring. Once the spring is gone, the tape measure itself has to be replaced. The only time it makes sense to buy a refill tape is when the original has been damaged prematurely.

Sliding T Bevel

The sliding T bevel has a steel blade held by a wood or metal handle. The blade is slotted at one end and

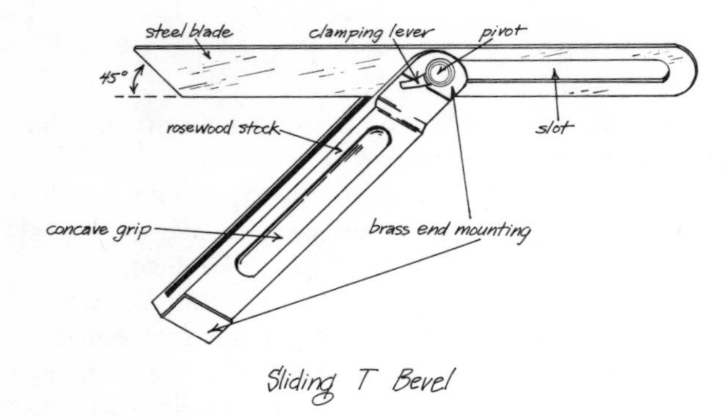

Sliding T Bevel

You can use your bevel *to take an angle from one piece of wood and transfer it to a second piece.* (Photos by Carol Newhouse)

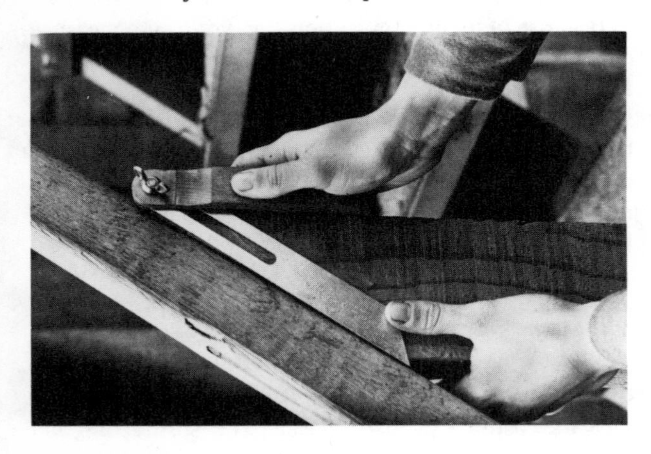

can be loosened and adjusted to make more or less of its blade usable. This tool is used to transfer angles and is essential to all sorts of carpentry and cabinet work. The bevel pictured has a rosewood handle and a steel blade. The blade is locked in position by means of the clamping lever (or, on some models, a wing nut). A second type of sliding bevel comes with a nickel-plated metal body and a steel blade. The blade is locked and adjusted by means of a screw mechanism located at the base of the body (or handle). This second bevel is slightly better designed for use in setting an angle on a circular saw, because the adjustment screw doesn't get in the way (the wing nut or clamping lever does). For sheer beauty, the first type of sliding bevel can't be beat!

The sliding bevel is a fairly delicate tool in that the blade can be bent, which destroys the accuracy of the tool. It should always be placed in a safe spot on the job and carried in a protected part of the toolbox. The safest way to set the bevel for storage or transporting is with the angled end of the blade slid fully into the handle and the slotted, curved end protruding.

Try Square, Miter Square, Combination Square

The *try square* has a steel blade (usually about 8″ long) fixed at a right (90°) angle to its hardwood handle. It can be used to mark, set, or check right angles.

The *miter square* has a slightly longer blade (usually 12″) attached to its hardwood handle to form a fixed 45° angle. The ends of the miter square are angled (45°), too. This tool is useful for laying out 45° miters.

A *combination square* can do the work of either try or miter square and more besides. This square has a metal handle (or body), which holds a steel blade, 12″ long and about 1″ wide. The blade can be fixed in place or loosened and adjusted so that it extends mostly in one direction. Certain combination squares are designed to enable the blade to be pivoted around and set absolutely flush with the straight surface of the handle; this allows the combination square to be

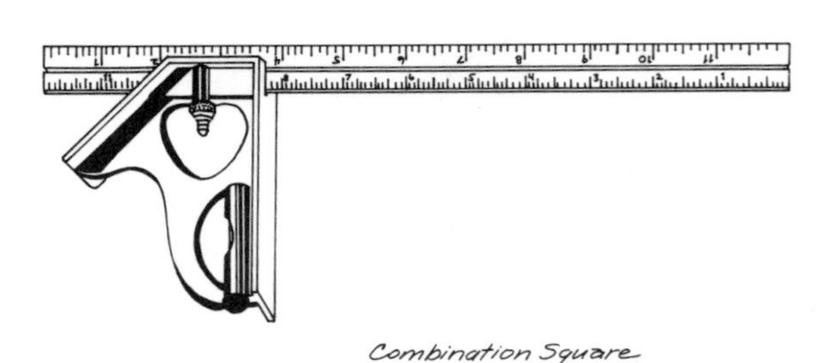

Combination Square

used as a try square and butted into a corner. The combination square has a 45°-angle face and a right-angle face, so it can be used to measure and mark these common angles. The blade is graduated and can be used as a ruler or tape. Some squares have a removable metal pin that functions as a tiny scratch awl or marker (this pin almost always gets lost if you use your square a lot!), and some have a built-in level vial.

The combination square is one of the most versatile and necessary of tools. When you buy yours, choose a good one. Stanley makes one of the best. Cheaper squares are either poorly designed or made with inferior materials; they wear out and never really give accurate readings. Your combination square can be ruined if it is bent. Keep it in a special section of your toolbox, and when you use it, be careful to put it down where it can't be stepped on or have a heavy piece of lumber dropped on it. With this minimal care, a combination square can last for years and years.

Steel Square or Framing Square

Whole books have been written about the use of this deceptively simple-looking tool; supposedly it can be used to calculate almost any problem in carpentry—including wages for a job! At its most basic, the steel square checks right angles in squaring up a building you're laying out or checks the corners of the window frame you've just put in. It is absolutely indispensable for laying out rafters and stair stringers. This square has a longer arm called the *body* and a shorter arm called the *tongue.* If you hold the square so that the body is in your left hand and the tongue in your right hand, the face, or face side, is up (toward you) and the back is away from you. The steel square is covered with markings, graduations, and information. For example, on the face side of the square are the roof framing tables, which show the length of a common rafter per foot of run according to the pitch of the roof. For most of your work with the steel square, you'll be using the more simple graduations on the 24″ body and the 16″ tongue.

In buying a steel square, don't be tempted by the "economy" models. These are usually not standard size, which makes working with them more difficult. They are more easily bent and just not the quality tool you need for precise work. You can buy a good aluminum or steel square (the aluminum one is called a *rafter square;* the steel square is the same tool, really). The aluminum has an advantage in that it won't rust, but I like the sturdiness and clear markings of the steel square. Along with your square you should probably invest in a pair of stair gauges or *staircase fixtures.* These are tiny brass attachments that come in pairs and can be fixed on the square for repetitive marking or gauging (necessary when you lay out a rafter or stair stringer).

The framing square *(or steel square) may be used to solve the complications of roof framing or stair building—but it is also handy for checking square corners.* (Photo by Carol Newhouse)

The steel square is another ultra-delicate tool that should be protected on the job and in the toolbox. If it gets bent it will lose its reliability, and if it gets rusty it will lose its necessary markings. The steel square is so commonly used that most toolboxes come with special slots meant for the tongue of the square (the body is carried inside; the tongue protrudes a little because most toolboxes aren't 16″ tall). If you're working with your square, place it carefully aside when you don't actually have it in hand.

Four-foot T Square

If you are working a lot with plywood, Sheetrock, or paneling, a 4′ T square is a handy tool. It functions as a lightweight, easily manageable straightedge and is marked in inches graduated down to eighths on both body and tongues. With this tool you can measure and mark large pieces of material or doors in one quick operation. You'll probably find so many uses for it that you'll wonder how you got along without it!

In shopping for this tool, take the time to choose one that is *exactly* square. This T is manufactured in two pieces that may not be joined together precisely, in which case the tongue and body do not form perfect right angles. Pick out one that looks square—and then check it with a combination square or a try square. Don't be surprised if you have to go through two or three before you find a really good one.

Once you have a perfect specimen, handle it like

the delicate tool that it is. Avoid putting it where it might be stepped on or have a heavy tool or board dropped on it. Carrying this tool to and from the job requires a little extra thought and caution—ours always rides in the cab of the truck, safe from sliding loads and other mishaps. At home, the T square should be hung safely on the wall where it won't be accidentally bent. With this defensive care, it should last you for many years.

Chalk Line

The chalk line is simply a string housed in a metal compartment that also holds powdered chalk. The string can be pulled out for use and then rewound into its housing. The chalk line is commonly used to mark cutting lines across distances too great for other measuring devices, to aid in lining up courses of shingles, and so forth.

To use your chalk line, you will want to stretch it taut between the two points you wish to connect. First give the chalk line a shake to distribute the chalk through the string. Flip up the arm on the housing to allow the string to feed out. The end of the string is attached to a small metal triangle with a little "foot." This foot can be slipped over a finishing nail driven at one of your points or hooked over the end of a board or sheet of plywood. As you pull the string out and move toward your second point, try to keep the string above the surface you are going to mark (this keeps the surface free of confusing extra marks). Stretch the string tightly down to your second point. Holding it securely against the mark with as much tension as you can get, use your other hand to reach out and give the string a pull upward, allowing it to snap down sharply *once.* This single snap will give you a clean, clear line to follow. Lift the chalk line so the string is again clear of the line you have marked and rewind. When you're done, press the arm back into the housing to hold the string in place. Periodically you will have to replenish the supply of chalk in the housing (chalk is available, in plastic bottles, at the hardware store). You can also buy replacement string if yours breaks or wears out.

Scratch Awl

A wooden handle and a long, thin metal blade with a sharply pointed tip make the scratch awl useful for marking wood and punching small starter holes for screws, drill bits, and so on.

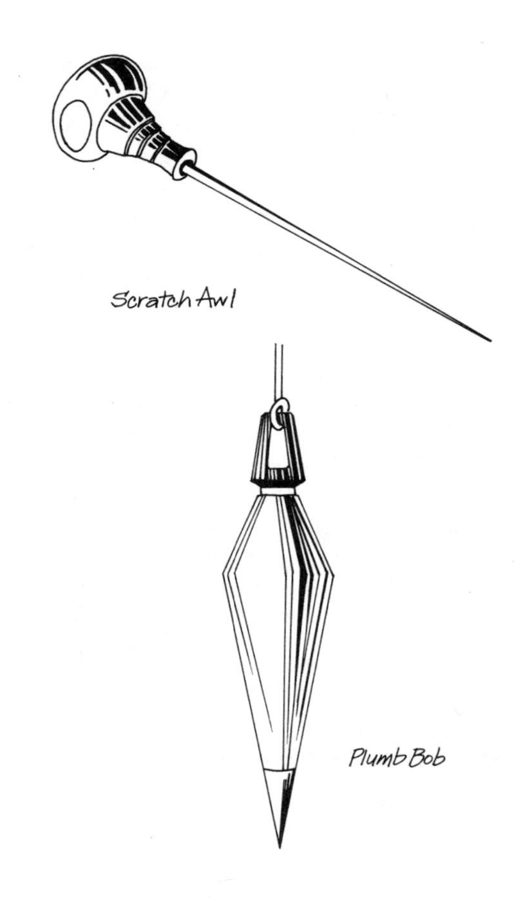

Plumb Bob

The plumb bob is a weighted metal body suspended by a string; it will help you locate a point exactly below another point (perpendicular). The best plumb bobs have a solid brass body, a finely pointed lead tip, and a strong string attached to the end opposite the tip. The finer the tip and the heavier the body of the plumb bob, the more precise the work it can do. To protect the tip of the plumb bob when it's not in use, a leather case should be used. If properly cared for, a plumb bob is one tool you should never have to replace. Even though you'll find multitudes of uses for it, it rarely wears out.

Butt Gauge and Contour Gauge

A *butt gauge* is used to mark the thickness and exact position of hinges on doors and hinged windows. It is sometimes used to help locate strike plates or parts of locksets. The butt gauge comes either in a fixed size (common sizes corresponding to the sizes of hinge are available) or in an adjustable type that will work for three or more sizes. The butt gauge is a specialty tool that you'll only really need if you're doing a lot of door hanging.

The *contour gauge* is an odd-looking item, like a cross between a comb and tool. It consists of a row of stainless-steel teeth held by a metal bar or band. When the bar is loosened the teeth can move in place. The contour gauge is used to make a duplicate of complicated moldings or any irregular surface 6″ wide or less. By loosening the bar, pushing the line of teeth against the surface or molding, and then tightening the bar, you obtain a temporary "copy" of the surface contour that you can then transfer to a piece of paneling or whatever. This is one of those tools that look intriguing and are fun to have—but are quite limited in their usefulness!

Wing Dividers

A wing divider is the simple tool you used in grade school math to scribe arcs and circles. It's a little metal device that holds a pencil in one "leg" while the other leg is fixed to a point or run along a surface. The wing divider is handy for measuring and marking off small distances. It is essential for one carpentry job at least: placing wood paneling or Sheetrock in a room with an uneven corner. For example, if you're finishing a basement room and want to match the last piece of corner paneling to the rough concrete of a retaining wall or foundation, you can trace the exact contour of the rough wall onto a piece of paneling by using a wing divider. The metal leg of the divider is run along the wall and the pencil leg, fixed about an inch or two out, scribes the line onto the paneling (which is butted as close to the corner as possible). You can then use a power jigsaw or saber saw to cut the paneling and have a nearly perfect fit.

TOOLS FOR CUTTING, SCORING, AND SHAPING

Handsaws: Rip and Crosscut

The saw used for most general carpentry work has a fairly wide blade, usually about 26″ long, with a row of cutting teeth. The blade is fastened to a wooden handle with brass rivets. A *crosscut saw* is designed to cut across the grain of the wood; the teeth are beveled or pointed and set at alternating angles to one another. This slight bending or setting of the teeth allows the saw to cut a path (called the *kerf*) for itself through the wood—otherwise the saw would bind up,

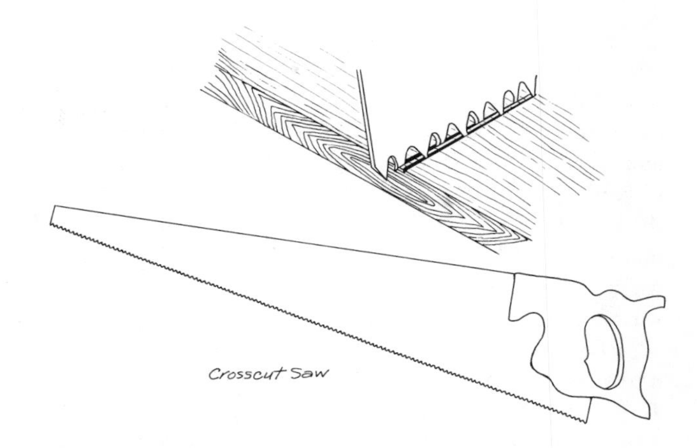
Crosscut Saw

unable to move. A *ripsaw* is designed to cut along, or with, the grain of the wood. Its teeth are chisel-shaped rather than pointed, but are set in much the same way for the free passage of the saw blade through the wood. Both ripsaw and crosscut saw are spoken of in terms of "points," which refer to the number of teeth per inch of saw blade. A saw with many, close teeth makes a finer cut than one with fewer, coarse teeth, but the cutting goes more slowly. Ripsaws have from 4½ to 6 points (teeth per inch) and crosscut saws from 7 to 12 points. For general carpentry, a 5½- or 6-point ripsaw and an 8- or 10-point crosscut saw will do most jobs efficiently and well. If you have a power circular saw, you'll probably use a hand ripsaw very infrequently—ripping is done quickly with the circular saw, and with none of the effort that ripping with a handsaw takes. But even with a circular saw, you'll probably find much use for your crosscut saw(s).

Learning to use a handsaw properly is one of the first steps in learning carpentry. To begin with, make sure that you have a sharp saw that is properly set. Sometimes saws are sold that are neither sharp nor properly set. If you hold the saw so that you're looking directly down on the cutting edge of the blade, you should be able to see the set of the saw: the teeth bent out slightly in an alternating fashion (one to the right, one to the left). The individual teeth should have bright, sharp edges. If you're in doubt about your saw, show it to a more experienced friend. Ask her to try it or give you an opinion.

Before you attempt a cut, make sure that the board you're going to saw is properly supported on a worktable or across two sawhorses, and that the end you're cutting off can fall free. Practice cutting off pieces that are a foot or so long—longer pieces are heavier and may need a second person to catch them or support them so that they don't bind the saw or crack the board you're left with.

Use your combination square to make a straight line across the face of the board (I'm assuming you'll start with a crosscut saw and cut across the grain of the board—the most common cut you'll make). Remember that the saw cuts a path for itself—the kerf. If you're doing precision work, you have to take the thickness of the kerf into account. If you're just learning, it's a good consciousness to begin developing. Always make your cut so that the kerf comes off the *waste* side of the wood. This means that if you want a piece 24″ long, you measure and mark your wood and then cut on the waste side (leaving the 24″ intact). If you try to cut right down the center of the line you've drawn, you will end up with a piece *slightly under* 24″; if you cut on the 24″ side of the line, you'll end up with a piece as much as ⅛″ too short. This may sound inconsequential, but when you're building something, it can throw off your work surprisingly much! I always try to make my cut so that the pencil line I've drawn is left intact right along the edge of the piece I'm saving. This is one way to be sure that I'll have the full length I need and won't be short.

All of the following is structured for the right-handed person. If you're left-handed, you'll have to adapt it to your own needs. The procedure will basically be the same. I like to hold the saw with one hand—a firm grip on the handle. I keep my second hand free, sometimes using it at the very beginning to guide the tip of the saw, and perhaps later using it to catch the waste piece of wood if I'm afraid it's going to crack the board or bind the saw. A handsaw does most of its work (cutting) on the downstroke and about one quarter on the upstroke. Sawing is a smooth up and down motion, but the beginning few strokes are different. Hold the saw so that the part of the blade closest to the handle touches the edge of your board at (just to the right of) your mark. Draw the saw up so that the teeth cut a little kerf. Your saw blade should make more or less of a right angle with the face of the board; if you tip the blade to the right or the left too much, you'll get a bevel cut on your board. This is perhaps the hardest discipline to learn in sawing. If you can't "see" the angle, try holding your combination square—or anything with a right angle—in place so that you can see how the saw should sit. This angle is different from the angle of the saw itself. The saw should be held at about a 45° angle to the work while crosscutting and about a 60° angle when ripping. This 45° or 60° angle should be between the face of the board and the tooth of the blade that's just below the handle.

It's common practice to saw with one knee on the board to help steady it. When you make your first few updrawn cuts with your saw, you might find it easier to steady the board with your free hand. Once you've begun your cut by drawing the saw up a couple of times, you can begin to saw with the regular up-down motion; the little kerf you started will guide the blade. Keep your eye on the line you've drawn to mark your cut and try to keep the saw (and cut) straight. Learning to use a handsaw well takes practice, so don't be discouraged if your initial attempts are wobbly. It will also take a while to get the proper angle, the right pressure on the saw, and so on. Sawing takes a bit of arm muscle, but you shouldn't really have to force the saw—the cut should be smooth, with a gentle pressure applied to the saw. If your saw won't work unless *forced,* it might be dull, or not properly set. Or you might be working with extra-hard wood. Some wood, Douglas fir for example, takes on a stone-hard quality when it ages (new fir is fairly soft and easy to cut). Oak is hard to begin with and gets *very hard* later. Experiment with different types of wood of different thicknesses to see how differently they act under your saw.

Held at the correct angle and used to cut across the grain of the wood, the crosscut saw *will make a smooth cut.* (Photo by Carol Newhouse)

Along the same don't-get-discouraged lines: don't let onlookers intimidate you with advice. Some advice is valuable and worth listening to, but some is intended (consciously or not?) to discourage you. When I was just learning carpentry, I asked a friend (a man, and a professional carpenter) to show me how to use a handsaw. He ran through the basics and then stood back to watch me. I had barely two seconds of cutting accomplished when he interrupted with: "No—no—that's all wrong. Women *always* hold the saw that way. That's wrong!" I was holding the saw at too sharp an angle to the wood, a common enough error for a beginner. But the message behind "Women *always* . . ." was a subtle and familiar one. I was left feeling that I was doing something vaguely wrong, wanting to learn carpentry—something that was not "natural." Learning around other women, whether they're teaching you or learning with you, is a decidedly different experience. I've never had a woman tell me I held the hammer wrong because it was my biological destiny.

But back to the gentle art of handsawing. . . . Once you feel comfortable with your crosscut saw, try a ripsaw. The work is very much the same, except that the saw handles a little differently—the teeth are coarser, so it cuts somewhat quicker. Your orientation to the board is different, as you're cutting down the center rather than cutting off an end. As before, having the board well supported is important. Don't try to cut from an end of the board that is extending too far beyond the supporting sawhorses or table; the board will be too springy and the cut hard to control. Position the board so that the part you're cutting is fairly close to the support and feed it out bit by bit. You can either stand to the left side of the projecting board or actually stand between the sawhorses (again, on the left side of the board) and lean over the board to rip it.

Keeping your handsaws sharp and well set will make working with them a pleasure rather than a chore. You may want to learn to do your own sharpening and setting; there are jigs and files available for sharpening and saw sets for setting. If you don't know anyone who can teach you the techniques, there's at least one good book out on the subject (Walton's, listed in "Resources"). Until you learn, you can usually have a saw sharpened and set in a local saw shop for a couple of dollars.

Keeping the saw sharp involves a certain amount of preventive care. First of all, buy a plastic or rubber saw guard for each saw you have. This is an inexpensive little piece that fits over the cutting edge of the blade when the saw is not being used, and protects the teeth from being chipped or dulled by other tools in your toolbox. Second, keep your saw dry so that it doesn't get rusty or pitted. This means keeping it in a dry place and periodically oiling the blade. If you're working outside and it starts to rain, make sure that you wipe the saw dry before putting it away. One final thing: don't cut into nails! One cut into a board with a hidden nail can take the edge off some teeth, if it doesn't actually chip the teeth off. Be extra-careful whenever working with old or recycled wood. You might even want to buy one cheap saw to use on such wood, saving your good saw from possible damage and wear.

I've left the buying of handsaws until last. As usual, every carpenter has her own preferences for certain brands, sizes, and styles of every tool. The best handsaws have blades made of high-grade steel. Sheffield steel, from England, is one of the best; Swedish steel is also very good. My favorite saw is an old Simonds that my partner owns, which has a beautiful springy blade that fairly sings in the wood. In choosing a saw, look for a flexible blade, preferably one that's "taper-ground" (each sawtooth diminished *gradually* to a fine point).

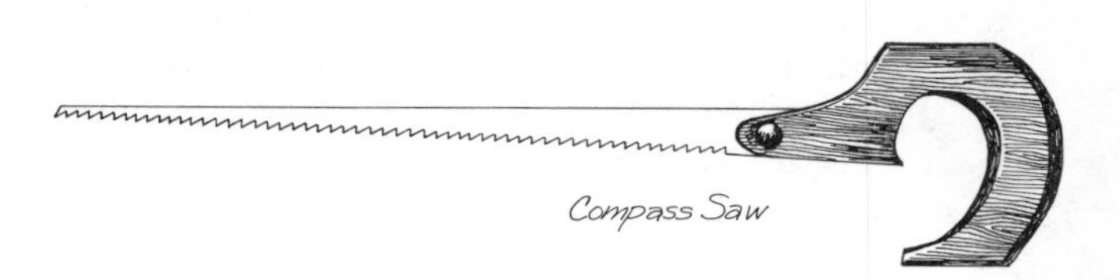
Compass Saw

Compass or Keyhole Saw

The compass or keyhole saw has a fairly short, tapered blade, which may be easily replaced so that you can use it as a crosscut saw and then as a ripsaw minutes later. Because of the design of this saw, it may be used to make cuts in the center of panels. You can start a hole with your drill, then slide the nose of the keyhole saw into the hole to continue the cut. A keyhole saw is useful for making difficult cuts in gypsum board (Sheetrock) panels: for example, mid-panel cuts for electrical outlets or for pipes. The keyhole saw blade has a bit of flexibility, which allows for some irregular cuts, too.

Backsaw and Dozuki Saw

The *backsaw* is a small saw with a fine blade that is reinforced or stiffened for controlled, precise cuts. You may choose a backsaw with up to 26 teeth per inch; many have 10- to 14-point blades, so the finer teeth should be specially sought out. The backsaw is used

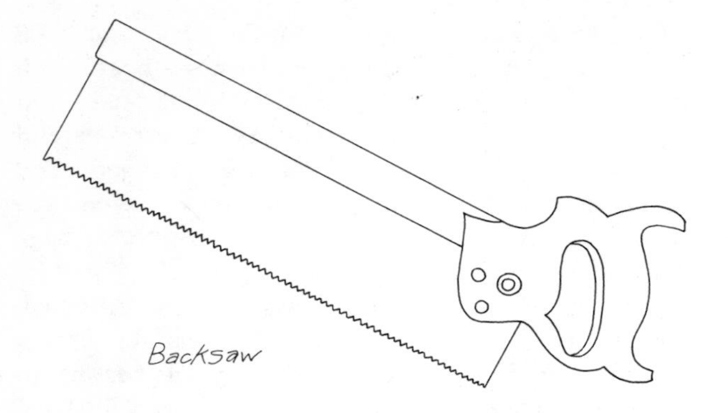

with a miter box for cutting trim, frames, etc.

Jan Zaitlin offers the following about the *dozuki saw:* "The dozuki saw is the Japanese version of the European dovetail saw (a small backsaw). Like all Japanese saws, it is designed to cut on the pull stroke and therefore can be made of thinner metal than the European push saw. It also has very little set on the teeth. This makes it excellent for cutting fine joints because the cut starts easily (i.e., the blade doesn't jump at the start of the cut) and leaves a thin kerf. This particular saw has 23 teeth per inch and is about 21″ long, including the 8¼″ cutting edge. Japanese hand tools are available through several of the larger mail order tool houses and at the Japan Woodworker in Alameda, California."

The dozuki saw *is the Japanese equivalent of the backsaw and is useful for making very fine, controlled cuts.* (Photo by Carol Newhouse)

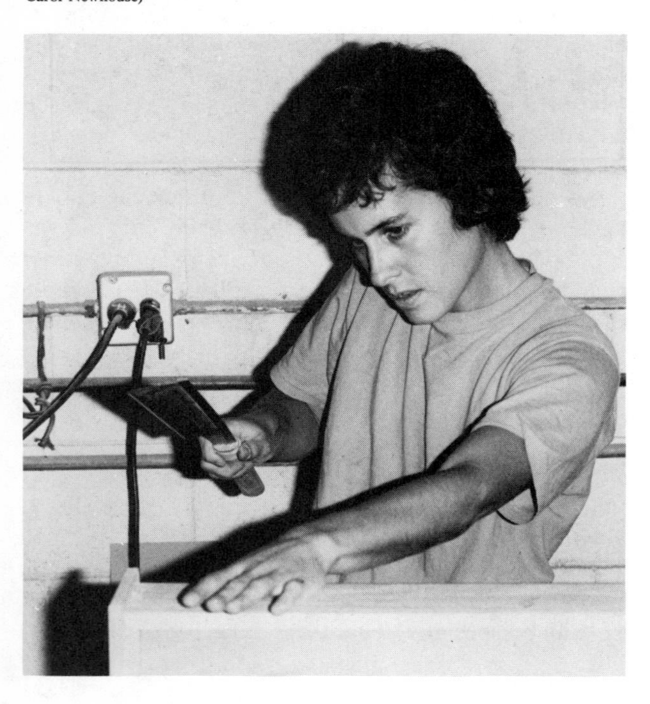

Miter Box

A miter box is actually a guide that holds the saw in a fixed position relative to the wood being cut and thus allows very precise cuts that may be exactly duplicated many times over. The miter box is used with a backsaw. The very simplest and cheapest miter box is one made of hardwood; it is slotted for right and left miters of 45° and for right-angle cuts. This miter box can be used for precision cutting of simple trim. More elaborate miter boxes allow you to cut a variety of angles, to use more varied thicknesses of wood, and to make more precise cuts.

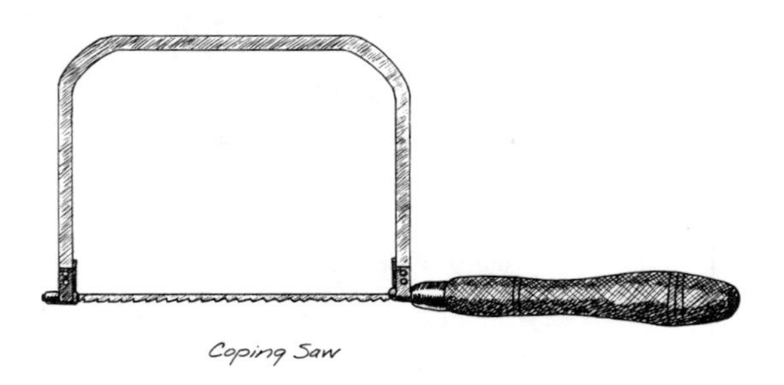

Coping Saw and Fretsaw

Both the coping saw and the fretsaw consist of a thin metal saw blade set in and held by a curved metal frame with a wood handle. The *fretsaw* usually has a blade with 32 points, which makes it suitable for intricate cuts in very delicate, thin materials (thin plastics, wood veneers, even thin brass). The *coping saw* has a 15-point blade and is generally used on thin wood. The coping saw may be used to cut curves, scrolls, etc., in (thin) material.

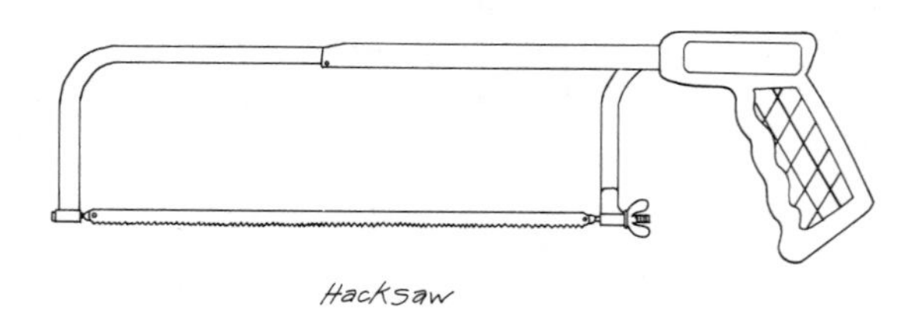

Hacksaw

This is a saw with a very fine-toothed blade made for cutting metal. A hacksaw can be used to cut pipe, rebar, bolts, nails, and other hardware. The blade of the hacksaw can be adjusted to cut in different positions (vertical or horizontal; left or right) and can be easily replaced if it wears out or breaks. There is a small close-quarter hacksaw that looks more like a knife (the blade extends straight out from the straight handle); this is useful for cutting in tight places where a regular hacksaw couldn't reach.

Wood Chisels

Wood chisels cut and pare away sections or levels of wood. They have so many uses that a set of three or four chisels in varying widths should be part of any basic tool set. An individual wood chisel consists of a steel blade attached by a thin projection (the tang) to its handle, which is made of wood or of specially hardened plastic. The blade of the chisel has a beveled cutting edge. The size of the chisel is determined by the width of the blade, which varies from ⅛″ to 2″. For general use, a set of four sizes should suffice. The most commonly used are ¼″, ½″, ¾″, and 1″. The handle of the wood chisel usually has a protective steel impact cap, which absorbs the blows when the chisel is driven with a hammer. Chisels come in a multitude of styles and special types, but for most general carpentry use a fairly stocky, short-handled, sturdy chisel is fine.

A wood chisel can cut out a mortise, make a groove, pare down an edge or corner, and more. It may be used to make a vertical or a horizontal cut. If you're working with soft wood, or want to shave away very minute amounts of material, the chisel can be guided with hand pressure. Most work is done with the bevel edge of the blade held up; for cutting a concave corner or for correcting a cut that is going too deep, reverse and hold the chisel bevel edge down. Experiment to see the difference in cutting each way. As much as possible, work *with* the grain of the wood. In general, the slighter the angle between the chisel itself and the surface being worked, the less wood is removed with each stroke and the smoother the work will be. If you have to make a fairly deep cut or are working with hard wood, use your hammer to strike the chisel handle and drive the chisel along. Your left hand holds the chisel and guides it while your right holds the hammer. Use short, gentle taps. Finish the cut using only hand pressure to guide the chisel. If a cut seems to be going too deep, reverse the chisel so that the bevel edge is down. One method for cutting out an area (for a mortise or a gain for hinges) is to hit a series of slashes in the wood (using your chisel and hammer) and then pare the wood out from the opposite direction. A vertical cut at either end of your work will check the cut: i.e., act as a stop; otherwise, the

In many cases, your wood chisel *is driven with a hammer; if the wood is very soft, you can use the chisel alone.* (Photo by Carol Newhouse)

wood tends to split along the grain past the area where you are actually working with the chisel.

A good-quality chisel will hold a sharp edge longer and perform better, so in choosing your chisels look for tempered steel, one-piece blade and tang, and an impact cap (or leather washer on wood-handled chisels) to protect the head. Storing your chisels in a leather roll will protect them and help keep them sharp, too. You may want to buy a couple of cheaper chisels to use for rough work on wood that might have embedded nails, thus saving your better chisels for finer work.

If you want to learn how to sharpen your own chisels, look at Walton's *Home and Workshop Guide to Sharpening* (see "Resources"). Companies such as Woodcraft (see "Resources") sell jigs and stones that will make your work simple and efficient.

Bricklayer's Chisel and Cold Chisels

These may seem odd tools to include in a carpenter's toolbox, but my experience has proven them very handy. A *bricklayer's chisel* is a wide, heavy-duty chisel made (as the name implies) for splitting bricks. It's

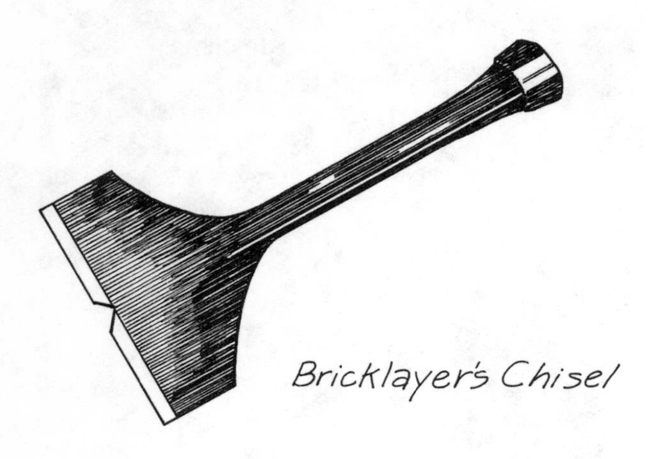

a useful tool for breaking up small areas of concrete. If, for example, you're adding a stairway or a deck to a house that has concrete walkways around it, the bricklayer's chisel will enable you to chip away areas and insert rebar, post anchors, or other connectors.

The *cold chisel* may be used similarly; it is a thick, stubby chisel actually made for work on metals. Either of these chisels is used with a sledgehammer. Always wear safety glasses while using either tool, so that flying bits of concrete don't end up in your eyes!

If you use either chisel a lot, the top will eventually begin to "mushroom" from the hammer blows. Take the time and precaution of filing the thin edges back so that slivers of metal aren't going to be flying along with the concrete chips.

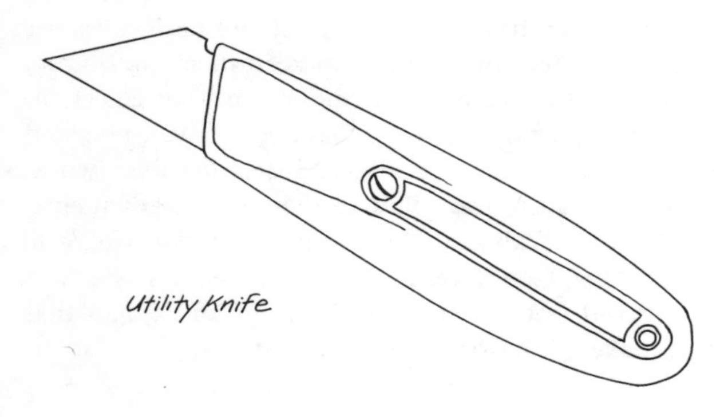

Mat or Utility Knife

This is a small knife with a razor-sharp replaceable blade. It has many uses in general carpentry, such as scoring Sheetrock, cutting felt paper, and trimming soft wood. There are two versions of this knife, one with a retractable blade and the second with a blade that rides inside the body of the knife and must be taken out and fastened in place for use. Despite the obvious advantages of the retractable blade, some say that it tends to break under pressure. I've always opted for the slower second type. Handle this little tool with a lot of respect—I've gotten some nasty cuts when my attention has slipped! Working with this knife usually means exerting quite a bit of pressure. If it misses its intended course, there's a lot of force behind it.

Planes

The plane is used to smooth and finish wood surfaces, and planing with a hand tool is one of the most pleasant tasks of carpentry. The hand plane consists of a metal cutting blade (much like a chisel) held in a metal or wood body. The blade can be adjusted to cut more or less wood with each stroke of the tool, and some planes have a variety of interchangeable blades for fancy and difficult cutting and shaping. A partial list of planes and their uses is below. For most general carpentry work, a block plane and a jack plane will be used.

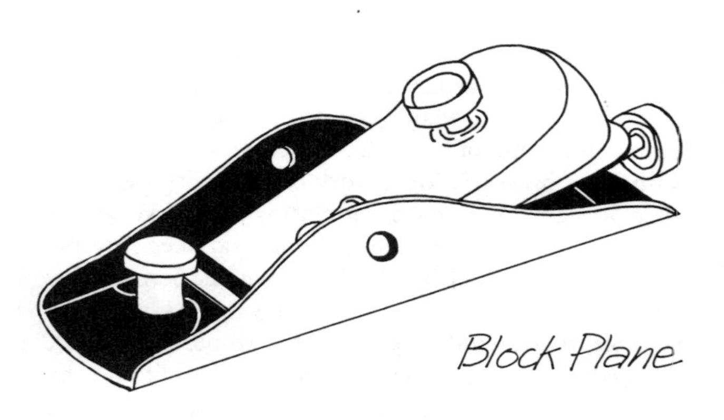

Block Plane This is a small plane, usually about 6″ long and about 1½″ wide. It is used on small pieces of wood and to cut across end grain. The cutting blade of a block plane is set at a very low angle (12° or so) to make end-grain work easy (the slighter the angle between a cutting blade and the wood it is working, the less the wood resists the cutting).

Jack Plane Perhaps the most useful plane for general carpentry, the jack plane is 11″ or 12″ long and suitable for planing the edges of doors and windows, smoothing flat surfaces, and so on. The jack plane is small enough to handle easily but has enough length of body to give you a fairly constant finished surface.

Smooth Plane This plane is used after a jack plane to further smooth a surface before sanding. It can be set to shave off very fine amounts of wood.

Fore Plane and Jointer Plane Either of these planes may be used to make a very straight edge on a board or to plane a large (or long) surface smooth. These planes are in the 18″-to-25″ size range and are quite heavy. They are more suitable for work in the woodshop than on a construction site or for around-the-house carpentry. If you are doing cabinetwork or hanging a lot of doors, the jointer plane is a good tool to have. For general work, the little jack plane can substitute for jointer or fore plane.

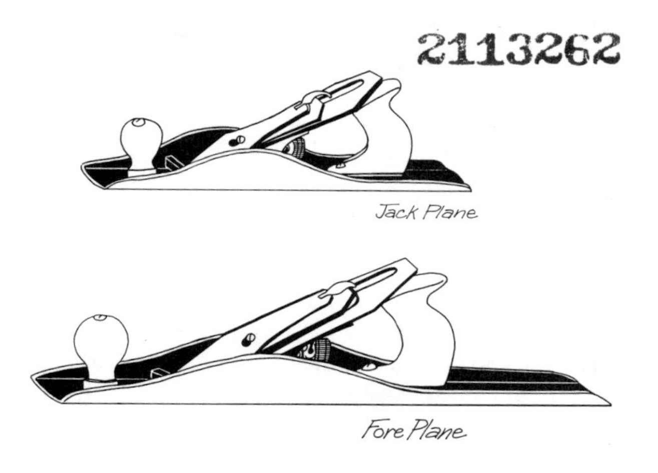

Router Plane This is a specialty plane designed for cutting grooves, dadoes, etc. If you are hand-making window frames, cabinets, shelves, and so forth, a router plane is a useful addition to your toolbox.

Spokeshave Spokeshaves smooth and finish rounded surfaces that other planes can't follow. They come in a variety of shapes with convex or concave bottoms as well as a straight-bottomed type.

Before using a plane, you should check and adjust the blade. The blade should sit squarely in place, with the distance between the edge of the blade and the edge of the opening (or throat) constant. If the blade is out of square, use the lateral adjustment lever to set it correctly. This lever is located close to the handle top on the upper surface of the plane. It moves to the left or to the right, changing the position of the plane blade accordingly. In small block planes, this adjusting must be done in a slightly different way. A lever on the upper body of the plane is swung to one side to release the plane blade. The blade is then adjusted manually, and the lever is turned back to clamp the blade in place. The distance the plane blade projects out through the throat will determine the amount of wood the plane shaves off with each stroke. Taking fairly thin shavings off each time will make the work go smoothly and finish nicely. There is an adjusting nut on each plane that can be turned to move the plane blade in or out. The opening between the plane blade and the front edge of the throat (the entire opening that the blade projects through) should also be adjusted; it determines how much of a shaving can run unbroken through the throat. This opening can be changed by moving the frog (the metal piece supporting the plane blade) back and forth. To do this on some planes, you have to loosen a screw that holds the frog in place; other planes have adjustment levers. When you have everything adjusted, give the plane a try. You may have to readjust after you see how it's working.

When planing, it is important to exert an equal pressure on both the knob and handle of the larger planes. Hold the plane square with the surface you are working. Experiment with the use of the plane on various types of wood until you can see clearly what to expect under different conditions. If you find that you're getting an uneven surface, try putting more pressure on the knob (or front) of the plane when you begin a stroke and more pressure on the handle (or back) of the plane when you finish the stroke. Check your work with a try square or combination square to make sure that you aren't creating high and low spots along the surface of the wood. Working with the grain of the wood will always be easier and smoother than cutting across or against the grain. Working with a long-bodied plane on a long surface will give a smoother overall finish.

If you are storing your plane in a toolbox or moving it about, adjust the plane blade so that it is retracted inside the body of the plane and can't be chipped or dulled. The plane is a delicate tool and should be treated carefully. Periodically the blade will need sharpening. You can buy a jig and stone and do this yourself, or have it done at any shop that does a good job sharpening chisels.

Surforms

Surforms are a family of shaping tools, something of a cross between a rasp and a plane. The cutting/shaping surfaces are peppered with tiny teeth not unlike those of a common grater, so that the wood shaves off in slivers. The larger plane-style Surform tool has a handle at each end so that you can guide it with equally applied pressure. A smaller version of this tool is shaped to fit comfortably into the palm of your hand; it is one of the most versatile of the Surforms, with a multiplicity of uses. The cylindrical Surform may be used to shape and enlarge holes (as when you install a roof jack—a metal device set into a roof when a stovepipe or vent is installed). Surforms have so many uses that it is impossible to suggest them all. They are suitable for shaping irregular pieces of wood, smoothing edges, fitting doors or windows, and so on. The blades are replaceable, and the Surform tools may be used on wood, soft metals, and plastics.

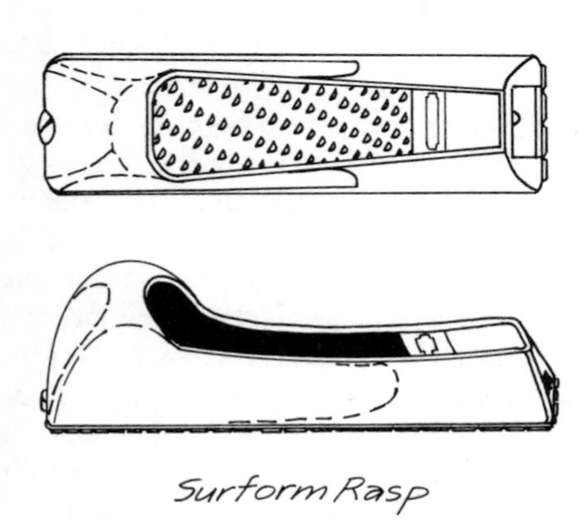

Surform Rasp

The Surform *may be used to smooth rough edges in a variety of situations.* (Photo by Carol Newhouse)

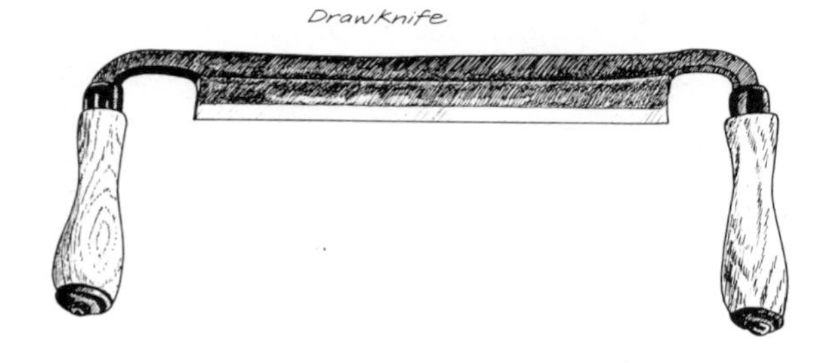

Drawknife

Drawknife

The drawknife is handy for furniture making, cabinetmaking, post and beam building, and wood carving. It consists of a steel blade with a wood handle at either end. With a firm grip on each handle, you can draw the blade along a surface, shaving off more or less wood according to the pressure you exert, the angle of the blade, and so on. This tool is useful for smoothing and shaping poles or posts and working curved surfaces in general.

Tin Snips

Tin snips come in straight-cut, right-cut, and left-cut types and various styles. They cut light sheet metal, plastic, rubber, and so on, and are especially helpful for cutting flashing and roof nosing. The fancier versions have molded plastic handles and locking devices to keep the snips closed when not in use and thus minimize damage to the cutting blades (and danger to you). Wiss is a good brand name to look for in choosing this tool.

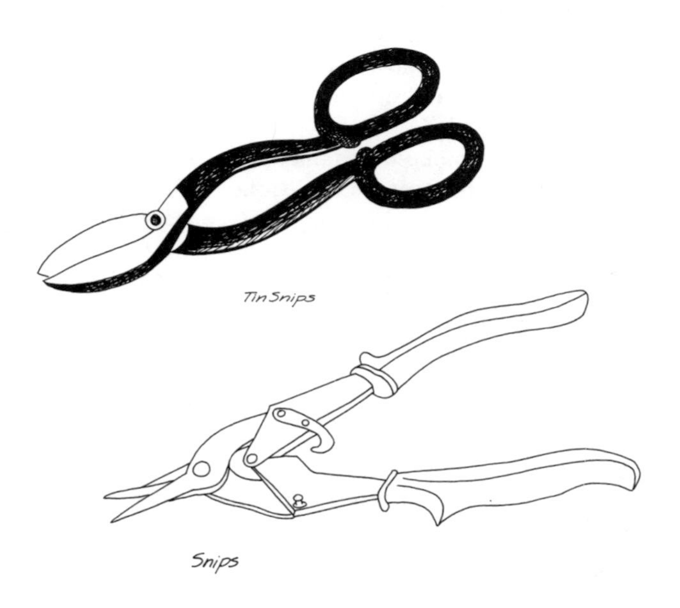

TOOLS FOR FASTENING THINGS TOGETHER

Hammers

Learning to use a hammer efficiently and well involves choosing a tool that fits you and the work you're doing as much as knowing the basics of hammering technique. Hammers range in size from tiny 5-ounce tack hammers to 10-pound sledges and come in a wonderful variety of styles for specialized work. For most general carpentry, a hammer in the medium weight range—16-ounce or 20-ounce—is used. If you are just learning, a 16-ounce hammer will provide you with sufficient weight and power to drive fairly large (16d) nails with ease but won't be so heavy that it will be tiring or hard to use. Once you've become comfortable with the 16-ounce hammer, you may want to exchange it (at least occasionally) for a 20-ounce or even a 24-ounce framing hammer for really heavy work (repetitive nailing of 2″-thick material, for instance). For nailing up trim, working on cabinets, and so on you'll probably want to switch to a 12-ounce hammer.

The general-purpose carpentry hammer comes in two basic styles: the *curved-claw* and the *straight-claw.* Since the claw of the hammer is used for pulling nails, the design of the claw is a secondary consideration in choosing your hammer. The curved-claw is slightly better for pulling nails; the straight-claw (or *rip hammer*) has an advantage when it comes to prying things apart. Most carpenters I know prefer the curved-claw (as I do). Of more importance to the actual working of the hammer is the material the handle is made of. Traditionally, hammers have been constructed with wooden handles joined to steel heads. The wooden handle provides the maximum shock absorbency—a feature you'll appreciate if you are hammering many hours a day with repetitive banging. The disadvantage of the wood-handled hammer is that the head may eventually work loose and the handle itself may become splintered and chipped.

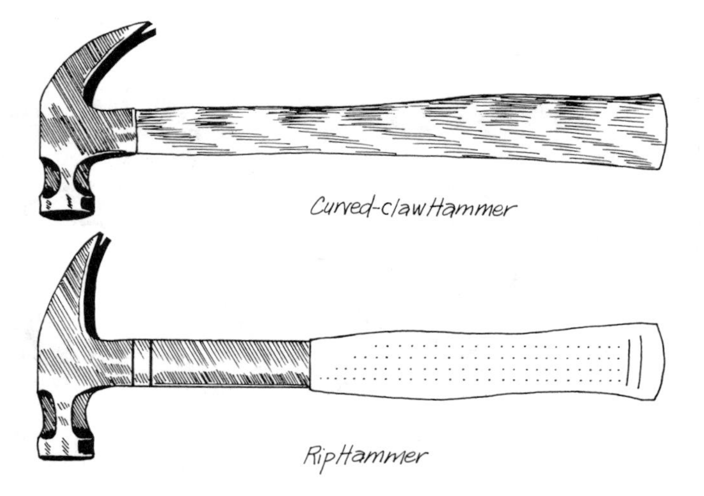

A correction of these problems was sought in the making of the steel-handled hammer, a single-piece tool whose handle is covered with leather or neoprene (a synthetic rubber). A substitute for the solid-steel hammer is the tubular-steel hammer, which has the advantage of being one-piece but absorbs more shock than the solid steel. Hammers with fiberglass handles bonded to steel heads are also now being manufactured.

Before choosing a hammer, try working with as many kinds as you can. Each hammer combines a different set of advantages and disadvantages. The weight and balance of the hammer are critical. Pick one that feels good in your hand, as your hammer will be functioning more or less as an extension of your hand and arm. My favorite hammer is the 20-ounce Estwing—tubular steel with neoprene handle. This is an exquisitely well balanced hammer with enough cushioning in the handle to make it a reasonable tool for all-day banging. Another good hammer is the 20-ounce wood-handled Plumb. Estwing makes a tubular-steel hammer with a leather-covered handle that feels wonderful to the hand (and is handsome!), but the leather loses its resiliency long before the hammer wears out and the shock-absorbing capacity this hammer initially has is lessened. Fiberglass-handled hammers always feel too light to me, but you may well find one you like.

Ask around to see what others think about particular models and brands before making your choice. Any individual hammer might have faults or features you are particularly concerned with. My Estwing has too-soft steel in the claws (my partner's Estwing doesn't, so this may well be a fault of my particular tool), but I like it so well otherwise that I'm willing to put up with this and would buy the same hammer again.

Learning to drive a nail quickly and cleanly takes a little practice; learning to drive nail after nail after nail takes a certain looseness in the arm and shoulder that you'll develop in time. "Hitting the nail on the head" is more than a cliché—it takes skill! To begin with, make sure you are working with reasonably soft wood; hard wood is much more difficult to nail into and beginning with a piece of aged oak is sure to discourage you! Use a sturdy nail—an 8d common is perfect, fairly short-shanked and strong. Hold the nail with the thumb and forefinger of your left hand (assuming you're right-handed) and position it so that it stands perpendicular on the surface of the board you want to nail into. Your best place to hold the nail is about

½" up from the tip. Hold your hammer close to the end of the handle. There's a tendency to grasp the hammer closer to the head for more control. This is called choking up on the hammer; it limits the ease of your swing even though it initially gives you a feeling of greater control over the tool. It's better to develop a habit of holding the hammer right (for maximum power output). Later on you may find that choking up on your hammer is sometimes desirable.

To begin, touch the face (the tip) of your hammer to the head of the nail. Lift the hammer slightly and tap down, using your wrist to drive the hammer. Repeat this procedure two or three times, being careful to strike the nail squarely on the head. If you hit it sideways, it will bend—especially if you deliver a more powerful blow. When the nail is set into the wood enough so that it stands on its own, withdraw your other hand. Now you can increase the strength of your blows. Try to feel the motion of striking with the hammer all the way up through your arm and into your shoulder. It should be one fluid, continuous movement from hammer to shoulder. The more liquid and loose you can make the motion of hammering, the easier it will be to drive the nail, and the more nails you can drive without becoming tired.

When you feel confident about driving 8d nails, try some other sizes. Remember that the hammer you use should be matched to the work: if you try to drive 20d nails with a 16-ounce hammer, it will take more work than driving them with a 20-ounce or a 24-ounce framing hammer. It is very hard to use a heavy hammer to drive small nails—they bend over easily. If you want to drive *very* tiny nails, use a small hammer—and a piece of cardboard to hold the nail rather than your finger and thumb. Take a scrap piece of cardboard and pierce the tip of the nail through it. This will provide you with a little handle to hold—or it can be set in place, with the nail sticking up and ready to hit.

If you strike a nail off center and bend it over, you can use the claw part of the hammer to bend it straight again or to pull it out entirely. A series of short, firm taps may work to drive the bent but straightened-up nail in. (See the "Nails" section of the "Materials" chapter in this book for more hints on nailing.)

In normal situations, your hammer is a sturdy enough tool to pretty much take care of itself. You won't necessarily have to be very careful about where you place it. It *is* a good habit to keep it on your tool belt so that it's handy (if you're working on the roof or on a ladder, it may be safer to keep the hammer in your hand). Never leave your hammer lying on a ledge, ladder, or any place it might fall from (and hit someone below!). Of course, you shouldn't leave your hammer out in the rain or in any damp place, as it will rust. If it gets wet on the job, dry it off well before packing it away.

Tack Hammer For very light hammering, a 5- or 6-ounce tack hammer is used. This hammer has a small, elongated head and is so light that it can drive very tiny finish nails or wire brads.

Wallboard Hammer This is a special hammer for nailing up gypsum wallboard (Sheetrock). It has a "checkered" head, slightly convex for setting the nail heads, and a hatchetlike blade with a nail-pulling indentation.

Ball Peen Hammer Varying in weight from 8 ounces to 32 ounces, the ball peen hammer has one rounded head and one flat head. It is used by machinists and designed primarily for metalwork.

Wooden Mallet The wooden mallet drives a chisel, but its use is mostly in the woodworking shop. If you are doing a lot of chiseling, consider trying the wooden mallet to save wear on the end of your chisel.

Nail Set

A small, cylindrical metal tool with a tapered head, the nail set (also called a punch) drives the head of a nail slightly beneath the surface of the wood. The set nail may then be covered with wood putty and more or less concealed. Nail sets come in various sizes, and for general carpentry work you'll need at least three or four to fit a range of nails (from small finishing nails to large casing and siding nails).

A nail set *is used with your hammer to drive the head of a nail just below the surface of the board; it should be matched in size to the nail being driven.* (Photo by Carol Newhouse)

Shingling Hatchet

A shingling hatchet, or shingler's ax, is a specialized tool used when applying wood shingles or shakes. It has a corrugated or waffled striking face that allows you to drive nails without splitting or otherwise damaging the soft wood of the shingles. Opposite this is a sharp, rather fine blade that can split a shingle to the width you need for a particular spot. Most shingling hatchets also have an adjustable weather exposure gauge, which is simply a device you can preset and use to check that the shingles you nail in place have a uniform exposure (I find that I never use this "aid"). Some shingling hatchets have a little V groove on the side of the blade. This is theoretically for pulling nails, but can really only work if the nailhead hasn't been fully driven in—or when you have to break a shingle out of its place and substitute another width.

A shingling hatchet is a surprisingly efficient tool, and if you are doing a lot of shingling you'll probably find it a worthwhile investment. It has very little use beyond this work, though. My favorite shingling hatchet is the one-piece Estwing with a rubber handle; it is sturdy and well balanced and holds up well.

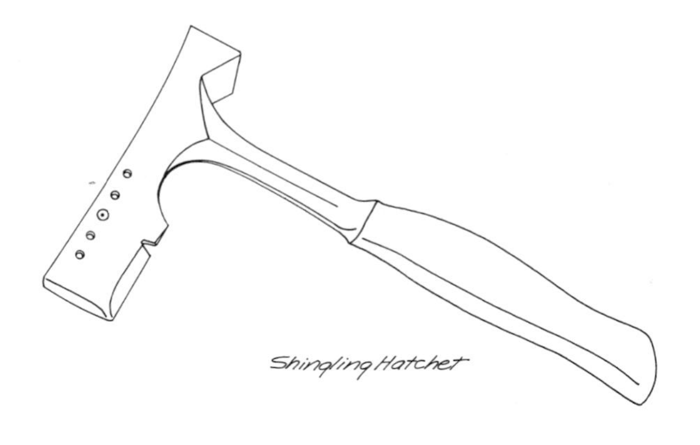

Shingling Hatchet

Steel Stud Driver

This is a special tool that fastens light shelving, furring strips, conduits, and so forth, to concrete, brick, or cinder blocks. The fastener itself is like a short, thick nail. It is put in place in the driver and the driver is held against the object or piece you are putting in place. A blow to the driver with a heavy hammer (2 pounds or more) will drive the fastener into place. This tool and its fasteners are limited to light materials; for heavier work, you will have to go to a masonry anchor or shield.

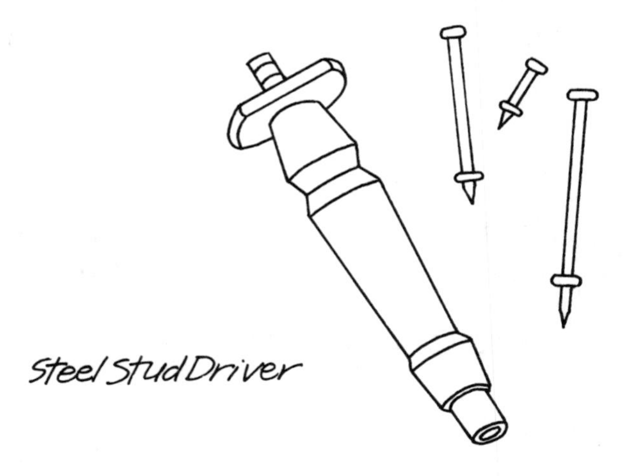

Steel Stud Driver

C-clamp

This is a simple clamp made of metal that joins and holds pieces and parts together temporarily. A C-clamp is helpful if you are gluing a joint or want to hold a straightedge in place while marking or making a cut. C-clamps come in a variety of sizes and are quite inexpensive—a couple of pairs in different sizes will come in handy for general carpentry work.

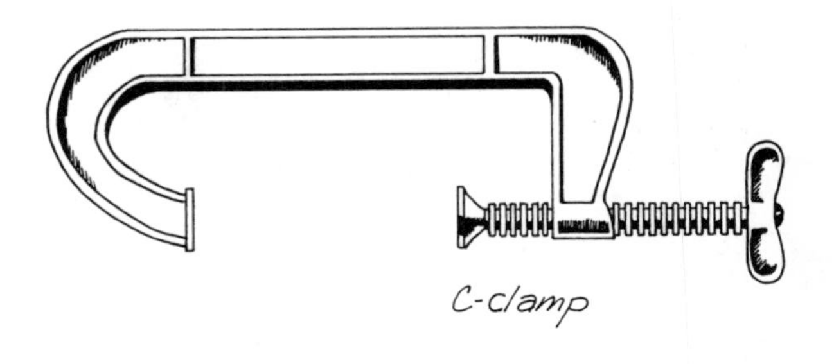

C-clamp

Brace and Bit and Eggbeater-style Drill

For boring holes in wood without a power drill, you can use a brace and bit or an eggbeater-style hand drill. The brace holds an auger bit and can make holes up to 2" in diameter. The eggbeater drill holds a twist bit and makes smaller holes for screws, nails, etc. Both of these tools function in the same way: the bit is held and guided by the drill and is fed into the wood by a turning mechanism on the drill and gentle hand pressure.

To put a bit in either drill, hold the drill firmly with one hand and use your second hand to twist the chuck (the part of the drill that actually holds the bit) until the hole in the center is large enough to

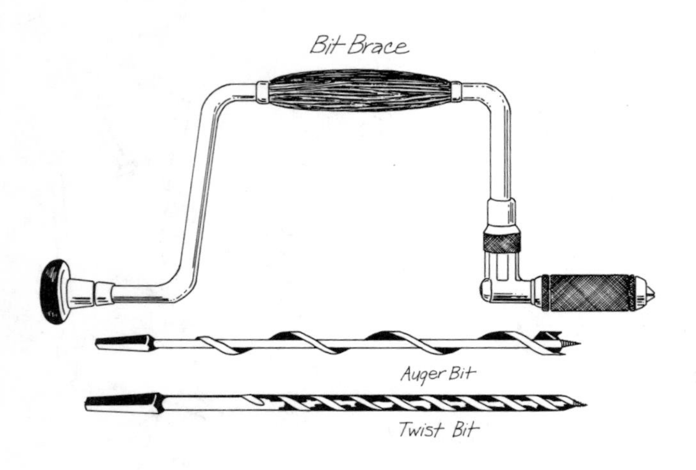

receive the shank of the bit. Usually twisting the chuck to the left will open it, to the right will close it. Set the bit in place and turn the chuck until it clamps the bit firmly. Either the brace and bit or the eggbeater drill should be held so that the bit is exactly perpendicular to the wood or surface being drilled. If you have problems "seeing" the right angle formed by the bit and the wood or surface, check yourself with a combination square or a try square. As you become more familiar with using either tool, you'll develop an eye for the angle or position you need to hold the tool in to drill a straight hole.

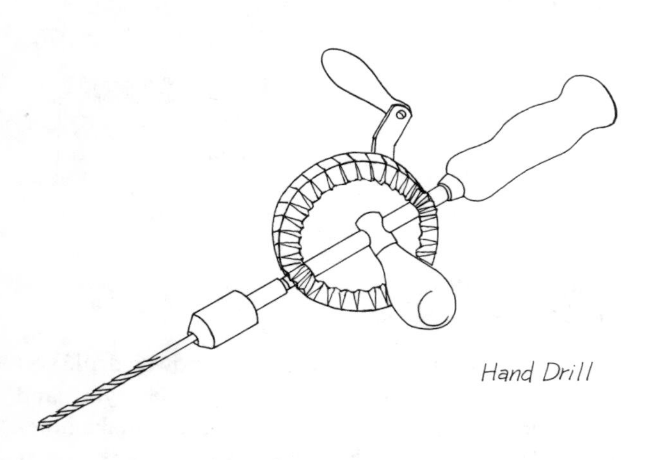

The *brace and bit* can be used in two different ways: it can be held by the head (the very end of the brace opposite the bit; it has a wooden or plastic cap, usually rounded), and the handle (between the bit and the head) can be rotated; or it can be held by the handle and the head rotated. Either motion will feed the bit into the wood. To drill a vertical hole, you usually hold the brace with your right hand firmly gripping the handle and turn the head with your left hand. Bear down just hard enough to feed the bit in; the tool should do most of the work. To drill a horizontal hole, you usually hold the brace by seating the head against your body (hold it cupped in your left hand, set against your stomach or chest) and turning the handle with your right hand. Again, use gentle pressure to feed the bit into the wood. Some braces have a ratchet mechanism that may be used when you are drilling in tight quarters and don't have enough room to fully rotate head or handle. There are special *corner braces* designed for work in awkward places, too.

The auger bits used in the brace range from ¼″ to 2″ in diameter and are usually graduated in sixteenths of an inch. The size of the bit is stamped on the shank and will appear as a "5" or an "11," indicating that the bit is 5⁄16″ or 11⁄16″ in diameter. These bits are fairly expensive (from $4 to $7 for the smaller diameters and up to $28 for the larger ones!), so they should be carefully used and kept in a protective box or case. Avoid drilling into wood that might have concealed nails! Or, if you have to, consider buying a couple of cheaper bits for this work. A brace will also hold an *expansive bit,* which takes the place of several sizes of large bits. The expansive bit can be set at different diameters by turning the setscrew (in the head of the bit) and then moving the cutter (a gauge shows you the positions for different diameters) and retightening the screw. An expansive bit will allow you to drill holes up to 3″ in diameter.

The *eggbeater drill* works much like the brace and bit, though making smaller holes. To drill a vertical hole, hold the drill by its top handle and turn the crank with your free hand. To drill a horizontal hole, seat the top handle against your body, hold it steady by the side handle (close to the wheel), and use your other hand to rotate the crank. As with the brace and bit, you need to exert only a gentle, steady pressure to feed the bit into the wood.

The *breast drill* is a larger version of the eggbeater-style drill. Its top handle has a plate that can be seated against the body for extra stability or pressure on the drill. This drill will take larger bits but is a cumbersome tool and not as popular as the brace and bit.

Bits for the small eggbeater-style drill are graduated in thirty-seconds of an inch and are commonly sold in sets of several sizes. The smaller bits break fairly easily. Most twist bits drill holes in light metals as well as in wood.

To make a smooth hole through a piece of wood with either the brace and bit or the eggbeater drill, clamp a piece of scrap wood to the opposite surface of the wood you are drilling. When the bit breaks through the opposite side, it will be less likely to splinter the wood. Another way to drill a clean hole is to drill through just enough so that the tip of the bit is protruding through the opposite side, and then to remove your drill and finish the hole by drilling from the opposite direction.

Drill bits may quickly become bent, dull, and/or chipped if they aren't stored carefully in a box or a cloth (or leather) roll. Carry them with your more delicate tools. Similarly, your drill should be stored and carried carefully so that the turning mechanism isn't damaged.

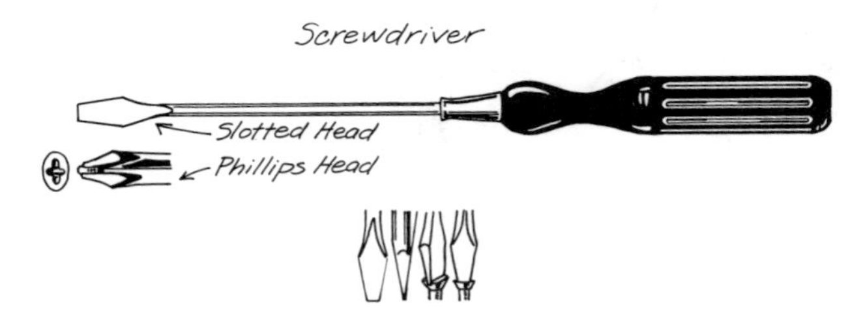

Screwdrivers

The simple screwdriver is said to be one of the most dangerous of all hand tools. This is a reputation you really can't take all that seriously until you're working late one afternoon and have that very last hinge to put on and the screw won't turn down into its hole. . . . Having already packed away your drill, you're *determined* to finish and bear down extra hard on your screwdriver—which slips as you give that last twist and bears down into your hand! A few gouges like that will build your respect for what you couldn't *believe* could be a "dangerous" tool. . . . But, its bad reputation aside, the screwdriver is an essential aid for hanging doors, building gates, installing locks, and more. There are some general habits you'll need to develop to use the tool well.

Unless you're working with extremely soft wood, it's always a good idea to predrill for a woodscrew or at least to use your awl (or even a nail) to start the hole. If you try to force a screw into wood that just won't give, the head of the screw will be damaged (or "stripped"). At the least, this damage will make your work look shoddy; at the worst, the screw might be fixed half in place and all but impossible to remove (*or* finish).

The second most important habit to develop is taking the time to choose a screwdriver that fits the screw you're working with. Screws and screwdrivers come in two basic types: the *standard screw* has a slotted head that fits a standard flat screwdriver; the *Phillips screw* has a +-shaped slot that requires a Phillips screwdriver with matching head. Beyond this basic style difference, you will need a screwdriver that is just big enough to slip into the slot on the screw head. If you use a screwdriver that is too small, it can twist and slip, damaging the head of the screw and making it very difficult to turn at all. I've demolished countless screwdrivers learning this principle, for the tool is damaged by this abuse as well as the screws! If you use a screwdriver that is the right thickness but is much wider than the screw, both screw and screwdriver will survive intact, but the wood surrounding the screw will be damaged.

Building up your collection of screwdrivers may take some time, but eventually you'll find yourself with exactly the right tool for each situation. In general, a long-handled screwdriver will give you more turning power. Short, stubby screwdrivers are best for working in some difficult, tight areas. You'll probably want an assortment of sizes and shapes. Most screwdrivers for carpenters are made with tough, transparent plastic handles and steel blades. I have a few with rubber-coated handles—these are a little easier on the hands and nice to have if you're using them over and over and over in a single day. Ratchet-type or "Yankee" screwdrivers work on a push-pull rhythm and come in regular and offset styles for special situations that you can't reach with your conventional screwdriver. There are also right-angle screwdrivers for working in certain situations.

The rule of thumb for drilling holes to receive screws is to drill them half the length of the threaded part of the screw in soft wood and two thirds the length of the threaded part in hard wood.

Wrenches, Pliers, and Socket Sets

Every toolbox probably has a totally different assortment of these tools, as they are so much a matter of individual need and preference. It helps to have a variety of sizes and shapes and "specialty" wrenches and pliers. The *Locking Plier-Wrench* or vise-grip is a versatile tool that can function as pliers, clamp, portable vise, and gripper. It has serrated jaws that will grip tightly and a locking device/release lever that allows you to clamp the tool in place and leave it if you need to. An *adjustable-end wrench* is another general-purpose tool that's handy to have; the jaws can be adjusted to fit exactly to the nut or bolt (or other fastener) you are working with. *Electrician's long-nosed pliers* are sometimes handy for getting into spaces

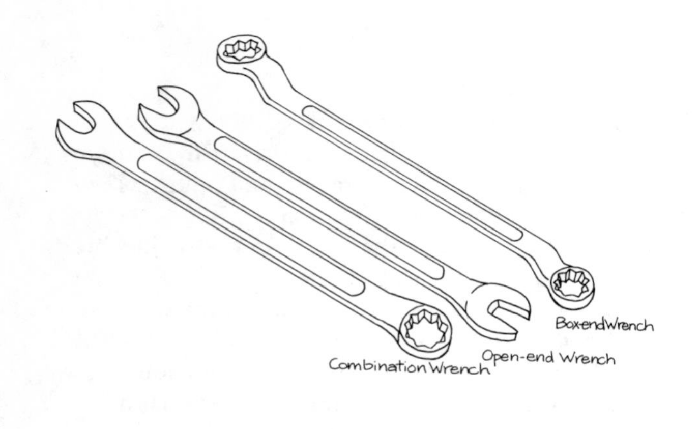

where other conventional pliers and wrenches can't reach. A *socket set* will also help you to reach those more difficult connections and is almost a necessity if you are doing a lot of work with bolts or lag screws.

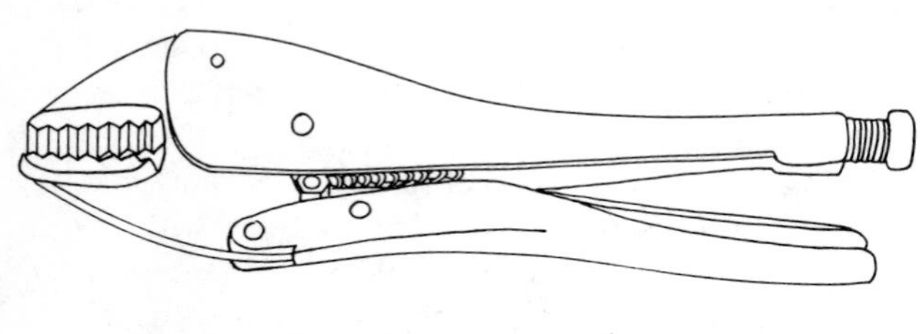

Locking Plier-Wrench

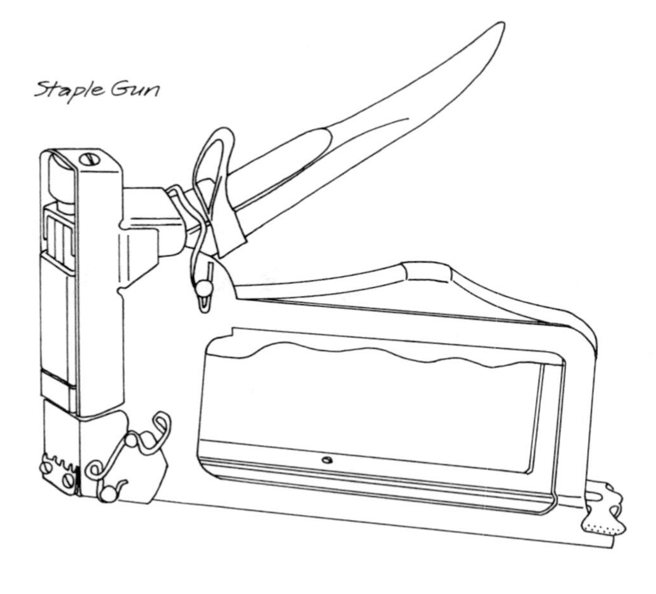

Staple Gun

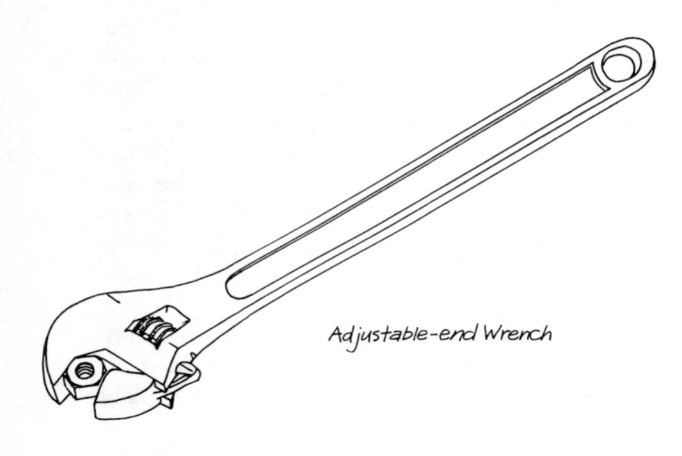

Adjustable-end Wrench

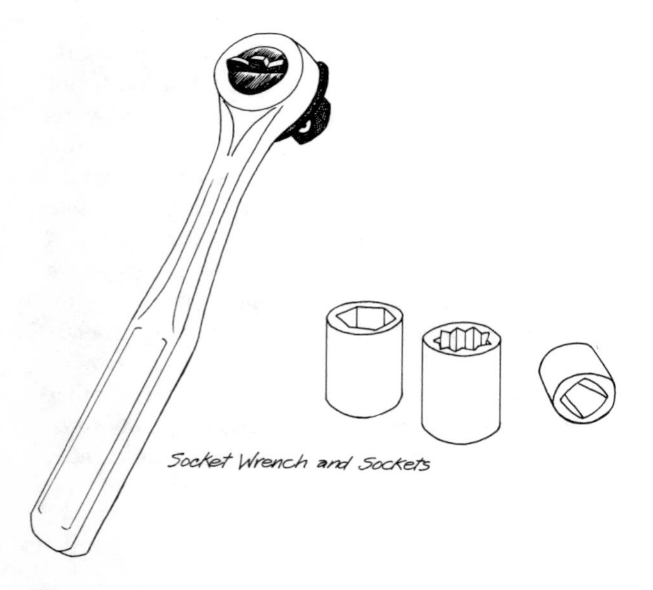

Socket Wrench and Sockets

Staple Gun

For installing insulation, putting up ceiling tiles, securing felt paper, and multitudes of other work, a staple gun is an indispensable tool. There are two styles of manual staple guns: the *common gun tacker* and the *hammer tacker*. The first of these is worked by holding the tacker firmly against the surface or material to be fastened and then compressing the handle. The staple is thus driven into the material. The hammer tacker is used as its name implies: by driving it against the surface as you would drive your hammer. I've always chosen the gun-type tacker because it seems easier to handle and offers more control; the hammer tacker might be best when you are fastening something above your head. Electric tackers are also available.

If you are going to be using your staple gun a lot, choose a good heavy-duty model that will hold up to the work. My choice is the Duo-fast gun tacker, which is beautifully made and functions smoothly and consistently. Other staple guns I've worked with tend to jam up a lot or are much harder to compress, which is tough on your hand. The Duo-fast also has an unconditional guarantee—the company will repair the gun free if it breaks under normal use. Make sure that the staple gun you choose is a fairly common make so that you can easily find staples for it. I was once given what seemed to be a good staple gun, but the make was so obscure that I couldn't find staples to fit it! Another point to consider in choosing your tool is the size of the staples it will handle—staples vary in width and in length. The better guns will take a variety of lengths from ¼″ to 9⁄16″ so that you can choose the right size staple for each job. The ⅜″ staple is a good all-purpose size. One final point to check is whether the gun has a locking device so that it won't accidentally discharge when not in use.

DEMOLITION TOOLS

Cat's-paw

This chunky little wrecking tool is used whenever a nailhead has been driven deep and can't be pulled with a hammer or regular crowbar. The cat's-paw itself can be driven into the wood and thus catch under the head of the nail and pry it up and out. This tool damages the wood and is strictly for rough framing or demolition work.

Wonder Bar

Most carpenters will readily list the Wonder Bar as a "basic" tool. It is a small pry bar with one curved end and one straight end; the entire tool is flattened. The thin body of this tool gives it wonderful advantages: it can be driven under edges, between joints, under nailheads, and so forth, with very little damage to the materials you are working on. Some of the uses of the Wonder Bar are to pry up shingles or roll roofing, to remove siding, to loosen window trim, and to dismantle temporary bracing. There is a tiny version of this tool (6″ or 7″ long) that works well in removing delicate trim, window beading, and so on. "Wonder Bar" is actually a brand name for one version of this tool, so look for it under other names, too; Estwing calls its a Handy Bar; Sears sells a Craftsman Pull-n-Pry Bar.

Wonder Bar and Cat's-paw

Nail Nippers

This rugged, simple tool resembles a pair of pliers with cutting edges rather than gripping edges. It can clip nails off or pull out nails that have lost their heads. It will also pull out nails that are impossible to remove with a hammer. The rounded surfaces of the nippers allow you to work nails out with minimum damage to the surrounding wood. For demolition work and general carpentry, this inexpensive, versatile, and hardy tool has no peer.

For pulling out difficult nails, few tools work as well as the nail nippers. (Photo by Dian)

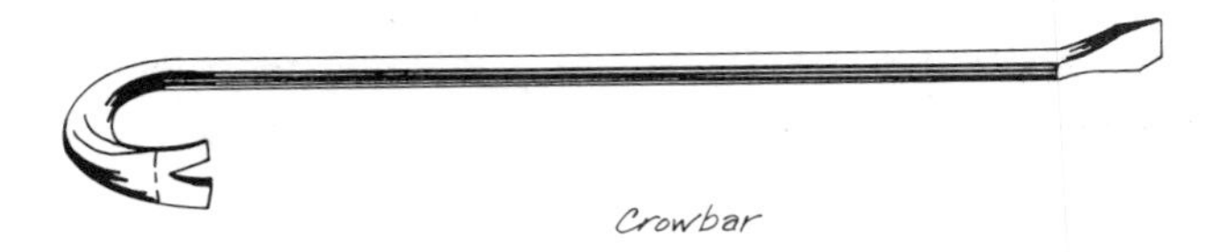

Crowbar

Crowbars and Ripping Bars

Crowbars, or wrecking bars, come in a variety of sizes and styles and are used for prying things apart. The length of the bar will give you more or less leverage. For general work, a 2′ bar and another about 3′ will handle most heavier jobs; you'll also want a smaller (12″ to 15″) pry bar or a Wonder Bar. A *ripping bar* (or rip bar) is similar to a crowbar but has a slimmer,

For chipping concrete, stucco, or other masonry products, the masonry hammer *should be used. Always wear protective glasses when using this tool!* (Photo by Carol Newhouse)

straightened end that enables you to drive it into joints more easily and with less damage to the material.

In choosing a crowbar, look for one with a fairly clean and tapered nail-pulling end, which will make it easier for you when you want to drive the end under a nailhead to pull it. Wrecking tools will damage your other tools all too easily, so it's a good idea to provide them with a toolbox or carrying case of their own.

Sledgehammer

Sledgehammers range in weight from 3 to 10 or more pounds, but for most carpentry work a 3- or 4-pound sledge is sufficient. A short-handled sledge will be easier to maneuver than one with a long handle, even though you lose a certain amount of power when you sacrifice on handle length. For general use, a 15″ handle is fine.

Sledges are useful for, among other things, moving walls into (or out of) position, adjusting posts to plumb, and driving form stakes. A sledge with a wooden handle is the very best for your hands—this handle has maximum shock absorbency. The disadvantage of the wood-handled sledge is that the head can work loose from its handle. The wood handle will also splinter if you use the sledge really roughly and/or carelessly. I have an Estwing sledge that's all one piece—head and handle one continuous piece of steel. The handle is coated with rubber, which gives a good grip and does a certain amount of shock absorbing. I like the balance and sturdiness of this tool well enough to overlook the slightly inferior absorbing power of the handle.

Mason's Hammer

This tool technically belongs to another trade but has its uses on certain carpentry jobs. It's a heavy hammer somewhat resembling a sledge, but with one tapered end. It can be used to break up concrete or to chip stucco away from exterior walls. Always protect your eyes by wearing safety glasses when you work with this tool!

MISCELLANEOUS TOOLS

Levels

A *spirit level* is a tool that shows whether the surface it is held to is exactly horizontal (level) or vertical (plumb). Smaller levels usually have a single transparent vial housed in a body of metal or wood; larger levels have sets of vials, which allow you multiple readings for extra accuracy over a given distance. A vial may be made of glass or plastic. It encloses a slightly arched tube partly filled with a liquid (usually alcohol), thus creating an air bubble, which is your guide to

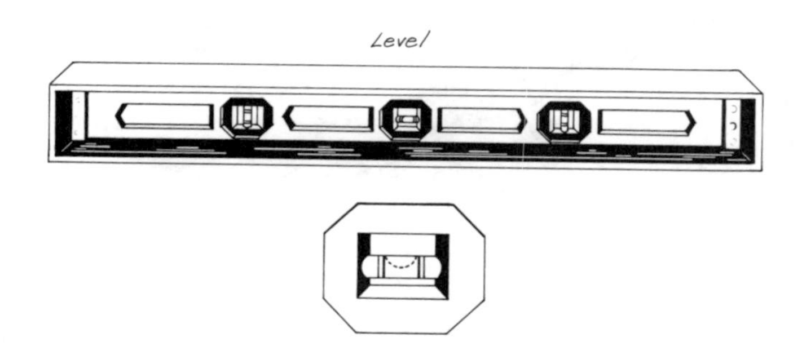

Level

level or plumb. The arched tube has markings to show its exact center, so that when the air bubble is directly between the lines the reading is perfectly level or plumb. Some vials will have two sets of lines on the tubes, and the perfect reading is exactly between the center lines.

Remember in using your level that any slight margin of error will compound itself over distance. Thus if you are using your level to place a long girder or to plumb up a wall, and the air bubble is touching one or the other of the lines, you should adjust your work until the bubble is perfectly centered. If you allow yourself to be even "slightly" imprecise at this point, you may well end up with an out-of-level floor or a leaning wall. If you are doing rougher work, such as setting a fence post, you can be a little more lenient. Another point to notice is that a level can give you an accurate reading only if the surface it is held to is smooth. A tiny nailhead protruding or a gouge in the lumber can throw the level off. A warp or bow or uneven thickness in the lumber can likewise change the reading. For these reasons, it is usually a good idea to check your level in more than one spot on a particular stud or beam or whatever. If you are trying to level a flat surface (such as a pier top), or checking the level of a floor section, place your level in a few spots and in a variety of directions to get the full picture.

Most accurate when used to check a relatively short surface, the torpedo level *is a surprisingly useful little tool.* (Photo by Carol Newhouse)

Levels come in sizes ranging from the tiny level to 4- and 6-footers. They may be made of wood (sometimes bound with metal) or aluminum, magnesium, or other metals. There are two basic "styles," one a solid rectangle or box type, the other more skeletal. I succumbed to aesthetics and bought a beautiful brass and mahogany level that is a delight to the senses. But even when I conquered my reluctance to subject this handsome tool to the inevitable scratches and mars of use on the job, I constantly found myself borrowing my partner's level. Hers is a sturdy metal skeletal-type and feels more comfortable and better balanced in the hand. It will probably outlast my beauty by years, even with both of us using it. . . .

There are three sizes of level particularly handy for general carpentry: the line level, torpedo level, and the two-foot level. Longer levels are nice to have for special work, but they are fairly expensive and harder to carry around safely. You can substitute for a longer level by using your two-foot level "extended" on the straightest (warp-free) board you have.

The little *line level* is suspended on a string stretched between two points and is commonly used in laying out batter boards or in any long-distance work. Because of the inevitable stretch in the string and because of the great distance it is usually asked to gauge, the line level gives you a pretty rough reading. A *torpedo level* (about 9″ long) is modestly priced and versatile. It can be used in leveling piers, setting windows and doors, putting up shelves, checking form walls, and so on. Its small size lets it fit where a larger level can't. The third commonly used level is the *two-foot level.* This, like the torpedo, has an abundance of uses and is just about indispensable for most carpentry work.

In choosing a level, look for the best quality you can afford. A well-made tool will greatly outlast its "bargain" version and is usually easier to work with and more reliable. Most good levels have replaceable vials, and the best vials are made of glass (the plastic ones scratch too easily; glass may break but it can be replaced). If you are looking at a long (4′ or 6′) level, your best bet is aluminum (light) or magnesium (even lighter), easiest to handle in such lengths.

Levels of any size should be handled with some care. Be careful not to leave them where they might be stepped on or have something dropped on them; don't carry them in with your heavy demolition-type tools; don't leave them on ladders or ledges where they might fall. A good-quality level gently cared for will be with you for as long as you care to do carpentry.

Toolboxes

Toolboxes for storing and transporting your growing collection of tools are a must! They come in endless variations, and choice is strictly a matter of individual taste and needs. It's a good idea to have two or three toolboxes so that delicate tools (combination square, plumb bob, levels) won't be battered or bent by rougher, heavier tools (wrecking bars, sledges). Smaller toolboxes (19″×7″×7″, for instance) usually have a removable tray that is an ideal place to keep your line level, combination square, pencils, and so on. Larger toolboxes (mine is 32″×8½″×9½″) may also have removable trays, but in addition have wooden slots for handsaws, a slot in one end for your framing square (it fits inside, with the blade or tongue sticking out through the slot), metal clips for levels, and so on. In choosing a larger toolbox, look for one with good locking devices; a box with three of these rather than two will distribute the weight and strain more. My box has a handle on the top plus a handle at each end—the end handles are nice if you've got a really heavy load of tools and can get someone to grab one end!

Handmade wooden toolboxes are nice to have, too. These can be any style, size, and shape you need. Most of them tend to be too heavy, so try to economize on weight by using ½″ plywood for the sides, with ¾″ for the bottom. Dowels make good handle material. A wooden toolbox, open at the top, is a handy place for extension cords, droplights, some wrecking tools (not too many, or your box will be too heavy to lift!), and other items.

Floor Jacks

Floor jacks are used to raise and support areas while you replace rotted or otherwise damaged floor joists, sections of a foundation, girders, and so forth. Sometimes a floor jack is left in place to substitute for a post, but usually this tool is used temporarily. Floor jacks may also support rafters or ceiling joists while you are putting in a new window or a doorway. You can buy or rent floor jacks to lift virtually any part of a structure.

The Little Giant (brand name—see "Resources") floor jack is an amazingly powerful jack for its size and very inexpensive. It is very versatile because it is designed to be used with a pipe or wood post, and thus can help hold up ceiling joists, etc. It is quite strong and dependable—my partner and I used sets of Little Giants to hold up a three-story house while replacing whole sections of the foundation. Other types of floor jacks are available at well-stocked lumber stores or building supply stores. They range in price from $20 to $40, according to size, quality, and so on. For some work you can use a good-quality hydraulic car jack in place of a conventional floor jack. Always be certain a jack is set on a solid base and that no one will tamper with it once it's in place.

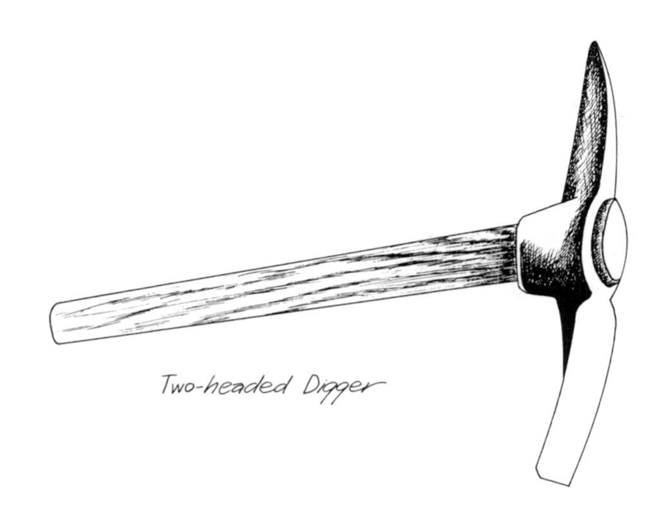
Two-headed Digger

Two-headed Digger

This tool is usually found tucked away in the gardening section of a general hardware store, but it's a worthwhile addition to your carpentry tools. I first discovered this little digger while replacing sections of foundation under an old house; the owner of the house lent me what looked like a miniature pickax, assuring me that it was the best tool in the world for digging foundation footings in awkward places. And indeed it was! Since then I've discovered that it's perfect for digging footing holes for piers, small trenches for water lines, and more. Whenever you need to make a fairly straight-walled hole or trench in cramped quarters, try this tool. It's handy in the garden, too. . . .

Bolt Cutters

These come in various sizes (14″, 17″, 24″, etc.), and are used to cut steel bolts, lag screws, and so forth.

The larger sizes will cut larger-diameter steel. The better cutters usually have rubber handles to absorb some of the shock of the cut and give you a better grip.

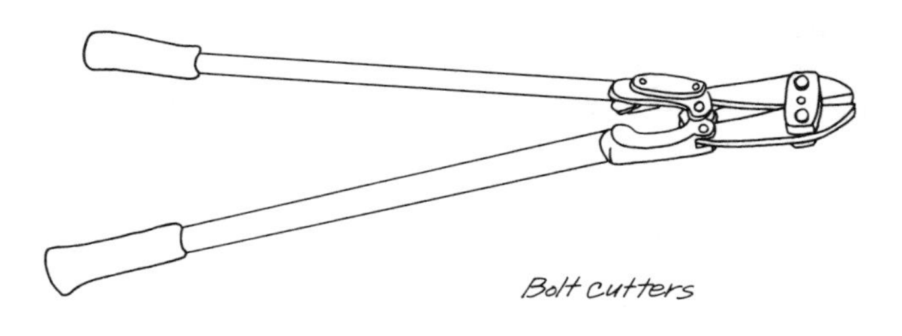
Bolt cutters

Droplights and Spotlights

For extra visibility where you're working, or for light in difficult areas (in the crawl space under the house, or under eaves in an attic), a droplight or spotlight is handy. The *droplight* is a mechanic's tool and has a hook that allows you to hang it from a nail or joist. The *spotlight* has a hood that directs the focus of the light. The best I've seen is sold by Woodcraft (see "Resources") and has a magnet base that will attach to any steel or ferrous metal surface (as well as a clip for attaching it to joists, studs, etc.).

TOOLS FOR BODY PROTECTION

Earmuffs/Hearing Protector

If you are working with certain power tools or in a shop or area where power tools are being used, you should consider protecting your hearing and reducing the noise-related stress of such working conditions. Read the chapter on "Noise and Vibration" in *Work Is Dangerous to Your Health* (see "Resources") for a full explanation of what is happening when you work all day with a wood planer or jointer, or find yourself hard of hearing after a few hours of ripping wood with your circular saw. Briefly, any continuous noise that is louder than about 80 decibels strains the nerves of the inner ear. This strain leads to a fatigue that results in a temporary decrease in your hearing ability and a general stress on your body. A wood jointer produces about 90 to 99 decibels; a wood planer or a wood lathe produces 100 to 109. Wearing ear protection while working with or around these tools is obviously a good idea.

A pair of good-quality rubber earplugs will help, but the best protection is gained when you wear a pair of special noise-reducing earmuffs. These have adjustable head straps and can be worn in a number of different positions. The muffs themselves are rimmed with soft rubber that will mold to the contours of your head. They are not uncomfortable to wear and can make your work much less straining. I bought a pair to wear while chain-sawing and now use them when ripping wood or using the circular saw for long periods of time. You can buy muff-type hearing protectors for high noise levels or for moderate noise levels, depending upon your work. They are not expensive ($8 to $12) and, if properly cared for, should last you a long time and help your hearing to last even longer!

Ear Protector

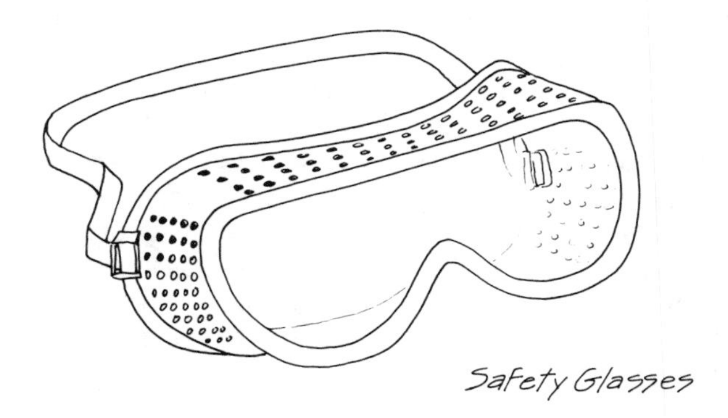
Safety Glasses

Safety Glasses and Goggles

Safety glasses are designed to protect your eyes from particles flying through the air when you're cutting or sanding material, hammering, and so on. These glasses have side shields of wire mesh; the individual lenses are treated to withstand impact. *Safety goggles* are one solid piece; they will fit over conventional glasses or simply give you an unobstructed view through the single wide lens. They are made of lightweight, strong plastic, with indirect ventilation through vents at the top and sides.

Make a habit of wearing safety glasses or goggles whenever you do work that releases a shower of tiny particles or might send a sliver of metal or concrete flying. They are also nice to have for overhead work in dirty or dusty areas: for example, when working under the house on floor joists or girders. If you throw either goggles or glasses loose into your toolbox, the lenses will get badly scratched. Try to be conscious of this and either keep them in a special bag or protective cover, or carry them outside of your toolbox.

Work Boots

A moment of carelessness in placing your circular saw down after a cut and the scare of a graze across the toe of your boot can make you think twice next time you put your saw down! It can also make you appreciate the protection of a sturdy work boot. A lighter shoe or sneaker, and that graze could have meant an ugly cut. A good pair of work boots should rate high on your list of priority "tools."

The sole of your boot is of special importance if you work professionally. Certain black rubber and synthetic soles can leave nasty black marks on flooring, particularly when you have to kneel to work and the toe and edges of your boot scrape on the floor. Crepe soles do not leave these marks. In addition, the crepe sole will have a long life and is a comfortable, well-cushioned base for long work hours.

Work boots come in a wide variety of styles, heights, quality, and prices. The more expensive boots, in my experience, do not necessarily outlast the less expensive ones, but they are usually made of more supple leather and break in much more easily. Choose a boot or shoe style that gives you good protection (a steel toe may be desirable) and a comfortable fit. When your boots begin to wear out, you will want to keep them for concrete pours and other rough work.

Respirators

If you are working with Sheetrock, insulation, and various adhesives, plastics, or glues, you should be considering a respirator as one of your indispensable tools. Dust, fibers, and fumes from certain of these and other materials can harm you in degrees ranging from mild (irritation) to severe (cancer). A simple, relatively inexpensive air-cleaning respirator can protect you from potential harm.

The type of respirator you will need depends, of course, upon the work you are doing. A large building supply store or hardware store usually has an array of respirators suitable for general needs. These respirators should be clearly labeled: i.e., they should state whether they offer protection against large particles, dust, fumes, and so forth. The type of respirator you will be looking at is usually made to fit over the nose and mouth and filters air through one or two disposable cartridges or filters. The very simplest and cheapest protection is a surgical-style mask, but this protects only against large particles.

The most difficult part of choosing the right respirator is finding one that actually fits your face. Manufacturers attempt to make a respirator to fit a multitude of facial types and sizes, but most respirators I've tried have been too large. Air can leak in around the edges, bringing in the particles, fumes, and such that you are trying to filter out. A respirator that fits this poorly is no protection. Fitting the respirator tightly can eliminate most of this leakage, but in an actual work situation you'll find the pressure of that tight fit painful, and you'll be tempted to abandon your respirator and its protection altogether. Take the time to search for one that fits you. Usually a generous edging of very pliable rubber or soft plastic is what's needed. I went through three or four styles before discovering one that fit me snugly and fairly comfortably. If you wear

When operating power tools, wear protective devices such as a good respirator *and* hearing protectors. (Photo by Carol Newhouse)

eyeglasses and/or will be wearing safety glasses while you work, make sure in advance that your respirator fits compatibly with these. It's a good idea to buy a supply of disposable filters at the same time that you buy your respirator so you won't find yourself in the middle of a job with a dirty filter and no replacement.

A respirator should *not* be kept loose in your toolbox where it can be crushed (making the original snug fit impossible to reestablish) or the filter contaminated. A special protective carrying case or box should be used. My partner and I found that a metal lunch box is a perfect solution for storing and carrying respirators and filters.

While working, you should check your respirator frequently to make certain that there are no particles *inside.* These would indicate that the fit is wrong, the filter improperly fitted, or the filter overused and possibly clogged. Keep your respirator clean by wiping it after use with a soft cloth and storing it in a safe place. As needed, give it a thorough washing in warm water and soap, then rinse and air-dry. Replace filters often. The rubber edges of a respirator are often the first areas to show wear, so check for signs of cracking and deteriorating. I've read that massaging the rubber parts will keep them pliable and extend the work life of the respirator. Also, protecting the respirator from exposure to high temperatures will make it last longer.

Working with a respirator on is *not* comfortable. It is harder to breathe, especially if you are doing heavy work like demolition. It is all but impossible to talk through a respirator, which means you can't shout a warning to your partner, or even exchange pleasantries as you work. The rubber makes you sweaty, and even the best fit gives you pressure pains after a while. The only justification for all of this is the *absolutely necessary protection!* Wear a respirator when you rough-sand any plastic-type coating (for example, between coatings of Varathene), when you work with Sheetrock (especially sanding joints, making cuts, and sweeping up afterward), or anytime you handle insulation (either installing, removing, or transporting). Protect yourself from dangerous vapors when working with adhesives and glues. Keep fine sawdust and particles of all sorts out of your respiratory system by the simple precaution of *using* your respirator. Discipline yourself to use this tool; you'll be safer and healthier for it.

HAND-HELD POWER TOOLS

For fast, efficient work, hand-held power tools are the step between hand tools and shop machinery or table-base tools. These portable tools save time and labor, make repetitive drilling or ripping of wood easy, and are inexpensive enough to be affordable even if you don't plan to use them "professionally." They are designed for safe use by anyone with basic knowledge of their workings and can be used wherever you can plug into a regular 110- or 115-volt power outlet. The power tools I've chosen to include in this section are those I've found most useful for general carpentry work. For information on using other portable power tools, such as the router or the disk or belt sander, see *How to Work with Tools and Wood* (listed in "Resources").

In buying any portable power tool, try to choose a good-quality tool that will work well and last. Commercial-duty tools are on the whole sturdier, better balanced, and made to *work.* If you're going to use a tool only occasionally, you can get by with a cheaper version, but if you expect to be doing a lot of carpentry (either on your own home or professionally), you'll probably be happier with a commercial-duty tool.

Within any brand name line of tools, you'll find some with excellent features and some with problems. It's a good idea to ask other opinions before choosing a tool because it has a well-known brand name. Within each tool line, too, are better models of each tool. Working with different models is very helpful if you have the opportunity: you may discover that one tool really outperforms the other and is only a few dollars more. You'll definitely also find tools that feel better to work with, are quieter, better balanced, and so on.

If you choose a good tool to begin with, and care for it, it should last you for many years. Keeping blades or bits *sharp* is especially important. Trying to make a tool work with a dull bit or a damaged, dull blade can put a strain on the motor and even make the tool unsafe. Develop a sensitivity to the performance sounds of each tool. A whine or "drag" in the motor is the clue that something isn't working properly. Don't try to push the tool if it's giving you this message. Each tool has certain maintenance needs, so follow the manufacturer's instructions for lubricating, etc. Avoid carrying any portable power tool by its cord, which can damage the connections inside. Store and carry your tools carefully so that the arms that hold blades don't get bent. Keep them dry to avoid rusting and corrosion. If you lend your tool to a friend, make sure that she understands the basic care and workings of the tool. This may sound paranoid, but I know of at least one electric drill that was almost burned out because someone tried to drill into a piece of wood with the "reverse" switch on—a simple enough mistake if you aren't familiar with the tool and its various controls.

One final hint in choosing a portable power tool: look for tools that are "all-insulated" or "double-insulated." They are designed so that if the insulation inside the tool fails, none of the exposed metal parts of the tool will become electrically live. Most tools come with a three-wire cord and plug, which includes a ground wire, but many electrical outlets are not properly grounded. Even if you take the time to use a proper adapter and even to ground the "pigtail" (third wire) on adapter or plug, the outlet may not be grounded correctly. Using a double-insulated tool is an extra safety step.

Any portable power tool is potentially dangerous and should always be handled with respectful attention. One especially important habit to develop is always to *unplug* a tool before doing things like changing blades or adjusting a base to angle-cut. This is a hard discipline to keep to because it always seems that it will just take a minute to make this or that change, but you are setting yourself up for a nasty accident every time you take that chance! If you're working with a partner or on a crew, always think twice when approaching someone who is operating a power tool—don't walk up behind a friend who has her circular saw going and put your hand on her shoulder!

Another habit to develop when working with power tools is to wear safety glasses or goggles to protect your eyes and a hearing protector for your ears. Every time you run a circular saw through a piece of wood, you are creating noise that is potentially harmful. If you're working a lot with power tools day after day, you can significantly damage your hearing. Safety glasses are *absolutely necessary* if you're ripping wood, drilling overhead, and so on.

And for one final caution: don't operate a power tool if you're standing in a wet spot! You are risking getting a bad shock, which might be harmful in itself—or might cause you to drop your tool or otherwise lose control of it. Wear rubber-sole boots if the ground is damp, but if it's really wet either move indoors or use a hand tool.

Precautions aside, portable power tools are a wonderful aid for all sorts of carpentry work. They *are* rather intimidating, so don't feel strange if you are a little hesitant to turn on your new circular saw or try out a friend's reciprocating saw. It takes a while to begin feeling at ease with a noisy, powerful, mechanical tool. Practice enough with any tool until you pass the "uneasy" stage. When you're sure of how the tool functions and are comfortable with your ability to handle it, you'll find that there's a great deal of pleasure in the efficiency and versatility a power tool offers.

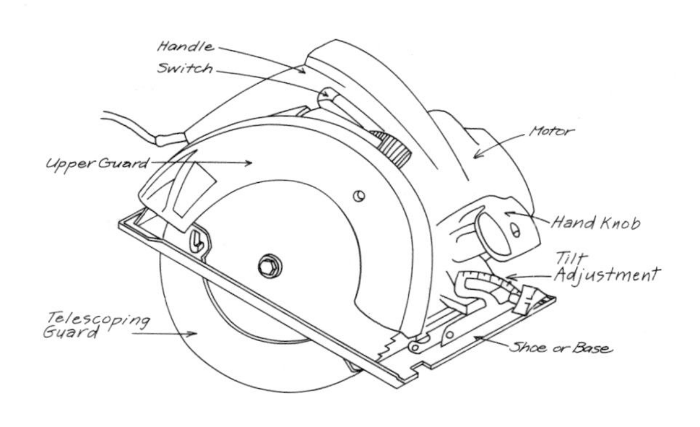

Circular Saw

The portable circular saw is probably one of the first power tools you'll learn to use because it is so efficient and fast for cutting and ripping wood, can duplicate angle cuts exactly and simply, and is relatively easy and safe to handle. Circular saws come in many models, which vary in power, in weight, and in the size of material they can cut through. These saws are sized according to the diameter of the blades they will accept; for example, a saw is spoken of as a 7¼″ model or a 6½″ model. For most general carpentry, the 7¼″ saw is a good choice. It can cut 2× (1½″) material either straight through or when set up to a 45° angle. Models range in size and may weigh from 9 to 13 or more pounds. Before you choose a saw, try to work with several different models to see how they feel. I like the weight of a heavier saw. Even though it is a little harder to hold and operate, it seems to ride more smoothly and is less prone to jump out of a cut. The heavier saws usually have more horsepower (2½ hp, compared with the 1½ or 1¾ hp of lighter models) and will hold up under sustained use. Always look for a double-insulated tool with basic accessories such as rip fences available. In general, the models advertised as "heavy-duty builder's" saws are the best constructed and will function smoothly and wear well.

Each model of circular saw is slightly different in terms of adjustment nuts, handle, and trigger design, and so on. All saws have a shoe or base, which is the plate the saw rests on when cutting. On most models, this shoe may be adjusted so that the saw can cut at different angles (up to a 45° angle). This adjustment mechanism (labeled "Tilt Adjustment" on the illustration) is usually toward the front of the saw. Most models have a telescoping safety guard controlled by a tension spring: when the saw is cutting, this guard is forced up in a position paralleling the body of the saw; when the cut is finished, the spring sends the guard down to cover the lower part of the blade. This allows you to set down a saw safely even though the blade is still spinning. If you have to feed the saw into a partially done cut, a lever on the safety guard allows you to lift the guard manually. All saws have a fixed guard over the upper part of the blade, and most saws have a depth adjustment that allows you to set the saw with more or less of the blade extending below the shoe for deeper or more shallow cuts. The depth adjustment actually moves the position of the shoe relative to the body of the saw. If, for instance, you are cutting very thin sheets of plywood, you can adjust the depth of cut so that the saw blade just clears the material. This feature also allows you to cut grooves, dadoes, etc. The depth adjustment mechanism is usually located behind the blade (toward the back of the saw). A rip guide or rip fence is available for some saw models. This attachment allows you to rip pieces of wood to uniform widths accurately and quickly. A protractor gauge for cutting off angles to the right or left (adjustable up to 75°) is also optional for some models. Some saws can be fitted with special abrasive wheels that may be used to cut plastic, nonferrous metals, or steel and iron.

A rip guide *enables you to remove a set amount of material with your circular saw.* (Photo by Carol Newhouse)

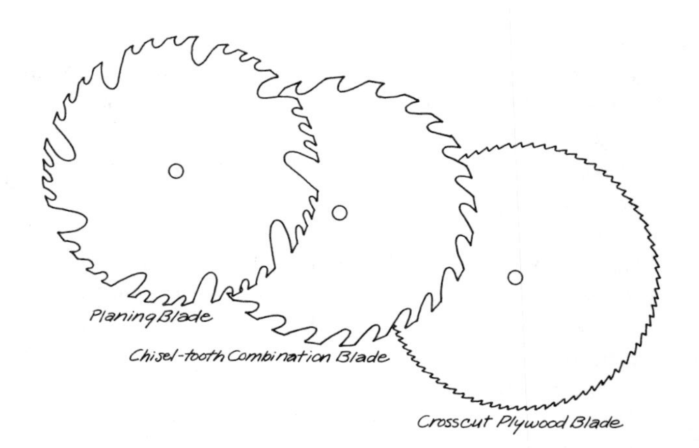

Wood-cutting blades for circular saws come in a variety of types with general or specific uses. A general ripping and crosscutting blade (or chisel-tooth combination blade) is adequate for most work; it is generally called just a combination blade. If you're doing strictly ripping, a rip blade with large, widely spaced teeth is best. If you're doing crosscutting exclusively, you can use a fine-tooth crosscut blade. Plywood cutting blades also have very fine teeth. A plywood blade can be used to cut other stock, but it does so very slowly and puts an unnecessary strain on the motor of the saw. A regular crosscut or rip blade, likewise, *can* cut plywood, but will make a very rough, splintery cut. A planing blade should be used for very fine work—when you want an extra-smooth, finished cut. There is a special blade for cutting fiberglass and light-gauge metals; it has slots rather than teeth. Carbide-tipped blades will stay sharp much longer than ordinary blades, but if you hit a nail with one, the tip might break off one tooth (and be shot like a bullet from the spinning blade). Once this happens, the blade is ruined for smooth, fast cutting. If you're working strictly with new lumber, carbide-tipped blades make sense.

All regular blades must be kept sharp; how often you sharpen them depends upon the use they're getting. If your saw starts to cut slowly, whines a lot, or burns the wood, chances are the blade is dull. Having a set of several blades is a good idea: you can put on a sharp blade and go on with your work. Blades can be sharpened at most saw shops, or you can buy jig and files and do your own. To give your blades an extra-long sharp period, keep them wrapped in newspaper or soft material when they aren't in use. If you put them unprotected into a toolbox, contact with metal tools such as your hammers or wrecking bars can dull or even chip teeth on the blades.

Always use your circular saw with absolute attention. Although it is well and carefully designed for safety, it is a very dangerous tool. A few saws have a power brake with which you can stop the spinning blade instantly, but most don't. The blade continues to spin with a lot of power even after you've stopped depressing the trigger. If you're feeling especially tired or preoccupied, don't use this tool!

Before beginning any cut, make sure the stock is securely resting on a table or sawhorses, with the end to be cut off extending well beyond the support. If you're cutting plywood or other large sheets of material, you can set your depth adjustment and cut the sheet by placing a couple of scrap boards under it but leaving it on a stack of material or even (if you're careful) on the ground. Always position the wood or stock you are cutting so that the body of the saw will ride on the portion of wood that stays on the supports—the portion that falls free will not then take the saw with it! Set your blade so that it clears the stock by about 1/8". Make sure that you *unplug* the saw whenever making any adjustments or changing blades. Wear safety glasses and a hearing protector for any sustained cutting.

To begin a cut, place the saw with the edge of the blade about an inch from the beginning of the line you want to cut. You can use the notch in the forward edge of the shoe as a guide in feeding your saw into the cut, or you may find it easier to do this by sighting more or less freehand. Remember to compensate for the thickness of your blade: position the saw so that you go right along the edge of your line and leave the full width or length of stock you want. Depress the trigger to start the saw and feed the blade into the mark. If your saw has a knob at the front, use one hand on it to guide the saw along. Your other hand will hold the handle and operate the trigger/motor switch. If your saw doesn't have that extra knob, you can guide the saw by holding the angle adjustment wing nut. Keep your second hand away from that spinning blade at all times! Never reach under the stock you're cutting. Make sure that the electrical cord doesn't get pulled under the saw, either. If you find yourself veering off your line, don't try to force the saw to go back in place; this will either make a messy cut or cause the saw to jump in the cut. Instead, back the saw up an inch or two and feed it back in the right place. Be prepared as you near the end of a cut to take the full weight of the saw (up until now, the stock should be supporting the saw). This is especially important when cutting 2×4's, which are cut through in seconds. Release the trigger as soon as the cut is finished. When the lower guard has swung into place, you can set the saw down. If you put the saw down too quickly and the guard hasn't had a chance to go into place, you can gouge a floor or deck or feed the blade into dirt and dull it. Watch out for your electrical cord or extension cords, too!

To set your blade for an angle cut, first ensure the saw is not plugged in; then loosen the adjustment nut that controls this setting. Most saws have a gauge right on the housing for setting angles: you can move the shoe to the angle you want and then tighten down the adjustment nut. Or you may want to use your sliding T bevel to duplicate an angle you need. Loosen the adjustment nut and set the bevel against the shoe of the saw. Move the blade into position against the bevel and tighten down the adjustment nut. Proceed with your cut as described above, making sure that you are placing the saw on the correct side of the stock for the angle you want.

A rip guide can be used when you want to remove a uniform amount of material from a board or panel. For example, you might want to rip 1" from a board to fit it in place. The rip guide can be set either to the left or to the right of the saw body. It slides into a pair of slots, one of which has a screw that can be loosened to receive the rip guide bar and then tightened down to hold it in place. Set the bar loosely in place, and then line up the blade of the saw exactly as you want it. The fence of the rip guide will travel along the edge of the wood; if the edge is uniformly straight (in most milled lumber it is), your cut will be uniform, too. To operate the saw with the rip guide in place, hold it with a slight tension pulling the guide tight to the edge of the board. Feed into the cut as usual. Make sure that you keep the guide in place against the edge as you cut.

To change blades on your saw, first unplug the saw. Use the blade wrench to loosen the nut at the center of the blade (the spindle that projects from the body of the saw). Remove the nut and hold the lower guard up, allowing you to free the blade. In placing your new blade, make sure that you put it on the spindle with the teeth of the blade facing in the right direction. Most saws have a little arrow engraved on the upper or lower guard to show you how the teeth should be pointing. This is very important! Otherwise, the saw can jump wildly when you turn it on and feed it into the stock. With the blade in place, replace the nut and tighten it securely. Plug the saw back in and go to work.

If you keep the blades sharp, follow the manufacturer's instructions for lubricating your saw, and in general treat your saw with respect, it can work years without needing any serious overhauling. Try to correct any problems before they cause you to overwork or abuse the saw. Once you've become familiar with the working of the saw, you'll be able to tell when anything is wrong—and take steps to correct it before

With your saw set at an angle or bevel, *you can make repeated, accurate bevel cuts.* (Photo by Carol Newhouse)

it does damage. The circular saw is a durable machine that should cause you very little worry and produce a lot of hard work.

Portable Electric Drill

The portable electric drill is simple to use and quite versatile; it will probably be one of the first power tools you'll want to buy. Besides its basic function of drilling holes in wood, metal, plastics, and other materials, the portable drill can be fitted with attachments that let you use it to sand, buff, grind, wire-brush, and more. The drill consists of a body, which houses the motor; a handle with trigger switch; the chuck, which holds a bit or attachment; and a cord to connect it to the power outlet.

The size of the drill is spoken of in terms of the chuck: the largest-diameter shank the chuck will accept is the "size" of the drill. Common sizes are ¼", ⅜", and ½". For most general work, the ¼" or ⅜" size drill is used. The ½" drill usually turns more slowly but has more power; it is used for repetitive drilling of large timbers. The actual power capacity of any drill is measured in horsepower and in rpm (revolutions per minute). A good general-purpose ¼" or ⅜" drill ranges from ⅛ to ¼ hp and about 1,200 to 1,700 rpm; the ½" drill should be up to ⅞ hp and about 500 rpm.

One feature to look for in a drill, regardless of size and hp, is a variable-speed motor. This means that the drill can be operated at a slower speed for really hard work and at the faster speeds for light work. Another feature to look for is a reverse drive, which is useful when you're drilling deep holes and have to pull the bit out repeatedly to help clean sawdust from the hole. A good-quality chuck is important, too. A chuck made with good metal will last a long time, loosen and tighten down quickly and easily, and hold your bit or attachment securely. Rigid and Jacobs are two companies that make very good chucks used on some of the best drills.

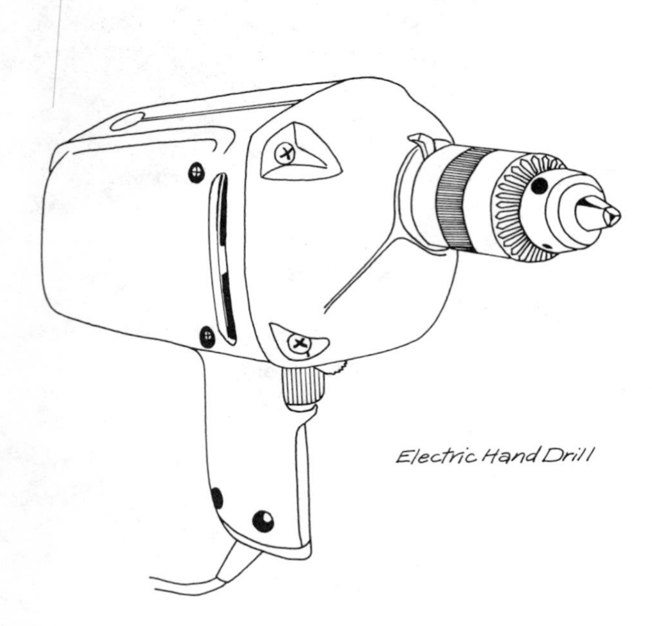

Electric Hand Drill

One final feature to consider is the housing of the drill, which may be metal, plastic, or a combination of the two. In general, the metal housing is more durable and can't be cracked or damaged. More and more drills are being manufactured with plastic housing, though, or a partially plastic housing. Be sure that the drill you choose is double-insulated, which means that if the insulation around the electrical parts inside the housing fails, the housing itself will not become live. The actual design of the drill is a matter of personal choice—try for whatever feels good in your hand and seems to be fairly compact.

There are various types or styles of bits available for electric drills. The common *twist bit* comes in sizes ranging from $\frac{1}{16}$" to ¼" and can cut into wood or metal. The twist bit is used, among other things, for making holes for screws and for pilot holes for nailing hardwoods. *Spade or flat bits* are used for drilling larger-diameter holes and come in sizes up to 1". They cut a very rough hole, tending to splinter the wood a lot. If you use these bits, keep them sharp. They will burn out your drill motor if allowed to get dull. *Power bore bits* range from ⅜" to 1" in diameter. They have a sharp tip that helps to center and begin the hole and a "spur" that cuts the wood fibers. They make a clean hole with little binding and consequently with little danger of damaging the drill by overheating. *Auger bits* come in larger diameters, ranging from 1⅛" to 1½", and have a twisted shank with a cutting spur. If you can't find auger-style bits for your electric drill (they are not common in some areas), you can buy the auger bit made for use with a brace and bit and hacksaw off the wedge-shaped end. The auger bit makes a clean hole and is durable. *Masonry bits* are made for drilling into concrete. They come in various diameters and will fit most drills—but use with caution. If you don't have a heavy-duty drill, trying to push your drill to cut into concrete can damage the motor. If the drill complains a lot and whines, consider borrowing or renting a heavier drill for the work. In some instances, you may actually need a *rotary hammer drill,* which works like a miniature jackhammer to drill into concrete, brick, etc. Another specialty bit is the *hole saw,* actually an attachment for drilling the large-diameter hole you need for a lockset. You can buy a "nest of saws," which is a set of hole saws of various diame-

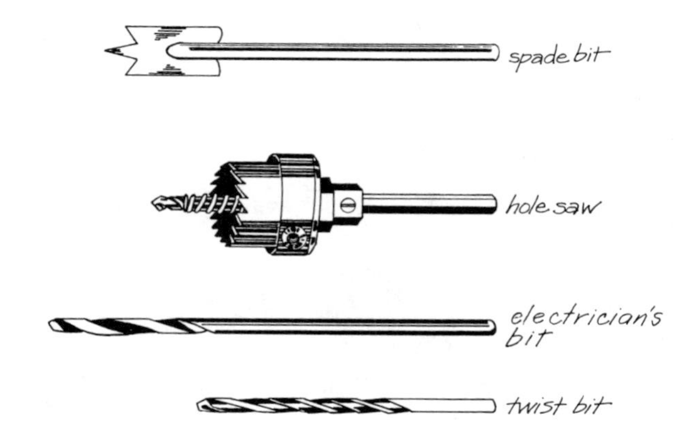

ters that all fit together for storage. The hole saw has a pilot bit called the mandrel, which centers and begins the cut. The *electrician's bit* is a standard ¾"-diameter (made for standard cable) with a tapered end that allows you to start drilling right on a set point. A *countersink bit* is another specialty bit; this drills a tapered hole so that a screw can be placed with its head sunk below the surface of the wood. A right-angle attachment is also available for drilling close to walls, between studs or joists, and in other difficult spots.

Besides all of the bits and specialty bits listed above, you can buy the following attachments for most drills: wire brush, disk sander, grinding wheel, buffing wheel, screwdriver attachment, stirrer (for mixing paint), and a surform drum (which acts like a rasp). You can also buy sets of odd-shaped burrs for special woodworking. These are sometimes called rotary cutters—they'll work on wood and plastic, but the work they do is of questionable usefulness!

To use your drill, first use your chuck key to loosen the chuck, insert the bit, and retighten the chuck. Run the drill for a second to make sure that the bit is centered and turning freely. With the switch off, place the point of the bit against the spot to be drilled. In some instances, you may want to use a scratch awl to punch a little hole to help you in centering and beginning. Turn on the drill and feed the bit in. The pressure you need to feed the drill into the hole should be medium-light; it will depend a lot upon the power of the drill, the hardness of the wood, and so forth, and is something you'll develop a feel for. Learning to drill a straight hole—i.e., holding the drill so the bit is perpendicular to the surface of the wood, forming close to a 90° angle—is something else you'll have to develop a feel (and an eye) for. If you're drilling a fairly deep hole, pull the drill out periodically to help clear the sawdust. You should never have to really force your drill. If you find yourself putting a lot of pressure on it, check to make sure that the bit is sharp, that you aren't hitting a nail or other metal attachment, or any other problem. If you drill into metal, wear safety glasses to protect your eyes. If you want to avoid splintering wood that you're drilling through, try clamping a piece of scrap wood to the underside of the piece—the splintering usually happens when the drill bit breaks through the opposite side, and having a piece of wood there will help. When you aren't using a bit, store it in a case or leather roll to protect the tip and help keep the bit in sharp working condition.

Saber Saw or Jigsaw

The saber saw or jigsaw is a lightweight, versatile power saw that can use a variety of blades to cut wood, metal, and plastics. It can be guided along irregular lines, can make circle cuts and curves, or can do simple straight cutting. Another special feature of this saw is its ability to make a plunge cut, which allows you to cut a hole in the center of a panel or board. The jigsaw can often be used for work you would normally do with your circular saw, such as cutting 2×4 material or making repetitive angle cuts, and is lighter and easier to use than the circular saw. It is perfect for cutting shingles to size while working up on a scaffolding, for making difficult cuts in trim, and other tasks. It is also a fairly inexpensive power tool, so that you can justify overusing your jigsaw and later replacing it. With this in mind, you can use the jigsaw to cut form wood, cut rebar, and so on.

In choosing a jigsaw, look for a variable speed or at least a two-speed model that will allow you to cut more slowly when working with larger material or cutting metals. The blade attachment is another important feature to consider. An attachment that opens up entirely (using a small Allen wrench) to receive the blade will tend to vibrate loose and wear out a lot faster than an attachment that opens just enough to slide the blade into place. (An Allen wrench has a square end that fits into a matched hole.) Look for a model offering a tilting base that can be set up to a 45° angle (right or left) for bevel and angle cutting. Some models can be outfitted with an optional rip guide for making straight cuts and a circle guide for cutting perfect round holes. Cutting up to a vertical wall with most jigsaws is difficult or impossible because the base plate or motor housing gets in the way. A few models have special (optional) flush-cutting blades or adjustable base plates that allow you to cut up flush to or nearly flush to a wall. In general, a jigsaw can be used to cut through 2× (1½") stock; some jigsaws take a standard blade that is 2" or 2½" long—the longer the blade, of course, the deeper the cut possible. Make sure your jigsaw is double-insulated.

To use a jigsaw, first—before you plug it in—put the blade in place and make sure that it is firmly and correctly seated. Blades come in a variety of sizes and types; a general-purpose blade can cut both wood and light metals. The number of teeth per inch of blade will determine the roughness or fineness of the cut and will also effect how hard the tool has to work to make a cut. The more teeth per inch, the finer the cut; the finer the cut, the more power the tool must put out. With the right blade in place, plug the jigsaw in and hold the tool with its base plate (or foot) firmly against the material to be cut. The jigsaw is light enough for you to do most cutting while holding the tool with one hand. The trigger that switches the motor on and off is located in the handle and easily accessible. Depress the trigger to start the motor, and feed the blade into the wood. If the saw seems to chatter or vibrate a lot, you are probably not bearing down hard enough to hold the base plate firmly against the wood. This will result in an uneven cut and in some instances may even snap the blade in two. The jigsaw blade moves up and down; it cuts on the upstroke and splinters the top surface of the wood. It is a good idea to position wood so that any good side (or surface you want to remain smooth) is *down.* For rough or general cutting, this makes no real difference, but if you cut paneling or trim, say, you should remember to place it good side down.

To make a plunge cut in the center of a panel, hold your saw tilted with the front edge of the base plate on the material just forward of the start of your hole (you can more or less line up the blade itself). Switch on the tool and slowly lower the blade into the material. As the blade feeds in, you can guide the saw down until the plate rests firmly on the wood. To cut circles and curves, you simply guide the saw with a light pressure. You'll have to experiment to see how tight a curve you can make with your individual saw. To make angle cuts, first unplug the saw and then use the adjusting wing nut or screw to tilt the base plate to whatever angle you want. You can use your sliding T bevel to set an exact angle. Plug the tool back in and proceed as for any regular cut.

Always store your jigsaw with its blade removed. Blades are easily broken if the tool is put in a toolbox intact. The blades themselves should be kept in a box or special container.

Reciprocating Saw

An invaluable tool for remodeling work is the reciprocating or "all-purpose" saw. This electrically powered saw is somewhat like the jigsaw or saber saw in the working motion of the blade, but its unique design allows it a different range of possible jobs. A reciprocating saw is perfect for rough demolition work, remodeling, and all sorts of repair. It has the flexibility to follow odd contours and work in difficult situations where a circular saw, jigsaw, or even a handsaw won't do. For example, you can use your reciprocating saw to cut along flush to a wall when you have to remove adjacent flooring sections to reach floor joists or repair flooring damage. No other power tool that I know of can do this job, and a handsaw that hits an embedded nail will be damaged (the reciprocating saw has blades that can cut through nails without being damaged). Another good use of the reciprocating saw is in remodeling: it can cut out wall sections, studs, and so on, with minimum damage to surrounding wall areas. This is especially helpful if you are adding a window or door to an existing wall. Blades for your reciprocating saw are available for cutting everything from tile to metal, from cast iron to wood.

The reciprocating saw *may be used for removing form wood, for demolition projects, or when cutting out for new window or door openings.* (Photo by Carol Newhouse)

One drawback I've found in working with a reciprocating saw is the expense of the blades—and regardless of their ability to cut through steel or tile, they do snap and break fairly frequently. The individual blade is not all that expensive, but when you go through three or four blades in an afternoon's work, the expense begins to add up to a significant amount! The ready justification for the expense is the versatility of this tool and its amazing speed on the job. Buying your blades by the box (if you can afford to) cuts down their price somewhat. In use, a reciprocating saw feels a little like a miniature jackhammer. Pressing the foot of the saw firmly against the material you are cutting reduces the vibration somewhat and, incidentally, helps keep your blades intact. Blades break frequently when the saw is too loosely held or when you keep the saw going while pulling it from a cut.

In choosing a reciprocating saw, look at and handle the different brand-name makes: they have quite different features in terms of body weight and flexibility. For general carpentry, the lighter-weight reciprocating saw is more comfortable to handle. If you are doing a lot of remodeling, demolition, and the like, a heavier commercial-duty saw will probably hold up better. Check out the foot of the saw and see how it will work in a tight situation. Some models are better designed to maneuver into corners; other models have an optional (added-expense) right-angle attachment for this work. Try to choose a saw that is fairly commonly used in your area so that replacement blades are readily available. A carrying case for your reciprocating saw and blades is a good protection for this valuable tool. Some models and brands come complete with carrying case.

WOODWORKING MACHINERY

This section is intended to give a brief introduction to several of the basic machines used in woodworking and to be used as a guide when buying these machines. For safe operation and more detailed instructions on techniques, uses, and maintenance of these machines, some instruction and/or woodworking textbooks are necessary. Two texts I recommend are:

Modern Woodworking, by Willis H. Wagner. Goodheart-Willcox Co., Inc., South Holland, Ill., 1970.
Cabinet Making and Millwork, by John L. Feirer. Chas. A. Bennet Co., Inc., Peoria, Ill., 1970.

In the sections on maintenance, I have noted the tasks unique to each particular machine and have excluded the maintenance procedures common to most machinery. These are such things as:

1. Keeping sawdust out of the motor (it can be blown out periodically with an air hose or vacuumed).
2. Oiling and greasing moving parts—90 weight automotive oil is good for most oil wells.
3. Light oil (e.g., "3-in-1" type) for places that need lubrication but are exposed to a lot of sawdust and would clog up if a heavy oil were used.
4. Blowing dust out of switch boxes periodically (or whenever the switch doesn't work).

In the sections on buying, costs, and brands I have used the most recent (at time of writing) price list on Rockwell machinery. (Prices include stand, motor, and electricals.) This is not meant to give more endorsement to their products but rather as one way of giving accurate current prices of new machinery. Some brands are less expensive than Rockwell and are quite good. As a rule, used machinery is the better buy, but it is often hard to find the used machine you want when you need it. I have purchased some new machinery and have found that dealers will often bargain a little. Sometimes they will give discounts to students and sometimes they can be convinced to throw some free accessories into the deal. When inquiring about prices be sure to find out if the price includes the motor, stand, and electricals (switch, cord, etc.). One good rule to follow when buying machinery is that the more power the machine has, the safer it is. Underpowered machinery causes the operator to force the wood against the blade, which increases the chances of being off balance and slipping. Sometimes also, instead of cutting, the blade will just throw the wood back at the operator. So don't be afraid of the more powerful machinery.

Table Saw

The table saw (also called the circular or bench saw) is one of the most useful machines in a woodshop. It can be used to *accurately* rip, crosscut, bevel, cut single and compound angles, taper, resaw, and make molding cuts. With different blades, jigs, and guides, it is possible to cut a number of joints including finger joints (also called box joints), spline joints, lock miter joints, and tenons, dadoes, rabbets, and grooves. The table saw can be used to cut many duplicate pieces accurately and, with the use of jigs, can be used to cut small pieces. If the blade is replaced by a sanding attachment, the saw can be used as a very accurate disk sander.

The table saw is basically a large table with a slit in the middle where a circular blade protrudes. Inspection under the table shows that the blade spins on a shaft (called the *arbor*), which is usually powered by belts and pulleys connected to the motor. On most modern machines the arbor (and motor) can be raised and lowered for cutting varying depths. On some older machines, the table tilts and the arbor is stationary. There should be positive stops at 45° and 90°. The blade spins toward the operator and wood is fed against the blade. The wood is usually guided either by an adjustable fence parallel to the blade, or by a miter gauge (sometimes a crosscut guide) that holds the wood at varying angles (45° to 135°) to the blade. The miter gauge is T-shaped (when set at 90°) and rides in a slot in the table parallel to the blade. The fence is most often used for ripping and the miter gauge for crosscutting and angled cuts.

The size of the saw is determined by the largest blade it will accommodate. For example, a 10″ saw (a good medium size) will hold a blade up to 10″ in diameter and will cut a maximum depth of 3⅛″.

A wide variety of blades and accessories are available for the table saw. Blades are sold by their diameter and arbor hole size (⅝″ arbor is standard on a 10″ saw). Some blades are designed for specific operations such as ripping, finish cuts, or crosscutting, and some are made to be used on specific materials such as plywood, particle board, plastic, or hardwood. Combination blades are also available. These save time because it is not necessary to change blades for different types of cuts. Blades are made of steel or of steel with carbide-tipped teeth. The steel blades are much less expensive (approximately $15) and can be resharpened in the shop with some practice. The carbide-tipped blades cost between $50 and $75, but last 10 to 50 times as long between sharpenings as their steel counterparts. It is necessary to have them sent out for resharpening (which costs about $13 for a 10″ blade with 60 teeth). The number of teeth on a saw blade is important to consider when shopping for blades. The more teeth, the finer the cut; blades with larger and fewer teeth are used for coarser cuts. A good blade with which to start a collection is a 10″ carbide-tipped, 60-tooth combination blade (for a 10″ saw, of course). The best blades have relief slits in the edge. These allow the blade to expand when it gets hot and thereby prevent it from warping.

Dado sets are used on the table saw to make grooves, dadoes, and rabbets. The set consists of two blades (similar to other types of blades) and chippers, which are only partial blades and have two teeth each. Chippers are usually ⅛″ and 1⁄16″ thick. Combinations of these chippers and the two outside blades are put together and mounted on the saw to make cuts of varying widths.

Another type of blade used to cut dadoes is called the *adjustable dado blade.* This is a single blade and is mounted on a tapered washer that causes the blade to swing sideways as it spins. The blade actually wobbles from side to side. This looks scary, but the blade is as safe as others and gives a clean cut. The distance the blade swings sideways is adjustable by means of a dial on the washer and determines the width of the cut. The adjustable dado blade comes with carbide-tipped teeth and is relatively inexpensive ($35). The standard dado set is available with steel or carbide teeth (steel, $40; carbide, $175). Both will cut a maximum width of 13⁄16″.

Molding cutters are another accessory available for the table saw. These come in sets consisting of a metal base (cutterhead) and a variety of cutters. There are three cutters of each shape and there are three slots in the cutterhead in which the cutters are mounted. Once the unit is assembled with the desired molding cutters, it fits on the arbor like a standard.

Another accessory available is the *sanding disk plate.* This mounts right on the arbor just like a blade and, of course, can be raised, lowered, and tilted. (This plate can be homemade with a round piece of plywood.)

Other accessories are:

Safety guards These are usually included in the price of a new saw. They protect against wood kicking back at the operator and catching fingers in the blade.

Extension bars The fence rides on two bars, one in front of the table and one on the back edge. The extension bars make it possible for the fence to move farther away from the blade, thereby allowing wider cuts. A must, especially for cutting plywood sheets. (Sometimes dealers throw them in for free.)

Tenoning jig This is used to hold a piece of wood on its end in order to cut a tenon. It is simply a large metal block that slides in the miter gauge slot on the table. It has a clamp that holds the wood in place. This makes tenoning a lot easier and safer. It is very expensive ($115) and can be homemade out of wood.

Extension tables These are homemade tables that can make cutting large pieces a one-person job. Build them so that they are the same height as the saw table. It's a good idea to make them slightly adjustable to allow for unevenness of the floor. One way to do that is to put a T nut, a bolt and two nuts, into the bottom of each leg. Bevel the front edge of the outfeed table to allow the piece being cut to slide easily onto the extension table.

The table saw requires a lot of space and is best located in the center of the room. It is a good idea to have at least 9′ of clearance in three directions (front, back, and left side—if you are right-handed) for cutting long boards and plywood. Table saws have big motors and therefore require a lot of power (at least 2 hp for a 10″ saw). Try to place the saw as close to an electrical outlet as possible. Good lighting is also important and should come from the left (if you're right-handed) and from above to avoid shadows on the blade and fence. This saw makes a fair amount of noise and a lot of dust. It is best to bolt it to the floor for stability, but this is often impractical in a small shop (sometimes the saw needs to be turned 90° to cut long or wide boards).

The only machine to which the table saw can be compared is the radial-arm saw. The radial-arm saw is a sorry replacement for a table saw in almost every category except crosscutting. The radial-arm saw gets out of adjustment too easily and is not as accurate as a table saw. The only other comparison that can be made is with hand tools and hand-held power tools. While these definitely have benefits, they cannot compare with the table saw for accuracy and efficiency.

When buying a new or used table saw:

1. The larger the better (if you have the wiring necessary, that is)—10″ is a good size. Be sure that it has enough horsepower; 3 hp is good, 2 hp adequate. You can get by with less, but cutting hardwood can be slow and the wood may burn.
2. The larger the table surface, the better. The larger surface makes cutting large boards and plywood much easier. If it's a used machine, take a straightedge to make sure that the surface is not warped.
3. A heavy base cuts vibration and increases accuracy.
4. Extension bars are important. Some older machines don't have them. Be sure to consider the maximum distance between the blade and fence on these.
5. Handwheels should move easily and tighten effectively. On some saws it is really hard to raise the motor or move it from 45° back to 90°. If the threads are stripped on the tightening knobs, they won't tighten, and with the vibration of the machine, the motor and arbor could slide out of the desired setting.
6. Cut a piece of hardwood. It should not burn or need forcing if the blade is reasonably sharp.
7. The heavier the fence, the better the accuracy of the saw.
8. Some miter gauge slots have a T-shaped cross section. A washer attached to the underside of the gauge bar rides in the wide part of the slot. This keeps the miter gauge from falling on the floor when it is not completely on the table. This is a good feature because it protects the gauge from unnecessary abuse and keeps it more accurate.
9. In most cases, the larger the saw, the better it is constructed.
10. Check that the fence is straight and square to the table surface.
11. Check for positive stops at 45° and 90° on the motor tilting mechanism. There should also be a scale showing the angle of the blade.
12. The tilting arbor-type saw is easier to use than the tilting table type. Cutting bevels on the latter is more difficult because the table must be tilted (it's heavy) and the wood has a tendency to slip. Also, it's difficult (though not impossible) to make an extension table for the latter because it must move up and down with the table.

Costs and Brands

10″ Rockwell/Delta Unisaw: $1,300 with a 3 hp motor (can buy these used for about $400 to $900, but they are getting hard to find); Rockwell also makes a 10″ Tilting Arbor Saw ($965)—lighter-duty, but adequate for some shops

Powermatic (comes with a better bench than the Rockwell Unisaw)

Inca (a small saw good for smaller, accurate work; it has a lot of good accessories)

Oliver, Tannewitz, and anything old and heavy may be good choices

Maintenance

Maintenance is fairly simple on the table saw. The fence and stops occasionally need adjusting. The table needs to be kept clean and slippery to make it easier to push wood across it. Use parafin or a silicon spray. The gears that tilt and raise the motor need to be kept clean and greased for ease of adjustment.

Radial-arm Saw

The radial-arm saw is designed primarily for crosscutting. It is especially useful for crosscutting large boards because, instead of moving the wood onto the blade (as with the table saw), you may hold the wood stationary while the blade is fed into it. This is a particularly good machine for cutting many duplicate parts to length. By making a variety of adjustments, you can use this machine for ripping, angle cuts (horizontally, across the width of the stock), and bevel cuts (vertically, through the thickness). In fact, with the necessary adjustments and attachments, it can be used to do just about everything a table saw can do. The problem with this saw, however, is that it has a tendency to be inaccurate because so many adjustments must be made in order to have it perform all of the functions of a table saw. The more adjustments, the greater the possibility of parts getting out of alignment. Greater accuracy is ensured if this saw is used only as a crosscut saw.

The radial-arm saw consists of a vertical column supporting a horizontal arm. This arm extends over

The radial-arm saw *is used for making repeated, perfectly accurate cuts in the shop or on the job.* (Photo by Carol Newhouse)

a stationary table on which the wood to be cut is held. The motor and blade ride on a track on the arm by means of a supporting bracket called the *yoke.* The handle on the yoke is used to pull the saw blade across the table and through the wood. The wood is held against a fence by the operator. When crosscutting (at 90°), the arm is perpendicular to the fence. To cut angles other than 90°, it is possible to pivot the arm on the vertical column so it is, for example, at a 30° angle to the fence. The arm can be locked in this position. Bevel cuts are made possible by tilting the yoke. It can be tilted at least 45° in either direction and on most models can be turned so that the arbor is pointing toward the floor. In this last position, the machine can be used for drum sanding and, with the proper attachments, may be used as a shaper.

So, by means of the pivoting arm and the swiveling yoke, the motor and blade can be adjusted in three ways: in and out, angled sideways, and tilted. The fourth adjustment possible is raising and lowering the blade, and is accomplished by cranking the arm up and down.

For rip cuts, the arm is set perpendicular to the fence and the blade (by means of the swiveling yoke) is turned parallel to the fence and locked in position. The yoke is positioned on the arm so that the blade will be the desired distance from the fence.

Although ripping on a radial-arm saw is similar to the same operation on a table saw, the differences between the two are worth noting. Ripping can be a little more dangerous with a radial-arm saw because most of the blade is above the wood. Of course, there are guards to lessen this hazard. Also, the radial-arm saw is limited by the distance between the blade and the table when the arm is at its highest point. For example, cutting open a box (a simple and standard technique on the table saw) is not possible if the box is a 3′ cube. There simply is not enough room to fit the box between the blade and the table. Another problem with ripping on a radial-arm saw is that it is a little more difficult to cut very narrow strips. This is because there is not room to get a push stick between the blade and the fence. (The push stick is used to feed pieces of wood into the machine, thus protecting the hands of the operator.)

The radial-arm saw can accommodate almost all of the attachments commonly used with a table saw (dado sets, molding cutters, sanding disks) and then some. It can be used as a shaper with the proper arbor adaptor and guards and can also accommodate sanding drums. Extension tables made with rollers are available and are a nice luxury. Tables can be made of wood, of course, and are extremely useful.

Good features to look for when buying a new or used radial-arm saw are:

1. Fence lock handles—these enable the operator to change the position of the fence to accommodate wider boards. To do this, you loosen the handles, slip the fence out, and put it closer to the column, then tighten handles.
2. Length of arm—the longer the arm, the wider the boards that can be crosscut (i.e., the blade can travel farther). If the saw is to be used for ripping, the longer arm will allow wider cuts.
3. Size of machine—a 12″ saw is satisfactory for most work. It will cut a 4×4 in one pass. A 10″ saw requires two passes to cut this size lumber. As usual, the larger machines are generally more sturdily constructed.
4. 1½ hp minimum for a 12″ machine. This depends upon how the saw is to be used. If it is to be used for ripping and shaping, more horsepower is recommended.
5. Play—if it is a used machine, lock the arm and the yoke and check for play.

Brands

Rockwell makes a nice 12″ radial-arm saw that costs about $841 new. Other sizes made by Rockwell are 14″, 16″, and 18″. DeWalt is another good brand.

Space Requirements

The best location for this saw is along a wall in your shop. It does not need clearance behind, but does require 12′ to one side (usually the left side) and at least 6′ to the other side. If space is limited, one end can be placed near a door that can be opened to clear long boards. This saw is a fairly noisy machine and produces a moderate amount of dust.

Maintenance

Maintenance on the radial-arm saw consists of making adjustments so that the saw will always be cutting square, and keeping the column and the track on the arm well lubricated. The table must also be checked periodically to make sure that it is parallel to the arbor along the entire travel of the arm. If the table is not parallel, cuts will be deeper at one point than at another. This is especially important when making dadoes because they must be a uniform depth all the way across.

With its thin, continuous blade, the band saw *allows you to make curved or straight cuts in a variety of material.* (Photo by Carol Newhouse)

Band Saw

The band saw has a variety of uses, the most common of which is cutting curved lines. It can also resaw thick materials, make cuts in pieces too small to be worked on the table saw, and make straight cuts where accuracy is not too important.

The band saw is a simple machine. The blade is a continuous band that rides around the circumference of two bicycle-type wheels positioned one above the other. Between the two wheels the blade straightens and goes down through a hole in the table. Wood is pushed along the table and fed into the blade, and can be turned to cut curves.

The alternatives to the band saw are the saber saw or certain handsaws. In most cases, the band saw is the best choice for the work described, but the band saw does have limits that the other types do not. The advantages of the band saw are that it can make deep cuts (it is possible to cut into a board 8″ thick), resaw thick boards to make two thinner boards, and cut faster and more smoothly than alternative tools. Unlike the reciprocating cutting action of the saber saw, the cutting action of the band saw will not chip out the wood on the top of the cut. The band saw is also more efficient than the saber saw because it is not necessary to spend time clamping down the work before making a cut.

There are only two disadvantages of the band saw. One is that it is not possible to start cuts in the middle of a piece of a wood as is possible with a saber saw or jigsaw. The second is that the distance between the blade and left side of the machine limits some cuts. For example, the saw has an approximate clearance of 15″ to 20″, so that it is not possible to crosscut a 48″×96″ sheet of plywood down the middle.

Blades come in a variety of sizes and types. The length of the blade is determined by the size of the wheels and the distance between their centers. To get this length, multiply the distance between the centers of the wheels by 2, and add the circumference of one wheel. (The circumference is 2 × 3.14 × radius—or $C = 2\pi r$.) Blades come in different widths for cutting different sizes of curve (narrow blades will turn tighter corners) and with different kinds and numbers of teeth per inch. Blades with more set in the teeth make wider kerfs; blades with more teeth per inch make smoother cuts. The cost of blades varies, of course, with the length and type. The cost of a 93″ blade ⅜″ wide with 4 teeth per inch is about $7.00. It is possible to buy steel blades that can be resharpened or disposable blades with hardened tips (these blades cost only slightly more than it costs to sharpen the other kind).

The size of a band saw is determined by the diameter of its wheels. The machines with the larger wheels have more clearance around the blade and therefore allow the cutting of larger pieces. The larger machines also tend to have larger motors and bigger tables, which make for easier and smoother cutting.

Here are some things to look for when shopping for a new or used band saw:

1. Excessive vibration. Smaller machines tend to vibrate a lot. Cut a piece of hardwood as a test.
2. Look for maximum distance between the blade and the left side of the machine and for maximum clearance between the table and the upper blade guide.
3. Better models have tilting tables. Some even tilt in two directions—left and right, that is. A degree scale is useful.
4. The bigger the table surface, the better.
5. On a used machine, look for badly worn rubber on the wheels (can be replaced) and worn guides, upper and lower (cannot be replaced).
6. Some band saws have a miter gauge slot and an adjustable fence. These aren't too useful, because this machine is not best used for accurate straight cuts.
7. On used machines, make sure that all of the adjustments work: i.e., upper wheel can be raised and lowered to adjust blade tension and tilted to track blade, and guides should move in all directions.

Brands

Old Rockwell/Delta, Powermatic, Crescent, Oliver.

Cost

A new Rockwell 14″ band saw costs about $840; a 20″ model, same brand, $2,000.

Space Requirements

The band saw does not require a lot of power (this depends upon the size, of course), nor is it particularly noisy, but it does produce a moderate amount of fine dust. It requires good lighting (some models come with lamps attached near the blade) and some space on three sides for long boards.

Maintenance

Maintenance of the band saw is as simple as the machine. Aside from the normal greasing and oiling, all that really needs to be done is cleaning the rubber on the wheels and occasionally adjusting the guides that steer the blade. The blades need to be removed and sharpened; removal is easy and blades can be sent to a saw sharpening shop. Depending on the length and type of blade, sharpening costs about $7 to $12.

Drill Press

The drill press is basically a power drill mounted on a large stand. It is primarily used to bore holes accurately and efficiently. In addition to this basic function, a variety of attachments enable the drill press to be used for mortising, planing, sanding, and routing.

The top of the drill press is called the *head* and houses the motor and pulleys, which are in turn connected to the spindle and chuck. The chuck accommodates at least a ½″ round shaft and is just like those found on power drills. The spindle speed can be adjusted by moving belts manually on the step pulleys, or by moving a lever on variable-speed models. The spindle and chuck can be moved up and down by means of a feed lever. The spindle unit is spring-loaded and therefore will always return the chuck to its uppermost position, so it is not necessary to pull the drill out of the wood—it springs up by itself. The spindle and chuck are centered over a table on which the wood to be bored is placed. The table is attached to a large column (as is the head) and can be adjusted to move up, down, and sideways, and can be tilted to the left and right. Drill presses are available in floor and bench models. On the floor models the column goes all the way to the floor; on the bench models the column is only about 3′ high and the entire machine is mounted on a stand. The bench model should not be confused with the small drill press stands that are made to mount a portable power drill. While these are much less expensive than a drill press, they also sacrifice much of the accuracy.

There are a number of situations where a drill press is more useful than a portable power drill. Almost anytime that duplicate holes must be drilled, it is possi-

The drill press *may be used to make accurate, repeated holes at various angles and depths.* (Photo by Carol Newhouse)

ble to make a jig or guide that can be attached to the drill press table to ensure speed and accuracy. And because the table tilts and locks in place, it is also possible to drill angled holes accurately. The drill press is especially useful for drilling large holes, and it has the added feature of an accurate depth control stop.

In addition, the attachments available for the drill press make it a versatile machine. The *mortising attachment* is the best accessory available. However, although using the drill press for sanding, routing, and planing by means of different cutters is possible, it is not highly recommended. These last three operations require the operator to apply sideways pressure on the chuck and spindle unit, for which the drill press was not really designed. Repeated sideways pressure can bend the spindle out of alignment, so care should be taken when using these attachments.

There are two main disadvantages to the drill press. First, it is limited to drilling holes no farther from the edge of the wood than the distance between the spindle and the column. This dimension—center of spindle to edge of column is multiplied by 2 to determine the size of the drill press. That is, a 15″ drill press has a clearance of 7½″ between the center of the spindle and the edge of the column. This limitation has been lessened somewhat by the development of the *radial-arm drill press.* On this machine, the entire head unit slides on an arm (much like a radial-arm saw) and greatly increases the distance between the spindle and the column. In addition, the head can be tilted 45 degrees to the left and right. The problem with this type of drill press is that, because it adjusts in so many directions, it can get out of adjustment quite easily and is therefore not exceptionally accurate.

Even the radial-arm drill press does not overcome the second disadvantage of the drill press, the fact that it is not portable and therefore cannot be used to drill inside of cabinets or in other awkward places—one of the jobs for which the portable drills are best suited.

When buying either a new or used drill press there are several things to look for:

1. Size—generally the bigger the better, because this machine does not take up much space.
2. Speed range of 400 to 1,800 rpm is good for wood boring, and at least 3,000 rpm for routing.
3. Check for play in the spindle. Does it chatter a lot when spinning? Take a *straight* ½″ piece of drill rod, and insert it in the chuck, and turn on the machine. Does the rod seem to wobble? If so, there's play, and you should be cautious about buying.
4. Length of stroke = distance that spindle will travel. Almost all drill presses have a 4″ or 6″ stroke. This determines the depth of holes possible.
5. How far the table tilts. Positive stops at 90° and 45° are good. Does it tilt to a vertical position? This is useful, too.
6. On some models the table is simply clamped to the column and raising the table involves unclamping and manually lifting the table. This gets tiresome. Better machines have gears and a crank to raise and lower the table.
7. Floor models are more useful for drilling holes in large pieces, but bench models are certainly adequate for most woodworking operations and are less expensive.

Brands and Costs

Jet—bench model about $300
Rockwell/Delta—$800 for a 15″ floor model
Powermatic

Space Requirements

The drill press does not take up much floor space. It can go against a wall but requires space to the sides and front for long boards. It has a small motor (½ to ¾ hp is common) so does not require a lot of power. It is quiet and relatively dust-free.

Maintenance

1. Oil quill (sleeve that houses spindle).
2. Keep column and gears clean and lubricated.
3. Do general motor and belt maintenance.

Jointer

The jointer is simply a large, stationary, motorized version of a hand plane turned upside down. Like a hand plane, the jointer is most commonly used for straightening and squaring boards and for flattening twists and warps. These operations are essential to cabinet and furniture making because wood must be flat and square in order to do edge joining, ripping, and crosscutting. The jointer can also be used for making chamfers, bevels, rabbets, and tapers.

Wood is removed on a jointer by feeding a board against rotating knives set in a solid metal cylinder, called a *cutterhead.* The cutterhead lies in a horizontal position between two tables (infeed and outfeed tables). On a hand plane, the bed is in a fixed position and the cutting edge is adjusted to determine the depth of the cut. On a jointer, the cutterhead is at a constant height and the level of the infeed table is adjusted to determine the depth of the cut. The outfeed table is set level with the peak of the cutterhead. The jointer tables work just as a plane bed does by guiding the knives so that the high spots on an edge or face are removed. One of the advantages of the jointer is that, since the tables are longer than a plane bed, it is capable of greater accuracy, especially on long boards.

Another advantage of the jointer is that it has a

fence that spans the two tables and is used to determine the bevel of the cut. The fence can be set at right angles to the table (to square an edge) or at any angle between 90° and 45° to make bevels and chamfers. The fence can also be adjusted from side to side, thereby utilizing the entire width of the knives. Knives can be 4″ to 18″ wide, depending on the jointer. Often woodworkers save one end of the knives for final clean cuts and use the rest of the width for cuts that can be more coarse. On the better models, it is also possible to set the fence diagonally across the tables so that the wood can be fed into the cutterhead on a slant. This produces the same shear effect that is acquired by holding a hand plane on a slant when planing. Because of all of these additional features, the jointer is faster, easier to use, and more accurate than a hand plane.

The jointer, of course, does require electrical power and a fair amount of space that a few hand planes do not. It is important to place the jointer as close as possible to an electrical outlet that has the necessary volts and amps for its motor (long extension cords often mean power loss). It is a good idea to have about 3′ of space on the operator side and at least 6′ both in front of and behind the machine. Leveling the jointer and fastening it to the floor ensures greater accuracy. With the larger jointers, especially, the tables may twist under their own weight if the machine is not level. Also, once the machine is leveled, it is easy to check for flatness of the tables. Once the tables are level, it is possible to set the knives and the fence. The jointer does not require any special type of floor, soundproofing, or ventilation. It does not produce much dust, only shavings and chips.

Aside from periodic oiling, waxing the tables, and checking for worn drive belts, the main maintenance job is replacing and honing the knives. Knives can be removed from the cutterhead and sent to a saw sharpening shop to be reground. (Some jointers come with a grinding attachment.) Putting the knives back in is time-consuming but not too difficult. Almost every comprehensive woodworking text has a detailed explanation of the knife setting and honing procedures.

When shopping for a jointer, the type of work and amount of shop space available must be considered. The size of a jointer is determined by the width of its knives (for example, a 6″ jointer has knives that are 6″ wide and tables that are slightly wider). The wider the knives, the wider the boards that can be face-jointed. The jointers with wider knives have longer tables, an advantage when working with long boards. It is difficult, but definitely possible, to joint 10′ or 12′ boards on a jointer with a total length of 27″. Larger jointers have larger motors and therefore require more power. It also costs more to have their knives ground because sharpening shops charge by the inch (about $.30/inch). Also, the wider the knives, the more time-consuming they are to set. It is a lot harder to set 18″ of knife at the right height than it is to set 4″ or 6″.

The prices on jointers vary depending upon their size, construction, and manufacturer. A new 6″ jointer made by Rockwell now costs about $750 with a stand and motor. This is a medium-quality jointer. A top-quality 12″ jointer made by Oliver costs several thousand dollars. New machinery is expensive. Used, older machinery is less expensive—and often better quality than its modern offspring. For example, three years ago my partner and I bought a trusty old 6″ Rockwell/Delta jointer for $150. Used jointers are fairly common and can be found with a little patience and legwork. Watch for ads in the newspaper classified sections or on bulletin boards in saw sharpening shops, lumberyards, and anyplace woodworkers congregate. Factory sales representatives often have used machinery as well as new for sale. Consider auctions, but be sure to get a good look during the preview before you decide to bid.

When looking at used jointers, there are several things to check for:

1. Make sure that it works. Bring a piece of hardwood and joint the edge and face.
2. Listen for obviously ominous sounds.
3. Make sure the fence is flat. Set it at 90° and joint a piece of wood. Check wood for squareness. Make sure the fence is square at both ends.
4. Bring a straightedge long enough to place diagonally across both tables. Check for twists and warping of the tables. If they are not flat, they need to be resurfaced at a machine shop.
5. Make sure there is a guard that swings over the cutterhead.
6. Make sure that everything that is supposed to tighten does.

Brands

Oliver (very expensive, but top-quality—often found in school woodshops)
Powermatic
Rockwell/Delta
Inca (Swiss machine available through mail order houses)
Crescent
Anything old with a heavy base is probably good quality

The wood lathe *is operated with calipers that gauge the depth of the cut; special gouges are hand-held against the turning wood to pare or cut away sections.* (Photo by Carol Newhouse)

Wood Lathe

The lathe is used to turn and shave square stock, thus making it cylindrical or round in order to create such things as bowls, knobs, and table legs. There is no other woodworking machine with which to compare it. Aside from its unique function of turning rounds from squares, it is the only machine on which a piece can be worked from start to finish. Once the piece is rounded and cut to the desired shape, it is possible to sand and apply the finish while the wood is still mounted and spinning on the lathe. It should be noted that while the lathe can be the sole machine required to work a piece, it is necessary to have a grinder with which to sharpen the lathe tools.

The wood to be turned is mounted on the lathe on a spinning point called the *headstock,* or *live center.* A cutting tool (a gouge) held by the operator and supported by a tool rest is fed into the spinning wood in order to remove the square corners and produce a round piece. Different-shaped cutting tools are used to make different shapes and contours in the wood.

Two basic operations are performed on the lathe. The first and easier of the two is called *turning between centers* or *spindle turning.* This is the method used to produce cylindrical pieces such as posts, legs, chopsticks, and baseball bats. In this operation, one end of the square stock is mounted on the headstock and the other end is supported by the tailstock, which in most cases does not spin. The tailstock is mounted to the bed of the lathe and slides along its length in order to accommodate different lengths of spindle. Thus the length of the bed determines the maximum length of spindle that can be turned. Standard lengths are 3′, 4′, and 6′ beds.

The second operation performed on a lathe is called *faceplate turning.* This is the method used to turn such things as stool seats, bowls, and platters. For this operation, the stock is usually first roughly band-sawed into a circle and mounted to a flat metal plate, which in turn fastens to the headstock. The tailstock is not used for faceplate turning. The largest diameter that can be turned is determined by the distance from the headstock to the lathe bed. If this distance is 6″, the lathe is said to have a 12″ swing. Some lathes have a gap in the bed that allows larger diameters to be turned. On most lathes, faceplate turning can also be done on the outboard side of the headstock, allowing larger diameters to be turned. This requires a very stable portable tool rest stand.

For both faceplate and spindle turning, it is necessary to be able to change the speed of the spinning headstock. Slower speeds are used for rough-cutting squares to rounds and for larger-diameter pieces; faster speeds are used for finishing cuts and smaller-diameter pieces. Two types of speed-changing mechanism are available. The variable-speed models simply have a lever that the operator moves while the machine is running. The other mechanism is not as convenient; it uses step pulleys with belts that can be moved up and down manually when the machine is turned off.

A wide variety of accessories are available for the lathe, almost all good. There are many different types of centers on the market, each with a special use (e.g., a screw center for turning small knobs and other small pieces, a chuck for drilling holes on the lathe). There are also sanding disks and drums, and grinding, wire, and buffing wheels that can be mounted on the headstock. (Wire wheels can be used to remove rust or paint from metal.) Turning tools (chisels, gouges, etc.) are of course necessary accessories. Some are designed specifically for faceplate turning, some for spindle turning; and some can be used for either. Turning tools are available in sets or individually. The best kind are long and heavy (approximately 18″, including handle). Sorby is a good brand. Sorby makes "long and strong" tools that really reduce the chattering characteristic of lighter, shorter tools. Calipers (both inside and outside types) are also a necessity for accurate lathework. It's a good idea to have several sizes.

That so many accessories are available for the lathe is evidence of the versatility of the machine. Entire books have been written on the art of wood turning (one good one is *The Craftsman Woodturner* by Peter Child). Public libraries are a good source for books on lathework.

Here are some things to consider when looking for a new or used lathe:

1. The heavier the better. The chatter and vibration of lightweight lathes make good turning difficult.
2. The more accessories the better. Basics are spur center, faceplates, several sizes of tool rest.
3. Variable models are better, but step pulley types are fine and less expensive.
4. Consider needs for distance between centers and swing (most table legs are under 36″).
5. Look for ease of adjustment on movable parts (e.g., tool rest, tailstock). Tightening bolts on movable parts gets tiresome. Look for hand levers instead.

Brands

Rockwell/Delta—12″ heavy-duty with variable speed control (costs $1,670 new)
Powermatic
Oliver

Space requirements

The lathe does not take up much space; it can be placed against a wall. It is a good idea to bolt it to the floor to reduce vibration. It does produce a lot of shavings and dust.

A Round House Is Something Special

I was born in Germany, and I remember quite well from my childhood a phrase that came up whenever something unusual happened in the United States or if someone became a millionaire overnight. While reading the newspaper reports my parents would exclaim, "America is the land of unlimited possibilities!" They were very poor, always living in rooming houses, yet they seemed not to conceive of the idea that they, too, had a choice to change and to grow. In those days I would never have dreamed that one day I would build and live in my own house in the United States.

It was after the Second World War that I began to realize that I had a choice. After all the suffering that I saw during that war, I entered the nursing profession. And in 1968, after twenty years of nursing, I felt again the need of change—this time, my entire life-style.

I was about forty when I quit nursing. My age didn't matter to me; I feel that I am ageless and secure enough to trust myself. I had moved to the United States and by some lucky circumstance I found a place in a rural area where I had access to two acres. I made the change. The losses were few; the gain was great.

While still a nurse, I rode a bicycle back and forth to work to save traveling expenses. I passed a lot of little square row houses and square backyards, and it was then that I set my mind on building a round house. After all, Africans have round compound houses, Eskimos used to live in round igloos, and these people lived so much in harmony with nature.

Ken Kern, author of many books including *The Owner-Built Home,* designed my house. My first enthusiasm dampened slightly, however, when I could not make head or tail of the building plans, and the mere word *pilaster* spelled disaster to me! I had no knowledge of all the intricate and awe-inspiring language of the building code; therefore my house was built somewhat by trial and error. When I began, I had little capital on hand—a mere savings of $2,300—and I reckoned that for a long time to come I would have to heat the house with a wood stove or fireplace. With that in mind, I chose a building site 120′ up on a hill, totally exposed to the sun, and reasoned that shade could always be had by growing trees.

My experience of building anything with wood was nil, but I began to read and to listen carefully whenever building was discussed. I soon learned the fundamentals, such as that most structured lumber is nailed 16″ on center, be it wall, roof, or floor. I soon became friends with neighbors who would prove to be of great help later on. One man in particular had a talent for gleaning building materials cheaply. A friend of his had bought some property on which stood an old country church that he wanted to have removed. So I could have all the lumber, provided that I would dismantle the church. And soon there I was, armed with crowbar and hammer,

(Photo by Dorothy Dobbyn)

walking about three miles each day to the site. I was pulling out nails, taking beams apart, and knocking down boards: somewhat heavy work. My friend completed the final stage of dismantling by pulling over all four corners of the building with a truck. With this used lumber, mostly oak, I built my first structure, a chicken house. The oak was very difficult to handle and I broke many nails. Nevertheless, the whole experience gave me more confidence and insight into the structure of a building. Unfortunately, any beautiful or decorative parts of the church had been removed previously.

It was about that time that I began to liken a house to something living and

(Photo by Dorothy Dobbyn)

began to build my own vocabulary. A house needs feet to stand on (footing or foundation), a rib cage and skeleton (the wall studs and rafters), something to walk and to dance on (floor joists), windows and doors to breathe with, some beauty to get dressed in (interior decoration), and, last of all, some people and animals to give it life. Once these ideas took shape, things appeared to be so much simpler. It was no longer a formidable task to build a house. I liked to do things with my hands and I knew that I would be able to do it.

Finally, I had the go-ahead, a building permit. The well had been completed previously, 120′ deep, and the percolation test was now done. Originally I wanted to build the walls of the house with cement, using stones as a facing, since there are a great many stones in the vicinity. I realized, however, that alone I could neither move nor cope with the slip form that had to be used for the walls, so I changed the plans slightly. Wood could always be knocked down and renailed, but that cement could not be pushed over if anything went wrong. Of course, Ken Kern's design was much more beautiful than my modified version. His wall was not of the same height, it was sort of spiraling, and the roof was not round. It had long straight lines that formed great overhangs. But then I saved a great deal of money in the end.

My friend was lucky enough to find some more free lumber. He got a contract with an industrial plant to clean up the lumberyard and, in return, receive some lumber that was unfit for sale because of slight warping. This plant manufactured prefabricated houses, and what a gold mine it was for me! I got all the small 2×4's and 2×6's that I needed for the building, plus all the 2×4 lumber for the outer walls and 2×8 and 2×10 boards that I used to hold the four main posts in place. I still remember the day. It poured all day and we were drenched and totally dirty, but that hardly mattered, I was so happy over all the lumber I had gleaned!

In my inexperience, I did not think of building the house in straight 4′-wide sections. Instead I built the wall all the way around. This first mistake affected the overall building process. Another serious mistake was in not telling the man with the bulldozer to grade and level the ground for at least 10 more feet surrounding the building site. He leveled only the area where the house was going to be, and later on I had

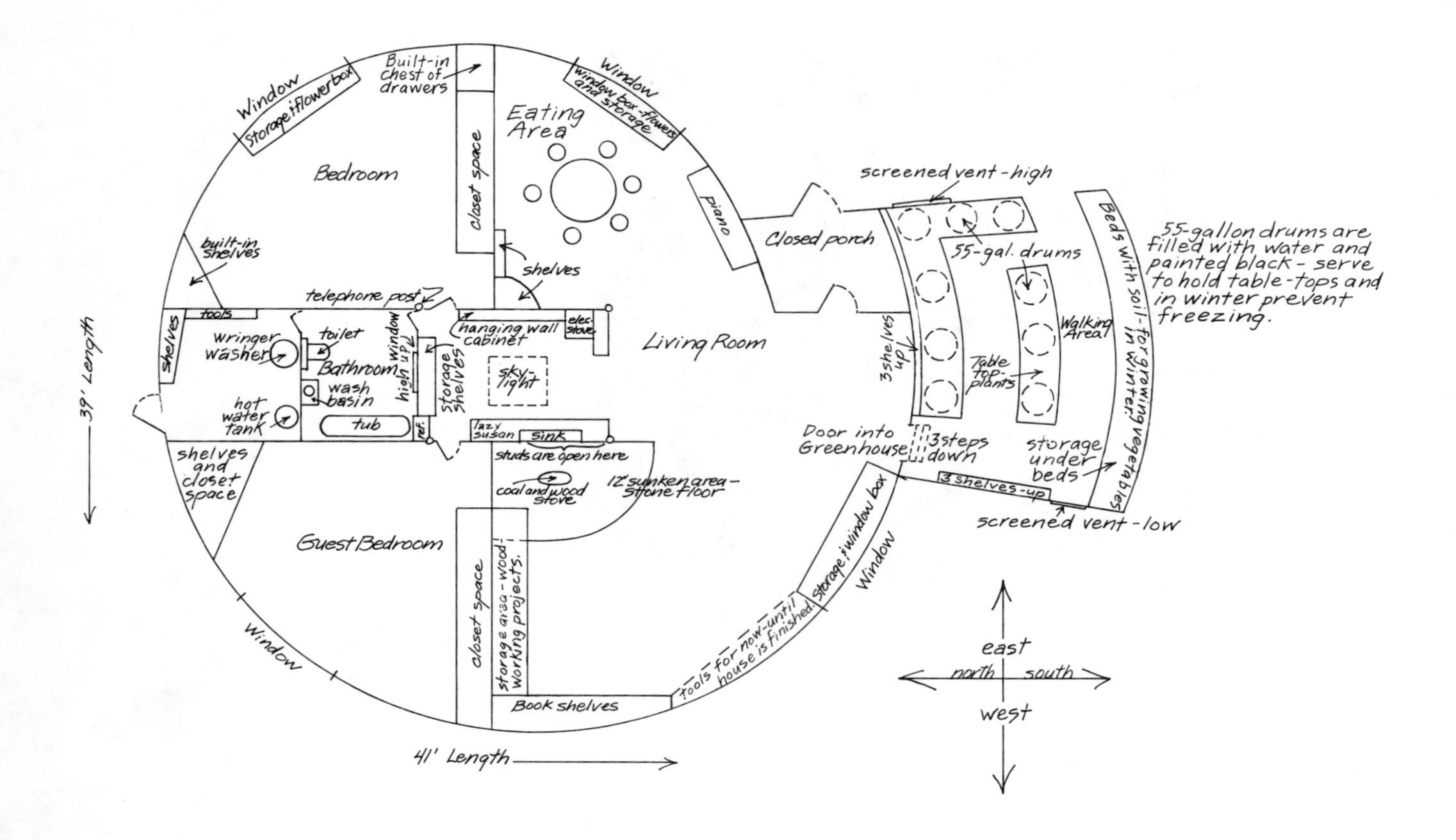

to remove hundreds of wheelbarrow loads of earth to avert flooding and erosion. I built terraces all the way around so that it looks almost like a Greek amphitheater. Eventually I am going to have a ditch dug and drainage tiles installed. I also learned that in leveling the site one should not rely on eyesight alone. When the operator was finished with the job he asked me if it was okay. We both went downhill and with hands cupped over our eyes gave it approval. Well, my friend, who happened to know a little more than I, checked it with a string level and it proved to be about 18″ off over the 41′ length east to west and 12″ off from north to south. Needless to say, the operator had to correct that mistake.

Another miscalculation was to allow a lapse of a few days between the excavation and the cement truck's arrival. After the foundation trench was dug, it poured with rain and much of the trench caved in. I spent days bailing out water and mud. When the cement truck finally made it up my hill, my careful estimation of the amount needed went out the window; there was just not enough to fill up the foundation to the desired 12″ all around. Fortunately I had a number of friends to help me spread the cement around. That is one job that could hardly be done alone. Later, I used some solid cement blocks and hand-mixed cement to build the foundation up to the desired height. By that time I had learned to use a string level and could make the trench reasonably level.

At the time I had no electricity, so my well was of no use to me. A well needs a pump and a pump needs power. Unfortunately, the rainy season started; it rained often, and water and mud covered the newly poured cement. It took me weeks to clean up that mess. Someone told me that the cement had to be free of dirt in order to adhere well with the mortar filling the space between cinder blocks and footing. I took that advice literally, and scrubbed it so spick-and-span that I could have used it as a dinner plate! When I finally laid my first cinder block, I danced around that work of art with excitement. However, that excitement faded as I tried

(Photo by Dorothy Dobbyn)

lifting the heavy cinder blocks. They also had to be laid down slowly and then tapped with a hammer to form a straight line. This is hard on one's back as well as elbows. I mixed all the cement by hand and soon learned the right consistency. Water for mixing had to be carried uphill, although often friends would drive up garbage cans of it. In the beginning I used my hands more than the trowel, until my skill increased.

It was autumn then, and I heard the birds and watched the small animals, as they watched me. My two dogs kept me company when they were not off catching the occasionally unlucky rabbit. The view around me was beautiful. Ten feet below the house grew a stand of white pines, and farther off was a valley where a small stream was winding its way. Bees from the nearby hives flew all around me to find their last food in masses of bright yellow goldenrod.

About 270 blocks were laid for the foundation, and when that phase was completed I had a "tennis elbow" for about a year. But I was happy and, besides, it was satisfying seeing all the blocks so neatly in line and the wall growing higher.

Many of my neighbors were friends who lived in a nearby community. They were always helpful, hauling the building supplies with me. One of the members was an architect who started to toenail the studs in place and showed me in detail how to frame a window opening. Then I was on my own again. I recall my apprehension at the thought of building the house wall—until I saw a few studs lined up and they remained standing even after a spring storm.

When the studs were nailed in place about halfway around the house, I wished I had electricity. Upon inquiring, I learned that in order to have a meter installed, there must be a waterproof roof first. So a roof I built. I nailed a waterproof box in between two studs and called an electrician. He was a middle-aged man and he did not say very much. All he did was lean against my studs, rocking and testing them. On his face was open disapproval that a woman should be building a house. Another electrician finally installed the meter and one electrical outlet. Now I could make use of my Skilsaw and other electrical tools, which made working so much faster.

Spring was well on the way, as I became more and more aware of the ever-changing nature around me. In the meantime, I found a new job. My neighbor, an arborist, taught me tree care, pruning shrubbery and trees, as well as landscaping. I liked the new work very much; it gave me the opportunity to be outside. As the days grew longer, so did my working hours (anywhere from ten to twelve or fourteen hours a day). But there was no one telling me when the house should be finished. I could sit down and enjoy the sunshine whenever it pleased me, and it was an exhilarating feeling of freedom, a freedom to choose to do meaningful work. Finally, after weeks had passed, the studs were all nailed in place, looking just like beckoning fingers against a darkening sky.

In the meantime, the building inspectors had paid me two visits. The first inspector asked where my building engineer was, and I replied that he had just gone to Florida for a short vacation. He seemed to believe me and left in a short time. The experience with the first electrician was still quite fresh in my mind, and I was afraid that I would encounter more prejudice. When months later another inspector came, he also asked for the building engineer, but he was not to be fooled. He looked straight in my face and asked, "Are you building this house by yourself?" I told him the truth then, that I could not afford to pay anyone to build the house. Despite my fears, the inspectors proved to be very helpful and understanding.

So far there had not been any work demanding great physical strength. The next phase was to build the stronghold of the house: the square in the middle of the house that would hold together roof, walls, and floor. That was heavy work. Four telephone posts had to be set in poured concrete blocks. I had help from some of my friends, which I greatly appreciated. The four posts are the corners of my kitchen, forming a square of about 10′×10′. They were then held together by 2×10 lumber at the bottom and at the top with large nails and spikes. Then came the question of the roof: how does one build a roof on a round house? I had no roofing plans whatsoever. My architect friend helped me out again. He drew some outlines on a sheet of paper, and there was the solution. My friends started to nail the first rafters in place, and another phase of the building process had begun.

This time it meant a great deal of climbing up and down ladders until I finally used a sheet of plywood to walk on. Since I worked mostly on evenings and weekends, the progress was rather slow. There were days when I nearly fell off the roof, which I took to be a warning of an accident-prone day—I usually stayed on solid ground when that happened a few times during the day! Each person's biorhythms differ, and I know that there are certain days when an individual is more accident-prone than on other days. A great deal of measuring and checking was necessary in order to have the rafters well aligned. At times I used rope slings for the suspension of one end of the rafter until the other end was nailed in place.

But not all was work. On hot summer days I would take a quick dip in a small creek. Under a bridge there was a natural waterfall. The waters came rushing down, cool and refreshing. Some snakes lived between the crevices under the bridge. They were nonpoisonous, of course, and while I was standing underneath the shower a few feet away, they would gracefully sway in the water, watching me. They never bothered anyone, except the trout that lived under some boulders. At times, when things did not go right, I would call my dogs and we would take a walk "around the block." That meant walking downhill, then along the road to have a short talk with some neighbors, and up the hill again.

I also learned how to deal with a thumb or finger that I hit with a hammer. I would take the affected part, squeeze it very tightly with the other hand, and hold it above my head for a few minutes. That usually stopped the injured vessels from bleeding too much into the tissue spaces, and also reduced the swelling. I never lost a fingernail, though quite a few were hit.

Some work was heavy: for instance, the roofing paper weighed 60 to 90 pounds.

I am slightly built and my shoulders are rather bony, so I could not carry the rolls up the ladder. It was frustrating, and I finally ended up rolling them slowly up the ladder before me and lifting them gingerly onto the roof, so as not to lose my balance. My greatest joy was to see a 4′×4′ picture window installed on the roof over the kitchen. I call it the "Highlight of the House." I found this window in a lumberyard. It had a slight crack in one corner, and the gas between the double panes had leaked out. They sold it to me for only $13.

Whenever there was a chance to get some discarded building materials, I went all out to get them. My friend the arborist found a source of discarded insulation material. Mostly small and at times large pieces of fiberboard ended up in a huge trash bin, and so did I. I was given a ride to the location and then spent hours on either Saturdays or Sundays throwing out these bits and pieces from among the trash. When I occasionally poked my head out of the huge bin, bypassers looked at me in a strange, startled way. Well, that material helped me quite a bit and I had some fun sawing and nailing these pieces together. It looked like a huge jigsaw puzzle. I used this fiberboard both underneath my floor and as part of the house wall. The floor was put down in the same fashion as the roof, except that I did not have an overhang. Actually, I finished the wall before I started the floor, using 3⁄8″ exterior tongue-and-groove plywood. It bent well to the round form of the house and I had no difficulty nailing it in place. Now, about five years later, the plywood (treated with creosote oil stain) shows no signs of splitting. With the outer wall finished, the house looked much more like a shelter.

When the floor joists were in place, it was time to have the plumbing installed. I knew a friend who was a plumber, and he agreed to a sort of "bartering" in terms of labor. We agreed that I would give him three hours of labor for every hour that he gave me. I finally had cold water inside my house! Days later a septic tank of about 1,000 cubic feet was installed; the connection from pipe to tank I did by myself. That was not difficult; all I used was plenty of cement between the opening of the tank and pipe. And so, in return, I helped the plumber's wife, who was very busy with a health food store and a large garden. The plumbing was, however, not acceptable to the inspectors, since the plumber lived "right around the corner" in another state. Consequently I had a hard time finding an agreeable plumber from the "right" state to okay the plumbing system my friend had installed. It cost me about $250 just to do that, plus ripping out some plastic piping that had to be replaced with copper. I thought I could get away with it, but the plumber had to put his signature on a piece of paper and it had to be "correct." I began to wonder about the United States and how "united" it is.

By now it must be apparent that I am no great teacher. I cannot tell how to build, but if I can instill into the reader's mind the hope that she also could build a dwelling, then that is just as important. When I think back on it now, I really got a kick out of using ingenuity to make do with what I had or improvising and inventing to make something serve as a tool. Whoever wants to build a house or other structure would be wise to look around for any sources of materials that could be useful and cheap. There are houses being torn down all the time, and often salvageable lumber with nails still in can be found for almost nothing. There are dumps and there are wrecking businesses that sell doors and windows and a great many other useful items much more cheaply than stores. All my doors are secondhand and they just look fine to me. Window sashes I bought new for $5.00 apiece. A carpenter showed me how to frame them, and I did the work as well as I could. Of course, the lumber that I had free for the framing was not straight, so now I live in a slightly crooked house, but that is part of its charm.

The other day I had a visitor who had heard of my house building. He wants to write a book about useful materials that could be found for building and interior decorating at very low prices. He found plenty in my place! I told him how I had found panels of real plywood in a lumberyard for $2.95 apiece. The cheapest wood panels here are now from $5.00 up. Apparently the staining of the panels was "irregular" and not evenly distributed. I found pieces of carpeting, which were free then but even now could be bought for low prices. I intend to nail them onto the floor of my bathroom. Since I am very much interested in all kinds of woodworking, I had visited a wood-veneer factory. Soon I was given permission to pick up all the discarded and torn pieces of different kinds of wood veneer. Again I was overjoyed, and I'll find many uses for it eventually. Leftover pieces of my wood paneling also found good use. The plywood panels came in 8′ heights and my exterior wall was only about 7′ high. I used many of the trimmed-off pieces on one of the interior living room walls in sort of a mosaic design.

I was fortunate enough to be able to live close to the building site while I was working on my house. The trailer I lived in used to be a construction trailer and was quite run down, with rats and mice living in among the insulation. This trailer had no water or plumbing facilities, but some of my neighbors let me use their facilities for the time being. The trailer was my home for about a year—until one night I felt something plump walk over my pillow. That was enough! I took my mattress and moved into my house and have lived there happily ever since. There was a pile of about forty sheets of Sheetrock waiting to be nailed to the interior walls. This pile formed my bed for quite a few days.

There was no grandiose moving in or housewarming, but there was a small celebration. I brought my little teakettle along, made myself the first cup of tea in the house, and had some cake with it. There was nothing that would get me out again; I felt at home at last!

With winter approaching, I first completed the kitchen square, which was to become my living quarters until spring. A little potbelly stove kept me quite warm. I had plenty of books to read, a radio, and visitors off and on who wanted to see the progress I had made. I have no recollection of the time it took me to build the house. My savings had long been used up, and I bought building materials whenever I had saved some money from pruning and landscaping jobs. There was no such work during the winter, and I began to take on a once-a-week housecleaning job that paid for my living expenses to some extent. Much of my food came from my garden and what I had canned in the summer. I also had chickens. Slowly the interior took shape. I built a cabinet for the sink, and some storage cabinets, and finally had sinks and the toilet installed. After a year-long search, I finally found an old iron bathtub big enough for me to immerse myself fully in it.

I like the decorating aspect much more than building, and that last phase might last another few years. The house plan is as follows: the living room forms a half circle with a sunken platform for a possible fireplace. The kitchen, in the middle of the house, has three doors leading to the bedrooms (on each side) and the bathroom, and a large opening with counter to the living room. Behind the bathroom is a small storage room. There is no basement.

When all the walls were finished, I began to do some remodeling. That is the joy of a homeowner. There is no landlord to ask for permission to cut some holes in the walls. Since the kitchen did not provide enough space for a refrigerator, I had to install it in the storage room. But soon that became a nuisance. Then I had the idea of making an opening between the kitchen and bathroom wall, where the refrigerator is now, fitting snugly halfway in between the rooms. The coils in the back of the refrigerator help to heat my bathroom a little, too.

Other openings in walls followed. I put a window between the kitchen and bathroom and can see the skylight from the bathroom now. The exposed studs and wall spaces I dressed up with aspenite wood, which looks very decorative. Many hanging flowerpots grace the open spaces between the studs. For my dishes and pots and pans I built a hanging wall cabinet. The first I built was a disaster. I had no plans and

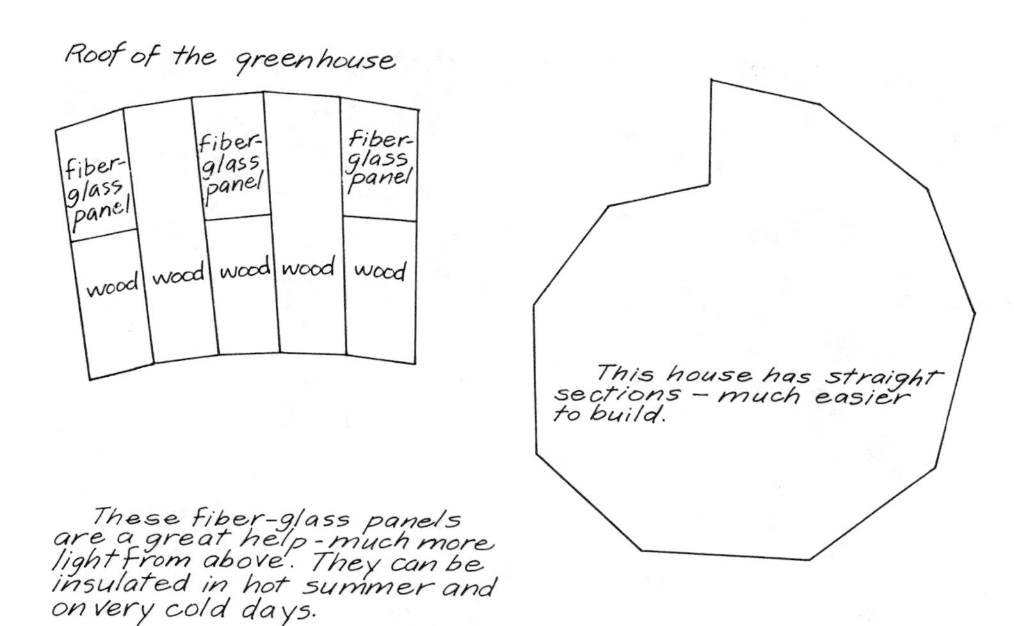

am not good at drawing and designing. So I just thought it over in my mind. The cabinet's dimensions turned out all right, but the lumber that I used was not strong enough. Off and on I have visitors who give me good advice and hints about my house building. This time, a woodworker pointed out that the cabinet I had just built would not hang for very long on the wall. It took me about one week to get over the shock, but I finally took the cabinet all apart and started anew—using stronger corner pieces and dowels to hold things together!

In the neighboring community there is quite a bit of activity going on; seminars are held periodically on alternative energy, gardening, beekeeping, and so on. Last summer's seminar dealt with building a greenhouse onto your house, so that it could distribute the collected solar heat to the house. A friend of mine, a designer, made plans to add a greenhouse to the south wall of my house. It was quite a task to design it, as it had to follow the round contour. In August of last year, during another seminar, we built this greenhouse in about a week's time. Altogether we were about ten people. Some of us were more or less experienced; others had no carpentry experience whatsoever. The outer wall of the greenhouse was built up with five courses of cinder blocks. At first these were laid freely, without cement between. When two courses were in place, we poured concrete into the holes between the blocks. Then we laid two more courses, poured the concrete, and so on. This way, the wall grew very fast and was up in no time. The blocks were waterproofed from the outside, and 1″ Styrofoam set against them. To ensure greater warmth, we heaped soil along the outer walls, creating a sloping mound to the top of the concrete blocks, then planted it to prevent erosion. The framing on the south and east sides was covered with Kalwal, a very durable material that permits a maximum of sunlight to enter. The south wall faces the sun at a 60° angle. The floor of the greenhouse is simply dirt covered with a 6″ layer of gravel. At the outer perimeter we laid drainage tiles. The roof has alternating panels of fiber glass and wood.

A built-on greenhouse not only functions as an additional heating source for your house; it provides greens for salads all winter and a place to start seedlings for spring planting. I had lettuce, endives, chives, celery, swiss chard, and marjoram all winter. As I sit here writing this article, the heat from the greenhouse warms my home. It is a sunny day in February. The outside temperature is about 20° F. (quite cold);

(Photo by Dorothy Dobbyn)

inside the greenhouse, the temperature is about 65°. On sunny days like this, I don't have to start a fire in my wood stove until late afternoon. In the greenhouse, the nine 55-gallon drums that serve as footings for the tables the plants grow on collect and distribute the heat of the incoming sunlight. They are painted black to aid in absorption of the sun's rays and are filled with water. These help prevent freeze-ups inside the greenhouse during the coldest nights. As there is still too much temperature variation in the greenhouse (as much as 30°), I plan to install a small wood stove for use in the worst weather. I am still experimenting with this addition to my house, and have learned now which plants can and cannot take the degree of cold. I saved all of my neighbor's and my own flowers from freezing by transplanting them into pots, and have seedlings growing and green vegetables as well.

I am still learning about my house and how to make it comfortable. At a conference on alternative energy I learned the value of insulation and talked with people who gave me ideas on how to make my house easier to heat. The floor of my house is very cold, so someone suggested that I place a sheet of plastic as a vapor barrier on the earth under the house, and then add Styrofoam insulation (2″ thick) under the flooring (attaching as well as can be done from under the house). A great many solar-oriented housebuilders now berm up the soil on the north and west sides of the house. Trees and shrubbery planted on these sides will help to serve as a windbreak. I am applying as much as I can of what I learn to my house. These seminars are also a place where I can meet with other women who are building and we can share experiences and learn together.

It seems a long time ago that I began to build my house. I've gained a lot of strength, much of which came from living alone during this time. I learned to rely on myself, to trust my intuition, to have the confidence to try. I know that if I'd been living with a man and learning to build, I would have given him most of the decision-making, and I would not have been so involved with the building processes.

I was also raised with the belief that men know and do better in certain areas than women do. I don't think I would have questioned that assumption—even if the man I was living or working with had known little or nothing about building.

I feel as though my house and I have grown together. If I've had times of tension and uncertainty, I've also had periods of joy as the building grew bit by bit. Now I have my round house—a warm shelter and a part of my life.

If the idea of a round house appeals to you but you have some questions on how to go about designing particulars, here's a lead that might be worth following: in *Low-Cost, Energy-Efficient Shelter* (see "Resources"), one of the USDA (United States Department of Agriculture) plans shown is for a round house. Unfortunately the plan number is not given. USDA plans are very cheap (only a few dollars) and pretty well detailed, so you might want to write and ask about this plan. Include a photocopy of page 367 of *Low-Cost. . . Shelter* for reference. USDA plans are ordered through Superintendent of Documents, U. S. Government Printing Office, Washington, D.C. 20402. You may not want to follow this plan exactly, but it will provide you with some good basics to build from.

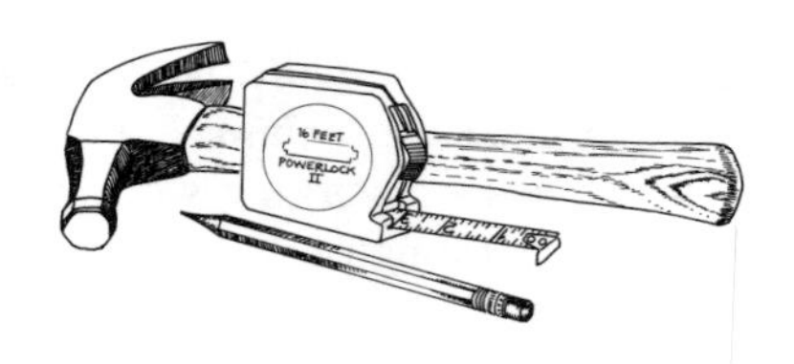

A Second Key: Materials

Learning about materials is as important a part of carpentry as learning to use tools or discovering how things work structurally. Knowing about materials will help to demystify carpentry for you, whether you are wondering what to use to patch a leaky roof or trying to find an alternative to conventional interior wall panelings. Much information about materials is woven throughout this book, as different women talk about choices they've made in house building, roofing, and so on. This section is intended to give you some basic information about common materials and the language to talk about them in.

One of the most intimidating situations for anyone who's decided she wants to do her own carpentry is to walk into a lumberyard or building supply store and find that the "common language" being spoken is entirely foreign. Far from being willing to explain or define what "S4S" or "W/P Sheetrock" means, most lumberyard personnel I've encountered seem to delight in muddling and mystifying one further. There are exceptions, of course, and that rare friend in the building supply store can be a treasure of information and help. Don't be overwhelmed by the terms, definitions, explanations, and such in this and other carpentry books. As you work with the materials and deal with different situations, the language will become more and more comfortable, direct, and meaningful. Don't be afraid to ask for an explanation if you don't understand the pricing or grading of some lumber you're interested in, or want to know what kind of caulking to use for a special job.

Another wonderful source of information about carpentry materials is the building supply store itself. Spend some time poking around, reading labels and looking at what's available. Once the store personnel are reassured that you aren't trying to shoplift the wet patch or rip off the weather stripping, they'll probably leave you alone to your discoveries.

This section is subdivided: "Lumber and Lumber Products"; "Nails, Screws, and Other Fasteners"; "Flashing and Caulking," and so forth. Other information about materials is included in specific sections and in "The Language of Carpentry" at the end of the book. New materials are constantly being manufactured and introduced, so if you are about to do a particular job you might want to do some investigating to see if any new or better materials or systems have been developed.

LUMBER AND LUMBER PRODUCTS

Lumber is the most common material used by the carpenter and probably the most diversified in terms of products. Basically, all lumber is milled from conifers (cone-bearing trees) or deciduous (broadleaf) trees. Wood from the conifers is called *softwood,* even though

---

Wood that is naturally resistant to decay includes the heartwood of bald cypress, black locust, black walnut, the cedars, and redwood.

---

it is not in all cases actually soft. Wood from the deciduous trees is called *hardwood,* even though it may in some instances or species be quite soft. The softwood trees include Douglas fir, western larch, spruce, the pines (southern, white and sugar), redwood, and cypress. The hardwood trees include: willow, walnut, white ash, birch, maple, oak, and cherry. The U. S. Government "Wood Handbook" lists some of the common uses of hardwoods and softwoods as follows:

Softwoods	**Used For**
Douglas fir	Framing members (joists, studs, rafters, braces, etc.), sheathing, subflooring
Western red cedar	Shingles, siding, cornice work
Western larch	Framing members (as above), sash and doors, interior trim
Sugar pine	Frames, drainboards, sash, doors
Redwood	Porch work, doors, frames, mudsills

Hardwoods	**Used For**
Walnut	Cabinetwork
Maple	Flooring, doors, interior trim
Oak	Sash and doors, interior trim, flooring, stair treads
Birch	Doors

From this list you can see that, in general, you'll be working with softwoods and using hardwoods only for special areas or uses (e.g. cabinets and flooring). What type of wood you actually use for a particular job or project will be largely determined by the area you live or work in. Certain types of lumber are cut, milled, and used only in certain regions; what you use for framing a wall or sheathing a roof in New England may be quite different from what is commonly used in Oregon. Learning what lumber is available in your area is a simple matter: talk with other builders, go into local lumber companies or mills, and ask questions.

Lumber is a general term for wood that is sawed into boards, planks, structural pieces, and timbers of particular lengths and sizes. *Boards* are 1″ thick and any width and length. *Dimension lumber* is material from 2″ to (and including) 4″ thick, 2″ or more wide, and any length. *Timbers* are pieces thicker than 4″. Lumber is spoken of in terms of its thickness (or depth) and its width, and the dimensions are commonly expressed in inches. A piece of material 1″ thick by 8″ wide (any length) is called *1×8* (1 inch by 8 inches). If you see a notation calling for 1× material, this means material that is 1″ thick by any width (and any length). Other common sizes are in multiples of two: 2×4, 4×6, 2×8, and so on. Lumber is cut in lengths that are multiples of two, and usually available in lengths up to 24′. The most common lengths available are 8′, 10′, 12′, 14′, and 16′. Other lengths may be special-ordered or cut at the lumber store if you ask. Hardwoods and certain fancy trim pieces may be sold in exact lengths because they are so expensive. For instance, you may be able to buy exactly 3′ of special molding for a project.

In talking about lumber, the term *board* is usually employed for 1× material. *Planks* are 2× material, usually in widths of 6″ or more, used horizontally as flooring, decking, and so on. *Beams* and *girders* are larger pieces (4× and up) used for structural supports. (A girder is the larger of the two, though the terms are often used interchangeably.)

Lumber is priced by the *board foot,* which is a unit equaling a piece 1″ thick, 12″ wide, and 12″ long. To convert material other than 1×12 to board feet, you have to use a formula:

$$\text{board feet} = T \times W \times L$$

(T = thickness of lumber in inches; W = width of lumber in feet or fractions of feet; L = length of lumber in feet). For example, say that you need 30 pieces of 2×6 material in 10′ lengths and want to know how many board feet you need. Using the formula, you set up the problem like this:

$$\text{board feet} = 2 \times 6/12 \times 300$$
$$\text{or board feet} = 2 \times \tfrac{1}{2} \times 300$$

Your solution: 300 board feet.

Usually a lumberyard will do this converting for you, but make sure that you double-check the figures. I've had the experience of running short by about one third of the materials needed to put down a new floor because the lumberyard made a mistake in its calculations. It was an expensive mistake in both time and materials!

Usually lumber is priced on a basis of 1,000 board feet (this may be expressed as /M—e.g., $400/M). To estimate the cost of the lumber you need, multiply the number of board feet by cost per 1,000 (/M), and divide that number by 1,000. To use the example above, if you need thirty 10′ lengths of 2×6, first convert this to board feet (300 b.f.). Now multiply 300 × 400 (lumber sold at $400/M) and divide that figure by 1,000. Your cost will be $120.

Sometimes lumber is sold by the linear foot, especially if it is very special hardwood or fancy finish material. In this case, you simply need to add up the number of linear feet of material you'll need and multiply that number by the linear-foot price. Whenever someone gives you a price for lumber, be sure to ask specifically whether this is a board-foot price or a linear-foot price. Plywood, particleboard, and hardboard are all sold by the sheet (usually a 4′×8′ sheet unless otherwise specified). Wood shingles are sold by the square (see "Adding the Roof" section). Some interior paneling and some flooring is sold in packages or units meant to cover a certain specified floor or wall area. Always be sure that you are absolutely clear about the material you are buying—how much it will cover, and how much it costs.

The price of lumber will depend primarily on its grade, as well as on the species of wood and the finish. There is a more or less uniform system of grading lumber according to its structural soundness, freedom from defects, and so on. The U. S. Department of Commerce sets up certain standards, which associations of lumber-producing companies then use to establish grades for their lumber. National associations such as the Southern Pine Association, the National Hardwood Lumber Association, and the Red Cedar Shingle and Handsplit Shake Bureau set up guidelines for all lumber mills to follow in producing and grading lumber of each species. Here are the most common grading systems for hardwoods and softwoods:

In working with lumber, you usually want to use the most economical grade that's sound enough to do the job. For most construction a medium grade (Construction or Standard) is the best choice. The Uniform Building Code spells out explicitly what grades of lumber are to be used for each phase of construction. Becoming familiar with general categories of use will allow you to make safe choices of lumber grades and keep the cost of your building within reason.

Lumber is sold either *surfaced* (planed smooth) or *rough.* In general, surfaced lumber is much easier to work with. Rough is used for certain special effects (e.g., a rough siding will give your house a "rustic" look). Rough lumber is a few cents per foot cheaper, so is often used as a minor economy item. Abbreviations have been developed within the lumber industry to tell you how many surfaces and/or edges of a piece of lumber are planed smooth and how many are left rough. Here are some common abbreviations:

S1S—surfaced one side
S2S—surfaced two sides
S4S—surfaced four sides
S1S2E—surfaced one side, two edges

Another commonly used abbreviation is "T + G," which means "tongue-and-groove." Still another is "D + M," meaning "dressed and matched." This is usually

Hardwoods (Oak, Maple, Etc.)	
top	FAS (firsts and seconds)—about 83⅓% clear
to	Selects
lower	#1 Common (66⅔% clear)
grades	#2 and #3 Common
	(*clear* means relatively free of defects)

Softwoods (Douglas Fir, Western Larch, Pine, Etc.)			
	Boards (1×)	*Dimension Lumber (2×)*	*Timbers (5×)*
top to lower grades	#1 Common or B & Better	Select Structural or Select Decking	Select Structural
	#2 Common or C Select	Construction or Commercial Decking	Construction
	#3 Common or D Select	Standard	Standard
	#4 Common	Utility	Utility
		Economy	

employed when referring to V-rustic or shiplapped siding, where boards are dressed (milled with special edges) to match (fit together in a particular way). By now you should be able to decipher what "200 b.f. 2×4 S4S Standard or Better" means—but there's more to come.

The actual and nominal sizes of any given piece of lumber are quite different because the piece loses width and thickness in the milling process. A modern 2×4 is actually 1½"×3½". It is important to know this difference in figuring out the dimensions of even the smallest carpentry job. For example, you may decide to put up a small area of wood paneling in a room and calculate your needs for 1×8 boards. The boards you get (if new) will be ¾"×7¼"—which will make a bit of a difference in your coverage! Rough lumber is usually the same in terms of actual and nominal dimensions, but most of the lumber you'll be using will be planed smooth (surfaced), and you'll have to take the nominal/actual difference into account. This difference can create problems when you're remodeling, too. For example, if you need to replace some wall studs, you may find that your new studs are ½" thinner than the old ones; you'll have to fur out the new studs to make an even nailing surface. ("Furring out" means adding wood or shims or other material to increase thickness.)

Yet another factor influencing lumber is whether it is green, seasoned, or kiln-dried. Freshly cut lumber has a high moisture content; as the lumber dries, the moisture leaves the wood, causing it to shrink. *Green* (freshly cut) lumber is wet, heavy, and subject to a lot of shrinkage. If you build something with green lumber, be prepared for changes; as the wood dries and shrinks, joints will pull apart, nails may pop out, and gaps will appear between previously tightly fitting boards. *Seasoned* lumber is lumber that has done most of its shrinking, though it will still shrink a little its first year. Lumber may be naturally seasoned if it is stacked with air space between the boards and left in a sheltered place. It must be carefully and evenly stacked to avoid warping as it dries. Another method of drying lumber is kiln-drying. In this process the lumber is actually dried in a carefully controlled oven. *Kiln-dried* lumber is more expensive than air-dried or green lumber and is used for fine work such as door jambs and cabinets.

Lumber is sometimes referred to as edge-grain or flat-grained. Softwood lumber that is cut so that the annual (growth) rings form an angle of 45° or more with the surface of the board is called *edge-grain* lumber. Hardwood lumber cut this way is called *quartersawed.* Edge-grain or quarter-sawed lumber will shrink less than flat-grained lumber, making it more expensive, and is harder to cut. *Flat-grained* softwood lumber is cut so that the annual rings form an angle of less than 45° with the board surface. Hardwood cut this way is called *plain-sawed.* Better-quality wood shingles are edge-grain; they will shrink and twist less than the flat-grained (lower-quality) shingles. Oak flooring advertised as quartersawed has this same advantage of less shrinkage. Most structural lumber and 1× is usually flat-grained.

One final thing to consider in wood is the defects. Some defects are present in lumber because of the area of the tree a piece comes from; a lot of knotholes means that the piece comes from a tree or part of a tree which had many branches. Some defects (twisted, bowed, or otherwise warped pieces) are caused by poor drying or seasoning of the cut lumber or by poor handling of the wood after it has been cut and seasoned. Even the best kiln-dried lumber can become warped if it is carelessly handled. Some defects that you'll probably encounter in the lumber you buy and work with are:

Warping A board or piece of dimension lumber can warp in a number of different ways. When a board *twists,* one end usually turns up. *Bowing* refers to a curving along the entire length of the piece, so that it resembles an archer's bow. Wide pieces of 1× are subject to *cupping,* which means that the edges curl up. One of the most common defects in structural 2× material is a *crook*—somewhat like a bow, but the curving happens along an edge. If you hold the material on edge and sight down along the top edge, you'll see a *crown* (rise) or a *hollow* (dip). Almost every piece of lumber will have a slight crook. If you are using a piece for a roof rafter or floor joist, place it with the crown *up;* the weight of roof sheathing or flooring will then force the crown downward, causing the board more or less to straighten out.

Splits and checks Here the wood fibers actually separate along the grain, creating cracks in the lumber; usually these begin at the ends of a piece. If you can't afford to cut off and discard the part of a board that is split or checked, be very careful when nailing it; if the board has a very straight grain, a nail placed too near to the check may send the check down the entire length of the board.

Knots These are found in pieces of wood milled from a portion of the tree that had limbs or branches. Knots vary in size and severity, but generally are considered to weaken the lumber if they are many in number or large in size. The wood surrounding a knot will be very dense and hard, so be ready for this when sawing. Pieces may chip away, too. Be especially alert for this if you are cutting with a power saw; a good precaution is to be sure to wear safety glasses.

Wane This happens along the edges of boards or 2× when a piece is milled close to the outer part of the tree. The wane is an irregular bevel along an edge of a piece of lumber; it is caused by bark, which may still be clinging in shreds along the edge that is marred. This isn't a serious defect, except that it gives you an uneven surface and slightly uneven width. It can be a problem if in a piece of flooring.

Blue stain A common moldlike fungus causes this familiar blue-black stain on wood. It won't harm the structural soundness of the wood.

Decay and/or dry rot Wood that has been subjected to insect, water, or moisture damage may be soft, crumbly, and/or spongy. Damaged boards should not be used if you want your work to last at all. You *can* cut off damaged portions and salvage the rest. Structural wood such as 2× that has areas of decay should likewise not be used; it is no longer structurally sound.

There are still more common terms used when talking about wood and lumber. The *sapwood* of a tree is the outer section of the tree between the center heartwood and the bark. Sapwood in the living tree is gradually turned into heartwood as the tree grows new layers. *Heartwood* is in most instances a darker color—it is the center wood of the tree. In general, the heartwood is the finer quality and lumber milled from heartwood is more durable than that milled from sapwood. When a tree is cut lengthwise, the annual growth rings of the tree form a pattern called the *grain* of the wood. If the tree has very small cells, tightly packed together, the grain is called *fine or close grain.* If the cells are large and open, the grain is called *coarse or open grain.* Oak and walnut are examples of open-grained woods; maple is an example of fine-grained. If the fibers and wood cells are fairly straight and parallel, the wood is called *straight-grained.* Twisted or slanted grain is called *cross-grained.*

Plywood

Plywood consists of built-up layers (plies) of wood glued together. It comes in sheets that are usually 4′×8′ and in thicknesses from ⅛″ to a full inch. When plywood is made, the grain of each layer is turned at a right angle to the layer next to it so that a very strong cross-hatching effect occurs. The inner layers of the plywood form the *core;* the outer layers are called the *faces.* Plywood may be softwood or hardwood. Generally, softwood plywood is used for sheathing, subfloors, and concrete form work. Hardwood plywood is used for paneling, cabinets, and work where a special finished surface is wanted. Plywood is made in interior, exterior, and interior/exterior types. These designations actually refer to the type of glue used to bond the layers together. Interior plywood is made with glues that are not waterproof; it is only to be used inside, where moisture is minimum. If used outside, this type will begin to peel and disintegrate. Exterior and interior/exterior plywoods are made with waterproof glues. They may be used unfinished outside, but will last longer is given a protective coat of paint or wood preservative. These exterior-glue plywoods can be used for almost any purpose.

Plywood comes in grades, which are stamped on each sheet. The grades range from A (top) to D (lowest) and are determined by the quality of the faces of the plywood. A grade is smooth and neatly finished; B and C grades may have knotholes and surface plugs; D grade has larger knotholes and plugs and may have splits. A sheet of plywood can have one face A grade and the second face C grade—or any combination of grades. Plywood is also called G1S (good one side) or G2S (good two sides). You can use G1S anywhere you need only one surface exposed; it is, of course, cheaper than G2S. Plywood is manufactured with a variety of special finished surfaces (such as textured, grooved, and corrugated) for use as exterior siding. Rough-sawn Douglas fir plywood siding, for example, makes an attractive, long-lasting exterior siding that is easy to apply and quite a bit cheaper than actual board siding. In general, plywood is an easy material to work with and is quite strong and durable. As our resources of lumber are not infinite, the use of plywood for siding, sheathing, and other aspects of construction becomes a reasonable alternative. Plywood produces less waste and more material from fewer trees.

Hardboard and Particleboard

Hardboard and particleboard are two more wood products, manufactured in sheets 4′ wide, and 8′ to 12′ or 14′ long. Either of these products may be used for interiors of cabinets, paneling, shelving, and so on. *Hardboard* is made by pressing wood fibers together to form dense, smooth sheets with a fairly hard surface. It comes in thicknesses from ⅛″ to ¼″. *Particleboard* is made with wood chips and flakes bonded together with adhesives or resins. Particleboard is usually available in thicker pieces than hardboard. It may be painted or stained. Although regular particleboard is not meant to be used outside, I've seen small buildings made entirely of particleboard weather through rainy coastal winters; this material is surprisingly durable. Particleboard is now also being manufactured with a special waterproof paper on both faces. The paper is smooth on one side and rough-textured on the other, to simulate wood siding. This type of particleboard can be used outside if it is painted.

Trims and Moldings

Specially shaped wood trims and moldings are used most frequently around windows and doors and to finish a room around the ceiling or floor. Some of these special pieces that you might find useful in your work include the following. *Quarter round* and *half round* pieces are used to hide joints; for instance, quarter round can be used where a wall surface butts against the ceiling. A *drip cap* is placed along the upper edge of the exterior of a window to help shed water. The *base, or baseboard,* is run along the floor to hide the joint of wall and floor; the *base shoe* is then set in front of the baseboard. *Coving, or cove,* is used to cover the ceiling-wall joint, for inside corners, and so on. *Stop material* is used around windows and doors. *Outside corner pieces* are usually placed where the name implies—to cover an outside corner joint.

There are many more shapes and types of trim and molding that you might discover and want to use. This is another instance where it helps to spend some time in a lumberyard or building supply store looking over their supplies so that when you come up against a special problem, you'll have some ideas to draw on. In general, trim and molding is sold in specified lengths. It's also good to know that you can order molding that might not be in stock, particularly if you're restoring an old house and need some very ornate or special pieces. Most lumber stores have books of moldings that are still being manufactured in limited supply.

More information about lumber, including types of siding and wood paneling, is in the practical section of this book.

One final matter that fits under general materials is the handling and storage of lumber. Wood should always be stacked carefully so that it won't twist, bend, or otherwise warp. Put it up off the ground, using lengths of scrap wood spaced at regular intervals and close enough so that the stacked wood won't bend because the span is too great. If your lumber will be stacked outside for any length of time before you use it, it should be protected with a tarp, plastic, or even a temporary shed. Kiln-dried lumber should be stored inside if at all possible; if you can't do this, try to buy this lumber just before you are ready to use it. Door jamb material and other fine finish material should likewise be purchased just before use and kept inside. If you have material delivered to a job site by the lumber company, be sure you are present when the delivery arrives. You'll want to check over the pieces delivered to make sure none is warped or of poor quality, and also to make sure that nothing is cracked or broken when it is dumped from the truck. You have the right to refuse an order if it isn't what you paid for: if 1× material cracks because it was unloaded improperly, or if 2× is warped or badly bowed. If you're there when the delivery is made, you can stack your lumber immediately and make sure that it doesn't have a chance to develop warps that will make working with it difficult.

NAILS, SCREWS, AND OTHER FASTENERS

Nails

Much of the work of carpentry involves fastening or joining pieces of wood to one another; this fastening is largely done with nails. Wire nails come in different sizes, gauges, and styles or types. There are general-purpose nails and nails made for very specific, limited uses. Nails are spoken of in terms of a unit called a *penny,* which is abbreviated as *d* (you will see "4d" or "20d" nails called for). In this system, the smaller the number, the smaller the nail. Nails range from 2d to 60d in size, but the most common are in the 2d to 20d range. To give you an idea of relative sizes, here are some of the nails you'll use a lot, with their sizes in inches:

2d = 1″	8d = 2½″	16d = 3½″
4d = 1½″	10d = 3″	20d = 4″
6d = 2″	12d = 3¼″	30d = 4½″

The *gauge* of a nail is a number indicating the thickness of the shank; the lower the number, the thicker the shank of the nail. For instance, you might see a requirement for a 12d, #8 nail: the "#8" refers to the gauge. Nails are usually sold by the pound, and a pound of smaller, lighter-gauge nails will of course contain more nails than a pound of heavier-gauge.

In choosing nails, you will also have to choose between galvanized and ungalvanized. The *galvanized* nails are treated to resist rusting and are meant to be used for outdoor work, or wherever the nails will be exposed to moisture. Galvanized nails usually have a rather dull finish, but some are "hot-dipped" and have a bright finish very similar to the finish of an ungalvanized nail. *Ungalvanized* nails are slightly cheaper per pound, so they should be used whenever possible.

Most nails that you will be using will have "regular" diamond points. There are special blunt-pointed nails that you might want to buy for working with hardwoods such as maple and oak. In general, a sharp-pointed nail is easier to drive but tends to split the fibers of the wood apart; the blunt-pointed nail spreads the fibers and won't split the wood as easily. You can blunt the tip of any nail by holding it with its head against a solid surface and giving the tip of the nail a quick tap with your hammer.

In deciding what size of nail to use for a particular situation, you can follow the general rule that a nail should be three times as long as the thickness of the wood it's intended to hold. Two thirds of the nail length will be driven through one piece of wood and one third into the other piece. Usually you'll want the nail embedded in the wood: i.e., you won't want the tip to protrude through the second piece.

With the size of nail you need in mind, you have a choice of several types or styles of nails. Below is a list of the most common types and their general uses:

Box nails A multipurpose nail used in general construction. Box nails have a slightly thinner shank than common nails and are not as strong. They are used whenever you want to minimize the danger of splitting the wood: for example, in nailing wood siding.

Common nails Another general-purpose nail used extensively for rough framing. Very strong.

Casing nails Come in the same weights and sizes as box nails but have a "finishing" (conical) head. Casing nails are used for putting on trim, window casings, etc.

Finishing nails Lighter than casing nails but with a similar conical head, finishing nails are used for cabinetwork, light trim, furniture, and anyplace where a small, thin nail is needed.

Wire brads Actually tiny finishing nails. They range in length from 3⁄16" to about 3" and are usually sold in a small box rather than by the pound. Used for fine work: for instance, tacking delicate wood bead or parting stops around a window.

Ring-shank, spiral-shank, helically ringed nails The shanks of these nails are specially formed to give extra holding power to the nails. Ring- and spiral-shank nails (also called annular or ringed) have barbed shanks. They are extremely hard to pull out once driven, and thus won't tend to work loose under strain. The helically ringed nail has a twisted shank that turns as the nail is driven in and is also very difficult to pull out (therefore won't work loose).

Duplex or double-headed nail Also known as a scaffolding nail. It has two heads separated by a short length of shank and comes in various sizes. This type of nail is used for any temporary work (concrete forms, scaffolding, temporary bracing). When it is driven in, the first head seats itself firmly against the wood and the second head remains protruding, making it easy to pull the nail out when you are ready to dismantle the work.

Spikes Actually huge nails, measured in lengths rather than pennies. Most of them range from 6" to 12". They are used for heavy construction.

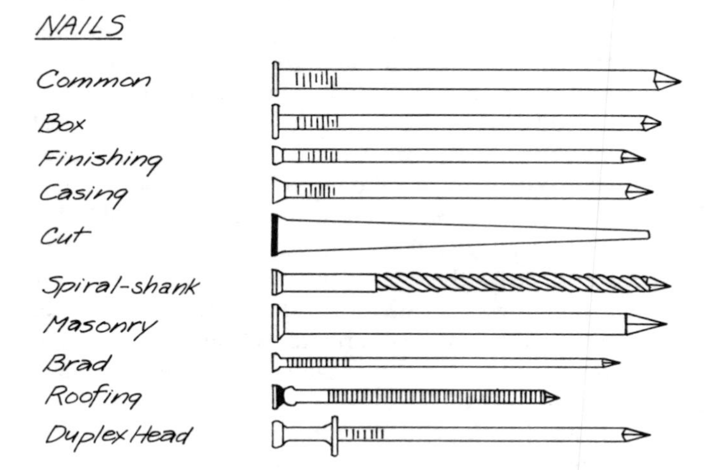

Rosin-coated and cement-coated nails These and certain other nails have a coating of adhesive or adhesive-like material that forms a bond with the wood when the nail is driven in.

Aluminum and stainless-steel nails Available for situations where you absolutely want to avoid rust and the resulting stains on surrounding wood. These nails are more expensive than galvanized nails, which may also be used for outdoor work. (Galvanized nails may rust after a long period of exposure; stainless and aluminum never will.)

Roofing nails Come in various styles and sizes for particular roofing jobs. Squat, thick-shanked nails with extra-large flat heads are used with roll roofing. These come in different lengths, to be chosen according to the type of sheathing you are nailing into and also whether or not you are reroofing and hence nailing through layers of old roofing. Asphalt shingle nails have a broad head, too, but have a twisted or a barbed shank. Some roofing nails have neoprene washers fixed just under their heads. When the nail is driven in, the washer is squeezed into place, forming a gasket-like seal. These nails are used with corrugated aluminum roofing, fiberglass panels, etc.

Cut nail A special squared (rather than rounded) nail used for flooring and for special effects.

Masonry nails Specially treated heavy nails that may be driven into concrete without bending or breaking.

Sinkers Plastic-coated nails now being used in framing. When one is driven into wood, the friction causes the plastic to melt and form a bond with the wood. Sinkers are supposedly easier to drive than conventional nails, and harder to pull out.

There are several ways to nail pieces of wood together, each with its own strengths and special uses. The "face" of a board is its width dimension, so that *face nailing* means nailing a board through its face (width) with the nail more or less perpendicular to the face. Face nailing is common wherever you join 1× material to studs or rafters, or splice together two pieces of lumber to make a built-up piece; for example, splicing together two 2×6's to make a header over a window opening.

End nailing means that one piece is nailed through its face (width) dimension into the end of the second piece. For example, if you are framing up a wall, you end-nail the bottom plate to the studs, in horizontal assembly.

Toenailing is a method used when you can't end-nail or face-nail. The nails are driven in at an angle, passing through the first piece of wood and into the second. An example of this type of nailing would be when you want to fasten a post to the wooden nailer block embedded in a concrete pier. When you are toenailing, the piece you are driving the nail into first will tend to move away from the angled blows of the hammer. You can deal with this in a number of ways. You can overcompensate in the beginning by placing the piece you are nailing well toward you so that it will be hit into the right place. You can brace the piece you're nailing with your foot, with a little block of wood tacked into place opposite where you are hammering, or with a tiny nail driven in temporarily from the opposite side. Another thing to watch for in toenailing is splitting the wood; this happens if you drive a nail in at too steep an angle too close to the edge of the wood. Experiment with different angles and positioning of nails until you get a feel for the "correct" placing of the nails.

Blind nailing is a technique used whenever you want the nailheads to be unexposed on the surface of the wood. It is common to blind-nail flooring and some wall paneling. The nail is driven in at an angle (toenailed) through the first piece of wood *through the edge.* A nail set may then be used to push the nail slightly below the surface of the wood. When the adjoining piece of wood is butted up to this piece, the nails are hidden and the face of each board will show no nailheads.

Every carpenter probably has her own techniques for driving nails. There are a few basics that might be helpful if you're just learning. First, it's important to use a hammer that is a reasonable size for the nails you are driving. A very light hammer will make driving 16d or 20d nails twice the chore it should be. For nails this size, use a 16-ounce hammer (minimum) or a 20-ounce (better). A very heavy hammer will bend or even demolish small nails, so for driving 4d or 6d nails you'll probably want something lighter than a 16-ounce hammer. My first hammer was a little 12-ounce "on sale special," which I used for years for all kinds of nails. I wasted a lot of energy pounding away with my "baby" hammer, but I was fond of it and became fairly proficient. My current hammer is a 20-ounce and, though leaning toward being a little heavy, is a good general-purpose size. You'll probably want to try two or three different-sized hammers with various-sized nails to see which match up in terms of the work you're doing and the energy it takes to do it.

The second important factor in nailing is the wood you're working with. A hard wood, whether it is classified as a hardwood or is actually a hard softwood, will be much more difficult to nail into. Very hard woods must sometimes be predrilled for nails. Driving a nail into old Douglas fir or into seasoned oak is like driving into stone or concrete. If you find yourself bending one nail after another, don't jump to the conclusion that you're a poor carpenter or incompetent! Practice on soft wood such as redwood, pine, or new Doulgas fir until you feel confident with your hammer and technique. If your nails go in but split the wood, make sure you aren't nailing too close to the edge of the wood. Try blunting the tip of your nail, also.

Holding the hammer so that it strikes the nailhead squarely, rather than at an angle, will help to drive the nail in without bending it over. Beginning your nailing with a couple of gentle blows rather than a hearty whack will let you see that the nail is beginning in the right position. A nail driven in at a slight angle will hold better than one driven in absolutely perpendicular to the wood. You'll have to practice to get the technique of hitting the slightly angled nail squarely on the head. To start, it might be best to drive your nails in straight until you have a good feel for the tool and the process of nailing.

One final factor influencing the ease of driving a nail is the nail itself. Certain styles or types are easier to drive; a box nail, for example, tends to bend much more quickly than a common nail of the same size. Casing nails are a little harder to drive than commons or box nails of the same size because of their small heads. The source (manufacturer) of the nail is important, too. Nails made in Belgium and in the United States are of very good quality and will drive without bending under ordinary circumstances. Other imported nails, sold as "bargains" and "specials," are often made with inferior steel and will bend and twist excessively. Buy good nails to work with; you'll not only save yourself some frustration and loss of self-esteem, you'll probably use fewer nails!

If, despite all precautions, the nail you're driving bends, you may still be able to save it by shortening and lightening your hammer blows. Sometimes a series of short, light blows will drive the nail in even though it has begun to bend—try it. If that fails and you have to pull the nail out, use a wood block under the head of your hammer, as shown. The block gives you extra leverage; it also protects the wood from being gouged by the hammer. If you've driven the nail into hard wood or the nail is a ring-shank or other hard-to-pull type, you can use nail nippers (see "Tools" section) to pull it. Place a wood block under the head of the nippers, too.

Another good tool for pulling difficult nails is a crowbar or pry bar, which will give greater leverage. The rule is: the longer the bar, the greater the leverage. Use a wood block under the head to protect the wood. If you need to pull a nail that has been driven flush with the surface of the wood, use a cat's paw (see "Tools" section) to lift the head enough so that you can pull with your hammer or crowbar. In some instances you might want to cut a nail off rather than try to pull it. For example, you've driven a ring-shank nail partially in and it has bent or hit some obstacle, and you want to avoid marring the wood you're working with. Here you can use your nail nippers to clip off the head and protruding shank of the nail. For extra leverage to clip larger nails, slip a length of pipe over each handle of the nippers; this will extend the handles and power of the tool.

How many nails you use in a particular situation and how you place them will make a difference in the strength of the joint and how it looks. In general, you should use the fewest nails that will secure the wood. A few well-placed nails can be as strong as or stronger than a lot of haphazardly placed ones. There are also certain features of nailing that you can adapt to your needs. One nail driven in to join two pieces of wood will hold them together, but will allow the pieces to pivot. This can be useful to know; for example, you may want to attach a temporary brace to a wall just before lifting it up. A brace with one nail can be swung into position, the wall plumbed, and the brace nailed more securely. Two nails driven parallel a few inches apart will hold a joint fairly well; this is a very common way to nail siding to sheathing or subfloor to joists. If you want an even more rigid joint, nail three nails in a triangle position. Staggering nails is another common pattern in nailing. Here the nails are positioned to form a zigzag pattern, which is very strong. This method is used when splicing pieces together, nailing on ledgers, nailing trim, and so on.

Sometimes nailing patterns are expressed in terms of the space between nails measured from the center of one to the center of the next (approximately). The pattern will be written using the abbreviation *o.c.* (on

Use a wood block under the hammer head to give yourself extra leverage in pulling out a difficult nail. (Photo by Carol Newhouse)

center), so that you may see a notation "6″ o.c. or 10″ o.c." The building code lists very specific requirements for nailing patterns and centers for all common situations: for example, how to place nails when attaching plywood subfloor to joists.

Clinching is one final technique to know about in nailing. Under most circumstances, you will want a nail to be entirely embedded in the wood it is joining, with no protruding tip on the side opposite the head. Sometimes, though, you can make use of the technique of clinching to give your nail extra holding power. To clinch a nail, you drive it through the wood so that an inch or more protrudes on the other side. This part is then smashed over and flattened against the wood. Clinching a nail can give it 30 or 40 percent more holding power. One situation in which you might want to use this technique is when building a gate with 1× material for both frame and siding. Clinching the nails will make the gate much stronger.

Screws

Wood screws will give you more holding power than nails under similar circumstances. They are more expensive and more time-consuming to use, so will be the choice only for certain jobs. Wood screws are normally used wherever you want a neat-appearing joint that can be undone without damage to the materials. They are used to fasten wood to wood or metal to wood: hinges to a gate, for example. Wood screws come in three "styles": flathead, oval-head, and roundhead. They range in size from as small as ¼″ up to 6″ and are made of brass, steel, aluminum, and even stainless steel. Wood screws for outdoor use are zinc-coated steel (rust-resistant); brass screws are commonly used for interior work. A wood screw may have a slotted head or a Phillips head, which has a "+" on the head and requires a special Phillips screwdriver.

Most wood screws are set into a predrilled hole. The hole should be made slightly smaller in diameter than the threaded part of the screw at its thickest part (wood screws taper to a point). The hole is usually drilled to a depth that equals one half to two thirds the length of the threaded part of the screw. If you are working with very hard wood, you may have to drill a deeper hole. In this case you may want to switch to a smaller bit on your drill for the deepest drilling, allowing the tapered point of the screw to be better seated in the smaller hole. In drilling a hole for a wood screw, be very careful to drill at right angles to the surface of the wood you're drilling into. Otherwise you will have an angled hole and the screw won't fit in properly: the angled head won't sit flush with the surface of the wood. It helps to train your eye to position the drill at a right angle to the wood. There are special jigs made for some drills to help with this, but they really aren't necessary. In some instances you can set a wood screw without predrilling a hole. If you're working with quite soft wood, it may be enough to "punch" a starter hole with an awl or even with a nail. If you find that you are having to *force* the screw to turn, though, it's better to drill, because the head of a wood screw is fairly soft and will be mangled if forcible turned into wood that won't accept it.

Be sure that you use a screwdriver that fits the screw. If the screwdriver tip is so small that it moves around freely in the slotted head of the screw, it can also mangle the soft metal and be impossible to turn. Choose a screwdriver that fits snugly into the slot. Phillips screws and screwdrivers likewise come in different sizes that have to be matched together. A long-handled screwdriver will give you more torque (twisting force) and is preferable except where you need a short, stubby screwdriver to fit into a tight spot. There are right-angle screwdrivers made for difficult work situations, and "Yankee" type ratchet-driven screwdrivers.

Sheet metal screws are similar to wood screws in appearance but have flat bottoms, rather than tapering to a point. They are used to join metal parts and come with oval, flat, and round heads. Drill a hole as deep as the length of this screw.

Lag screws are heavier and larger than wood screws and are used in similar ways to make strong joints (wood to wood or wood to metal) that can be easily undone. Lag screws have square or hexagonal heads and come in a range of lengths. Holes must be drilled for lag screws; they are then tightened in place with a wrench or a socket set. Lag screws are used to secure ledgers to walls, fasten metal post anchors to wood posts, and so on. They are much stronger than nails for this type of work.

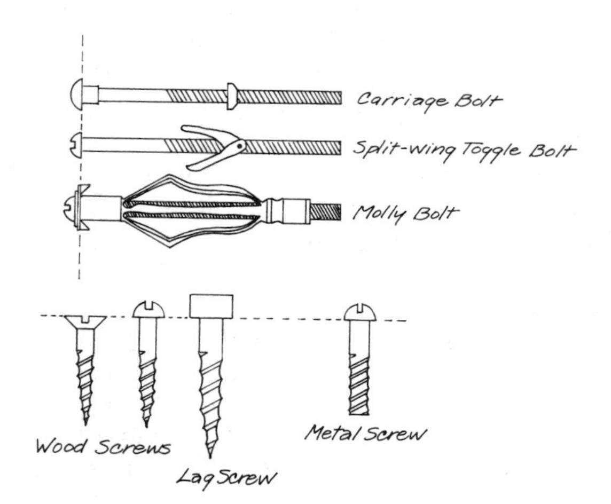

Bolts

Bolts are used when you can or want to drill through the two pieces being fastened together and want a connection that is extremely strong and can be disconnected or tightened as needed. Bolts come in sizes from ¼″ to 1¼″ in diameter and up to 30″ in length. They are of two basic types: the carriage bolt and the machine bolt.

The *carriage bolt* is primarily a wood-to-wood fastener. It has a smooth, rounded head that makes an attractive joint. Just below the head is an enlarged portion of the shank that is designed to catch and hold its place in the wood. The carriage bolt is tightened in place by means of a nut, with washer, which is turned onto the threaded end opposite the head. The hole drilled for the carriage bolt should have the same diameter as the bolt's shank. The bolt is driven into the hole with a hammer, and the washer and nut put on and tightened down. One disadvantage of the carriage bolt is that it can wear loose its "seat" in the wood if subjected to a lot of jostling or strain, or if the joint is disconnected and reconnected a lot. The wearing happens when the enlarged portion of the shank rubs or wears down the edges of the hole where it is seated. Once this has happened, the carriage bolt can't really be tightened down and the joint becomes loose.

The second type of bolt is the *machine bolt,* which has a square or a hexagonal head. Like the carriage bolt, this one is driven with a hammer into a predrilled hole. You may choose to put a washer under the head to minimize damage to the wood and to keep the bolt from being drawn into the wood when it is tightened down. When the bolt has been driven into place, a second washer is added to the threaded end and a nut is tightened on after it. To tighten the nut, you'll have to hold the head of the bolt with a wrench or socket set. Then use a second wrench or socket set to turn the nut until tight. The advantage of this type of bolt is that it can always be tightened down; it can't wear out its seat and become loose permanently. Bolts of this type are widely used to fasten big timbers in place.

In placing either type of bolt, it's important to drill a hole that is straight. If you drill at an angle, you won't be able to tighten the bolt properly. In extreme cases, a bolt set in at an angle might shear off (break) if under a lot of stress.

Countersinking a bolt, wood screw, lag screw, and so on, is a means of dropping the head of the screw or bolt just below the surface of the wood. To do this, a second hole is drilled after the first has been made for the shank of the screw or the length of the bolt. This second hole should be slightly larger in diameter than the head of the bolt or screw and should be deep enough to accept the head so that the head will be indented or positioned just below the surface of the wood. There are special countersink bits you can buy for your drill if you're doing a lot of this type of work and want it to go fast and look extra-neat. Sometimes a wood plug is placed in the hole over the countersunk head and glued there to totally hide the metal head.

Lock washers are made for use with joints that will be subjected to movement that might work the bolt loose. The lock washer is split and slightly bent; when it is tightened in place, it produces a pressure against the head of the bolt that helps to keep the bolt from working loose.

Stove bolts are small bolts used for light assembling of wood/wood, metal/metal, and wood/metal. They are used with square nuts and differ in appearance from other bolts in size and in having the entire body of the bolt threaded. Stove bolts may have flat or round heads and may be either countersunk or left flush with the surface.

Other Fasteners

Other fasteners include screw eyes and screw hooks, cup hooks, corrugated fasteners, and L-shaped screw hooks. *Corrugated fasteners* are used to join or splice fairly small pieces together in situations where screws or nails would split the wood. They are made of sheet metal and are corrugated (with alternating ridges and grooves). One end has a beveled edge; the other end is square-cut and is the surface you strike your hammer against to drive the fastener in. Corrugated fasteners are used in pairs or sets of three; in both instances, they are set parallel to one another. This type of fastener repairs wood window sash that is splitting open in corner joints and joins mitered pieces of wood, among other uses. *Screw eyes and hooks, cup hooks* and, *L-shaped screw hooks* all have threaded, pointed shanks, which are twisted into a predrilled or punched hole. The ends are specially shaped for different uses or situations.

Toggle bolts and molly screws (or molly bolts) are used to fasten things to wall surfaces that are hollow behind; gypsum wallboard, for example, or particleboard. *Toggle bolts* are long, slender bolts with screw heads and pairs of "wings." A hole is predrilled into the wall surface equal in diameter (roughly) to the toggle bolt with its wings folded against its body. The bolt is then pushed into the hole. The wings spring open and grip against the wall as the screw head is turned down. The toggle bolt is fairly strong. If you want to attach something under the screwhead, you must unscrew the wings beforehand and slip it on, then replace the wings and slip the bolt into place in the wall. If you want to remove a toggle bolt, you can unscrew it by turning it until the wings fall free. The disadvantage of this fastener is that you'll need to replace the wings if you want to reuse the bolt.

Molly screws or bolts are similar in appearance and in use. They are long, slender bolts with a cylindrical sleeve in place of the wings of the toggle bolt. The molly bolt requires a smaller hole than the toggle bolt but is about equal in holding strength. When the molly bolt is slipped into its hole and the screwhead turned, the cylindrical sleeve expands and grips the inside surface of the wall. The molly bolt can be unscrewed and the sleeve will stay in place. Either of these fasteners can be used to place pictures and other light hangings on a hollow wall surface.

A variety of metal fasteners are used to join pieces of wood together under special circumstances. The simplest of these fasteners is a *mending plate,* which is a small strip of thick metal with holes drilled in it. This plate may be screwed in place over two pieces of wood that butt together, to join and hold the pieces. A *tie strap* works the same way, but comes in lengths up to 48" and in light to heavy weights. Tie straps are used to join trusses to rafters, to tie walls together, and so on. *Corner irons* are similar to mending plates but are L-shaped. An *L strap* is a heavy-duty corner iron and a *T strap* another heavy-duty version of the same kind of fastener. A T strap or L strap may be used to strengthen the joining of a beam or girder to the wood post it sits on. Another general-purpose fastener is the metal *reinforcing angle,* which comes in lengths from 3" to 9" (see photograph). *Post caps* are used to connect posts to girders or beams. They come in a number of different styles and must be bought in the exact size to fit the post. *Post anchors* are embedded in concrete piers or foundation walls and form a connecting point between the concrete and wood posts seated upon it. Again, these come in specific sizes for different-sized posts. *Joist hangers* are used to connect floor or ceiling joists to header joists or to beams. They vary in size and strength as well as in style. There are other special metal fasteners, but those listed are the most commonly used.

A metal joist hanger *allows you to set a joist flush with the girder or beam—useful if you have a limited space heightwise.* (Photo by Carol Newhouse)

Fastening things to concrete presents some special problems that can be solved with the right fasteners. For very light materials, concrete nails may be driven in with a *steel stud driver* (see "Tools" section). More commonly used are the slightly heavier *masonry shields and anchors.* These are similar to one another in appearance and use; the shield is of malleable iron and the anchor of lead. To use either, you have to drill a hole into the concrete. Use a carbide-tipped (masonry) drill bit. If you're using a shield, the hole should be about $\frac{1}{16}$" smaller than the diameter of the shield; if you're using an anchor, drill the hole exactly the diameter of the anchor. You can put a small piece of tape on your drill bit to show you when the hole is deep enough. Drive the shield or anchor into the hole, using a wood block to protect the shield or anchor from your hammer. When the shield or anchor is in place, turn a machine screw or bolt of the proper size through the material you want fastened and into the shield or anchor. As the screw goes in, it draws up on an expansion nut, which causes the shield or anchor to push out against the surrounding concrete, thus wedging itself in place.

Another way to fasten things to concrete is to use a *powder-actuated tool,* which is a gunlike tool that uses a small explosive charge to drive a special case-hardened (extra-strong) nail into concrete. This tool can be rented from some tool rental shops and is useful if you have to fasten furring strips to concrete walls or in any instance where you have a lot of fasteners to deal with.

At the opposite end of the fastening scale is the *hand-powered staple gun,* which drives small wire staples into a variety of materials where a lot of strength is not necessary: for example, to fasten felt paper to wall or roof sheathing. There are a couple of styles of staple guns, and most of these will take more than one size of staple for different types of work. (See "Tools" section.)

FLASHING AND CAULKING

Caulking

These two seemingly quite different materials are often used in conjunction with one another or to solve similar problems in construction. Basically, both are used to waterproof and weatherproof a building. Caulking is used to fill joints and cracks between like and unlike materials. Most caulking comes in tubes and is in a flexible form when applied. A caulking gun is a tool designed to hold tubes of caulking and to force out a bead of caulking as pressure is applied to the handle of the gun. Architectural-grade caulking compound is one of the most commonly used caulkings. It comes in different colors (red, white, etc.) that may be matched to the job. Its widest use is to seal joints in wood siding and trim. Once applied, this caulking will harden to a fairly solid form. It may be lightly sanded and can be painted over.

Silicone-based caulking compounds are more expensive but form an excellent, durable, waterproof bond between glass and wood, metal and wood, or plastic and wood; any of these juxtapositions might arise if you're installing a skylight. Caulkings of this type usually remain slightly flexible and cannot be sanded or painted. They are more often used to form a gasket-like seal between unlike substances than to fill in joints or cracks in wood. Like the architectural caulking,

(Photo by Carol Newhouse)

There are two new products that you may want to locate and try; these are called "sealants" rather than "caulking," but they do the same work. One is Polycel One, a new foam sealant that can be used around window and door frames, around plumbing and other pipes, and so on. It is supposedly much more effective (and cheaper) than normal caulking. Unfortunately it comes only in aerosol cans in small quantity (18 ounces); larger amounts may be specially ordered in refillable cylinders weighing from 20 to 100 or more pounds. Polycel One is code-approved and works for all sorts of problem areas. The manufacturer is Coplanar Corporation.

The second sealant is one highly praised in *Low-Cost Energy-Efficient Shelter* (see "Resources"). Manufactured by the Tremco Company of Clifton, New Jersey, it is called Mono. It comes in tubes and is said to be excellent for glazing greenhouses, sealing skylights, and repairing rusted metal flashing. Suggested sources for this product are any large glass-storefront installer or businesses listed under "Caulking" in the Yellow Pages of large cities.

Wet Patch or similar plastic-type compound is handy for patching holes in a leaking roof. (Photo by Dian)

these come in tubes which are used in a caulking gun, and in certain colors or in clear (translucent). In choosing the kind of caulking you need for a particular job, be sure you pick one that will bond with the surfaces you're working with and that will have the properties you need. Decide if you will paint the area, or if you need a joint that will remain flexible.

There are a number of caulking-type materials that have special uses beyond sealing joints and forming gaskets. *Wet Patch* is the brand name of a very-good-quality plastic-type roofing cement that can be used as caulking material around vents, chimneys, and roof jacks. It is also an excellent material for patching leaks in asphalt shingles or roll roofing and can be applied in the rain (what better time to fix a leaking roof?).

Glazing compound is another caulk-like substance, this one used to fix and seal window glass in wood or metal sash. Old-fashioned window putty is linseed oil–based glazing compound. The newer glazing compounds are more plastic in consistency and don't dry out entirely but remain slightly flexible. *Spackling compound* is a thick white paste-like substance used to patch plaster, fill in holes and cracks in gypsum wallboard, fill in nail holes, and so forth. Glazing and spackling compounds come in cans and are applied with a stiff putty knife or a small Sheetrock knife. Spackling compound will dry to a hard finish and can be sanded and painted. If you are filling a large crack, you'll have to apply several layers or coats of spackling, waiting a day or two between coats to allow the compound to dry entirely and "shrink" as much as it needs to.

One final caulking material is *"tub and tile" caulk;* this is used around showers and tubs, between sinks and walls or counters, and so on. It comes in a small tube and is squeezed out in a bead; it will dry hard but isn't painted or sanded. You can use it to repair tiles and waterproof joints in kitchens and bathrooms.

A General Guide to Caulking: Types and Uses

Latex and acrylic latex caulking Used where caulking is to be painted, interior or exterior. Adheres to most common construction materials. Can be painted within 30 minutes of application in most instances.

Butyl caulking A Butyl-rubber-base caulking compound used to seal aluminum windows and doors, gutters, flashing, etc. Bonds to most metals and to wood and glass. May be painted over.

Polysulfide caulking Used to fill joints that expand and contract a lot. Forms a more or less permanent seal. Comes in white, aluminum (silver color), gray.

Silicone rubber sealant A clear material that adheres to glass, wood, metal, many plastics, etc. Cannot be painted over.

Architectural-grade caulking An economical, general-purpose caulking compound. Use around doors and windows, to fill cracks in masonry and wood, etc. Comes in a variety of colors that may be matched to the job.

Gutter and lap seal Another butyl-rubber caulking. This one bonds to aluminum and to asphalt products, so is excellent for use on roofing joints, around gutters and vents, etc.

Plastic roof patch A thick black caulking that comes in a can or in a tube. Used to seal around chimneys, to seal down loose asphalt shingles, cover nailheads exposed on a roof, etc.

Most caulking compounds should be applied at a temperature of 40° F. or warmer. Most should be applied to dry, clean surfaces (plastic roof patch is an exception—it can be used in the rain).

Flashing

This consists of strips of sheet metal (quite thin) or of flexible asphalt roofing, both of which are used to weatherproof joints. Formerly, the best metal flashing available was copper. This has become too expensive, so aluminum flashing has taken its place. This comes in rolls of varying widths. Flashing is applied using the same principle as for shingles: the purpose is to shed water off from and away from an area. The upper part of the flashing is always tucked *under* the adjoining surface and the lower part of the flashing laps *over* the adjoining surface. Caulking compound is often used to seal horizontal joints when flashing is used. Most flashing on a building will be found on the roof. A metal edge of L-shaped flashing called *roof nosing* or a *metal drip edge* is placed along the edges of a roof under the roofing material. This flashing prevents water from being pulled up under the roofing at this point. Metal flashing is used around vents and chimneys and at points where roof lines intersect. In roof valleys, metal flashing may be used. A less expensive choice is to substitute mineral-surfaced roll roofing.

Flashing may also be used around the foundation perimeter of the house; here it acts to help channel water past the concrete foundation as it sheds off the walls. Flashing is used around window openings for the same reason.

In applying flashing, use nails that are suitable to the material (aluminum nails with aluminum flashing) and use caulking wherever needed. Metal flashing can be cut with tin snips and can be bent into shape for special areas. More on the use and application of flashing appears in the roofing section of this book.

ADHESIVES

An adhesive substance is one that will cause materials to bond or join together. Glues, mastics, cement, and resins are among the adhesive substances that you might work with or make use of in doing carpentry. The joining of wood using glue is a whole art in itself;

there are hot and cold glues of various types that are used in very specific ways to join wood, and each kind of wood has particular properties when it comes to glue. For general use, a product such as Elmer's Professional Carpenter's Wood Glue is fine; this is a fast-setting, very strong glue with a broad range of uses. It is inexpensive and easy to use, can be cleaned up with warm water, and will bond with most woods. Once this glue has set, it is heat- and water-resistant and can be sanded and painted. If you want to bond other materials to wood or to one another, you have a wide choice of glues, epoxy resins, and others. One good general-purpose adhesive I've used is Wilhold Glu-on Fixture Adhesive. This comes in a tube and can be used to bond wood, metal, rubber, glass, concrete, and tile. Like most modern adhesive materials, this one carries a bold *warning* on the label: "Danger: extremely flammable . . . toxic—vapors harmful." I have yet to figure out how you are supposed to use materials like this without breathing the vapors, but I suppose caution is the only help.

Mastics are thick, paste-like adhesive substances used to fasten wood parquet floor tiles to a plywood floor base, to hold acoustical tiles in place, and in certain instances to fasten gypsum wallboard in place. To use these mastics, you simply have to follow the manufacturer's instructions and try not to breathe too much while you're doing it. The worst I've worked with is the mastic used for the parquet floor tiles; it has an *unbelievably* strong odor, and the can it comes in is covered with warnings about breathing or touching the stuff. It's very sobering to be on your hands and knees spreading a gooey mass all around you that you know is *toxic.* Wearing a good respirator helps, but probably not enough. My only advice is: don't make a habit of working with these materials!

INSULATION AND WEATHER STRIPPING

Insulation

Insulating your house and sealing air leakage areas around windows and doors will cut down your fuel consumption—whether you are heating with fossil fuels, wood, solar energy, or some combination of these—and help maintain an even, comfortable temperature inside. Basically, almost any material has some insulating value; this means that the material helps prevent the passage of heat from a warmer area to a cooler area. In cold weather, the insulating value of a material will keep warm air from passing through the walls and ceiling/roof of your house to the outdoors; in hot weather, it will help to keep warm outside air from passing through walls and roof and heating up the inside of your house.

The insulating value of a material is expressed in terms of "R," or its *resistance* to the passage of heat. "R" is usually compared in units that equal a rate of heat flow through one inch of any given material. To compare the insulating value of several materials, you must know the R values *per inch.* The higher the R value, the more insulating the material will do. For comparison, according to *Architectural Graphics Standards,* the R value for oak, maple, and other hardwoods is 0.91 (per inch thickness); the R value for fiberglass insulation is 3.85 (per inch thickness). Adding 5½" of fiberglass insulation between the floor joists in your house or putting 3½" of this same material between wall studs could increase the R value of your floors or walls by over 21 in the first case and almost 13½ in the second. With the rising costs of fuel, you can readily see why insulating your house is an important consideration! Until recently, insulating materials with a total value of R9 or 11 were considered adequate for walls and ceilings; now an R19 or even 20 or 21 is considered desirable.

In planning the insulation of your home, you'll want to add as much good-quality insulation as is economically feasible. The initial cost will be balanced out over a number of years in fuel savings. How much insulation you'll want in your house will depend upon the area you live in, of course, plus the importance you place upon having a more or less constant temperature in your house. If you live in a very temperate area, or build with materials such as adobe or logs that have excellent natural insulating value, you may be able to skip this part of building altogether.

There are several types and styles of insulating materials available. New products are being created, so that your choices may be expanding. Basically, what's available now includes:

Fiberglass and Mineral Wool These are the most commonly used insulation materials, but you might do well to think twice about using them. Both fibrous glass and mineral wool have, as is pointed out in *Work Is Dangerous to Your Health* (see "Resources"), "properties similar to asbestos, which makes them both useful and harmful." Sufficient studies on the results of breathing in fibers of either of these insulating materials haven't really been done, but there is enough early evidence that lung damage (scarring and quite possibly cancer) can occur. Looking at what has happened with asbestos, asbestos products, and workers associated with this material should be a good warning. It has taken fifteen or twenty years for the asbestos industries to admit that their basic material is in fact deadly to the people who work with it. Like the asbestos fiber, fibers of mineral wool or fibrous glass are virtually indestructible—and it takes very little common sense to come to the conclusion that these fibers won't do much good if they get in your lungs.

If you want to work with these materials, wear a top-quality respirator, and wear gloves and an extra loose shirt over your regular clothing. Remember that you'll be covered with tiny particles from just handling these materials—and you'll bring them home on your clothes if you aren't already bringing them home to install! Don't encourage friends to stroll around your work area and don't let children "help out" with the stapling or handling of these materials.

If this hasn't discouraged you sufficiently and you're still thinking of working with these materials, you have a choice of styles. Fiberglass and mineral wool (or rock wool, which is similar) come in *blankets,* which are rolls of standard widths and thicknesses, or in *batts,* which are like rolls in thicknesses and composition but come in 4′ or 8′ lengths. Blankets and batts are made to be placed between studs, floor joists, ceiling joists, or rafters. You should choose a width and thickness that fits your exact needs. The R value for these materials is usually about 3 per inch of thickness (for instance, a 3½″ thick batt will have an R value of 10½). These materials are fire-resistant, rodent-resistant, and readily available—pluses that don't, in my mind, add up against their possible hazards to the health of people working with them!

Rigid Insulation This material is usually a polystyrene or other plastic foam that has been permanently bonded and produced in the form of a board or panel. It comes in a couple of different sizes, thicknesses, and types. The R value of a panel may vary from 4 to 5.5 (or more). This insulation is less dangerous to work with, as there are no free-floating fibers to contaminate your lungs. It may be used over roof sheathing (under the roof covering) to insulate an exposed or cathedral-type ceiling, or it may be applied outside the perimeter foundation to insulate foundation walls. Polystyrene and similar products have one serious disadvantage: they are a fire hazard and give off highly toxic gases when they burn. For this reason, they should not be used on the walls of your house.

Insulating Wallboard and Insulating Sheathing The first of these products is intended for your interior walls. It is simply gypsum wallboard with a regular paper face but with a backing of aluminum foil. The foil acts as insulation and as a vapor barrier, preventing the movement of moisture occurring in the house from cooking, bathing, and so on, from moving through and possibly condensing within the walls.

Insulating sheathing is a wood fiber material used on the outside walls of the house under the regular siding. It comes in ½″ and 1″ thicknesses and has an insulating (R) value of about 2 per inch.

Plastic Foams In addition to the plastic foam rigid insulation, you can use one of the new foams that are blown into place behind walls with pneumatic equipment, usually by specially trained contractors. One of these, polyurethane, has an R value of about 6.5 per inch, which is quite high. Like polystyrene, though, this foam is somewhat hazardous if your house catches on fire. Urea formaldehyde, another foam, has a lower R value (about 4.5 per inch) but is fireproof. You may want to do some research in your area to see what is available and compare costs, R values, fire safety, and so on.

Perlite and Vermiculite These two materials, once widely used, are largely being abandoned as insulating options because they present several problems. They come in a loose fill that can be poured into walls, between attic joists, and so on. They tend to be low in insulating value (about 2.5 per inch), will settle down in walls and leak through tiny holes, and are not fireproof.

Cellulose This is a relatively new form of insulation that deserves your attention and consideration. Made with specially treated low-grade cellulose fiber such as recycled newspapers and telephone books, this insulation offers a good R value (about 3.7 per inch), is fire-retardant, rodent-resistant, safe to work with, and ecological. It may be poured into place behind walls and between ceiling joists, or blown into place with special equipment, which you can usually borrow from the store you buy your insulation from. Cellulose insulation comes in a loose form in bags, and also in blankets or panels. Some of the manufacturers of this new product also make a type that is sprayed in place, using a machine that mixes the dry material with a liquid adhesive. If you do carpentry for a living and get a lot of calls for insulation work, it may be worth learning how to apply this insulation; contact one of the companies below to see what's involved in becoming franchised and licensed. If you use cellulose insulation, make sure you buy a brand that carries the UL (Underwriters' Laboratories) seal on the package—which means it is approved as truly fire-retardant—or is on an approved list put out by your state fire marshal. Any product that is not officially tested and approved can be a fire hazard. Two companies producing quality cellulose insulation products are:

National Cellulose Corporation, 12315 Robin Blvd., Houston, Tex. 77045
Mono-Therm Insulation Systems, Inc., P.O. Box 934, 551 S. Yosemite Ave., Oakdale, Calif. 95361

Applying Insulation

To install any form of insulation, follow the manufacturer's directions, which will vary somewhat from product to product. The following are some general procedures you may make use of. Fiberglass and mineral or rock wool in blankets or batts must be applied between studs, joists, or rafters, and this work should be done when your walls and ceilings are still open or exposed; the exception may be in the attic, but if it has an unfinished floor, you can apply the insulation directly. These insulations usually come in two distinct forms: with and without a foil vapor barrier. As explained before, the vapor barrier prevents the passage of moisture into the walls; it also acts to reflect warm air back into the room if installed carefully. Always place this type of insulation so that the foil faces the room: if applying between rafters, then, place foil down; if applying between floor joists, place foil up. The width of the insulation should be chosen so that it will fit snugly into the space you want to fill. If, for example, you have to use two lengths of insulation in an area, the joints should be butted tightly together. Don't try to compress the insulation; it should be loose and fluffy when put into place. Staple the flanges along each side of the blanket or batt directly to the wall stud, rafter, or joist. Ideally, you want an air space of 1″ or so between the foil face of the insulation and the finished wall; this allows the foil to reflect heat back toward the room. Staples should be placed about every 8″. If you are insulating an attic floor, it's best to use a blanket or batt without any foil vapor barrier as the foil might cause condensation, which would mat the insulation or damage the ceiling covering. If you have only foil-faced insulation, slit the foil in several places to allow moisture to pass through freely. Be careful not to block attic vents. Cracks around windows and doors should be stuffed with loose chunks of insulation and then covered with a vapor barrier; 4 mil polyethylene sheeting does this job nicely. If you are stapling insulation up between floor joists over a crawl space, use chicken wire or a crisscross pattern of thin wire to hold the insulation in place (remember to put foil face up toward the room). You may want to insulate a crawl space further by covering the ground with a vapor barrier of roll roofing or 4 or 6 mil polyethylene sheeting.

Loose-fill insulation of any type, such as cellulose or vermiculite, is usually poured into place behind already existing walls and between attic joists, so there are few general instructions. You'll have to get access to these areas by drilling holes through your exterior siding and sheathing between studs, or by removing a top board of your interior paneling, if possible. Before pouring loose insulation in your attic, check to make sure all wiring is in good condition. Be careful not to cover attic vents. Also keep the insulation back from chimneys, roof jacks, recessed lighting fixtures, or any such area that could create a fire hazard if insulation were piled against it. You may have to construct little boxes to help "control" the insulation in these areas.

Weather Stripping

A great deal of what you gain by insulating your house may be lost if you fail to seal around windows and doors. The first step is carefully to caulk any cracks around window and door frames. Then add weather stripping to any areas likely to leak air. Typically, these include under the door (between the door and the threshold), along the top and sides of the door, cracks between the two halves of double-hung windows (the meeting rail where top and bottom sash come together), and along the sides of casement windows.

To solve problems of drafty doors, there are some products specifically intended for particular areas. You can use a *threshold and door-sweep combination* on the bottom of your door. This has a layer of crushable vinyl, which is attached along the bottom of the door (it is compressed in place when the door is closed, sealing off the area between the door and threshold), and a rain drip cap, which is attached to the lower outside edge of the door. You may have to plane or cut your door down slightly to get the right clearance, but otherwise this addition is easily made (it is screwed into place). The *door bottom sweep* is simpler yet; it is applied to the lower surface of the exterior of the door and is like a little lip which closes the gap between door and threshold. Either of these devices will cost a few dollars and take less than half an hour to apply—and may end an annoying influx of cold air.

Along the door edges, you can use the old-fashioned strip of spring metal (this looks like a folded metal ribbon), which is nailed or screwed in place and is compressed when the door closes. You can also use the newer aluminum "retainer" with a vinyl lip; this is screwed in place on the existing doorstop and again compresses when the door is closed, forming a tight seal. There are several types of self-adhesive strips of neoprene, polyurethane, or vinyl foam that can be applied directly to edges of doors or windows; these form a gasket-like seal when the door or window is closed. The old-fashioned solution for these areas was a strip of matted felt, which was nailed in place with little wire brads; this is still a good choice, but hard to find. Windows may also be weather-stripped with the spring metal strip described above.

You might do well to spend some time poking around several hardware and building supply stores to see what is available, compare prices, and see what new options have appeared. For very little money and effort, you can make your home much more weather-tight and cozy by adding weather stripping all around!

Log Cabin: Building Together

Two summers ago, Kathy and I decided to open our home (a large women-built A-frame) and land (a few acres of pastureland) to other women. Many women came to share this space, to enjoy the country and the company of other women. One afternoon, two women who had visited us several times stopped by with a gift of several finished pine logs, which they thought would make fine seats for gatherings around the campfire. They had spotted the logs at a local campground where they'd been staying, asked where more could be located, and after some searching successfully located the site. These women gave Kathy and me directions to the "woodpile," as they referred to it, and their enthusiasm about all of the free wood available there was enough to incite our curiosity. We decided to set a day to take the thirty-minute drive to see for ourselves what was there. Little did we know what our curiosity would lead us into!

Traveling backcountry roads as instructed, we found the ten-acre wood lot at the end of a dirt road. Piles and piles of discarded, fresh lumber surrounded us. The sight and the smells of this place were overwhelming. Most of the wood consisted of log pieces from 1″ to 5′ long; it was the waste of a lumber company that manufactured precut log cabins. This was their "garbage": freshly milled seasoned pine, flattened on both sides, any size you want, ready-to-use logs for building. We felt like kids at a huge playground, with grown-up-sized mounds of Lincoln Logs! Immediately our vision began to expand beyond simple seats for campfires to other kinds of furniture, barns, fences, and so on, although on this first day we didn't yet know what we would end up doing with our discovery. However, we loaded some pieces into our van and took them home.

The company had also discarded long pieces of slabwood and various lengths and pieces of other size lumber, plus piles of sawdust. Eventually we hauled out of that lot enough logs to build two cabins, at least a cord of supplementary wood to burn in our stoves for the winter, three of the windows used in one of the cabins, enough slabwood for a cabin roof, 2″ to 3″ log ends to lay as flooring, bags of sawdust for the outhouse, and truckloads of stripped bark for mulching!

To create needed space to dump their garbage, the lumber company sent a bulldozer in every week to push the wood over cliffs. More scrap wood would be dumped every day during the summer, and if we'd had the time, energy, and equipment to handle it, we could have had enough free lumber to build a hundred cabins! We shared our find with friends—two neighboring women salvaged slabwood to build a barn for their horse and goat. Others gathered what they wanted for firewood and furniture. This dumping ground is a sad but familiar example of wasting resources, but such places can become a valuable resource to women living in mill areas. Asking

(Photo by Dorothy Dobbyn)

around at lumber and building companies in your locale may lead you to a local "wood dump" and a free lumber or scrap wood supply.

A couple of weeks after our find, a friend of ours was spending a few days with us, as she often did whenever she could manage to get away from her work obligations in the city. Upon observing our new campfire seats, Polly, who has the soul and spirit of a backwoodswoman, asked if we could get more logs. When she heard that the supply seemed endless, she gave a shout of joy that still rings in my ears: "Let's build a cabin! That's always been my dream, to build a log cabin in the woods!"

With much other work to be done, we weren't exactly thrilled by her suggestion. Besides, we really didn't need another structure on the land—the A-frame provided sufficient indoor living space. But Polly's enthusiasm began to spread as we thought of all that wood just lying there, free for the taking. Polly and Kathy and I decided we would build the cabin, fully encouraged by Batya, another woman who'd been visiting and who offered her van for hauling logs. She turned out to be an endless source of encouragement and help, often cooking meals for us also when we didn't want to stop work on the cabin. Polly's sense of optimism and cheer lasted for the entire project.

We didn't have any plans: just decided to build a simple rectangular 9′×12′ cabin. Kathy directed the project, as she had a lot of building experience behind her, including the large A-frame she and Jackie (who also lived with us but was away for a short time) had built. Kathy and I had worked on several projects during the past two years, including our own and other people's interior finishing and remodeling of houses, adding a one-room addition and a house trailer, and building a 9′×12′ half-in-the-ground, rock-walled tool shed. Polly was familiar with many aspects of building and carpentry work, and Batya had almost no previous experience but was willing to learn. We all grew very close to one another as we worked together. We spent many evenings relaxing from a hard day's work by playing cards and laughing while we smoked our corncob pipes. We knew we were strong women and were enjoying feeling that strength.

As to the actual building process, we began by selecting a level area, measuring with a string and line level for the dimensions we wanted. We then dug a 12″ trench, using the line as a guide for straightness, and filled the trench with about 6″ of concrete. The first row of logs was laid in the wet concrete and given time to set. We continued to lay one row after another, nailing the logs together with 16-penny nails driven in at an angle from the new log to the top of the row beneath it, as well as to the logs in front and behind it.

Transporting the logs from the dump to the land was more time-consuming than actually laying them, and the walls were completed in about four days. By now Jackie had returned, and her fresh energy was welcome, as the slower, more tedious work was ahead of us. Various women who were visiting helped for a few hours here and there, nailing the logs and working at the tedious job of chinking the cracks (which were filled with cement).

We built a gable roof whose pitch was determined to a large extent by Kathy's accurate carpenter's eye. We cut dead trees from the woods to use as roof support poles, inserting them into the front and back walls and nailing them together at the angle we had determined for the roof pitch we wanted. We covered them with 1×12 and 1×14 sawmill-cut pine boards, and covered these boards with tar paper and light-colored rolled roofing to reflect the sun's rays and keep the cabin cooler, as we expected to use it mainly during summers.

It was a fun project, and we laughed at ourselves for building a cabin for the enjoyment of the experience rather than true necessity—it was a fantasy come to life. One afternoon Batya sat down on a pile of logs, fixed us with a serious stare, and pronounced: "The secret of youth is immaturity." It was true; our hard work wasn't really "work"; it was "play." The cabin was completed about three weeks later: bed built, door and curtains and herbs hung, and Batya named it "Radclyffe Hall"!

One of the rewards of building with logs is that there is no work of building inside walls. Aesthetically, the logs make beautiful inside as well as outside walls and are self-insulating if properly chinked. There is at least 6″ to 8″ of solid wood between the inside of the cabin and the outside weather elements. When I'm inside the cabin, it feels like a more natural place for people to live. It feels more earthy and satisfying to look at wood walls rather than at plaster or Sheetrock. The smell of the pine lingers for months, which adds to the natural quality of the cabin. It's a snug little house, good for all kinds of weather if a small wood-burning stove were added. It's also a structure that could be built by one or two strong women.

With the knowledge of a second summer approaching, and lots of women using the space in the A-frame, Kathy and I began to muse about living in our own cabin. Radclyffe Hall was already reserved for two other women, and we felt the need to have some quiet, private space for ourselves. Since many women had expressed a desire to build a cabin after seeing the existing one, we decided perhaps we should build a second cabin, making the project a workshop for those women who wanted to learn or improve their carpentry skills, and for those who wanted to learn to build a log cabin in particular. We'd all learn something, and Kathy and I would have our private space a lot faster than if we built it by ourselves.

In early spring, we broke ground for the second log cabin. Before it was finished, a month later, at least thirty or forty women had had some hand in the project. It was a highly charged emotional experience and an adventure from start to finish. Before we actually started building, it was necessary to make several trips to the woodpile to collect a supply of logs and slabwood. It was much easier this time, as we'd gotten an old truck to use for hauling. Different small groups of women made these trips for the wood, with Kathy and/or me usually along too.

About twenty-five women assembled the first weekend, and Kathy began the task of directing a large group of women. Most of them had little or no previous experience in skilled carpentry work and none had built a cabin of any kind before (except for me, working on the first one). It was quite an experience, as we tried to keep up with everyone's questions and suggestions (which often came simultaneously!). Everyone was anxious to see that cabin go up, and at one point one "team" was starting the first, second, and third row of logs for the first wall while another was still trying to get the foundation set for the last wall!

We worked all day with hammers ringing. We took beer, coffee, and lunch breaks, then different women would take up the shovels of cement, hammers, and picks, and the work and talking and laughter continued. Everyone was doing something. Two women were even filming the event as others lifted the logs and set them in place and still others toenailed them together. At times it seemed like mass confusion to Kathy and me, as so many hands made it nearly impossible for us to keep up with directions of how to cut out spaces for the windows and door, where to put supports, what was square or level enough, and so on, and so on. At the end of the day we were all tired. Muscles hurt that weren't accustomed to being used. Sunburns and cramps were a big topic of discussion, too. Nobody stayed up late, but all fell into bed sharing a feeling of a day's work done well. We had completely set the foundation and had gotten a good start on the construction of the walls.

Early the next morning we went back to work, and the learning and sharing continued. During one of our sit-down breaks, the two women who had been doing the filming of our work wanted to tape an interview with the group. They asked why the women present were working here without pay, what they expected to get out of it, and so forth. Some women wanted simply to learn to build a cabin or improve their carpentry skills. Others expressed a personal desire to help Kathy and me get a "house." Some were delighted with the idea of utilizing the free wood, unable to believe that all of this material would otherwise be wasted. One woman expressed her primary desire: to just be with and work with a large group of women. I think that everyone worked for these reasons initially, but that most stayed with the project for the mood that was being created. It made us imagine pioneer days, when the community rallied together for the support of the individuals within it, with everyone gaining in the experience of giving and taking. This communal aura was beautiful, and we felt our strength in working together as our confidence grew.

By the end of the second day, the walls of what was to be a 10′×14′, one-room log cabin, with a 6′×10′ sleeping loft, were over halfway up. The door and two of the three windows were framed in. After an exhausting weekend, we shared food, massage, baths, and a general sense of weariness along with a deep sense of satisfaction and pride in our accomplishments.

I remember at one point looking at the cabin in dismay. Because there were so many of us working on it (and at such a fast pace), this cabin wasn't quite square or level. At some places it seemed as though the walls might not meet, and one small section over the doorway bellied out in an alarming fashion. These problems arose from the walls' being built before the first row of logs had set properly in the cement foundation. Also, some hurried choices were made for certain logs that turned out to be too large or bowed for their particular niche. In the end, these errors were almost imperceptible (from inside especially) and the cabin became at least as beautiful as the first, enhanced especially by the higher log wall and the added dimension of the sleeping loft.

The following weekend was reserved to ready the ground for our vegetable garden and to put a fence around the garden area, but many of the original group planned to return the weekend after that to resume and, we hoped, complete the log cabin we'd started. We expected to finish building the walls, set the windowpanes in place, build the sleeping loft, and finally build the roof for the cabin. We figured the floor could be laid, door hung, and cracks filled in at a slower pace, after the roof was on to keep it somewhat protected.

Spirits were high as women began to arrive on the Friday evening of the third weekend in May. Our energies were renewed and we were anxious to resume our work. The plan was to start early Saturday morning and have the cabin basically finished by Sunday evening. Later that night, however, after everyone was in bed, a telephone call came informing us that Helene, Kathy's mother, had died of a heart attack. We left early the next morning to begin the sad task of caring for Helene's affairs, and women who two weeks earlier knew nothing about building took over the direction of the work and continued to build the cabin in our absence.

Helene originally found this land, and to me that cabin is a kind of memorial to a strong, vital woman who shared her strength and knowledge with her daughter and others. In a period of time when women were frowned upon and discouraged from doing such "unfeminine" work as carpentry, plumbing, and repairs, Helene had taken over the job of maintenance in her childhood household after her father's death. She became very handy and adept at these tasks; it was work she enjoyed and found far more satisfying than many other things. Teaching Kathy, not just these skills, but the natural attitude that she could and should be able to do this work (or any), Helene kindled in her daughter a sense of independence and strength. Kathy, in turn, furthered her reach and learning and shared what she could do with other women. So this heritage of Helene's continues, mother to daughter, sister to sister.

When Kathy and I returned home a day later, the walls of the cabin were completed. The next morning we all joined hands, pausing quietly in thought of the mother, daughter, and sister just passed from our lives, and then we went to work and put the roof on that cabin.

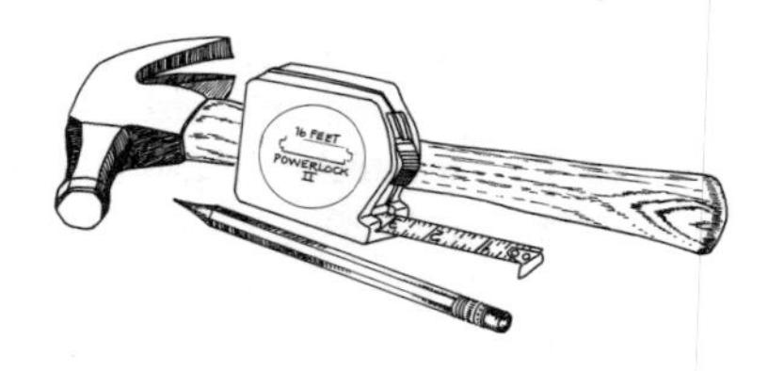

What Did We Miss in High School Shop?

Like most women I know of various ages, backgrounds, and present life-styles and persuasions, I grew up within a culture and an educational system that quite clearly delineated the "natural" spheres of women and men. In the realm of carpentry and building, this delineation took its usual gross and subtle forms. Although I was accepted as somewhat of a "tomboy" (our language doesn't even contain a real word to denote a young female who is physically active, prefers being outdoors to playing inside with dollhouses, and likes experimenting with tools or motors), I was never given the hammer I yearned after or a tool set to practice with. Somehow my friends and I copped or conned cast-off and lesser tools, pooled them, and learned on our own how to use them.

When I was about seven, my mother went off to the hospital for several days to deliver her new baby. I was given a wonderful new doll carriage, which was wonderful to me only by virtue of its enormous rubber-coated wheels! By the time my mother came home, I'd hidden the remains of the carriage out in the backyard and was coasting down the street on a smooth-rolling, rubber-wheeled "gig" (a crude, hand-built sort of go-cart). My aspirations as a carpenter and creator were not stymied by the lack of "junior tool boxes," much as I hankered for one. Years later, I'd cast a furtive look through the half-opened "shop" door on my way to some other class in high school. There would be rows of handsome machines, shining and fascinating—and the smell of fresh sawdust. The door to the shop was effectively closed, though; there was no question about it, only boys could take the courses in woodworking, carpentry, construction. There was no conning your way out of home economics. And there was no women's movement to raise questions (or press lawsuits) about this. My interest in building things gradually languished and atrophied. To be sure, there was no one in career counseling who suggested that I might like to take up carpentry after high school!

Years later, living on a farm in the country with another woman, I began to rediscover how to handle a hammer, saw a board, put things together. Neither of us knew how to build the gates and sheds and feeders we needed for our quickly multiplying livestock, so we improvised and blundered along. We worked with inferior tools (no one ever suggested to me that my hammer was about 20 ounces too light!), and few of them at that (owning neither level nor plumb bob, we somehow "sighted" our little buildings up into place). It was at about this point in my life that I began to recognize what I *had* missed in high school shop. By missing out on any formal instruction in the use of tools or materials, I'd reached my mid-twenties feeling like a child in the realm of building. I had no idea how to go about planning a project, what materials to consider working with, or the basics of building. Most important, I didn't feel comfortable or confident about this kind of work. Without any basic precepts to work from, I had a hard time visualizing solutions to problems. Without a basic understanding of the jobs tools can do, I was mystified by carpentry.

Fortunately, I was working in a place and under conditions where no one looked over my shoulder or paralyzed me by criticizing my attempts as amateurish (which they were) or even entirely wrong (which they often were). I was free to learn on my own and at my own pace. My tools were a source of discovery, and if it took me a while to acquire them and learn how to use them, there was neither judgment nor pressure involved. Being able to work alone, on my own, also allowed me to figure out solutions to problems. Sometimes I'd discover a basic on my own—with surprise I recognized one afternoon that if I just turned the roof rafters on edge, the roof wouldn't bow in when it was done and I was up walking around on it! Often I'd just make the mistake and puzzle over it later. I was pretty much working in the dark—but enjoying it!

Another step forward came when my friend Harriet decided to hold a sort of show and tell session at the site of the new house her husband and friends were building. She had been slowly involving herself in this project, initially taking care of all ordering of lumber and materials but now getting more and more interested in the actual construction. She decided that her consciousness-raising group (of which I was part) would benefit from a lecture-look at this house in progress. About half of us showed up at the site one morning as arranged, and Harriet conducted us from foundation through framing, carefully naming parts and showing how and why it all fit together. At the time, it seemed a casual enough episode—but it was to fire all sorts of things! For the first time we were experiencing something of what was happening in the cities, where the earliest visible results of women's consciousness-raising groups included women's classes in practical skills. Evolving a feminist politics involved pragmatic changes, too.

A few weeks after this house session, Harriet decided to start learning carpentry. Her first project was building a bookshelf for her daughter's room. Lacking her own tools but not resourcefulness, she used a cookie sheet for a square and went to work. We were all invited to see the new bookshelf—as monumental as a house to someone who has never attempted to build anything before and succeeds!

Now, many years past Harriet's house lecture, I still find it a point of reference: my first real look at a building in its component parts, my first vision of a house not as something that "just happened" (entire, mystifying, and intimidating) but as something carefully put together in certain ways and stages.

So what, after all, *did* we miss in high school shop? Perhaps most important of all, we missed growing up with a sense that tools and materials are approachable, accessible, and not all that mysterious. Because as girl children and adolescents we were rarely, if ever, encouraged to experiment with tools, as grown women we tend to be discouraged or mystified by them. How many of us try once to use some tool, find it difficult, and stop at that stage? How many say "I just can't hammer a nail straight" or "I can't saw a line with its becoming all crooked" and go no further? And how many men reinforce our feelings of awkwardness or uncertainty by maintaining that "women just aren't meant to do this kind of work" or by the here-let-me-show-you-how method, which usually means that the tool is taken out of your hand and you are "shown how" by having the work done *for* you? To be sure, not everyone has the temperament, physical coordination, or even the desire to become a skilled carpenter—as with any other trade or craft or art—but it's important to realize that you may not be able to hit that nail straight at first because you've never had any training, not because you are naturally talentless. Along the same lines, you needn't remain mystified by the processes of carpentry, whether these are as simple as fixing a window or as complicated as building a house. Men have no more "natural aptitude" for these matters than women; they have simply grown up being taught about them and being talked to in the language. As you will discover, things fit together in quite logical and often simple ways, and the house (like the bookshelf or the handmade gate) is built step by step by step. Carpentry may or may not become a passion for you—it may remain a simple practicality—but, whatever the outcome, learning to repair your house or build the things you need, imagine, and design has its own heady rewards.

Besides the basic information in the "Tools" and "Materials" sections of this book (which include tips on *using* tools and materials as well as choosing them), there are many little methods, hints, and processes that you might have picked up years back if anyone had thought to share them with you. Like my "discovery" that a roof rafter is stronger if it's placed on edge, these things are often assumed to be "givens" by carpentry books and even those involved in teaching carpentry. I sometimes feel that there is a great conspiracy to keep these matters secret, so that those who know them can remain in the rather powerful position they are used to occupying. Fortunately, women who learn carpentry have a quite different sensibility about it and are willing to share what (and as) they learn. The following is a somewhat random section of things which don't really fit elsewhere but which are as important as learning to frame up a wall or hang a door. Certainly you'll add hundreds of similar discoveries as you work; we hope you'll find a means of sharing them with other women. By learning, teaching, and sharing, we can keep that "shop door" wide open.

PUTTING PIECES TOGETHER

Whether you're attempting to build a house or a bookshelf, you'll be faced with the necessity of joining pieces of wood together in certain ways. The ways in which you cut the pieces and then fasten them together may vary from simple joints and splices to quite complicated joinery: the art of making joints elevated by the craft and skill of cabinetmakers and woodworkers. Basically, *splices* are connections made when two pieces of wood join end to end. They may be overlapped and then fastened together with nails, glue, wood screws, or even bolts, or they may be placed with their ends butted together and then fastened by means of a plate (metal or wood) that spans the joint and is nailed or otherwise secured to each piece of wood. In carpentry and woodworking, a *joint* is defined as a connection between two pieces of lumber that come together at an angle. There are many types of joints, but for most of your work, five basic types will suffice: the butt joint, the dado, the half lap, the mortise and tenon, and the miter.

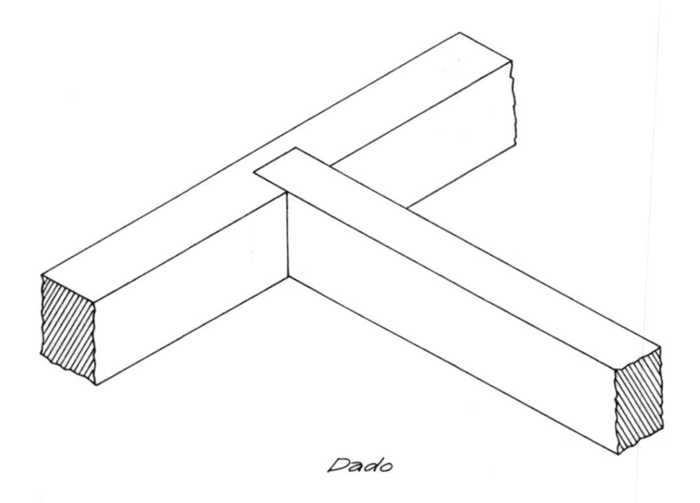

Dado

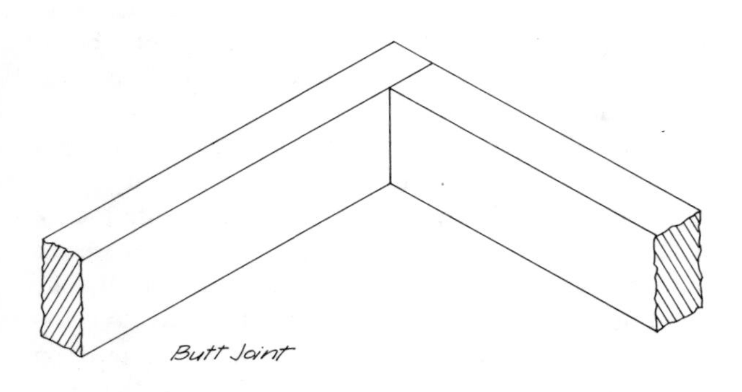

Butt Joint

The *butt joint* is one of the most commonly used and perhaps the simplest; it is also considered the weakest joint. Butt joints are used extensively in rough framing, in making gates and shelves and other simple projects, and even in constructing frames for windows and doors. The word *butt* is an adaptation of *abut,* which means to lie adjacent or to border upon. When two abutting (adjacent) pieces of wood meet at an angle (usually 90°) and are fastened, you have a butt joint. This joint is usually quickly secured by end-nailing one piece through to the other. In some cases you may have to toenail the pieces together. What is most important to remember in making a butt joint is to make your cuts absolutely square. If the cut is angled at all, the joint will be *very* weak because the pieces can't really butt smoothly together. You can use a block plane or surform to try to square up your cut, but this is hard because you are working with end grain, which tends to split and shatter if you try to plane it. Fortunately, most butt joints are found in rough work. By the time you get to places where the joint might show, you'll probably have had enough practice to make a really good one!

A *dado* is actually a groove or slot cut into a piece of lumber; it is commonly cut across the grain (across the face) of the board. The second piece of wood is then fitted into the dado for a very strong though simple joint. Dadoes are used for shelves, drawers, closed stair stringers, and so forth. To cut a simple dado, mark the location and width you want on the face of the board. Turn the piece and mark the thickness or depth of the dado on each edge. There are many ways to cut the dado. If you're working with simple hand tools, you can make several saw cuts (kerfs) on the face of the board: one to define each side of the dado, and then several more between the side lines. Make these cuts as close to the depth of the dado as you can. Then use your chisel to chip out all of the wood and to make the dado, which is simply a slot. Remember in making all cuts to keep *inside* your original lines. If you cut on or beyond the lines, your dado will end up too wide or too deep.

There are many variations of dadoes; the one I have described is the simplest. You can make your dado at an angle if, for instance, you're making a window frame and want to slip the beveled sill into the side jambs. For more details on making and using dadoes, see *Against the Grain, How to Work with Tools and Wood,* or other texts listed in "Resources" at the end of this book.

The *half lap joint* is a nice one to use if you're making a door or gate or cage and you want it to have a narrow, smooth frame. In this joint equal amounts of material are taken from each piece of wood and the pieces are then lapped together as shown and fastened with nails, glue, or screws. A half lap joint may be made at the end of two pieces, which may join at a right angle to make a corner, or it may even be made midway on one piece, with the second piece joining perpendicularly. The important thing is to remove

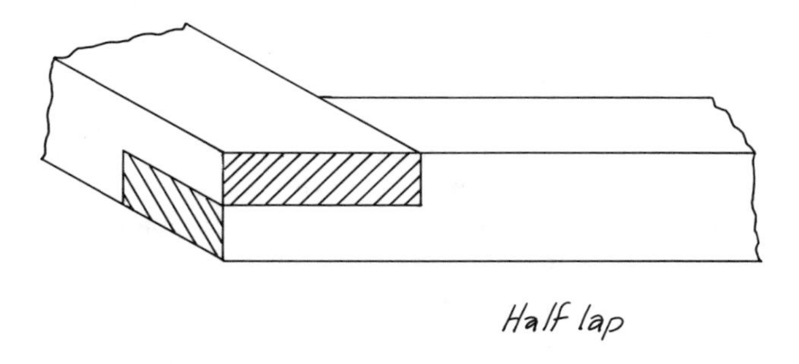

Half lap

exactly the right amount of material from each piece, so that when they are joined together their faces and edges will meet smoothly and in the same planes. If you're making a rough half lap (for example, if you're building with poles), you can mark each piece across its face and down the two edges to establish the perimeter of the cut. Use your handsaw to cut across the face, going just as deep as you want: in other words, the depth of your cut should equal the amount of material you want to remove from this piece. Using a hatchet, shingling ax, or even a wide chisel, you can usually split out the entire chunk of waste wood. Hold the cutting edge of your tool against the end of the piece, directly opposite the cut you've just made. If you give your cutting tool a series of taps with your hammer, the waste section will often split away quite smoothly. You can then use your chisel to smooth away any splinters or rough areas. Duplicate your work with the second piece of material, and you can join the two pieces and fasten them.

A *mortise and tenon* is a strong, versatile joint that can be used to hold a bookshelf or an entire timber-framed house together! The basic mortise and tenon consists of a tongue or projection on one piece of wood that fits into a recess in the second piece. The tongue is called the tenon and the recess is the mortise. Exact instructions for laying out and cutting mortise and tenons is given in *The Timber Framing Book* and in *How to Work with Tools and Wood* (see "Resources"). Each piece must be very carefully cut to fit the other. You can cut a mortise by using your drill (hand or power) to make a series of holes within the perimeter you have marked out. The depth of the holes and consequently of the mortise may be controlled if you make a small mark on your drill bit equivalent to the depth you want—a felt-tip pen or crayon will make a mark that will last through a few drillings. Use your wood chisel to chip out the waste wood remaining in the recess or mortise. To make the tenon, first mark the end of your second piece of wood carefully. You'll want a line running continuously around the piece where you want the tenon to end: that is, marking the length of the tenon, the same as the depth of your

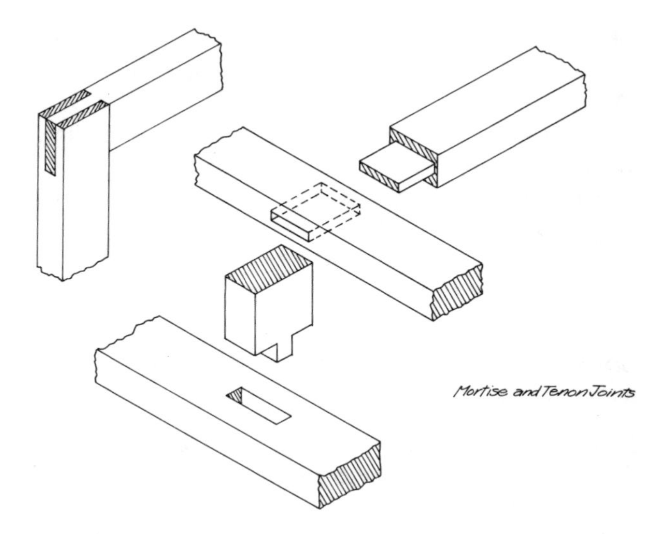

Mortise and Tenon Joints

mortise. A second set of lines will mark the thickness of the tenon. Holding your piece of wood securely in a clamp or vise, use your handsaw (a stiff backsaw is the best) to cut down from the end of the piece of wood—one cut to each side of the tenon, which will be the wood remaining in the middle. Next turn the wood first on one face and then the other, using your saw to cut away the waste wood from each side of the tenon. Now you can mark and cut away the tenon to the desired width. Use your chisel to smooth and finish the work. Mortise and tenon joints may be further secured with dowels, wedges, and glue.

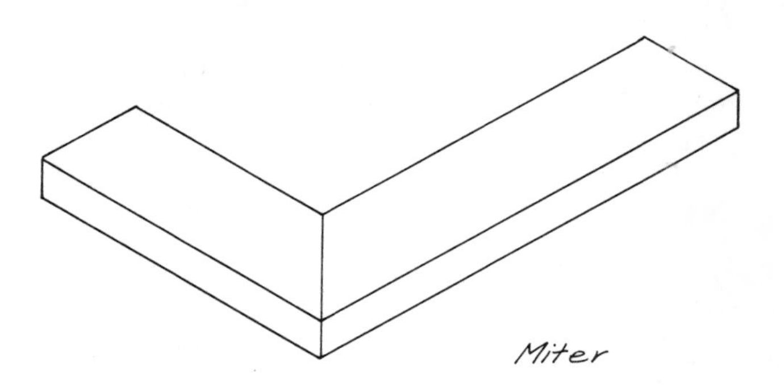

Miter

The *miter joint* is another common one that you'll use for such items as trim and frames. *Miter* means angle, and a miter joint is one formed when two pieces are each cut to one-half the angle created when they come together. For example, the common right-angle (90°) miter joint is made up of two pieces cut at 45° angles and joined as shown. To make a good miter joint, it's usually wise to use a miter box. With this tool, you can cut an exact angle ranging from 30° to 90°. Metal miter boxes are adjustable to this or a broader range; wooden miter boxes come with slots for 45° and 90° angles only, but you can add your own range by sawing cuts into them at the required angles. Once you've cut the two pieces, you can join them with nails, glue, dowels, or corrugated fasteners.

Some Principles and Terms of Working with Wood

Besides the knowledge of the basic joints and how to make them, there are some principles behind the joining and placing of wood that you should understand. A piece of lumber is always stronger when it is placed on edge, which means that it is placed so that its wider dimension is running vertically. For example, a 2×6 should be placed so that the 6″ dimension runs vertically. You can easily demonstrate the truth of this principle by taking a length of lumber—a 2×6 or a 2×4—and placing it with a brick or scrap block of wood or other material under each end, making sure the blocks you use are about the same size. Have a friend steady the piece of lumber on edge while you stand on it about midway. Now turn the piece the other way and stand on it. Whenever you use a piece of lumber for a floor joist, a roof rafter, or a girder or beam, keep in mind that it is strongest on edge.

Along the same lines, a piece of lumber is always stronger when it is placed in a *plumb* (level in the vertical plane—or, more simply, straight up and down) or a *level* (level in the horizontal plane—or straight across) position. This is important to consider if you're placing posts that you intend to have support your house or a gate or if you're making legs for scaffolding. The reason is that a piece of wood can take an enormous pressure as long as it receives it equally; when it receives it tilted on an angle or all at one point, the wood tends to topple over or crack, split, or tear. Always place posts carefully and come back to recheck them before you nail bracing on to "fix" them in place.

Joining pieces of wood over a common nailer such as a stud, a joist, or a rafter involves some thought beforehand because you have to leave room (nailing surface) for each piece. Thus, if you're going to nail sheets of plywood onto walls as exterior sheathing or if you're laying down plywood as a subfloor, you have to make sure that each piece takes up about half of each stud or floor joist in terms of nailing surface. This is important, too, when you're nailing on horizontal paneling and roof sheathing, among other things. All of your measurements and cuts should take it into account. It's a habit you'll soon develop, but one that takes some attention at first.

Blocking is extra material added usually for nailing surface, sometimes to stiffen and strengthen an area. For example, if you're putting up vertical boards inside to finish off your walls, you'll find that you don't have anything to nail to midway up the wall. You may need to add blocking—short lengths of material equal in thickness to the wall studs, toenailed and/or endnailed in between them—at the height you want. A second use of blocking, for stiffening, is seen in the use of blocking between floor joists. Blocking may also refer to pieces of wood set in to close up open spaces; for instance, blocking is added at the eaves of a house in between the rafters. Otherwise there may be an open gap between the siding of the house (which extends up to the top plate) and the roof sheathing and covering (which is suspended above a distance equal to the thickness of the rafters).

Bracing is added in many cases to stiffen a structure and prevent lateral movement. If you build a small square or rectangular box with scrap 2× material, you can see why bracing is necessary. By pushing this box from various points, you can cause movement, twisting, and distortion of the original shape. No number of nails, no matter how well placed, can stop this from happening. Now add a number of little braces running diagonally from corner to corner. Try to push the box out of shape again—you won't be able to! A house is subject to a lot of force exerted by winds, movement of earth, and natural settling. Without diagonal bracing, it can be in serious trouble. By the same reasoning, many temporary constructions (such as scaffolding) and smaller projects (such as sheds, cages for animals, even rough cabinets) need to be braced if they are to remain square and true.

In making out a plan for even the simplest project, don't forget to figure in the *thickness of the material.* Forgetting this is one of the most common mistakes in carpentry. Two or three ¾″ thicknesses forgotten can total up to quite a disaster at the end of your day's work!

If you need to make a *vertical measurement* in "open space," you can use a straightedge and your level: for example, say you want to add a deck to the sloping side of your house and need to see approximately how high in the air you'll be if you run it 10′ out. Stand the straightedge (any piece of fairly straight lumber—2×4, 1×—whatever you have) at the point you want to measure. Get it plumb, using your level, and measure up to any height you think you want. You can secure the straightedge temporarily with braces and make measurements back to the house, or run a line and a line level across from this temporary "post" to get more ideas on your deck plans.

Shims are an indispensable help in many situations. You should not only know about them, but know that everyone uses them for correcting some problem or other! A shim is simply a wedge-shaped or tapered piece of material (usually wood, often a shingle), which is driven into place to fill up a gap, level something, or equalize pressures. Whenever you need to bring a surface out slightly, consider whether you can make use of a shim. Driving shims in pairs from opposite directions will create a perfectly level "lift" under a post or a girder or between a door frame and the rough studs of a door opening.

When a piece of lumber is left to *run wild,* it extends past a given point by any random amount. This is done when you aren't sure how much material you'll need past that point—or when you're in a hurry and can come back later to cut things back evenly. It's common practice to let decking run wild at the edges, for example, rather than taking the time to cut each piece to exact length. When all the decking is down, you can snap a chalk line along the edge and then cut off all the randomly protruding pieces at once.

LOCATING STUDS BEHIND THE WALL

If you want to add shelves, closets, or a new partition wall to a room, you'll have to locate the studs (vertical framing pieces) behind the walls, assuming, of course, that your house has standard construction rather than post and beam. If your house is framed post and beam, the posts will probably show inside. If nothing shows, and your walls seem to be a regular thickness and the house is more or less standard in other ways, you can pretty well assume that studs are to be found.

First take a close look at your interior wall covering to see if you can find regularly spaced nails. Studs are usually spaced 16″ o.c. (16″ from the center of one to the center of the next one) or 24″ o.c. There are extra studs around doors and windows, but the regular spacing is usually followed as much as possible. If you can't find any visible signs of nails, try tapping along the wall with your fist. There should be a hollow sound until you tap over a stud: then the sound should be more solid. You can almost feel the difference, especially if your wall covering is thin. When you've located one stud (or think you have), measure over about 16″ and try tapping there. If no stud "sounds," try a 24″ measuring. You can sometimes begin in a corner and measure over about 14″ and locate the first stud (this difference is due to the thickness of adjoining walls, not to an odd spacing of the studs).

There is an inexpensive little gadget sold in most hardware stores called a *stud finder.* It has a magnetized needle that fluctuates when passing over a nail, so that if you run this gadget along your walls it might help you to locate studs. But you'll have to run it along the floor (about 1½″ up) or just below the ceiling, and because there are nails holding the bottom plate to the subfloor, and so on, this gadget doesn't always locate studs—it just identifies nails rather randomly!

Whichever method you use to locate the studs and their o.c. spacings, you can double-check yourself by driving a small finishing nail or drilling a tiny hole where you think the stud is located. Make these holes in some easy-to-hide location—just above the baseboard, for example. Later you can spackle or putty over them.

DETERMINING A BEARING WALL

A bearing wall is one carrying weight beyond its own, such as the exterior walls of the house, which bear the weight of the roof. Usually there are interior walls that are bearing, too; they usually run parallel to the exterior bearing walls. You might need to identify a bearing wall if you are planning to put in a new window, add a room, or change your interior space by tearing out an existing wall. With most houses, you can tell rather quickly which are the exterior bearing walls by looking at the slope of the roof. The walls the rafters bear on will be those the roof slopes down to; there will be an overhang, quite possibly with visible rafters. To identify an interior bearing wall, try going up in the attic or attic crawl space. If your ceiling joists have a broken span and are butt-jointed or lapped over a particular beam (which runs perpendicular to them), chances are there's a bearing wall below at this point. The basement or crawl space will likewise often reveal a bearing wall: it should have support below, as it takes weight that the normal floor and floor joists shouldn't bear without extra support. One final point to look for is that ceiling joists and floor joists will usually be running perpendicular to a bearing wall.

If you're remodeling your house or doing this work for someone else, you may want to consider removing a bearing wall. While this is a serious project, it *can* be done. The usual procedure is to replace the wall with a girder that spans across where the wall has stood and takes the weight the wall has been bearing. This girder is placed just under the ceiling joists, running perpendicular to them. Steel or wood girders may be used; they may be concealed within the ceiling, or a wood girder might be left exposed. At each end the girder will have to be supported by a wood or steel post, which may be concealed within the wall or left exposed. In doing this work, you'll have to set up a temporary support system for the period of time between tearing out the wall and replacing it with the girder and posts. A temporary girder can be raised and supported on posts with or without floor jacks. This girder should be strong enough to take the full weight that the wall is bearing; it should be placed running parallel to the wall and as close to it as possible, leaving yourself a reasonable work space. Once you have your wall removed and the permanent girder and posts in, your temporary girder can be removed and used elsewhere or kept for just this type of work. Posts supporting a bearing girder should be placed so that they have support below; if they don't naturally fall over a portion of concrete foundation, a girder, or a pier, you may have to add piers, footings, and posts below to support this concentrated pressure.

ADDING A PARTITION WALL

Before tackling this project, read through the section on "Rough-framing the Walls"; much of the information and procedure will be helpful to you here. Although every situation will demand some creative response and special considerations, there are some basic steps that you'll probably need to follow in putting up a partition wall. To begin, locate and mark where you want the wall to run. You may choose to build a full wall, running floor to ceiling, or a partial wall serving more to break the space rather than make an entirely separate room. Mark the location of your wall-to-be on the ceiling, measuring out from the corners and/or walls of the room and snapping a chalk line across the exact location. If possible, try to situate the wall so that at least one end can be nailed to an existing stud in an adjacent wall. If you're working in a room with concrete walls, you'll have to use special fasteners to secure your new wall, and you may want to add furring strips or nailers at this point. (*Furring* is thin strips of wood used to bring the thickness of a wall out or to provide a nailing surface for paneling, Sheetrock, and so on. A *nailer* is any piece of wood added in any area or space to allow you to fasten something such as paneling or flooring to it.) The chalk line along your ceiling will locate the top of your new wall, and, using a plumb bob, you can quickly transfer this line to the floor. Make a series of X marks to the side of each line where you want the new wall to stand, and double-check yourself to make sure that your X's are on the same side or your wall will be a "leaner"!

Now determine how you are going to attach your wall to the ceiling above and the floor below. If your wall runs perpendicular to existing ceiling joists, you'll be able to nail the top plate of your wall directly to these joists. If it runs parallel to the joists but doesn't sit directly below one, you'll have to add blocking between the joists and nail your top plate to that. If the wall runs directly below a particular joist, of course, you can nail to it. You may have to do some searching and marking to locate your joists (follow the same general procedures explained earlier for locating wall studs). If you have to add blocking, you'll have to remove a portion of the finished ceiling. If your floor is wood or is tile or linoleum over a wood subfloor, you can nail the bottom plate of your new wall into the flooring itself. If your floor is concrete, you'll have to use special fasteners. With the location of your wall ready and nailing surfaces prepared, you can begin on the wall itself.

There are two easy ways to frame up a partition wall: you can construct the wall entire and lift it into place, or you can put the top and bottom plates up first and toenail your studs into place. Partition walls are usually framed up with studs 24″ o.c.; doorways may have simple 2×4 headers over them as they aren't taking any displaced weight. If you have the work space, it is faster to construct the wall on the floor and lift it up into place. You can avoid a space or squeeze problem if you plan your wall to have a double top plate. Construct it with a single top plate, so that when you go to lift it up, this extra 1½″ will give you just enough room to stand it up. You can then slip the second or double top plate into place and nail up through both plates into the ceiling. If you decide to place your top and bottom plates first and then add studs to your wall, you'll avoid this problem altogether.

Before nailing your plates to ceiling or floor, make sure everything is perfectly located and, in the case of the preconstructed wall, perfectly plumb. Use 10d nails placed 24″ o.c. to secure your plates. Remember that the bottom plate in doorway areas should not be nailed; this section will be cut away when the wall is completed. Your wall should be nailed to adjacent walls and should have some diagonal bracing if it is a long wall. Fire blocking may also be added to stiffen the wall. Finish the wall by paneling, sheetrocking, or what you decide on, and hang your door(s). Trim in the form of baseboards, ceiling molding, and so forth, should complement that on adjoining walls.

DEALING WITH A STICKY DOOR

If you've been living with a chronically hard-to-shut door, a few minutes of investigation might reveal that it's a simple matter to fix. Many doors that swing freely while the weather is dry tend to swell in damp or rainy weather and become sticky, difficult to shut. A door should always close easily, with a gentle push; otherwise it is straining its hinges, hanging unevenly, or somehow in trouble. Begin by closing the door very slowly but firmly, observing where it seems to be catching. Now take a look at the hinges and the screws holding them in place. Quite often an older door may simply have worn away the wood that had been holding the hinge screws in the doorjamb. If you can turn the screws easily with a screwdriver (or with a twist of your fingers), the solution to your door problem might be as simple as replacing the screws with slightly larger ones (in length, not diameter).

While you're looking at the hinges, take a careful look at the doorjamb itself. If the wood looks soft or pitted, check it with your awl or pocketknife; if it yields to a light pressure, the jamb might be rotted and may need to be replaced. (See sections on "Replacing Your Door" and "Hanging a Door: New Construction" for procedures). At this point you might also want to check your doorjamb to see if it is plumb and level. Houses often settle as they age and throw doors and windows slightly out of level. If this is your problem, you may want to remove the trim around the door (exterior and interior) and replace the shims or add new ones to readjust the jamb. You might opt for cutting or planing the door to fit if this procedure seems too complicated.

If hinges and jamb seem to be in good shape and you can see where your door is sticking, you may simply need to plane or cut it down a little to alleviate the problem. Most doors are hung with loose-pin butt hinges, which are actually hinges consisting of two separate pieces or sections held together by a removable pin. The pin has a rounded top and is removed as follows: with the door open, take your hammer and a broad-tipped screwdriver; place the tip of the screwdriver just under the round top of the pin and give it several taps upward with your hammer. The pin will be driven up and out of its seat. If your door has been painted, you may have to chip away the paint to free the pin, or do some extra or more vigorous tapping to loosen it. Now remove the pins from the remaining hinges. Usually the door will hang in position even with the hinge pins removed; you can loosen it by taking hold of it from the doorknob or latch side and rocking it gently, then catching it as it falls free. It helps to have a partner to do this part of it—and watch out for your fingers! With the door removed, you can plane down the areas that seemed to be sticking. Remove any hardware from areas you are planing. Be very cautious in the amount of planing or cutting you do; it may take a little longer to replace the door and check your progress, but it's better to underdo this work than end up with large, unnecessary gaps around the door.

Replacing the door is, again, better done with a partner. One of you can lift the door from the latch side and the other can help to guide it onto the hinges and replace the pins. Pins should be slipped into place and then tapped down with a hammer. You may have to replace and remove your door several times before you get it planed down just enough, but if this small amount of work means your door closes easily from now on, it will be worth it.

REPLACING A FOUNDATION POST

If you live in a house that is built with a pole foundation or is up on piers and posts, it's a good idea to get underneath periodically to check the condition of your posts. The posts are vertical supports spaced regularly under the girders, which in turn hold up your floor joists and the rest of your house. These posts or poles may be embedded in the ground or set on concrete pads or piers. Embedded posts/poles that have been pressure-treated with a good preservative will last up to thirty years or longer, but if they are merely soaked in preservative or painted with it, they may begin to rot after a season or two in the wet ground. Posts that are up on piers or pads usually have a long life, but if dirt and debris have been allowed to pile up around them, they too may begin to rot prematurely.

You can check the soundness of your posts by pushing an awl or pocketknife into them. Sound wood will be hard, even though the outer inch or so may feel soft; wood that has begun to decay will be soft and spongy. It's a good idea to replace any post showing signs of damage. If you wait too long, the pressure bearing down on the rotting post can cause it to give, letting your house drop at that point. Even a subtle drop can throw windows and doors out of alignment, so it's best to head off this type of damage before it happens. You should likewise keep an eye on posts/poles under decks, porches, exterior stairs, and so on.

Another problem to watch for is posts that are out of plumb. Go under the house or deck with your level and check any that look as if they are tilting or leaning. In some instances you can straighten out a post in place by simply detaching it from the pier or pad it sits on (use your nail nipper and/or cat's-paw for this work and try to do as little damage as possible to post and pier) and then using your sledgehammer to hit it back into a plumb position. Then you can refasten it in place. If the post is badly out of alignment or is embedded and can't be hit over, you'll have to replace it.

Replacing a post is a simple enough project. Basically, what you need to do is set up some sort of temporary situation to take the pressure of the structure off the post so that you can remove and replace it without disturbing anything above. Depending upon the distance from the girder (which is resting on the post) to the ground, you may be able to use a jack and a pier or wood block, or may need to use a post with pier or block. The Little Giant jack mentioned in the "Building with Piers" section and listed in the "Resources" section is the ideal tool for this job, as it can be used with a post of any length and can take enormous amounts of weight. Without the help of a Little Giant, here are two basic approaches to take:

If you're working with a short post, you can use a regular floor jack or a good hydraulic car jack and place it directly under the girder. Set the jack up about a foot or a foot and a half from the post you want to replace. Stand it on a pier or a solid wood block or even a concrete block—anything that will give it a strong base. Setting it directly on the ground will create problems because any pressure on the jack can drive it into the relatively soft ground. You can use lengths of 2× or even 4× material as "shims" to make up for height differences if necessary, but try to keep everything very solid and not wobbly. Use the jack to lift the girder *a hair;* this will release all pressure from the post so that you can remove it easily. If you're going to replace the post immediately, the jack can hold the girder while you're doing this; if you're going to come back later to replace the post, or if you're casting a new footing or pier below the new post, you may want to set a temporary post and pier under the girder to hold the weight while you work. Usually the jack is stable and strong enough to do this work. Once you've placed and secured your new post, you can let the jack down and remove it, or you can use the jack to lift the girder slightly while you remove a temporary post, then let it down onto your new permanent post. Instructions for setting or casting piers, casting footings, placing posts, and so on, are all found in the "Building with Piers" section of this book.

If you're working with a tall post, you'll have to use a post of equal length to replace it temporarily while you remove it and set up a new permanent post. Again, locate your temporary post about a foot or so from the one you want to work on. The temporary post should rest on a solid base—a precast pier or wood block. Cut it to the exact length (from girder to top of pier or block) and stand it in place. Now you can use this post to lift the girder slightly by driving wood shims between the post and the girder and/or between the post and its base of pier or block. Drive the shims in pairs from opposite directions and they'll lift the girder evenly and with very little effort. You may want to tack this whole affair (post, shims, pier or block) in place to make sure that nothing moves when you take out the post you're replacing. It's important, of course, to set even this temporary post up in a plumb position so that it can take any downward-bearing pressures evenly. Now you can knock out (or dig out) your old post and replace it. To remove your temporary post, you can either drive the shims loose by hitting them sideways (assuming you have a bit left protruding from under the girder or above the pier/block) or knock the post loose by hitting it at the base with your sledge.

The main thing to keep in mind when working under your house in this way is to set up all temporary supports on solid bases and to keep things plumb. With these precautions, the work is easily and safely accomplished and your house will have a stable, sturdy base for years to come.

REPLACING SASH WEIGHTS AND CORDS IN DOUBLE-HUNG WINDOWS

Double-hung windows, often found in older houses particularly, come in sets of two windows positioned one above the other. The top window can slide down or the bottom window can be pushed up for ventilation. A common malady in the older house with double-hung windows is the broken sash cord. This is a simple cotton cord that travels on a pulley inside the window jamb; it has a weight on one end that acts to stabilize the window in an open position. When the cord breaks or comes loose from its weight, the window won't stay open without a prop (stick or book shoved under it). Replacing or retying the sash cord is a simple matter once you know about the existence of the pockets in the side jambs of your window.

To investigate your window, first take your Wonder Bar (or any flat pry bar) and *gently* remove the stops from around the window. These are located on the inside and consist of three small pieces, one vertical at each side of the window and one horizontal running across the top. They are nailed into the jambs with finishing nails, which will usually pry loose quite easily. With the stops removed, you can take the bottom half of the window partially out of place or remove it altogether. If your window works with sash cords, you'll see the groove cut down either side; each groove has a small round spot for the knotted end of the sash cord to nestle in. You'll probably find a piece of broken cord in place. Now look carefully at the side jambs. Somewhere an inch or two up from the sill you should see a little "door," which may be very well concealed, particularly if the window has been painted. Two or three wood screws may give away the location even if you can barely see the outlines of the door. Remove the wood screws and you'll be able to tip the little door out—and there will be the hidden pocket! Inside should be the sash weight, which is an iron bar or cylinder with a hole at the upper end. If the weight is missing but everything else is obviously in working order (a pulley set in the upper portion of the jamb, plus the grooves on the window itself), you may have to go to a junkyard or salvage store to get a new (old) weight—new weights are almost impossible to find in hardware stores. Sash card may be purchased at most hardware stores. It is sold as such and comes in an inexpensive hank.

To put in a new cord, feed one end down through the pulley (upper jamb) and catch the end below. Tie it to the sash weight. Then hold the window approximately in place and cut the cord so that you have enough to knot the end and slip it into the groove in the window. Now set the window back in place; the cord will be sandwiched in securely. You should adjust the knot at the weight end so that the weight is suspended inside the pocket. Experiment a bit with this to see how high it should hang to make the window work right. When you're satisfied that your window is once more functioning, close up the pocket, slip the window into place, and replace the stop on that side. You can then do the other side and replace the other two stops. Your window should now be restored to working order!

Some double-hung windows work with a tension spring rather than a sash cord. This, too, may break and need replacing. You can buy new springs in the hardware store and replace them. Again, a careful look at the window and its workings should show you what you need and how to proceed.

A sticky window that is difficult to slide up and down may need other attention, varying from planing to resetting the actual jamb pieces. First, take your level to the jambs and see if they've settled out of place. If they're only slightly out, the easiest procedure is to plane the window a tiny bit to fit. If they're badly out, you may want to remove the trim from inside and outside and redo the shims between jamb and rough frame. Often a sticky window is just layered with paint and you'll have to scrape it down or strip it to get it working again. Sometimes the parting bead is damaged and should be replaced. The parting bead is the thin strip of wood that actually separates the two windows (you'll see it running along each side of the jamb between the windows and across the top in between the two). Layering it with paint is another quick way to make a double-hung window ornery! You can pry the parting bead out with a lightweight pry bar or a good stiff putty knife and buy a replacement piece at your local building supply store.

HIGH SCHOOL SHOP WAS NEVER LIKE THIS

Admittedly, most of the preceding material would never have been revealed behind those closed doors anyway, but it describes the kind of problem or procedure you'll be facing often enough, whether you are working on your own house or going to work for someone else. Whatever the project you decide to undertake, try to look at it as the logical situation it is: if you know the basics of how a building goes together, how joints are made, why things are structured the way they are, you should be able to come up with a workable solution to any problem. Try to balance your day so that you're making plans or doing the most demanding part of the work when your mind and body is fresh and rested. Working when you're tired, preoccupied, or pressured will make the difficulties of a project doubly hard; if you come back to it the next morning, you might find it not so monumental! Whether you become enchanted with the work or simply accept it as a practical, necessary skill, learning carpentry will definitely make you feel more at home in your house.

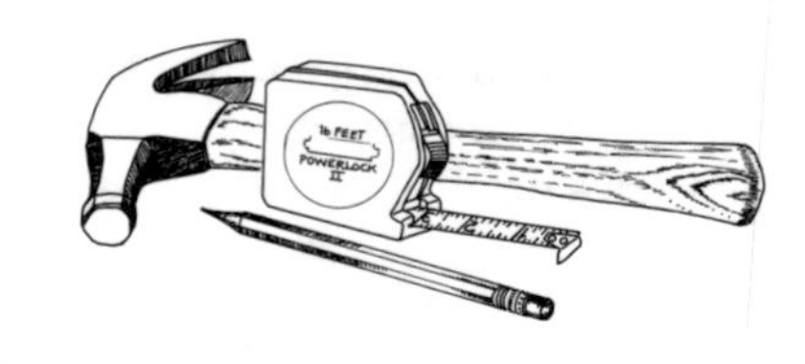

Some Simple Projects

SAWHORSES

Sawhorses are an almost indispensable tool for any carpentry work, and making a pair of them is not a difficult project even if you are just learning how to use your tools and work with wood. There are many designs for sawhorses, and your preferences and needs may lead you to want a heavier construction or different structural setup.

Sawhorses are used to hold materials that you are going to measure, cut, plane, drill, and so forth. They should be strong enough so that they won't break under a reasonably heavy load (a 10′ 4×6 girder, for example, or a couple of sheets of ½″ plywood). They should also be light enough so that moving them isn't a two-woman job! The sawhorse pictured shows the design my partner and I use. It is made entirely with 1× material but is surprisingly strong. It's so light that one person can carry it—or a pair—around easily. There are some other nice features, too. The two top pieces are held apart by the upright piece at each end, and the 1″ space (we used actual, not nominal, 1× material) between them makes a perfect slot for storing a handsaw between cuts. The piece that runs below and ties the two sets of legs together serves as a "shelf" where you can safely set your combination square, chalk line, and so on, while you're working. The material for this sawhorse is very cheap. We used rough redwood 1×4 throughout, and needed only a handful of 8d galvanized nails plus a few 6d nails.

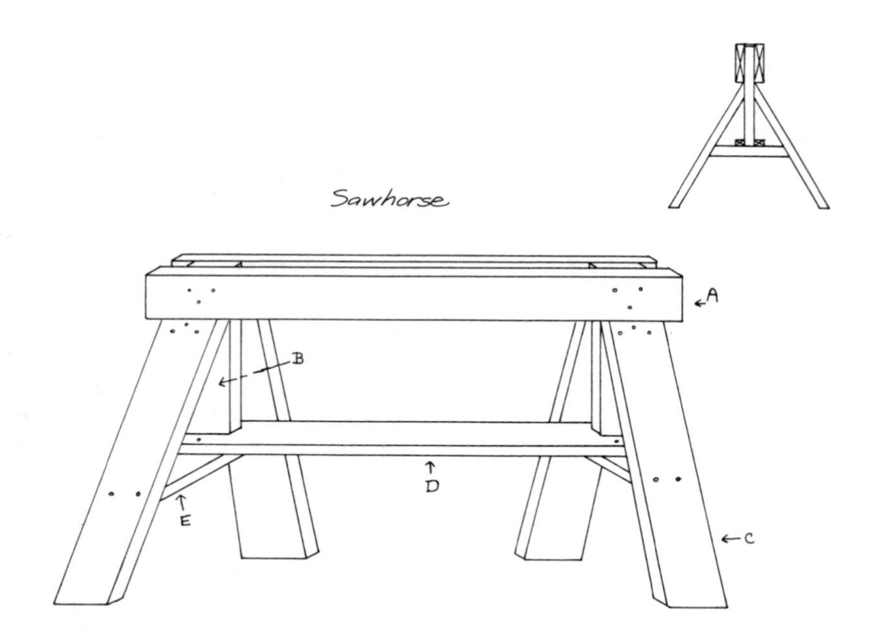

Sawhorse

To duplicate this design, you'll need:

2 ledgers (A)—3′ each
2 uprights (B)—1′2″ each
4 legs (C)—2′ each (plan on a piece about 2′4″ long for each, as you'll have to angle-cut each end)
1 bottom crosspiece (D)—slightly less than 3′
2 crosspieces (E) for legs—9½″ each (again, plan on slightly longer pieces, as they must be angle-cut)
some 8d galvanized nails and a few 6d or 4d galvanized nails

You may choose a different way to make this sawhorse, but here's a step-by-step method to follow. First, mark and cut your two 3′ ledgers (A). Next, mark and cut the two 1′2″ uprights (B). You can now nail the uprights in place between the two ledgers. The ledgers should extend about 2″ or more beyond each upright piece (look at drawing). Use your combination square to make sure that the pieces line up squarely at the top before you nail them. To do this, place your square so that the body of the square butts against the side of the ledger piece and the blade of the square runs across the tops (or ends) of the ledger, upright, and second ledger. When all three pieces meet smoothly against the blade of the square, you can drive in your nails. Placing three nails in a triangle shape, as shown, will make a sturdy joint that has very little to no movement. If you're worried about splitting the wood, blunt the tip of each nail before driving it in.

Next, cut your four legs (C). The most crucial angle is the one at the top of each leg; it should be between 30° and 45° to give enough (but not too much) spread to the legs. You can use your bevel to make an angle under 45°. If you want a 45° angle, use your combination square. Trace the angle on the edge of your piece of 1×4 (the 1″ or ¾″ dimension). Then continue this line across the face of the board with your combination square. Repeat the angle line on the other edge, and complete the face line so that the entire board is marked. This may seem unnecessary, but it's a help when making an angle cut, especially if you're working with a handsaw and if you're just learning. To make the actual cut, you'll want to hold the board on edge and brace it against something to keep it from vibrating and moving while you saw. Remember to start your cut by drawing *back* with the saw a couple of times. As you cut, you can use the lines drawn on the faces of your board to make sure that you are cutting evenly and squarely. Without these guidelines, the cut can easily go off and result in a curved end rather than a neat angle cut.

When all four legs are angle-cut at the top, you'll have to do a little experimenting to get the angle for the bottom of the legs. This angle will be only slightly different from a right angle, just enough to give the sawhorse a stable set of feet. The angle will depend upon the one you've chosen for the top of the legs and the size of the material you're using. One way to get the angle is to tack the legs in place. Butt them up under the ledgers and tack each with a single nail; leave a part of its shank extending so that you can easily pull it out. With the legs tacked on, stand the sawhorse on level ground and use your bevel to get the approximate angle you need. To angle-cut the bottoms of the legs, untack them and use the same procedure as for the top angle cuts.

Now you can nail the legs back in place, face-driving three nails (triangle pattern) per leg, as shown. Next, cut the long crosspiece (D). This piece will keep the legs of the sawhorse from spreading out in the long direction (away from one another) and will further strengthen and stiffen the sawhorse. Measure from the outer edge of one upright to the outer edge of the second upright and cut the crosspiece to this length. Now you'll have to make a notch or slot in each end of the crosspiece so that it can slip into place. Lay the crosspiece flat and mark the center of the face (2″ if you have an actual 4″ wide piece). Now measure ½″ to either side of this mark (if you're using actual 1″ material) and make two more marks. This is a way to center your slot or notch. Each slot should be cut to a length (or depth) equal to the width of the upright; if you're using actual 1×4, the slot should be 1″ wide and 4″ deep or long. You can set your combination square to make the lines for your cuts. (If all this seems too complicated for a simple procedure, take a look at the illustration and you can see exactly how the crosspiece fits and needs to be cut.) Once you've marked and cut your two parallel lines, you can chip out the wood between by placing a 1″ wood chisel across at the end of each cut. Using your hammer, tap down on the chisel until the piece of wood breaks loose. If it comes loose in pieces, use your chisel to smooth the end of the slot.

When the crosspiece has been notched at both ends, put it aside while you cut the two smaller crosspieces (E) that go between the legs of the sawhorse. These pieces keep the legs from spreading out when heavy weights are placed on the sawhorse. Each piece must be angle-cut at either end. To get these angles, you can hold a length of board in place so that its top edge butts just up against the upright and it extends past the legs at either end. The smaller drawing shows where this piece has to be held to give you the angle cuts for E. Now you can scribe the two angles exactly as you need them. Mark and cut each crosspiece. Before you nail them in place, though, slip the long crosspiece on. Now you can nail the two shorter pieces. Blunt the nails you're using as they'll be going into the ends of the crosspieces and might split them otherwise.

You can also turn the sawhorse upside down and place two more (blunted) nails into the end of each upright. Now the long crosspiece can be tacked to the two smaller crosspieces; use your smaller (6d or 4d) nails and place one nail into each "tail" of the long crosspiece, as shown. Now stand your sawhorse up and make sure that it feels sturdy. If there seems to be movement in any joint, add a nail or two until it feels stiffer and won't move. Your second sawhorse should be twice as easy!

BUILDING A GATE

Gates are simple to build and are a good introduction to carpentry. Whether you are fencing your garden or your yard, a gate can be beautifully made as well as sturdy and functional. A carefully constructed gate should last many years, amply rewarding an afternoon's labor and a relatively small investment.

The first step in gate building is to plan out the entire process, deciding what materials—lumber and hardware—you will need and what type of gate best suits the situation. Any gate that you build will consist of two basic member types: the *framing* or "rails," and the *siding*. The frame is usually built of wood and the siding may be wood or wire. The wood-frame and wire gate is quickly built, cheap, and very light in weight. It works well for a garden or poultry yard, but will not afford any privacy if you are considering a gate for your yard. The wood-sided gate will be stronger and more durable, with more general uses. You will generally want to use 1× material for siding, and you have a choice of 1× or 2× for the framing. (See "Materials—Lumber and Lumber Products" for information about dimensions.)

Smaller gates (3′ or so in width and 3′ to 4′ in height) may be built entirely of 1× material; larger gates (over 3′ in width; over 4′ in height) will usually need a stronger frame of 2× material. A frame built of 2×4 will be sufficiently strong for most gates. Larger framing material (2×6, 2×8) may be used in special cases, but it adds a considerable amount of weight, which will strain hinges and post. You might want to use 2×3 for an intermediate-weight frame. In addition to framing and siding, most gates will require at least one diagonal brace, which should be of the same size material as the frame—1× or 2×. If you are building a particularly wide gate for a driveway or roadway, you may want to use a set of diagonal braces intersecting one another. Or you may choose to build a pair of gates, running your bracing to form a pattern.

In calculating the amount of material you will need, it helps to sketch out your gate to approximate scale so that you can see all of the framing, as well as how the gate will look when it is put together.

If you are planning to run your siding vertically (generally the most pleasing to the eye), you should plan to extend the siding an inch or two above the top rail and below the bottom rail. How closely you space your siding is a matter of gate purpose and aesthetics. If your gate is part of a fence intended to give you privacy in your yard, you will want a solid or nearly solid siding, with little spacing between the boards. If your gate leads to an animal pen or pasture, be careful to space the siding so that an animal can't get its hoof caught. When building a garden gate, you have to beware of leaving too much space and having a rabbit or ground squirrel squeeze through.

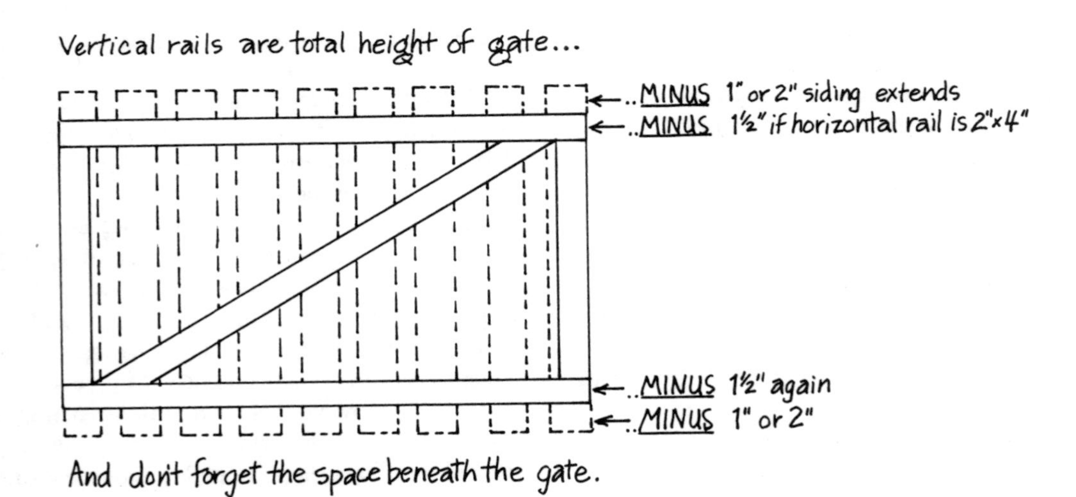

You can use boards of varying widths for nice visual effects, or run your siding in a diagonal or herringbone pattern. Be cautious about choosing very wide (1×12 or bigger) boards for your siding, since these have a tendency to cup (curl up from the edges); 1×6 and 1×8 are good sizes for most gate siding. Narrower 1×4's or 1×2's or even lath or furring strips may be used for a more delicate-appearing gate. Any of these smaller sizes will, of course, entail a lot more time in cutting, laying out, and nailing. Other choices for gate siding include poles, grape stakes, or milled lumber such as shiplap and V-rustic (these boards interlock for a tight, solid surface). Your least expensive choice is probably 1× material in 6″ and 8″ widths. If you're using recycled materials, your choices will be predetermined by what you have available.

If you are buying materials, you have a choice of rough or surfaced lumber. Rough lumber is a little thicker, therefore slightly sturdier, than smooth, or surfaced, lumber. It is splintery, so it's less desirable to work with or live around. Usually it sells for a few cents per foot less than surfaced lumber of the same size, grade, and type. If you plan to paint or stain your gate, get surfaced lumber. The type of lumber (i.e., species of wood) you choose for your gate will depend upon the area you live in. Choose a wood that weathers well and is known for its durability outside. Some types tend to rot, crack, or warp more quickly than others. Douglas fir, cypress, pine, redwood, and oak are recommended for gate building; in my area, redwood or cedar is used for unpainted gates and Douglas fir for gates that are to be painted. If you are buying wood for your gate, choose a fairly inexpensive grade. "Common" rather than "select" is what you need. Avoid the lowest-grade, really cheap "bargain" lumber, as these "specials" are often virtually unusable because of splits, warping, cracks, and so on. If you have a local lumber store that you trust, ask them to recommend the most suitable grade and type of lumber; if you have no one to consult, find out what's available at local lumberyards and check out your choices in the charts of "wood use" in this and other carpentry books.

Next, you should consider the nails and hardware that you'll be needing. All nails that you use should be galvanized. These have a rust-resistant coating for outdoor use. They are almost twice as expensive as ungalvanized, interior-use nails, but are absolutely necessary if your gate is to last in the weather. Box-type nails with their thinner shanks may be used on surfaced wood to reduce the chance of splitting; common nails are sturdier and well suited to rough lumber. Generally commons are used for all rough framing and box for finer work. Use 8d or 10d nails for fastening on your siding (smaller if you are using lath or 1×2), and 16d nails for constructing 2× frames.

Hinges for outdoor gates come in two basic styles, the *strap hinge,* with two equal leaves, and the *T hinge,* with one long leaf and one smaller, rectangular leaf. Generally I like to use the T hinge. The rectangular leaf is fastened to the post and fits perfectly on a 4×4 or 4×6 post; the longer leaf is fastened to the gate. A strap hinge is normally used when you fasten a gate to a broader surface—to a house or barn or a very large post. The elongated leaves of the strap hinge distribute the pull of the gate over a greater area, but if used on common-sized gateposts they often must be bent around to fit.

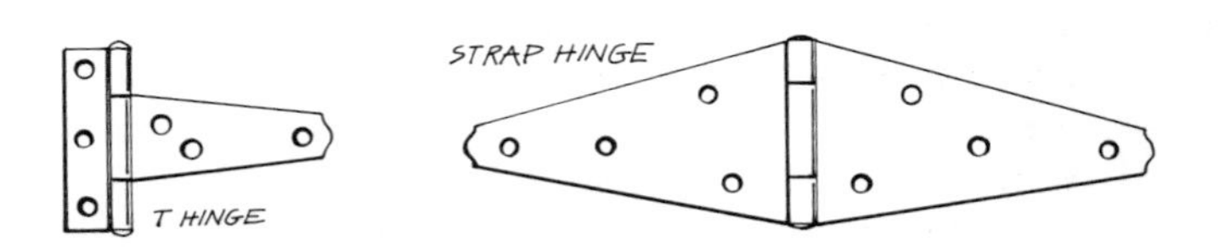

Hinges come in various sizes, weights, and lengths. They will be stamped or package-marked "light" or "heavy." Unless you are constructing a very delicate or extra-small gate, choose the "heavy"-weight hinge for wooden gates. A 6″ hinge will support a wood-frame, wood-sided gate in the 3′-wide and 3′- or 4′-high range. You will usually have a choice of buying your hinges prepackaged with wood screws or buying hinges and screws separately. It's better to buy them separately so that you can choose the length of screw that will fit the material you are building with. Normally your hinges should be attached to your gate where the horizontal rails run. Choose a screw size that will not extend all the way through your siding and rail, using the longest screw that will fit for maximum holding power. You can go to bolts or lag screws if you want an extra-strong hinge/gate connection, but these are usually not necessary unless it's a *very* heavy gate. The average wood gate in the 3′ size range will take a wood screw of 1″ or 1½″ length. Always choose zinc-plated or other guaranteed rust-resistant hinges for your gates.

If you want a gate that will automatically swing shut after you, you can buy spring-loaded hinges. They are a bit more expensive and sometimes a little harder to find, but are very useful where a gate gets a lot of traffic and not enough attention to closing. You can create your own self-closing gate with an expensive spring or homemade weight and pulley system.

If you'd like a gate that can be easily temporarily removed for some particular purpose, there is a type of strap hinge made for this. The strap is bolted to the gate and the screw hook is attached to the post. The gate may be lifted on and off the hook attachments. These hinges are particularly good for very wide, heavy gates, as they can take a lot of pressure and pulling without letting the gate sag. The bolt-type is especially strong and can be periodically tightened up.

Ornamental hinges in black or brass are available for very fancy or special gates. These usually run two to three times the price of a similar-size "plain" hinge, but as hinges are priced in the $2.00-to-$5.00 bracket to begin with, you might like to splurge and dress up your gate.

The final matter in gate materials is the latch. There are many types to choose from, ranging from the simple hook and eye to elaborate ornamental handles. Most latches are surprisingly inexpensive. Look over what's available and choose what seems most suitable to your needs. Ring- and lever-type latches may be opened from either side of the gate. There is a style of thumb latch that can be installed only when the top of the gate is flush with the top of the post. Barrel bolts (which slide across and flip to lock in position) can be opened only from one side of the gate or by reaching over, so that they are limited in use to low gates or inside a yard for a lock. Hasps come in various sizes and a few styles; some must be closed with a toggle switch or padlock; others have a turning lock device that will accommodate a padlock or toggle switch, too. A simple hook and eye, or a spring-loaded hook and eye (usually in light, small sizes only) may be sufficient. Latches range in price from around $.39 for a tiny hook and eye to $3.00 or $4.00 for a heavy barrel bolt. Safety hasps are $1.00 to $3.00; ring and lever latches are about $3.00. After looking over all the latch possibilities, you might prefer hand-fashioning an old-style wooden one. These are simple to make, following the barrel-bolt sliding principle, and look very handsome on a handmade wooden gate.

With lumber, latch, hinges, and nails gathered or bought, you are ready to build and hang your gate. The tools you will need will probably be: a hammer, a saw (hand or power), a combination square, a tape measure, a pencil, and a level (for checking plumb of posts). Optional tools may be: a bevel for setting angle cuts, a coping saw or jigsaw for making curved gate tops, a surform for rounding post tops or smoothing cuts, and a drill or brace and bit for setting screws or bolts. Sawhorses or a workbench and a clear, fairly level work area should complete requirements. You will probably find that building and hanging your gate takes less time than gathering materials!

Before you begin, check your proposed gateposts for plumb: they should be level or nearly level vertically. See they are well set; a post that gives when you pull on it should be reset. If your posts are badly out of plumb, you should dig them up and reset them. The post that the gate will be hinged to is subject to a lot of strain and should be particularly strong. Both posts should be firmly tamped in place with gravel and earth, or set in concrete. They should sit squarely opposite one another. You may want to run diagonal braces to the next posts in line for additional support. Usually you will want your posts to be flush with the top of your gate or to run 12″ or 8″ above the gate. If you are planning to hang an extra-heavy gate, you might want to put in an extra-tall gatepost so that you can run a supporting diagonal cable or wire to take strain off the hinges. In this case, it is a good idea also to brace the post in the opposite direction. Cables or wires with plate-type attachments and turn-down tightening devices are available for this purpose.

A simple gate to construct is one made entirely of 1× material. It is lightweight but sturdy and can be used in many places. The vertical boards at either edge of the gate serve a dual purpose as siding and vertical rails. This type of gate looks a little rough on the back as nails come through the 1× rails and are clinched over. It will look better with age, as the nails and boards seem to blend together a little.

To begin, cut the four boards that will serve as your frame. The vertical rails should be cut equivalent to the height of your gate. Remember that you have to leave ample space beneath the gate to let it swing—4″ to 6″ will usually do (more if the ground is uneven; less if you are penning small creatures in or out). Take each vertical rail and mark a line 1″ to 2″ or so from each end; this is arbitrary—choose whatever looks good to your eye. The two lines drawn on each board will help you to place your horizontal rails. Make all of your cuts and lines square, using the combination square, so that things will fit together well. You will need ½″ to 1″ on each side of your gate to allow it to swing. Your two horizontal rails, then, should be cut 1″ to 2″ shorter than the distance between your two posts. Lay out your framing boards as shown in the drawing at the beginning of this section. Square everything up by placing your combination square in each interior corner, and place one small nail in the center of each joint. At this point your frame is roughly held together, but everything can still pivot.

Turn the frame over. You will now be nailing from the front of the gate and can nail each corner in place. Check your square as you go, using the combination square. Face-nail three 8d or 10d nails per corner in a triangle or use five nails in an H pattern. You can drive these nails clear through both pieces of wood (into the ground, or allow space if you're working on a table); later you will clinch the protruding ends down.

With your frame nailed together, you can fill in with your vertical siding. Use a template to keep an equal space between the boards as you set them. If you have to rip some of your boards to fit the width/spacing of your gate, plan ahead to place them in a symmetrical pattern. When all of your boards are nailed in place, turn the gate over and clinch your nails. Clinching (banging them over flat) greatly increases the holding power of the nails and hence the sturdiness of the gate.

Next you should measure and mark a board to serve as a diagonal brace. Always run your brace from the bottom of the side the hinges will go on to the top of the side opposite the hinges. This will help to counter the gate's tendency to sag and pull off its hinges. Use your bevel to get the correct angles on your diagonal, or simply lay the board in place as shown and mark at points indicated. Connect points to give your cutting line, and you should have a perfectly angled, snug fit.

Tack your brace in place on the back of the gate with a couple of small nails, then turn the gate over. Nail the brace firmly in place with one or two 8d or 10d nails through each piece of siding and into the diagonal. Turn the gate and clinch over these nails. This way, the clinched-over nails will not mar the front or face of the gate.

The next step is to attach your hinges. They should be attached to the front side of the gate, and should be positioned so that they are fastened through into the top and bottom rails. Now you can stand your gate in place, putting it up on blocks or scraps of wood to allow swing space below. Remember to leave ½" or so between your gate and your post. Make sure this space is consistent from top to bottom so that the gate hangs fairly level and plumb. Fasten your hinges to the post and remove the blocks—your gate should swing freely. The final step is to install your latch and, if you like, put on a stop for the gate. A stop is a strip of wood fastened to the post opposite the hinges. It runs vertically just behind the edge of the gate and must be wide enough to keep the gate

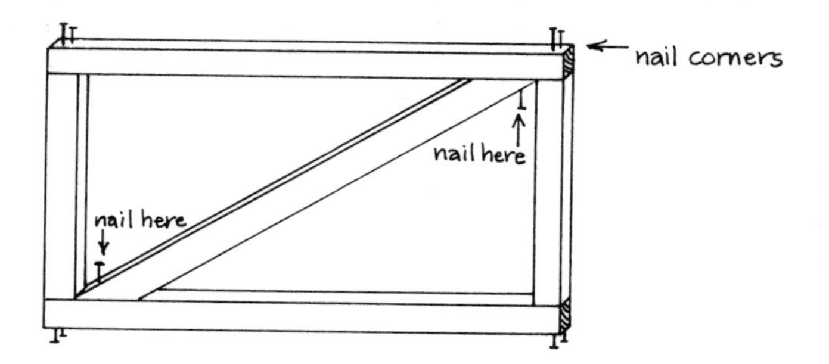

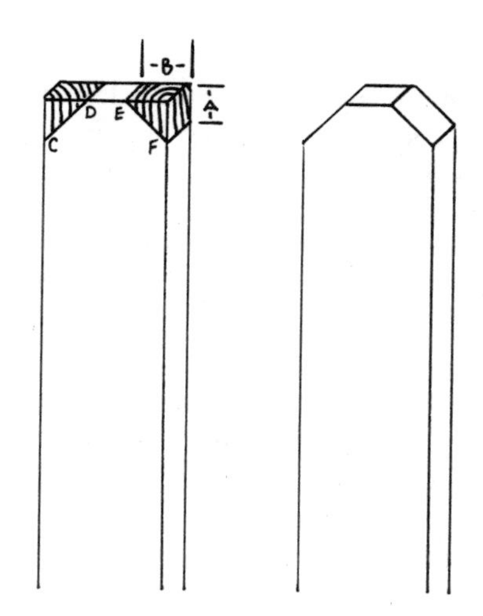

from swinging in past the post and pulling off its hinges.

A second type of gate is built with a heavier frame, and is more versatile and sturdier. You will probably want to use 2×4 for your frame and 1× for siding. You can run your siding diagonally or in a fancy pattern. If you want a simple wire and wood gate, follow frame building instructions and attach wire instead of siding. You can substitute light poles or 2×3 material for your framing. The 2× frame gate is cleaner appearing than the previously discussed gate because the nails that hold the siding do not extend through beyond the rails. The procedure in building is basically the same as that just described, with a few changes.

First, cut your horizontal rails to the desired gate width. The vertical rails will fit between these, and the siding will extend a few inches above and below, so that the length of your vertical rails plus the thickness of the horizontals and the extensions of the siding will be the total height of your gate.

Compute out and cut your two vertical rails. If all of your cuts are square, you can lay out your frame and nail it together (with 16d nails); two nails per corner should be enough. The joint you have created is a simple butt joint. You could instead make a half lap joint here, which is harder to make but is considered much stronger. (See "Putting Pieces Together"; also Wilson's carpentry book in "Resources" or you can have a carpenter friend or builder show you.) I've always found the butt joint to be plenty strong except when I've used poles. Here a half lap joint is really necessary because the curved poles can't be nailed tightly.

The next step is to lay your diagonal in place, mark, and cut it. It may be nailed on with 8d or 10d nails placed as shown. Now you are ready to cut and attach your siding. You might like to make the gate a little fancier by shaping the tops (as illustrated). These are called *dog-eared tops* and are easily created. With your pencil, freehand-draw the angles that look best to your eye. The more severe the angles, the more picket-like the boards will appear. Measure distances A and B and duplicate these on the tops of each board. The lines from points C to D and E to F in the illustration will become your cutting lines. You can make gently curved tops by cutting and then smoothing with a surform.

Another way to make this simple gate more attractive is to plan to curve the entire top. Cut all of your siding pieces extra long by 1′ or so and lay them in place. Draw an arc from one side of the gate to the other. If you aren't particularly talented at freehand circle drawing, make it easy for yourself, as follows. Make the center of your middle board the high point of your arc. In progressive steps, drop down ½″ or so for points A–D and E–H. Connect all points with a nice relaxed line and you'll have your arc. You can either nail all of the siding pieces on first and cut with a jigsaw or coping saw, or cut each piece separately and then nail on. Once all of your siding is nailed in place, your gate is ready to be hung. This procedure is exactly the same as described for the first gate.

Making a simple door is much the same as building a gate, so that you can proceed from one to the next, learning as you go. Much of what you learn building gates can be a foundation for your carpentry skills: choosing materials, handling tools, being patient and accurate, and planning things out. Your gate might be an opening to a whole new (pre)occupation!

HANDMADE DUTCH DOOR

My house *had* to have a dutch door! For me it symbolized the essence of a country house. And I wanted it to be big: no more squeezing furniture through narrow doorways for me! I had found a piece of thick glass from an old Jeep and I knew my dutch door would have a window. Then one rainy afternoon I gathered my tools and materials and built a beautiful dutch door.

These plans are for my dutch door; you can scale them up or down or change them all around to suit you and your door framing. Once you know how to make a simple door and set a window into it, you can make it any size that suits your whim.

I began by measuring the rough framing of the door opening. At this stage, my house was unfinished. On the side of the door opening opposite the side I intended to hinge, I nailed a finished (surfaced) board. The rough-cut lumber I'd used in framing is not pleasant to brush against when going in and out. You might prefer to build yourself a complete door jamb with surfaced, dry lumber and then follow some of my ideas on door building. I also made my doorsill out of finished lumber, largely for the sake of my bare feet! This sill extended out from the rough framing about 2″ on the exterior and lay flush with the framing inside. A ¼″ shim under the inside edge of the sill tilted it enough to cause rainwater to flow away from, not into, the house.

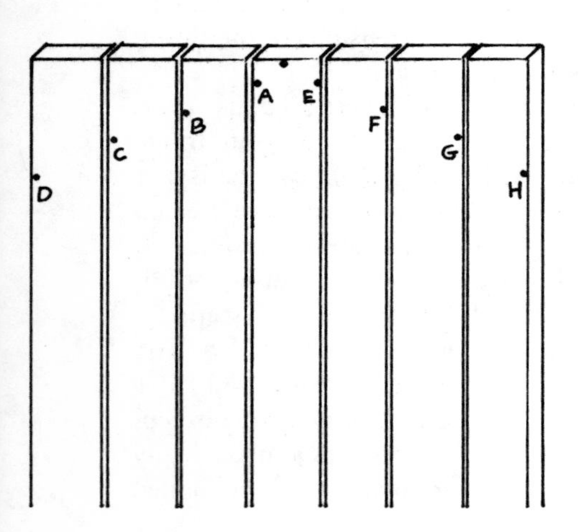

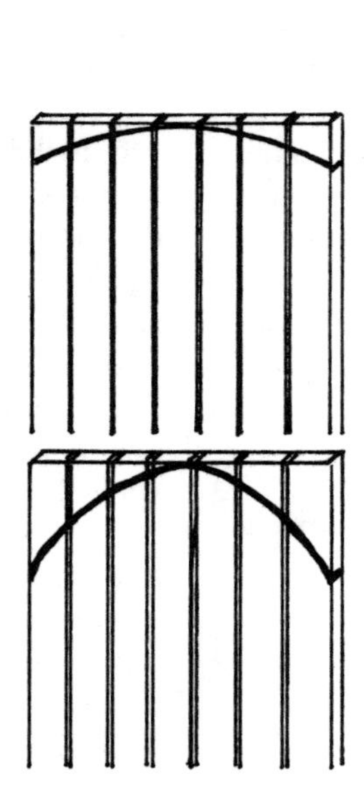

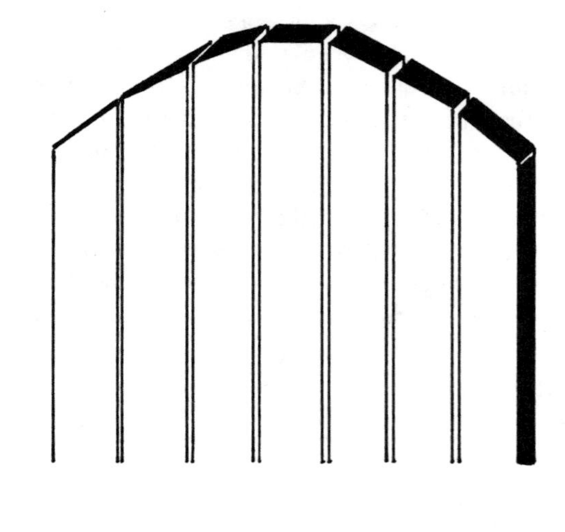

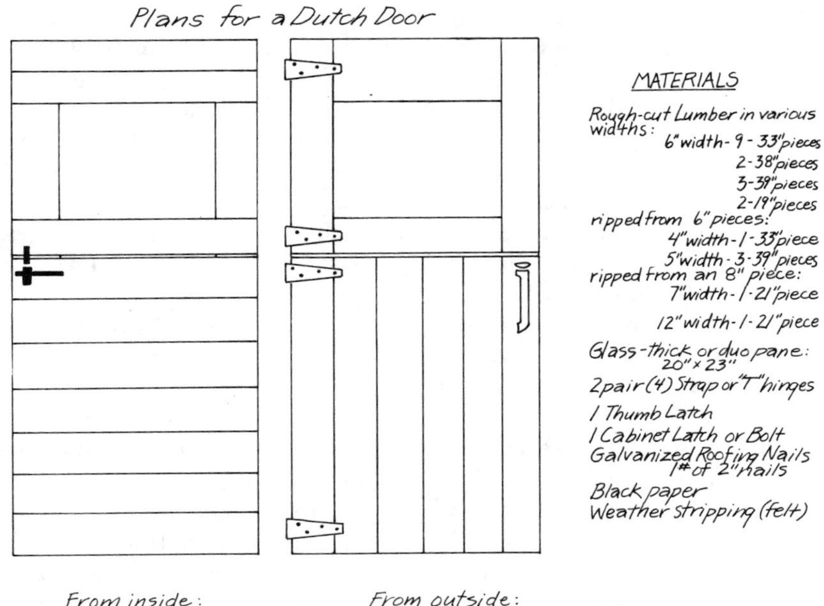

From inside: Top - 33" x 37" Bottom 33" x 40" From outside: Top - 33" x 38" Bottom - 33" x 39"

With these steps done, I measured the door opening again. I subtracted 1″ from the length and 1″ from the width to determine the actual size of the door I was about to make. This allows a generous ½″ all around for the door to open and close. If you fit your door too exactly to its frame, you won't be able to move it! When the door is done and hung in place, this ½″ gap can be covered by nailing 1″ stops all around the inside of the doorframe and stapling weather stripping onto the edges of the door. My doorframe was 78″ by 34″, so my door had to be 77″ by 33″. Both halves of the door would be 33″ wide; the bottom half would be 39″ high and the top half would be 38″ long. Since I did not have access to a planer and long clamps, I couldn't glue boards together to achieve a 33″-wide solid piece. Instead, I made a sandwich of boards: one layer running vertically and one layer running horizontally, with black felt paper in between. Thus I would have a sturdy, heavy, windproof door that I could nail together from individual boards.

I decided to have the door open out, away from the house, and I wanted to be able to open the top half while keeping the bottom half shut. The meeting of the two halves would be a simple and functionally elegant lapped joint, which would also prevent drafts. This meant that the bottom inside boards would be 1″ longer than the bottom outside boards, and that the top inside boards would be 1″ shorter than the top outside boards. (Building a good door is really easy—but figuring out all of the numbers isn't!)

Since the bottom half seemed easier, I tackled it first. My numbers were: inside boards, 40″ long by 33″ wide; outside boards 39″ long by 33″ wide. The outside boards were to run vertically, so I cut three 5″ and three 6″ wide boards to the desired length. I was working with full-dimension lumber, so these boards combined gave me my 33″ width. I decided to run the inside boards horizontally, so I cut six of my 6″ wide boards and one 4″ board (all 33″ long). Next I laid the inside boards down on the floor and aligned them neatly. I covered them with a piece of black felt paper, then placed the outside boards on top. After lining all the edges up carefully, I nailed the "sandwich" together with galvanized roofing nails—because I had a lot of them left over from roofing the house. I like the effect the big heads make regularly spaced on the door, and they don't leave rust stains running down the door. I nailed each board on the outside onto each board on the inside, using extra nails at the edges and corners. Then I turned the whole thing over and nailed each inside board onto each outside board. After I trimmed off the excess black paper, the bottom half of my dutch door was done.

I didn't know how to make the top half and wasn't able to find a book to explain how to set in a thick piece of glass between two pieces of wood. After some pondering and figuring, this is how I decided to do it. I laid out the pattern for the top half of the door on graph paper. For the inside I cut three 6″ boards 33″ long, and two 6″ boards 19″ long (giving me a total 37″). I then laid them out on the floor so that I could visualize the rest more easily. For the outside boards I cut two 6″ pieces 38″ long—these were to be the uprights. Instead of sawing several smaller planks to 12″ lengths to fit into the space between the uprights, I used a piece of 12″ plank I had for the crosspiece at the top. The bottom of the outside top half of the door had to be 1″ below the bottom of the inside top half (to make my lap joint); I ripped an 8″ plank down to 7″ for the bottom crosspiece. I laid all of these pieces out neatly on the floor, too.

Now I had to chisel out a space for the glass to nest into so that the two halves of the top could be nailed together. More numbers! Figuring out how much to chisel out of the uprights was easy: the door is 33″ wide, the glass is 23″ wide, leaving a difference of 10″. Since two 6″ planks equal 12″, I saw that I had 1″ of extra wood on each upright (12 minus 10 equals 2: 1″ off each side). I chiseled out a space 1″ wide, the thickness of the glass (½″) deep, and the height of the glass (20″) long on each of the uprights.

To determine how much to chisel out of the crosspieces, I subtracted the height of the glass (20″) from the height of the door (top, outside half—38″) and got a difference of 18″. I chiseled out a space ½″ wide and the thickness of the glass deep across the entire width of each of the crosspieces. As I chiseled, I frequently laid the glass in place to check for high or low spots. I finally dropped the glass into place and it fit snugly! I fit the boards of Part A over the boards and glass of Part B (no felt paper in this sandwich) and nailed it all together, first from the front and then from the back. I was careful not to nail where the glass was hidden in those chiseled spaces. Now I had the top half of my dutch door!

All that remained to do was to hang the door and put on the latches. I screwed the hinges into place on the door halves while my friend River got the ladder into position. Two hours and two tries later, we finally got the whole door hung so that the halves met and the whole door hung plumb and opened and closed freely. I can't explain how we hung that dutch door, but perseverance, strength, a long screwdriver, and pieces of wax help—and rain, wobbly ladders, and screws that bend over rather than go in don't! A few nails held the door shut until I found another afternoon to install latches, molding around the inside of the window, weather stripping, and stops. Now I have an amazing, incredible dutch door for my house. As soon as I figure out how to build steps, we'll be able to use it!

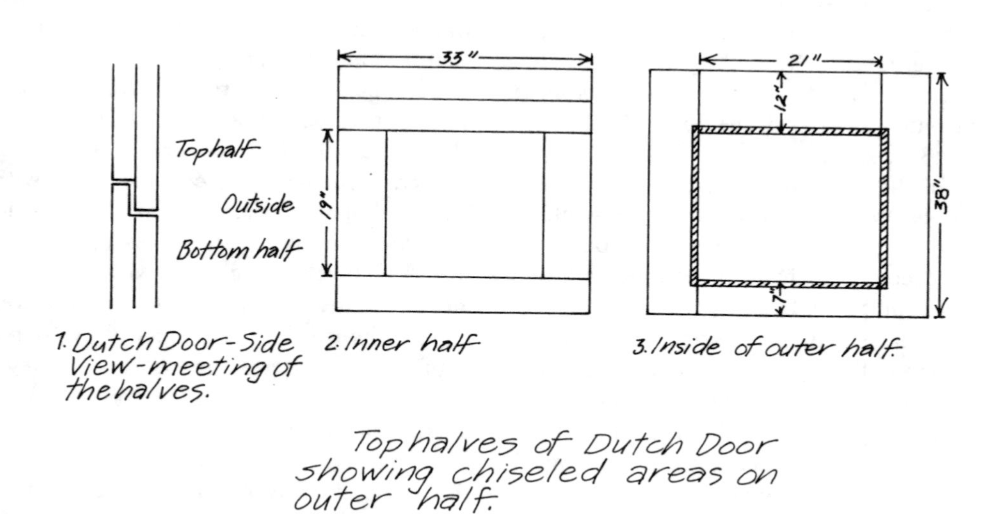

Top halves of Dutch Door showing chiseled areas on outer half.

Imagining the Real House

The first building on our land was a simple kitchen house. None of us (three women and two men, two girl children) was an experienced carpenter, so we all did it together. Yet nothing about the experience of "helping the men" build that kitchen led me to imagine myself building my own house. It was always a question of which man would do it for me. Not until two years later, when I met women making a living as carpenters, could I even begin to dream of being an equal in the process of building my own house. Anyway, I was so busy being mother, cook, and nurturer-therapist, I didn't consider having my house built for years. Yet I couldn't relate to my temporary pyramid with plastic sides, which was built for me. It stirred up an anguish that went beyond the knowing that it came from years of moving as a child, not having a room of my own even as an adult, and countless nightmares of being in unfamiliar buildings with hidden levels, huge gaping holes, dark mysteries. Now it seems that all buildings, all dwellings were at that time a mystery to me, part of my anguish at being in a world I felt no part in creating.

There came a time when it seemed appropriate to seek out a designer/carpenter for my house. That's the only way I can say it because in fact I wasn't ready to imagine the reality of a house. I acted as if I were in a great hurry, drew a simple design, and asked the male carpenter in the neighborhood with a reputation for fast, good framing to build it. All was set in motion, except I felt crazy. An enormous pressure stirred inside me until I was able to go to this man and tell him I wasn't ready to build the house . . . I just thought I was supposed to be ready. I was able to tell him that I had related to him as an authority—trying to please him with my efficiency, even designing a house that I knew he'd like to build (since it looked like all of his other houses). Then I went away, greatly relieved that it was over for a while. Somewhere deep inside I knew that I wanted more from the process than to have a man build for me yet I couldn't imagine what I did want. The dream of women building was buried too deep. Five years ago during an est process, we were supposed to build a sanctuary in our minds. All I could do was panic that I didn't know what to do, even in make-believe. Years later, during my first exposure to a woman's spirituality group, I was again asked to create a sanctuary in my mind. The difference was that I no longer felt panic and shame. Still, I couldn't imagine the structure beyond a vague circular feeling that kept going in and out of focus like the memory of someone you know well but can't visualize.

Then I met a woman carpenter who was willing to go on the journey of my house building with me—through its labor pains and my resistance to taking responsibility for my own visions. Before we even talked of building my own house, I started working with her remodeling a basement in the house where I lived. I remember

(Photo by Kathryn Biglow)

the filthy chaos of the basement and the respect I felt that inside of her was the capacity to see and create beauty out of it. After the demolition there came a moment of truth. She said that I was to construct a wall where there was none. I made an excuse to run upstairs, curled up in the fetal position, and fell asleep for five minutes. Coming back downstairs, laying out the wall, breathing deep through the pain of not knowing, the mystification of building . . . I could not have done it without this woman standing there so quietly imagining the real.

Gradually, as I built more walls, decks, fences, my dreams of houses changed. They became less amorphous. I saw structural details and there were no giant holes, jagged edges, dark places. I was "ready" to build my own house, even as I had been ready to give birth to my child, yet not able to imagine what the birth process would be like.

With each hammer stroke, saw cut, and decision to be made, I became more attentive to the present moment. At first, however, I still relied on my friend, Jean, to make all of the decisions. I could barely look at plans with her. Yet when she was planning with one of the experienced women who had come to help, I got angry. This anger was my way of covering the pain and frustration of not yet being able to take responsibility, of thinking that I had to know things before I had the experiences that would teach me. That episode was a turning point for me. My carpenter friends accepted me through my angry confusion. One of them showed me a poem she had written about the Sufi way of teaching, in which seekers of enlightenment are instructed not to study philosophy but to apprentice themselves to a worldly trade. I realized that none of the spiritual trainings I'd undertaken had begun to fill the deep hole of inadequacy in the world of "things and doing" that was daily being exposed by the process of building. In the past that inadequacy had always come out as resentment and rage. This time I chose to build my own house, to take an equal part, and to accept the differences of skill between myself and my co-workers as a challenge rather than as an indictment of my own inadequacy.

As the work progressed, my commitment to aesthetic perfection grew until I wanted perfection no matter how long it took. Joy in work and in myself grew with that constant commitment to excellence inspired by Jean's attitude and example. Every woman who worked on my house did so with a love and dedication that I cherish and feel permeates the structure itself.

During the building, I had initially felt some conflict between what my hands

(Photo by Sharon Marks)

were doing and paying attention to the feelings of those around me. I had to learn to shift from a predominantly people-directed consciousness to an attitude I'd always associated with and expected of men: an active, total focus on work. At times I couldn't handle the conflict between being a mother and a builder—and then there were the magic moments of play when we were all like children together. I especially cherish the day Jean, Chivvis, and three-year-old Rhama played store with the wood shavings on the second floor. We laughed and laughed.

When I had decided that yes, I wanted to build an octagon, I had only a vague idea of its complexity. It wasn't until we got to the roof that I understood what we had undertaken. Yet all along we kept choosing a more complex structure, always guided by what would be beautiful or fun to live in, not what was easy to build. We followed this method of choosing and building even though time and money pressures were constantly with us: yes on hand tools; yes on the loft, the dormers in the roof, the tower (still to be built); yes on the skylights at the two levels of the roof where the 5/12 and 10/12 pitches meet, and yes on cedar shakes with no paper underneath to hide the wood and cover any mistakes in waterproofing. Choosing an octagon shape for my house was, for me, asking for the stars and receiving them. The first nights I spent in the octagon after the roof was completed were like riding a wild horse—the molecules of wood were still spinning.

(Photo by Sharon Marks)

(Photo by Kathryn Biglow)

In a circular house, the energy flows around in a unique way, changing psychic walls and corners left from childhood years in square rooms. My house reminds me of nowhere I've ever been—only where I am. From within, it feels like an elf's cathedral; from without, it looks like a tiny hut from somewhere between Africa and Polynesia. It belongs with the trees. As we built, the guava trees fed us their fruit, and as we finished, it seemed that we had deserved their bounty: the house is like a geometric tree, part of the natural wonder.

(Photo by Kathryn Biglow)

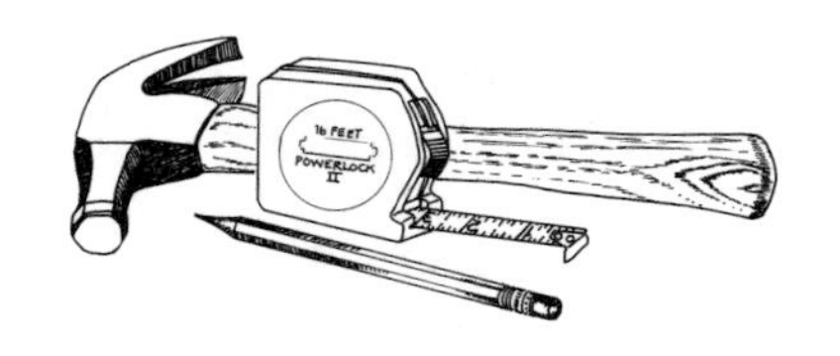

Changing Your Space -in a Day or Two

REPLACING YOUR DOOR

Replacing one of your house's exterior doors may be a matter of aesthetics or necessity, or a combination of both. You have a few options for your new door: you can make your own, buy one ready-made, search for an old door and recycle it, or have a door custom-made for you. Of these options, the second and third are the most reasonable in terms of time and money; the first and last can produce that unique door that adds a special feeling to your home.

Whichever option you choose, careful attention to the details of your existing door is the first necessary step. It's a good idea to take a level and check the plumb of your doorjamb. As mentioned earlier in "Dealing with a Sticky Door," older buildings shift and settle, commonly throwing the jambs of doors and windows out of plumb and creating some problems (doors that bind; windows that stick). Sometimes you'll find that someone before you has cut the door to match the out-of-plumb jamb: the door is not cut square, and attempting to replace it with a carefully squared new door can be frustrating! If you find this to be the case and are feeling really ambitious, you may want to go ahead and work on the jamb to make it plumb—or even replace it. If you haven't the time or inclination for this work and don't mind hanging another "custom-cut" door to match the situation, continue on.

Take the measurement of your existing door from top to bottom. If the door is beveled at the bottom, measure to the longer edge of the bevel. Measure the width of the door, checking top and bottom to see that these are equal. Finally, and of absolute importance, measure the thickness of your door. In choosing a new door, you will want one of the exact same thickness: otherwise it may not hang properly in the existing jamb. You will need a replacement door of exactly the same width if you can find it; or cut a wider one down with a handsaw or a circular saw with rip guide and fine cutting blade. The height of your new door likewise should be the same as your old, and any variance must be larger rather than smaller. A door that is even ½″ too short can create weatherproofing problems; a door that is a couple of inches too large can be easily cut down to size.

While you are looking over your existing door, decide whether or not the hardware on it is still functional. If you want to reuse it and plan to search for an old door to recycle, you will have to disassemble the hardware and take *careful* measurements (more information on these problems later). At this time, too, you should take a critical and careful look at how your door is working; does it swing freely—bind—close easily—hang right? Is the threshold still in good condition? Is the door a source of drafts, or is it fairly weather-tight? All of these details will affect how your new door hangs and acts, so it's a good idea to look them over ahead of time, try to foresee any problems, and think about how you might deal with them. Take an awl and drive it into the jamb itself in a couple of places to make sure it is still sound. If the awl goes into spongy, soft wood, you may need to replace the jamb rather than buy yourself a new door (or do both). Make sure that your door hinges are still in good condition, too, if you plan to reuse them. You may need to buy new, slightly longer wood screws to give the hinges a firm connection with the jamb.

If you are having a door custom-made or planning to make one yourself, your basic groundwork is over. Likewise, if you plan to buy a ready-made door, you can begin shopping around for the one you want. Lumber stores, hardware stores, and some furniture stores usually have ready-made doors in stock. Prices vary considerably from place to place, and you should be careful in making comparisons to price exactly the same style and quality of door.

Finding a suitable old door for recycling is a bit of a treasure hunt, but it may turn out to be well worth your time. Old doors may be found in antique or "junk" stores, but are usually most reasonable at places specializing in salvaged building materials. Armed with your measurements, a tape, and a pocketknife, you should search around as much as possible. If the salvage yard is well organized, doors will be stacked according to size. Interior and exterior doors are commonly stacked together, so you'll have to determine which are your exterior doors by weight, style, wood type, and tapping. A hollow-core interior-type door *sounds* hollow when tapped; it's also obviously light in weight. Solid-core doors are heavier and a little harder to manage, but generally your best choice for an exterior door. Most old doors are buried in layers of paint, and its's important to remember that paint can hide potential beauty—or flaws!

When you find a door of the right dimensions and of a style you like, look it over critically. A scratch or chip, a slightly uneven edge, or a broken pane can all be dealt with; joints that are opening, substantial splits in the wood, and areas that feel spongy or soft are all serious faults. If a door is heavily painted, you can use your pocketknife to scrape away a little paint along an edge, revealing the wood underneath. The beauty of an old door is often in the wood used: stripped of its paint and given a little oiling, the door may reward you with rich tones and perfect grains all but impossible to find in modern counterparts. Don't be shy about conferring with the shopkeeper about the type of wood the door is made of—often the doors are priced according to this.

Another consideration is the hardware an old door can accommodate. Some doors will still have their hardware intact and functional; others have sets of obsolete holes that may not take the type of lock or latch or knob you want, or have only part of the hardware set. In some cases you can adapt what is there, but more often you will be very limited. You may have to create some sort of cover-up for the old holes, while trying to install your new set of hardware above or below. This can easily destroy the aesthetics of your door, not to mention creating all sorts of problems when you try to match up the strike plate and other details. If luck is with you and you find a door that you like, that's a close match in size, that is sound and will actually take a standard hardware set (or has its old set intact), your final task is to check out the hinges. It is fairly common for old doors still to have their hinges—and underneath a layer or two of paint, they may be perfectly functional. Until it is removed, the paint may hold the hinges locked into position. If the hinges are missing, you'll have to replace them. With luck, the mortises or seats for the old hinges will accommodate new standard sizes. If not, you'll have to do some work to make *them* work. Remember that the hinges must fit snugly in place, so slightly larger hinges may be needed—you can chisel out room for them as long as they aren't *too* large. If you are planning to cut the door down width-wise, you may simply be able to cut the hinge edge off and freely install new hinges of whatever size you wish. Don't forget that you will have to match the hinges—old or new—against the hinge mortises already in your doorjamb; this may take a bit of important thinking and juggling. An extra-special old door may be worth every second of this extra work in the end—or you may opt instead for a "ready-made" and its simpler installation!

With the business of choosing or making your door behind you, you're ready for the work of hanging it. Hanging a door and installing hardware can easily be a full day's work, so if you find yourself well into the afternoon hours and think you must be awfully slow, take heart! This job requires utmost patience and total attention to detail. You are much better off to take your time and make each step perfect than to rush yourself and risk what may be an irreversible mistake.

To avoid scratching the finish or paint on your new door, put some cushioning on your sawhorses. If you are using a power saw, remember that some blades will splinter the wood along the cut, so choose a planer blade or another that gives a really clean, fine cut. Removing the existing door is a simple matter of holding a broad-tipped screwdriver up under the head of the pin of each hinge in turn and giving the screwdriver a few good taps with your hammer. When all three pins are removed, you can lift the door free and set it aside. If the old door was cut really oddly, you may want to use it as a pattern on your new or replacement door. Remove all hardware to avoid scratching the new one, then lie the old door on top of the new and trace carefully with your awl. This may sound extreme, but I've run into at least one situation where it was necessary. Usually, however, you can cut the new door to size using exact measurements for the old. Don't forget to duplicate the bevel if the old door has one. Be really careful to check and double-check before every cut (and later when you chisel or drill); you can't go back once you've begun, and the door is too easily ruined by a wrong cut or improperly placed mortise.

If you have to cut the door down width-wise, try

to do it symmetrically. If you have to take off only 1/2" or so, it won't matter which side you cut (unless it's a recycled door and you have to make allowances for hardware). If you have to take off more, it's better to take from both edges, leaving the door symmetrical, especially if the door has panels or panes or detailing of any sort. If you are cutting the door down in its height, you must similarly pay attention to aesthetics. Usually the door should be cut down at the bottom; the bottom rail is much broader than the top, giving you more leeway.

The next step is to set the halves of the hinges that belong on the door. Remember that the normal exterior door opens inward (screen doors and storm doors can then open outward), so that the hinges are set on the inside edge. You can take the measurements for locating your hinges from the old/outgoing door. If you don't have a butt gauge of the right size, use your combination square to trace out the exact mortise (seat) for the hinge. You will want the hinge to extend slightly (1/8" or 1/16") beyond the door. One procedure for setting a hinge:

Measure down from the top of the door and mark where you want the hinge top to be.

Set your combination square for the width of the hinge minus 1/8" or 1/16" and use it as a straightedge to mark the location of the hinge.

Chisel out this hinge mortise very carefully, going just deep enough to make the hinge sit flush with the surface of the door (you can use a hand chisel, a router, or a router plane for this work).

Set the hinge in place and mark with a pencil where the screws will go.

Remove hinge and drill for screws, then replace the hinge and screw in place.

With your hinges set in place on the door, you can now put the door in its place, rocking it until each hinge meets with its opposite half. Slip the pins in each hinge. If you've had to replace the jamb or you're installing a new door entirely, you should stand the door in place in an open position, giving you access to the jamb. It helps to have a partner here either to hold the door or to mark for the hinges. Put shims under the door to hold it up enough to give you free swing later. You should now be able to mark on the jamb where each hinge hits. Remove the door and chisel out the mortises for the hinges in the jamb. Set all three, then put the door up again and slip it in place, putting the pins in to join the hinges. Now give your door an experimental swing. By some miracle, it may work perfectly! More likely, you'll have to do some adjusting. A sticky corner or edge can

(Photo by Carol Newhouse)

(Photo by Carol Newhouse)

(Photo by Carol Newhouse)

be planed down. You'll have to remove the door for each attempt, but it's better to plane off cautiously, working in slow stages, than to take off too much at one time.

You can adjust the plumb of your door in two ways: by using shims in between the hinge and the doorjamb (cardboard is the old standby for this—funky but functional!), or by chiseling one or the other of the hinge mortises deeper. For example, if the door seems to hang out of plumb and this causes it to bind or just looks bad, you can chisel one mortise slightly deeper and see if that helps. It may not be quite enough, but you don't want to set your hinge in too deep, so try shimming out the other hinge instead. Keep in mind as you work that a door will swell slightly in damp weather, so that you must overcompensate a little all around by giving a bit more space for swing than you think you need.

A door should always close easily and go to the closed position naturally when given a gentle push. Getting the door hung exactly so that it closes without strain is one of the most critical steps in door hanging. If the door seems to close with a little jump back, it is not hanging properly. It will probably close when forced (even gently), but it is straining on its hinges and sooner or later this will cause trouble. This is the time for shimming, chiseling out your mortises a bit, checking plumb, planing, or whatever series of efforts, compromises, and balances will relieve the strain on the hinges. Look carefully at all the possibilities and make any adjustments you think you need *very cautiously.* You don't want to chisel too deeply or plane off too much. This can definitely be the most trying part of your day, but when you finally get it just right and the door swings nicely closed with no binding, no grabbing, no dragging—you'll have the satisfaction of a job well done!

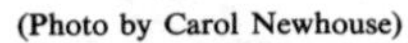
(Photo by Carol Newhouse)

The final step is to install your hardware. Fortunately the companies that manufacture locksets and hardware for doors include very carefully detailed instructions with their products. If you follow the instructions exactly and carefully, step by step, you can install a lockset without too much trouble. I can't overemphasize *carefully* enough! Proceed very thoughtfully and double-check all of your measurements and positionings. Once you've made a chisel mark or drilled a hole, you can't undo it, and there's very little that can be done to hide or alter a mistake.

Make sure that you buy a good-quality, brand-name lockset; cheaper ones are not only harder to install (critical parts break, jam, or won't fit), but they don't last as long or work as smoothly. The lockset must be chosen according to the thickness of your door. This is very important, so read the fine print if the set you're looking at doesn't state very clearly and boldly what doors it will fit. Some locks are also made to fit according to how a door opens. These may be coded with initials such as "LH" and "RHR." A door that is hinged on the left and opens inward takes a lock coded "LH"; if it opens outward, the code is "LHR." A door hinged on the right and opening inward takes a "RH" lock; opening outward, it takes the "RHR" lock. But not all lock companies use this system of coding. The lock you buy may instead have a template printed on one side for doors hinged at the left and on the other side for doors hinged at the right.

The tools you'll need will be a drill or brace and bit, and a special bit that will make the large (1¼" or so) hole required for the lockset. You can buy (or borrow) a relatively inexpensive "nest of bits" which includes a series of graduated bits made for drilling large holes; these have a smaller pilot bit at center to guide them into place. If you're going to do a lot of lock setting, you may want to buy a special boring jig for door hardware, or some good-quality large bits in appropriate sizes. You'll need your wood chisels

More on Doors

Panel doors consist of stiles (verticals) and rails (horizontals), which are usually solid, and panels of plywood, hardboard, or solid wood. The design of the panels and the number used in a door vary.

Flush doors consist of a wood frame with thin sheets of material applied to both faces. The core of the flush door may be solid (filled with wood blocks or other material) or hollow. Hollow-core doors are usually used for interiors.

Standard thickness of an exterior door usually = 1¾".
Standard thickness of an interior door usually = 1⅜".

Interior doors are usually 2′ to 2′6″ wide.
Exterior doors are usually 2′8″ to 3′0″ wide.
Both interior and exterior doors are usually 6′8″ in height.

Exterior *solid-core* doors are usually hung with three hinges. The top hinge is located about 7″ from the top edge of the door (to the top edge of the hinge). The bottom hinge is placed about 11″ up from the bottom edge of the door (to the bottom edge of the hinge). The third or middle hinge is centered between the other two hinges.

Interior *hollow-core* doors and other light doors require only two hinges, placed top and bottom at the positions indicated above.

The clearance you need around a door is usually about 1/16″ on top, 1/32″ on the hinge side, 1/16 to 3/32″ on the knob side, and 5/8″ on bottom (more if the room has thick carpets).

The lock/knob side of a door may be slightly beveled to allow it to close more easily.

for cutting out the strike plate gain (seat), a screwdriver with the proper head, and a pencil.

The procedure for installing a lockset will vary from set to set. There should be a template included with the instructions; the template is folded and set against the door at the height you want the lock and shows you exactly where to drill. Usually you mark and drill the large hole first and then mark and drill at the right distance from the edge of the door for the latch. Slip the latch into place and trace its outline so that you can chisel out a gain, or seat, for it, enabling it to sit flush with the edge of the door. Now you can slip the lock cylinder and latch into place and assemble them; exactly how you do this will vary from set to set. The final step is to mark for and set the strike plate on the jamb. You can get a center mark by putting a little chalk or crayon on the tip of the latch, closing the door with the latch held in, and then letting the latch out slowly; it will hit against the jamb and mark it. The strike plate should be set into the jamb so that it is flush. You should chisel out a gain for it. Once you've put the plate in, you may have to do a bit of adjusting to position it exactly right. Be careful not to set it in too deep; if you have, you can shim it out again with a strategically placed piece of cardboard. Make sure that the holes drilled for the screws that hold the strike plate are deep enough for the screws to fit flush with the plate; otherwise they will give you trouble, causing your door to stick.

HANGING A DOOR: NEW CONSTRUCTION

If you're building a new structure or a room addition and have come to door hanging, much of the previous material can be followed once you've set the door frame securely in place. In rough framing, you've constructed a rough door opening that allows you the proper space for setting in the frame and sill and hanging the door. A door frame may be bought entire from many building supply yards, or it may be constructed on the site.

Most simply, it consists of two vertical pieces (the side jambs) and one horizontal piece (the head jamb), which runs across the side jambs at the top. The frame forms a secondary "box" within the rough frame that is carefully plumbed and fixed; the door is hung within this frame. Stock door jambs are made with 5/4 material, which is finger-jointed (has a special multiple-edged joint) for maximum strength and stability. This material is clear (has no defects) and dry (kiln-dried). If you are making your own frame, you can buy lengths of jamb material or, alternatively, use a good grade of dry 1× material (1×6 or 1×8, depending upon your walls).

Calculations for your door should be made from the top surface of the finished floor within. If you're going to live with a rough subfloor for a while and add the finished floor when you have more time and/or money, take this into account in planning your door frame and hanging the door. You may want to hang the door as though the finished floor were already in and tack a temporary extension strip on the bottom of the door to block drafts until the floor is actually in place. Usually the side jambs are cut to a height roughly ½″ shorter than the height of the rough opening. They are dadoed to receive the head jamb, which runs between them. The head jamb should be cut about ⅛″ longer than the actual width of your door, and if you dado the side jambs you have to add an amount equal to the depths of the two dadoes. The head jamb should be located so that when the door frame is in position there will be ½″ to ⅝″ of clearance for the door itself to swing. To position your dadoes, add ½″ or ⅝″ to the actual height of your door; this measurement will give you the *lower edge* of the dado, measured from the bottom of each side jamb. With the dadoes made, you can slip the head jamb into place between the two sides and nail it with 8d finishing or casing nails. If you've elected not to use dadoes, and make a frame with the head jamb resting over or between the sides in a simple butt joint box, you'll have to adjust your measuring and cutting to make similar allowances for the door; give at least ¼″ between the door and the side jambs—⅛″ per side—and allow for ½″ to ⅝″ from top to bottom for swing.

With your frame nailed together, stand it in position in the rough opening. Use a level to check the frame; it should stand so that it is plumb on both sides and level across the head jamb. If your head jamb reads out of level and your floor *is* level, recheck the measurements of your side jambs (and dadoes) and make whatever adjustments are necessary. To secure the frame in a plumb position, the usual procedure is to drive pairs of shim shingles between the side jambs and the rough frame: one pair toward the top of the frame and one pair toward the bottom on each side. Check the jambs carefully as you drive the shingles into place; the more exacting you are when you do this, the easier it will be to hang your door and have it swing smoothly. When the entire frame reads perfectly plumb and level, secure it by driving 8d casing nails in through the side jambs at the locations of all shims; these nails should go through the jambs and shims and into the rough framing. Now place two more sets of shims per side at intermediate spaces. These will further strengthen the frame and keep it from warping or changing later. Nail the frame all around with 8d casing nails placed 16″ o.c.

With the door frame in place, you can now set your sill and threshold. Depending upon what material you use for your sill, you may need to cut back flooring and even part of the floor joists below the door opening. The doorsill should be slanted/beveled so that any water hitting it is channeled away outside. Standard doorsill material is cut to a bevel and comes in set lengths that correspond to standard door openings. The sill should sit snugly between the side jambs of the door frame, and the upper face should be level with the finished floor. Two projections similar to the sill horns of the windowsill are usually cut to extend beyond the door frame on the outside. If you are improvising rather than using standard materials for your doorsill, you should still try to make the sill beveled, level with the floor, and secure. After the sill is in place, you can put the threshold on; this is a standard piece of oak or other hardwood cut to fit over the joint of sill and finish floor. Again, you can improvise, but in this case the milled threshold is quite inexpensive and very durable. You can buy metal thresholds, which wear well but lack some of the charm of oak. Thresholds, wood or metal, come in several standard sizes, so match one to your door.

The next step is to put a set of trim or casing around the interior and exterior of your door. This functions both aesthetically—"finishing" the door—and practically—keeping out wind and weather. Casing material and design may be standard or inventive. As with window trimming, door trimming can add to the appeal, inner and outer, of your house. The usual trim or casing consists of two vertical side pieces and a horizontal head piece that rests over the sides. Clear (without defects) 1×4—and sometimes 1×6—is used and, depending upon the material itself and the mood and fancies of your house, this material can be left natural, stained, painted, or even ornately carved. The trim or casing will not only cover the joint (space) between the door frame and walls, but will help to hold the frame in place and thus keep your door swinging nicely for years. The trim pieces should be nailed into each side jamb they cover and into the rough-framing studs. Use small nails (4d or 6d) into the jambs and larger (8d) into the studs. Finishing or casing nails are the best choice.

With all of this work accomplished, you are ready to hang your door, install locksets, and add weather stripping. Door hanging has been described earlier; for some ideas on weather stripping, see the "Materials" section. There is one final step while hanging your door, and that is to put a piece of wood stripping called a *door stop* around the side of the door when it is hung. This strip will keep the door from being pushed in and straining the hinges. It can be made of 1× material (1″× 1¼″, actual dimensions, is standard) and should be cut and nailed in place when the door is in a closed position (on the *inside!*). The door stop pieces should mimic the jambs—two side or vertical pieces and a top or horizontal piece between them.

ADDING A WINDOW

Adding a new window to an existing wall in your house is not an overly difficult job and can make a dramatic difference in a room or in the overall feeling of the house. Your first step is to make the practical and aesthetic decisions about the window: How big should it be? Exactly where should it be positioned? How will that change the light and space and privacy of the room? Should it be fixed or openable? In choosing the actual type and size of the window, you should consider how it will affect the outside appearance of the house as well as the interior. Try to balance it with the other windows and doors of the house, and use a style of window that is more or less similar to those already there.

The next thing to consider is whether you are going to use an old, recycled window or buy a new one. If you choose an old one, you'll have to make a set of jambs and a sill for it. If you're buying a new one, you may want to buy one prehung: that is, complete with jambs and sill and ready to go in. Whatever your choice, you'll need some other materials; a header to span over the window, a stud for each side, and so on. Look over the chapter on rough framing and also the section on windows for an idea of the work and exact materials involved. You'll want to have everything there when you set to work so that your house doesn't have to stand too long with an open hole in the wall.

If the wall where you plan your window is a bearing wall, you'll have to make some special preparations before you can cut out for the window. The weight bearing down on the wall in the immediate area will have to be supported. A wall is considered a bearing wall when it is taking the weight of the roof rafters, so if the rafters come down or rest upon the wall you've chosen, you have to deal with this. The easiest way

(Photo by Carol Newhouse)

is to put a temporary header in place which will take the weight while you work and which can be removed once you've done all the rough framing around your new window (and your new header is taking the weight). The same problem occurs if the wall you've chosen is supporting the weight of ceiling joists and a floor (or floors) above. Again, you'll have to set up some system of temporarily deflecting the weight in the area of your work. What you need in either case is a header that will tuck up under the joists or rafters in the area, and a means of lifting that header so that it slightly displaces the bearing members above it while you work. This will release the downward-bearing pressure from the wall you're working on and make it possible for you to safely cut out your rough hole and then do the framing and installing of your window.

Your header should be placed so that it runs perpendicular to the joists or rafters it is going to temporarily support. If you're working with an open ceiling, this is easy. If there's any question in your mind (if the wall is bearing, the rafters or joists should be running perpendicular to the wall, anyway), you can go up in the attic or attic crawl space and take a look. If there's another floor above and the ceiling you're working under is also covered, you can usually tell by the nailing patterns of ceiling or floor (above) materials which way joists run: panels of sheetrock or pieces of wood paneling will be nailed to the joists at regular intervals.

How you go about setting up your header/deflection system will depend upon the exact circumstances you're working in. You'll want to take precautions to protect floor and ceiling coverings, so that you might want to use scraps of burlap or old rug pieces to cover the top of your header and the bottom ends of posts that support the header. If you're working on a rough floor surface that you plan to redo, you can make a header/post system as follows. Measure the distance from floor to ceiling (or rafters or joists), using a straightedge and level to make sure that you are measuring correctly. Subtract the width of your header from this measurment but add about 1⁄16" to it. If you are using any cushioning over the header or under posts, figure this in, too. Cut two posts to the length you've figured. You can use 4×4's (even two 2×4's spliced together) for posts. Your header should be at least a 4×6 (set on edge) and should be long enough to extend past your proposed window opening at least a couple of feet on either side. To wedge the header in place, you'll need the help of a friend (or two), a sledgehammer, and a couple of stepladders. Basically, you'll have to lift the header up into place (stand on your ladders for this part) and then wedge a post under it at either end. If you've cut the posts correctly, one of you will then have to go down to the base of the post and sledge it into place under the header. Make sure that both posts are standing plumb, and then tack everything together securely enough so that it won't fall on you or move while you're working. If your cuts were off a little and the posts weren't quite long enough to push the header up and lift the joists or rafters slightly, you can drive shims into place for the same effect.

A second way to set up a weight-displacing system, and one that will be easier on your floor, is to use a pair of house jacks (or good hydraulic car jacks) to support and lift the posts that hold the header. In this case, you can set each jack on a piece of cushioned 1× or 2× material, get an approximate measurement from the top of the jack (turned fully down, of course) to the ceiling. Cut your posts to this length. Again, you'll have to work with a partner or two at this stage; to lift the header into position and then place the posts and turn up the jacks will take two or three people. Little Giant jacks (see "Tools" section) are perfect for this work as they are made to be used with posts and can be put in place easily and quite securely. Using jacks to lift the posts (and header) is quick and minimizes damage to your floor, so it's a method you might well wish to try.

With your header in place, you're ready to remove interior siding or paneling and exterior sheathing and siding. Measure and mark an approximate rough opening for your window. It's a good idea to make the opening fairly close in size to the window—the less you tear up your walls, inside and out, the less trouble it will be to replace siding and patch up the walls. You will have to cut down below the rough window opening inside in order to be able to place studs and trimmers and you may have to cut above it for the same reason. The perfect tool for this work is a reciprocating saw. With this tool you can make a very neat cut with minimum damage to siding, especially exterior siding. You can also cut out studs and other framing very quickly and easily. If you don't have a reciprocating saw and can't borrow one from a friend, you might consider renting one for the day or for a few hours; they are easy to use and invaluable for this particular job. If you don't use a reciprocating saw, you can do the work with hand tools: a drill to start holes and a keyhole saw. You can remove studs by pulling nails with a cat's-paw and nail nippers or curved claw hammer. I've heard of people using a chain saw for cutting rough openings, but it makes a messy cut and is dangerous. The chain saw is *not* meant to cut into metal, and the chances of your hitting nails in the wall are quite high. I used my chain saw once to cut a rough window opening in my barn, but this was a single-wall building and I could see where the nails were. Besides, I wasn't too concerned with making a neat, waterproof job of it.

If your house is wired and you aren't sure where the electrical wires are in the wall, a good precaution is to make a small hole in the beginning and look in the wall with your flashlight. If you're dealing with an insulated wall, take care in removing the interior siding; be as neat as possible, and try to avoid tearing up the insulation if you can. Use gloves and a respirator when you remove the insulation.

With your rough opening cut out from both the inside and the outside, you can remove any studs that are in the way of your new window. If your wall has continuous bracing at this point, you'll have to reroute it somehow. You may want to make K-type bracing for the area, and this may involve cutting back more of the interior siding. Now frame in your window as explained in the "Rough-framing the Walls" section. If you are working on an old house that has full (actual) 2×4 studs and you're using new material, you'll have to fur out your framing pieces to make an even surface for interior siding and trim. Make sure that you fur out your header, too. Pieces of plywood or strips of lath can be used for the furring. The next step is to cut and place strips of felt paper (15-pound roofing felt or builder's paper) around the window opening. These should be slipped behind the exterior siding that adjoins the opening and wrap around the 2×4's and header and be stapled in place inside. Basically, what you are doing is covering the joint of the window opening and exterior sheathing, and also protecting the material that frames the window. Now you're ready to install your window.

A prehung or prefabricated window will come complete with its beveled sill, exterior casing, and the rest. Make a quick check to see that your rough opening is right and that your exterior siding/sheathing is cut back far enough to accommodate the entire window and its casing or trim. This is another part of the work that is best done with a partner. If the window is large, you'll both want to be outside to lift it up and slide it into place. The casing or trim will act as a stop, so that one of you can hold the window in place while the other one goes inside. Inside, you'll see that the sill is suspended a little above the rough sill because it is beveled.

The usual procedure in installing a window is to cut three (or more) wedges, which are driven, from inside, into this space between the sill and rough sill. The wedges help you to level the sill up, but they also strengthen the sill. When our cabin was still in the unfinished stage, we had a visitor who casually strolled over and took a seat on the sill of one of our windows. We hadn't put the wedges in place when we should have, and her weight (*quickly* removed!) pulled the sill partly loose from the side jambs. It was a lesson in finishing these details at the time of installation! Anyway, this is the time to cut and place the wedges.

You can drive the wedges in until the sill reads level. Place a wedge toward each end of the sill and one or several between, depending upon how large the window is. Now the outside woman can drive a couple of 16d casing nails into the bottom edges of the two side casings; these will hold the sill (and window) in place in the level position. Next you'll want to plumb up the side jambs, using pairs of shims driven between jamb and trimmer stud on each side, and checking by holding your level against the jamb. At this point you may also want to use your framing square to check your progress. If your sill is level and your two side jambs are plumb, a framing square should fit quite snugly in any corner of the window. If the fit isn't snug, recheck and adjust all of your leveling. When you're satisfied that the window is fitting in squarely and that the jambs and the rest are all perfect, you can place a nail (16d casing) through the top of each side casing. If you're putting in a window that opens, now is the time to check that it swings open and shut easily. If so, you can finish nailing the casing (use 16d casing nails placed about 16″ o.c.).

If you're using a recycled window or using a new one but making your own jambs and sill, you can follow the same general procedure. You may want to nail your window frame together without the exterior casing/trim and place it. To make this frame, you have to cut your sill piece first. The sill should be cut to equal the length of the window, but should have a small extension on each end called a *sill horn;* this is an L-shaped extension that will go beyond the exterior casing or trim on each side about 1″. Look at some exterior windows and you'll see how this extension works. It isn't absolutely necessary, but it makes a neat finish—a sort of base for the exterior casing to rest on.

Sills may be made of special 2× sill material, which is milled for this purpose and will give you a properly angled surface when in place. Or you may choose to use a piece of 2× or even 1× tilted to an angle you think is sufficient to shed water. The side jambs must be angle-cut at the bottoms to match the sill when it is set at its angle. If you are making your own, you'll have to experiment until you find the angle you need. A 30° angle is commonly used here. You may also want to make dadoes in each side jamb and slip the sill into these rather than angle-cutting the jambs and joining them to the sill in a butt joint. These dadoes will provide a tighter, more easily weatherproofed frame for your window. Remember to cut the dadoes at the angle you want your sill to be tilted at; in cutting your sill, remember to add the extra length to allow for the portions inserted into the dadoes.

The head jamb extends over the two side jambs, so consider this in cutting it to length. Jamb material should be very straight, dry, and preferably clear (defect-free), but you can improvise and use what you have and still make a fine window frame. Place your frame in its rough opening and level and plumb it as explained. In this case, you'll want to drive casing nails through the side jambs into the trimmer studs to secure the window as you level and plumb it. Then you can put your window in and hinge it or secure it in place with stops. Your final step will be to cut and place your exterior trim or casing and weatherproof the window. Then, of course, you have the interior to finish up.

Weatherproofing the window—whether you've used a prehung unit or made up your own—is one of the most critical steps in this whole process. If you've made a really neat hole and your exterior casing or trim meets the adjoining siding very closely, a thin bead of caulking can be applied all around the casing/siding joint. You may have designed your casing to fit over the sheathing, or you may be working with a shingled wall. In these cases, use caulking wherever you think water might be pulled in by capillary action or leak down and get behind the wall, between siding and sheathing, and anywhere else. Over the window is probably your most serious area. Here you have to devise a method of channeling water over and away from the joint between the top trim or casing piece and the siding. A drip cap is a piece of wooden molding used here; you should place flashing and/or caulking behind the drip cap to seal the joint. You may choose to use a thin piece of metal flashing embedded in a good flexible butyl rubber caulking to seal this area; the flashing will act as a drip cap, shedding water shingle-style away from the area.

With the exterior finished and sealed, you can go to work inside. Replace interior siding as needed. If you're working with gypsum wallboard, this process will take several days as you wait for the joint compound to dry between applications. Adding interior trim is the final step. Here you can be creative and inventive if you like. The standard procedure is to use a piece of trim or casing along each side of the window; this is sometimes set in about ¼" so that the jamb shows. A piece called the *stool* may be set over the sill—or you might prefer to have the sill showing. If the stool piece is used, it is usually cut with extensions similar to the sill horns outside, so that the two pieces of side trim or casing can rest on the extensions. Under the sill, a piece called the *apron* is used. It butts up under the sill and hides the joint where you've placed your wedges. The *head casing* or *head trim* extends over the two side pieces and finishes the window. You can use mitered joints for interior window trim, or cut your head trim to a fanciful shape. All trim should be nailed in place with finishing nails, which may be set in with a nail set and puttied over if you're going to paint the trim. If you're leaving the trim natural, you'll still want to set these nails slightly. When the last nail is set, you can collect the rewards of your work: a new flow of light into the room and the different feeling given to the space.

Casa Adobe

When I went to New Mexico in the winter of 1969, my sole purpose was to visit my level-headed, long-haired niece, whose youthful wisdom I believed might supply new direction for the head-changing journey I had been on for the preceding two years.

In wildest flights of fancy, I would not have dreamed of owning a farm on which I would build a home. However, with passing weeks, I fell under the spell of high meadows, sprawling mountains, and the prodigious blue splendor of the cloud-embellished skies of northern New Mexico. Then I met a woman of my own age, who was not only farming but busily engaged in building herself a home. I was inspired by her gallant, self-reliant spirit.

On the crest of a precipitous hill, north of the village of Arroyo Hondo, the 7½-acre farm I bought was traversed by a grassy-banked, willow-lined, burbling irrigation ditch and boasted a rickety fenced area, dignified by the appellation "barnyard." Huddled near the north bank of the curving ditch was a funky, weathered old adobe house of six small rooms.

At first the house seemed strange. Its irregular, foot-thick walls and oddly shaped windows and doorways were in sharp contrast to the linoleum-covered floors, papered walls, and fiberboard ceilings. I felt somewhat like an archaeologist when, after levering up six layers of musty linoleum, I found a handsome floor of wide, age-silvered planks, and the ripped-down ceiling cover revealed patinaed, hand-hewn vigas (roof beams). Laborious scraping away of many layers of faded, garish wallpaper bared the plush sheen of seasoned adobe plaster.

My private dig took me back to simpler, pre-plastic days. Here was I exposing the original, organic materials under the numerous, artificial coverings applied by those who inhabited the house after the builder. An ex-urbanite, I was enthusiastic about my return to the land and cherished this almost living structure, which I knew someday would collapse and, decaying, return without polluting it to the earth from whence it had come.

The house was rather dark, and some interior and exterior walls needed remudding. Since I felt incapable of knocking out bricks for larger windows or mudding walls, I hired a neighbor, a man of Spanish descent. Intimidated by forty-eight years of city life in the stereotyped role of a "helpless little woman," I was inhibited for months from attempting any tasks considered part of the male domain. The liberating influence of the women's movement had yet to reach me in the high country north of Taos, so I struggled along, dependent on the alleged expertise of men, before daring to strike out on my own.

When I seriously began to consider building a house for myself, I wisely decided

(Photo by Lisa Law)

first to experiment with a structure of more modest nature, a chicken house, for which I purchased three hundred adobe bricks. I hired a backhoe to excavate the side of a slope, so that only one wall need be constructed of bricks since, on the other three sides, the earth, lined with dry rock, would form walls.

Adobe bricks, made two or three at a time in wooden forms from a mixture of straw and extremely sticky adobe mud, are dried in the sun. Dried, they become sufficiently strong to build with, but, of course, are not as durable as kiln-fired bricks or cement blocks.

As the hen house grew, I learned to mortar the bricks with mud, and to fashion and place window and door frames, along with the onerous task of setting vigas in place and constructing a waterproof roof. With this experience, I felt ready to start. But construction of my house had to wait until after the passage of winter. Mortar, after being frozen, crumbles upon drying; work on adobe houses should not be done during the cold weather.

The following spring and summer were spent farming; hence ground was not broken for the house foundation until after the harvest, in early autumn. By that time my young friends who were working and living with me furnished welcome assistance with the many chores involved in construction.

The site I chose for the house was across the barnyard from the original adobe and farther from the highway. It was close to the soothing murmur of the irrigation ditch with its clumps of pliant willow, and near a great silver poplar. By this stage I had become familiar with the spot, aware of its shifting moods during the passing hours of the varying seasons.

(Photo by Lisa Law)

In the two years of planning the house, I had absorbed a lot of know-how, picking up ideas from home-building neighbors, as well as profiting from their mistakes.

Although I knew nothing of the technicalities of architectural plans, I began by sketching a rectangle, representing a 24′×30′ structure, and modifying it to suit my needs and artistic whims. I decided to narrow my bedroom on the northeast corner, to provide space for an entrance portale or porch, which would also serve as protection for stacked firewood. A bowed effect was given the south wall by angling it in three sections. The bedroom's east windows would ensure a ringside seat for sunrises and let me keep an eye on the goats in the barnyard. My plan placed the kitchen and dining area on the west, with large windows to frame the magnificent skyscape, extravagant sunsets, and the view across the broad gorge of the Rio Grande to the serrated peaks of the San Juans. A pair of wide, floor-to-ceiling french windows centered in the south wall would welcome winter sunshine. Also on this side, I planned a patio, overlooking, through the ditch's willows, the valley called Arroyo Hondo.

With the exception of three wine-bottle portholes, the north wall would have no windows. This solid wall would offer extra protection from the bitter winter winds blasting down from Colorado. Additional insulation from winter cold and summer heat was provided by setting the base of the house several feet into the ground. I planned on a depth of 3′, but the soil was so hard and rocky that, after much hacking toil, I settled for 1½′. Since the house site was on a slight slope, I fudged a bit, ending up with a split-level floor, a fortuitous situation, inasmuch as the location of the fireplace and wood stove on the lower level led to more efficient heating of the kitchen and bedroom on the upper.

I preferred that all interior walls should be functional, made of rough lumber. A full wall between kitchen and bedroom would provide for a closet and cupboards, while a waist-high wall between the upper and lower levels would provide counters and bookshelves. This was a happy decision, since in an adobe structure wooden supports must be embedded in the walls to furnish support for shelves or cupboards.

The foundation of cement and rocks was built 1½′ below ground level and a foot above, to prevent the house being washed away during summer flash floods. I discovered inflation had doubled the price of adobe bricks since I had built my chicken house, and the prospect of making, drying, and stacking my own was forbidding, so I opted for the method used by a friend to build her weaving workshop. This procedure utilized forms in the manner employed by early Spaniards to pour walls before they began making bricks.

The forms were 1′ wide and 1′ deep, of any convenient length, held together by hooks and eyes for easy removal. I borrowed two 3′ forms from my friend, constructed another 7′ long, and was ready to pitch in.

The forms, placed on the foundation, were filled with a mucilaginous mixture of adobe mud and straw. Just enough straw, chopped into 2″ lengths, was added to give body to the adobe. Only sufficient moisture was added for the mass to become workable. To do my bit for ecology and incidently reducing the amount of mud to be mixed, I placed rows of empty cans along the center of the forms. The resulting dead air provided additional insulation.

On warm days the forms could be removed within half an hour of being packed with mud. By the time a layer around the entire building had been completed, the adobe had dried enough to support the next tier. Placement of the forms was alternated at the corners, and no reinforcement was necessary. The end seams of the alternately placed forms resulted in a random pattern as the walls rose.

Frames for doors and windows were set into the rising walls. Projecting pieces of wood, nailed to the frames, were embedded in the adobe to hold the frames securely in place. As the adobe dried, it shrank away from these frames, so, after completing the walls, it was necessary to chink the resulting cracks with adobe mortar. Window

(Photo by Lisa Law)

and door trim may be dispensed with, but, if desired, extra pieces of wood must be included in the frame for a base on which to nail the trim.

Load-bearing lintels, wide timbers, 4″ thick, were placed over doors and windows, projecting 1′ or so on either side into the walls, to distribute the weight of the bricks above the openings. I placed additional lintels where, at some future time, I might wish to open entrances to additional rooms.

Framing of doors and windows is not vital. Without frames, however, making airtight doors is difficult, and window glass fixed in adobe affords no ventilation. I did make one small, unframed, arched window in the west wall and fashioned several portholes with gallon wine jugs in the bedroom and half-gallon jugs in the living/dining room walls. One of the joys of working with adobe is the possibility that, with little artistic talent, one can sculpt small shelves in walls or make windows of any size and shape.

Constructing a fireplace of adobe bricks is fun and affords an outlet for creative urge. Two women friends built one for me in a corner of the living area, mortaring the bricks into an attractive free-form shape, which I finished off with a hand-smoothed coat of mud.

The walls, upon completion, were capped with 1″-thick, 1′-wide boards, upon which the vigas rested, ensuring equal distribution of the weight of the roof. Instead of the traditional flat Spanish roof, with fire walls and canales, I chose a simpler, slanted roof, with a pitch of 1″ per 1′ as more secure, providing for runoff of rain or melting snow.

It would have been gratifying to cut and trim my own vigas, but I was totally lacking in experience with chain saws and wary of all those whirling teeth. The ones I bought were straight, well seasoned, and, stripped of bark, a mellow, mottled gray. Placed on 1½′ centers, twenty vigas, ten on each side, sloped from the hefty center beam to the north and south walls. This 10″×10″ beam stretched the width of the house, resting on the east and west walls, supported by four 10″×10″ 9′ uprights. The vigas were spiked to the center beam, with the other ends set solidly in the walls. Inch-thick boards of random lengths and widths, nailed to the vigas, formed the base of the roof. Over them two layers of Celotex were secured for insulation and covered by roofing paper and tar. The eaves were extended a foot beyond the walls to afford some shelter for the walls from eroding rain.

I had considered floors of adobe, treated with a mixture of linseed oil and kerosene. Since, however, this required repeated saturation of the adobe, with prolonged intervals for drying, and my friend Ruth and I had moved into the house, this idea was abandoned.

My problem was solved when a man arrived one morning peddling a truckload of local flagstone. Our initial effort, laying stones in the kitchen on an adobe and linseed oil base was a viscid disaster, for we failed to saturate and dry the base properly. We laid the flagstone successfully in the living/dining room over a 3″ base of cement. We never did get around to finishing the floor of the bedroom.

The final step in construction was plastering the walls with a coating of muddy adobe. A portion of cement was included in that used on the outside wall to make it resistant to the erosion of rain.

Plastering with mud is a simple process, learned best by doing it. The adobe must be wet enough—about the consistency of good mayonnaise—to smear easily into cracks, and thick enough to cover imperfections in bricks. If eaves overhang walls by 3′, plastering exterior walls may be postponed by as much as three years. Interior walls may crack after the first coat of plaster, but it is a simple matter to slap on another coat. Rough wall surfaces with a few minor imperfections are part of the character of an adobe house. It is even possible to tint adobe with the powdered colors used for cement, if this should be your fancy.

Adobe may be used for construction in any area warm enough for the bricks to dry. In more rainy climates broad eaves are mandatory, and the mixture of cement in exterior plaster is indispensable.

The building of my house took over six months. I started in September, but could work only until the end of October because of the cold. I made the mistake of failing to cover the last completed tier of bricks with plastic sheets to protect them from rain. As a result, they became wet and, when freezing weather set in, they froze; when I resumed building in warmer weather they were crumbling apart. The complete layer of faulty bricks had to be removed. I also discovered that one window was too high for any view from a sitting position, and had to knock out several tiers of brick in order to lower the window frame. I couldn't believe how hard those bricks had become! It took a sledgehammer, a chisel, and plenty of muscle to remove them.

The cost of materials (sand, cement, lumber, and roofing) totaled roughly $300. I began to install the electrical system, but chickened out when I realized I didn't know what I was doing and that fires could result from the mistakes I might make. For only $200 I got wiring properly installed, with plugs galore in every section of the house. Although I didn't install plumbing, to do so at a later date would pose no problem. One had merely to chisel holes through walls, place pipes, and cover them with mud.

I never worried about precise measurements, plumbed walls, blueprints, or anything else of a technical nature—another endearing aspect of building with adobe.

Anyone who wishes to work with adobe should and can do so. The primary requirements are motivation and muscle. Minimal skill in carpentry for framing and roofing and a little knowledge of handling the materials are all that are needed. It was a

(Photo by Lisa Law)

simple matter on the farm for some of the women who wanted to help for an hour or so to pitch in, mix mud, and shovel it into the forms.

Having lived now the past three years in a post-and-beam house in southern Oregon, I have come to appreciate fully the ecological benefits of adobe. Conservation of wood is a factor in its favor. The difficulties I encountered in adding a room to this wooden structure made me really appreciate my house in New Mexico, where an additional room entailed only a sledgehammer, chisel, mud, straw, and muscle. The feeling of warmth and security is greater within the more primitive, simple structure of earthen walls. Something atavistic within one makes appealing the insulation provided by materials from the earth.

House Thoughts

DESIGNING AN ENERGY-EFFICIENT HOUSE

A house, like a biological organism, is a complex body with the capability of doing much more than simply sheltering. It can be as responsive as your garden, which means that nothing stands in isolation: energy systems are tied to waste systems, which connect up with food production. In the house, these interdependent systems should be considered in the initial designing stages. If the house already exists, you can consider changing present systems and/or adding to what you have.

While this discussion will center around construction of new homes, the principles are applicable to fit existing situations in both urban and rural areas. The essential factor is that a home needs to be designed from a holistic standpoint: it should be looked upon as a whole consisting of interdependent parts. This perspective calls for an emphasis on natural systems, a wise and conscious use of resources. To live lightly on the earth, to move from a consumer to a conserver society, starts with simple acts. We have it within ourselves to change our culture: let's begin by redirecting our thinking vis-à-vis our most obvious statements—the structures we chose to build for our homes.

In order to begin to gain perspective, run down a stringent checklist for your house (proposed or existing):

Does it utilize natural energy systems?

Does it minimize waste through conservation, recycling, and good design?

Does it minimize and optimize materials usage? Does it need to be so large?

Is your home appropriate to the area? Are you using indigenous materials?

Is your house put up to last? Will future users thank you?

Has your new house minimized disruption of existing natural systems? Are you building in a particularly delicate area such as a flood plain or watershed? Why? Couldn't your house be elsewhere?

Are the systems within the house understandable to you? Are they needlessly complex? Did you construct them or could you fix them?

In the following, we will consider some major house systems in terms that make ecological and economic sense. The energy-efficient house should make minimum demands upon such natural resources as water and fossil fuels. In approaching systems for heating and cooling your house and for dealing with water and waste, or in adding a small food production unit to the house, alternatives such as solar heating units and composting toilets offer us some sane solutions to old problems.

Solar Systems

The American dream claims technology as the panacea; and the higher and more complicated the system, the more absolute is our belief in its powers. Since we're not generally schooled in the concept of alternatives, we've repeatedly let this attitude dictate a narrow vision of possibilities. There is a danger, with our greatly increased awareness of the sanity of solar systems, to look toward the more complex systems as being the most appropriate to most situations. But heating your home and domestic water with the sun is pretty basic—it needn't be a complicated, expensive proposition.

The definitions of passive and active solar heating

systems tend to get a bit hazy, but generally *passive systems* utilize the house and natural processes such as convection to collect, store, and transport heat. *Active systems* are commonly defined as systems that require a certain amount of pumping, blowing, and transporting of a fluid in order to heat or cool specific areas of a structure. These are the most common type of heating system but there are several problems with active systems: they're expensive; the user is removed from the process and often unclear about what is going on; and they break down. So if you're lucky and have the luxury of starting your home from scratch, why not give consideration to the possibilities available in low-cost—often owner-built—passive space-heating and water-heating systems?

To take space-heating systems first: the temperature of your home will vary according to the amount of solar energy received or shed and the ability of your house to store heat. Several things affect this. Windows are the biggest culprits for heat loss, and it's a good policy to minimize or eliminate them on the north side of your house (north light is lovely but it's a net loss in terms of heat—doesn't trap any and loses lots!). West and east windows should be included, but in moderation; west light, particularly, is difficult to deal with because the sun comes in at an awkward and disturbing angle in the late afternoon. West and east windows do, however, afford some heat gain to offset the disadvantage of heat loss through the glass. Concentrating windows on the south side (the solar system in my house is simply a wall of glass on the south side) allows for penetration of winter sun but, with forethought, keeps out unwanted summer sun. Different types of glazing will also have an effect on the amount of heat gained or lost. Double glazing is usually worth the extra dollars. Single-paned windows lose approximately twenty times the heat that a standard insulated stud wall does; double-glazed windows lose only about ten times as much. Reducing air leaks through the cracks around windows and doors is also critical. Caulk and weather-strip and make sure windows and doors are built to fit.

Insulating walls and ceilings—and in some climates, floors—has, in many states, become mandatory. Insulation is a function of *resistance* to heat flow and therefore keeps a house warmer in the winter as well as cooler in the summer. This resistance (indicated by an R value) is defined as the tendency of a material to retard heat flow. It is the opposite of *conductance,* or the heat-transmission properties of a substance (the U value). The higher the R value of a substance, the more effective it is in insulating a space. The most common insulating material, fiberglass, is rated R11 for 3½″-thick wall insulation and R19 for 6″-thick ceiling insulation. To put this into perspective, window glass has an R value of only 0.9. There are many types of insulation (see "Materials" section), but whatever type you choose, insulating your house is essential to conserving energy.

Insulating windows can also greatly decrease the overall heat loss of your house and can be inexpensive and easy to build. Insulating devices range from shades and curtains to more complicated and mechanized beadwalls and skylids. (The *beadwall* consists of a two-piece glass or plastic window with an air space between the pieces that is automatically filled with insulating beads at night or during cloudy weather. When the weather warms and the sun shines, a small vacuum sucks the beads out of the air space, allowing the sun to shine through. *Skylids* are insulated louvers that are placed inside the house behind vertical windows or skylights. These louvers automatically pivot closed during bad weather and at night. Both can be obtained from Zomeworks, Albuquerque, New Mexico—see "Resources": "Sources.") All systems require at least a modicum of participation on the part of the homeowner.

The orientation of the house is also critical. If possible, the house should be situated to provide the maximum amount of solar gain during the winter months, and the minimum gain during the summer. Different latitudes and climates require different orientations, and you'll need to do some specific research to find out which is best for you. The *ASHRAE Handbook of Fundamentals* (American Society of Heating, Refrigerating and Air-Conditioning Engineers, New York, 1972) and *Design with Climate,* by V. Olgyay (Princeton University Press, 1963) are both good references for this research work.

The preceding guidelines are basic to good design. If you choose to include a solar collection system to provide a percentage of your heating requirements, as opposed to simply minimizing heat loss, you'll need to do some investigative research into the different types of systems to find out how they work and which would be the best for you. There are three basic, generic types of passive and passive-hybrid types of collection systems. The simplest is that exemplified by an adobe house or, as in the case of my home, a well-insulated, tight wood frame structure utilizing south-facing glass as a collector; this is coupled with a system for storing the heat. These are called *direct gain systems* and function in a sun-space-mass mode: the solar radiation enters the space and from there is transported (naturally through convection or with a small blower or fan) to a storage medium, or mass (usually rock or water). An *indirect gain system* couples the sun to the collector (mass) and from there it is transported to the space as needed. A more involved system couples the sun to a collector (often a flat-plate hot-air collector), whence it thermosyphons into a storage medium for later retrieval. These are a few of the many types and varieties of system you might consider. See the "Resources" section for further readings.

The single most important aspect to solarizing is to provide your house with mass: that means providing a place for the solar (heat) energy to be stored. All materials will absorb heat, but their ability to do so—the *heat storage capacity*—varies. It is this capacity to accept heat which is critical to an effective storage system. A typical stud wall, for example, has a much lower heat storage capacity than a masonry wall, which means that it will reach its maximum capacity sooner, resulting in excess heat remaining in the air and overheating the space. When the sun sets or temperatures drop, the house will begin to lose heat. If adequate mass has been provided, this heat will be replaced by heat stored during sunny periods, and the house will remain comfortable. So a house with a lot of thermal mass will not fluctuate rapidly with outside temperature variations, but will consistently remain comfortable. Massive houses will also work for summer cooling, moderating temperatures inside the house and keeping it cool.

The process of adding thermal mass to your house need not be complicated, expensive, or unattractive. Five of the solar houses that are being monitored at the Farallones Rural Center in Occidental, California, have utilized the cavity inside the perimeter foundation walls as storage vaults for river rock. This rock provides thermal mass in an otherwise unused section of the house. Other methods might be massive fireplaces, containers of water, brick interior partitions, and, in some cases, concrete floors. No matter which type of space collection-storage-retrieval system you choose for your home, be prepared to spend some time with alternative, backup heating (or added sweaters). Even the most carefully designed solar systems are usually not 100 percent effective. But certainly with forethought and a little extra money you can greatly reduce your dependence upon outside energy sources.

Utilizing solar energy for water heating is not a new phenomenon. In the 1930s, in California and Florida particularly, simple solar water heaters were common and many of the designs now in use date from that time. Today the small-scale solar water-heating system in most cases is within the price range that makes its installation feasible. Because one requires hot water all year long, payback times are shorter than with space-heating systems. Extended cloudy days are a definite problem with solar space-heating systems as storage can usually take care of only two or three days with no recharging. It's not critical to go a couple of days without hot water, but if you deem the inconvenience too great, a conventional water heater (even a wood stove) can be plumbed into the system to provide the added boost to the water temperature. The types of solar water-heating system are about as numerous as space systems and range from black plastic bags filled with water, to tanks painted black and placed within an insulated box (breadbox collectors), to more complicated flat-plate systems. Again, as with space heating, the less complicated your system, the less trouble you'll have with it.

Water and Waste

There is a growing realization among many people that our present waste disposal systems are perhaps more damaging than helpful. The average American uses ten thousand gallons of purified water every year flushing away wastes which, when dried, could be lugged around in a five-gallon can! Our present sewage treatment facilities use one resource to destroy another and cost billions of dollars in the process. Furthermore, we are denying our soil a rich nitrogen source. Perhaps we should start thinking of our wastes as resources out of place and not as something we simply flush and forget.

Experiments are being run right now at various locations around this country and abroad to determine if dry, composting toilets can be a viable option to current on-site waste disposal practices. These toilets require no liquid medium to transport wastes, and they convert the wastes into usable soil fertilizer. The most historically common dry toilet is the backyard variety of pit privy. These are fine, but only in selected, sparsely settled rural areas. The pit privy may produce problems with odors, often the by-product of the anerobic decomposition; many people use a lime additive, which reduces odors but retards good decomposition. Composting toilets are fairly recent; they are most common in the Scandinavian countries and in the United States in areas where more traditional disposal systems are not feasible because of low soil permeability (inability of the soil to absorb waste water). Composting toilets operate much the way your garden compost does. They require a certain carbon-to-nitrogen ratio (30 to 1, obtained by augmenting the nitrogen in human wastes with carbon found in such things as grass clippings, leaf mold, and chopped straw), a moderate amount of moisture (too much urine causes excess ammonia buildup and odors as well as delaying decomposition), and aeration (through venting or manually turning the pile). There are a variety of types of composting and dry toilets from which to choose, some commercially available and some owner-built. You'll want to investigate them all in order to choose the best for your needs. Some of the alternatives are:

Biological Toilets These include the Clivus Multrum, Toa Throne, Farallones Vault Toilet, Drum privy, and Ecolet. These toilets rely on bacteria and microorganisms to decompose the feces. They require additives in the form of cut grass, leaf mold, or sawdust (*not* redwood or cedar, which do not readily decompose), which are added following each use to provide additional carbon as well as to reduce the moisture content of the pile (feces are about 60–80 percent moisture). Biological toilets rely on oxygen-loving organisms to do the work, so proper aeration is essential or the pile may go anerobic and smell unpleasant. Air is supplied through venting (a series of baffles runs through the Clivus and Toa Throne) or by manually turning the pile with a pitchfork (as with the early Farallones Vault Toilet). While the biological toilets may accept urine, it is essential that the pile not become soggy and overly moist. (Many users syphon off urine for later use on compost piles; other composting toilets use small electrical heating units to evaporate excess moisture). The composting process itself takes anywhere from six months (the Ecolet) to two years (the Clivus and Toa Throne). Finished compost, or humus, should be buried under 12″ of soil around fruit trees or ornamentals, not in your vegetable garden; tests are inconclusive regarding pathogen and parasite content following decomposition.

If you are just building your house and think you might want to make use of a biological toilet rather than the conventional flush type, take some time to research the various types available. Some, like the Clivus Multrum, are quite large and require a specially designed area to accommodate them. Others (the Ecolet, for instance) are more compact and so may be installed in a smaller space. Carol Stoner's book *Goodbye to the Flush Toilet* (see "Resources") is an excellent reference with clear comparisons of the alternatives now available, giving costs, advantages and disadvantages of each model, size, and other information.

Chemical Toilets These toilets rely on chemicals to do the decomposition work and are most suitable for temporary use. Wastes, instead of being composted and reused, are trucked to central sewage treatment facilities.

Oil Toilets The most complex of all, these toilets require a very complicated system of pumps, valves, coalescers, and holding tanks. The final product, unusable because it is covered with a film of oil, is pumped and trucked to land fills or central treatment facilities.

Low-water-use Toilets These are flush toilets that rely on a compressor to greatly reduce water use. Some have the capability of variable water settings, depending upon whether you're flushing urine or feces. These toilets aren't too expensive and can be plumbed directly into your existing system—a good alternative if composting toilets are, for one reason or another, not appropriate.

As you can imagine, the single biggest obstacle to dotting the landscape with compost toilets is the local health department. Currently the most widely used on-site disposal system is the septic tank/leach field combination, and most sanitarians are more than a bit hesitant about approving systems that deviate too much from this standard. Several counties in a few states allow compost toilets provided the owner also installs a traditional septic tank (at $1,500 to $3,000) sized to accept a design load of water that would far exceed that used by dry toileters. In most cases, you'll be facing a classic dilemma: the health department won't accept your permit for a dry toilet until it's been constructed and tested; and the building department won't allow you to put one up to test without the health department's okay. Getting around this problem isn't so easy and there are no clear solutions. The most likely course at this time is to submit your toilet under an "experimental" permit and demonstrate that a conventional septic system could be installed if needed. It helps to read everything you can get your hands on about composting and dry toilets so that you won't be totally at the mercy of reluctant county officials!

Used waste water that has not been contaminated by human excrement is called graywater and includes all water coming from showers, sinks, washing machines, and so on. The science of graywater disposal is a new one; there are comparatively few tried and true systems in operation. Several issues are clear: graywater recycling systems are not trouble-free and simple. The inexpensive and less complicated systems require continual and regular maintenance, which even the most dedicated recyclists might tire of. Even then they're prone to distribution problems at the least. More esoteric systems such as septic tanks with dosing syphons (which regulate flow) and filtration systems are expensive and still require a certain amount of conscientiousness. The state of the art is still quite a distance from safe, inexpensive, idiot-proof systems. If you live in a drought-affected area, or your water situation isn't what it should be, or you're simply dedicated to the principle of recycling water and you'd like to take on the task of graywater recycling, your choice of which type of system to go with will be determined by a number of factors. Some of these are:

Soil Types Fractured rock, for example, will require one type of distribution, while loam can handle quite another approach. Soils can accept only so much graywater per day per square foot. Soil permeability will determine the size of your disposal system and, in fact, whether or not your soil can even *have* a more traditional leaching field system.

Water Table If your water table is very high, you'll have to design your graywater system accordingly; it's not advisable to allow graywater leach lines within 24″ of a water table.

Slope Conditions You may be able to employ a gravity feed system if your site has adequate slope. If this isn't the case, you may find that you'll have to include a pump (preferably wind-powered!) in your system.

Use Rate Approximately 40 percent of domestic water use comes from flush toilet use. Other high-use areas are showers, tubs, laundries, kitchens, and lavatories. You'll want to contact your local health officer to find out what your design flow will be. Some states, such as Maine, will permit households with dry toilets to reduce their field and tank sizes accordingly; other states allow no reduction for compost toilets. It's clear, however, that a critical factor in keeping your system functioning properly is to minimize your consumption through flow restrictors and other conservation methods.

After analyzing your specific site and use rates, you will want to study carefully the types of graywater system available for your particular situation. Assuming you have good percolative soil, there are several ways you can go with recycling your graywater:

1. The simplest graywater system is, obviously, a bucket under the sink. This system is temporary at best; it may cause odors if the graywater is allowed to sit and ferment; and it's a nuisance to keep hauling graywater-laden buckets out to your garden. But as a stopgap measure, particularly in time of drought, this system is simple, easy, and cheap. Don't use graywater to flush your toilet by putting it into the tank—it will clog up the works. You *can* pour it directly into the toilet bowl. The graywater collected in an under-the-sink bucket may be used in the garden on nonleafy vegetables such as tomatoes or beans, or on ornamental plants. While graywater is not grossly contaminated with human wastes, it may contain high bacterial counts and strong chemicals. These bacteria will, in all probability, be filtered out as they pass through the soil (the best filter of all), but if you are suspicious of your waste water, don't put it in your vegetable garden. Instead, you may pour it around fruit trees.

2. The next step is to attach a graden hose to your house drains. This hose should be set up to direct the graywater into what's called a buffer tank. This is a 55-gallon drum set horizontally with an inlet at the top and an outlet at the bottom, controlled by a valve. This tank allows a mixing of the more grimy waste water, from the kitchen, for example, with the cleaner water from showers and bathroom or laundry room sinks, to provide a more homogeneous effluent. The hose from the buffer tank may be a standard ¾″ garden-type hose and should be perforated or have some sort of filtering device on the end (such as a small cloth sack) to trap large particles. These filter sacks must be regularly removed, cleaned, dried, and replaced. Graywater hoses should be rotated around your garden to prevent flooding, scum, and sodium buildup.

Two major problems may develop with this buffer tank disposal system: sodium buildup in the soil, which causes root and leaf burn and destroys soil aggregation; and clogging of the distribution lines due to particulate materials causing totally restricted flow and/or uneven distribution.

Soaps and detergents both contain sodium. Detergents have a higher content than soaps, and detergents with softeners have even more sodium. So while the phosphates in detergents do not have the same negative effects when deposited in soils as they do in rivers and streams, soaps are the better choice. The milder the soap, the less destructive it is for your soil. Even with filtration, graywater contains dissolved salts which, over time, build up. You will want to alternate graywater with fresh water in order to allow these concentrations to leach out.

Waste from the kitchen sink is another culprit, as it contains grease and food particles. It is absolutely essential that plates be thoroughly scraped into your compost bucket before washing, and that a mesh screen or trap be put in the sink drain to catch anything that has escaped your watchful eye.

You'll also want to remember with this system to turn the valve off and on to restrict flow to the garden (overflow problems can be avoided by plumbing the tank into your sewer or septic system).

3. Another method for graywater distribution is to hook your drains into a commercial or owner-built septic or settling tank system, making sure, of course, that no water from a flush toilet can find its way into the same tank from which you'll be drawing irrigation water. After the water has gone through a septic tank, it will pass through a dosing syphon (a method for regulating flow) and from there to a filtration system of some sort (slow sand filter or pea gravel roughing filter) and then to a holding tank from which it can be distributed to the garden. Some of these systems can be pretty complex, requiring extensive engineering and plumbing knowledge.

4. If you have a greenhouse (solar) system on your house, you might want to consider the utilization of graywater for the plants. This requires good sandy, well-drained soil. Greenhouses that are contained (that have barriers between the interior and exterior earth) will not tolerate the use of graywater due to salt and sodium buildup and the inability to leach these substances out.

No matter what your needs and choice of graywater disposal system, try to read as much as possible on the subject before making any actual installations. There are several individuals and organizations doing investigation into graywater systems with the hope of developing safe, inexpensive, effective treatment systems. The "Resources" section should help you in locating further resources.

Food Production

Lettuce cost $0.95 a head in our local supermarket this past winter. If price were the only criterion, the situation would be bad enough, but there were other vital issues involved: quality of food; the dysfunctioning caused by massive decentralized food growing, processing, and distribution systems; loss of soil fertility and upset of biosystems from extended use of chemicals; and cultural and social upheavals brought about by the rapid disappearance of the family farm. Providing for part or all of your basic food needs is as integral to a whole-home concept as heating with the sun. We're reminded of Barry Commoner's adage that "everything is connected to everything else" when we realize that, in fact, both processes (heating and growing food) work within the same context: that is, transforming light into other forms of energy. The ideal situation is to provide one space within which both situations can be performed.

Certainly your choices of food-raising method and scale are to a certain extent determined by your site—as well as your inclination. Not everyone has the desire or opportunity to live the rural life-style naturally suited to extensive food production. But even in high-density urban areas it is possible to utilize window greenhouses for growing some vegetables. If you're designing your new home with solar heating, think about including a solar greenhouse—functionally, aesthetically, and economically, it's difficult to do better.

Solar greenhouses, unlike traditional greenhouses, are insulated on all those portions of the building—the north and part of the east and west walls—that do not get direct solar radiation. The traditional greenhouse, usually not insulated in this manner, thus suffers heat losses that require large infusions of energy to maintain a constant temperature. In the solar greenhouse, a heat storage area (generally masonry or water in containers) is required to store the heat received as natural solar radiation for later retrieval. The heat storage area(s) provides enough thermal mass to counteract extreme temperature fluctuations within the space. Vents to the outside (on the lower south and upper north sides), which are operated either manually or automatically, are required to prevent overheating. In the attached greenhouse, vents along the north wall into the house itself provide for the passage of cool air into the greenhouse and warm air back into the living space. Other special features for attached solar greenhouses include double glazing (if you live in a cold climate), seals between the house and the greenhouse to prevent leaks, and proper drainage away from the greenhouse (if you plan on using graywater to water your plants inside the greenhouse, this is particularly important). And, of course, special attention will have to be paid to the interior arrangement and planting to maximize use of space, provide adequate light for plant growth, and give you the most food for the least space.

LET THE LIGHT SHINE IN

During the Middle Ages people believed that light streaming into a room was God entering that building. Thus they designed regal churches and intricate stained-glass windows to better "enlighten" the space created within the four walls. It is said that the eyes are windows to the soul. If this be true, it is no less true that windows are the "eyes" that reveal both the soul of a house and the outside world to those that dwell within.

Of major importance in designing a home that you might want to live in is the placement of windows. Size, shape, style (single- or multi-paned), direction the building is facing, ventilation, sun pattern, and position of the windows on the wall are a few important considerations. Because windows break up the large expanse of a wall, they become a focal point in a room. Extra detail in trim can greatly affect the aesthetics of a building.

When I first began building my house, I wanted to incorporate stained glass as well as a variety of window shapes and sizes. A weekend course in stained-glass making taught me the basic skills: the rest was practice. I have found that in my five years of building since then, being able to cut glass has greatly increased my options in designing a cabin or renovating an existing structure. The carpenter who chooses to leave glass cutting to the glaziers limits her possibilities. I'd like

(Photo by Carol Newhouse)

to share some of my discoveries about glass and to pass on enough information for you to feel comfortable cutting glass.

Placement

If you are designing a new house, try to spend time camping at your building site. Notice where the sun is at seven in the morning. How do you want it to enter your bedroom when you first awaken: in large screaming splashes of light, softly dappled ribbons, or just a cheerful quiet presence? Maybe you don't want to see the sun at all until ten o'clock. Check again in the afternoon and at sunset. Consider what you'll be doing where and at what time. Don't forget that the sun drops much lower in the winter. If you cannot actually observe the sun during that time, the moon in summer follows the same pattern the sun will take six months hence.

Is your building surrounded by trees or in the middle of a meadow? Southern light is strongest and hottest. If you have scorching sun and no shade, you might want to minimize your southern windows and opt for the cooler northern light. Consider using your southern exposure for some sort of solar heating plan or attached greenhouse. A roof overhang on the south side will also block the hottest summer rays but allow for the winter sun to enter. Placing windows on more than one wall will give a better light distribution throughout the house, as well as allowing for cross-ventilation patterns. The closer the top of a window is to the ceiling, the farther into the room light will penetrate. If the sun's rays are at a 45° angle, the illustration shows how the variously shaped windows would let in the diagramed amount of light. If you plan ahead and place the tops of all windows 6′6″ high, they will all be even with the top of a standard-size door as well as one another. This will prevent a restless, disjointed feeling in the room.

Size, Shapes, and Types

There are many traditional window types and patterns that an observant eye can see in various homes in any town or city. In general, one large area of light works more effectively than several smaller scattered windows because it prevents dark areas and provides a stronger focus.

Some common types of window include double-hung, casement, hopper, and fixed. In a *double-hung window,* the bottom and top halves slide up and down independently in the frame and are balanced by a system of pulleys and cast-iron weights, which sit inside a special pocket built into the frame. Newer double-hung windows now work with a spring balance system rather than these pulleys and weights, or use compression weather stripping for the tension that provides a counterbalance to the sash. Double-hung windows may be opened without any protrusion inside or outside of the building. *Casement windows* swing in or out. They are hinged on the sides with a double set of

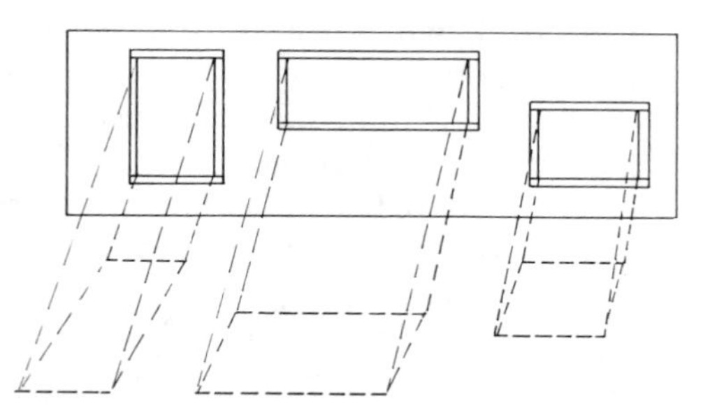

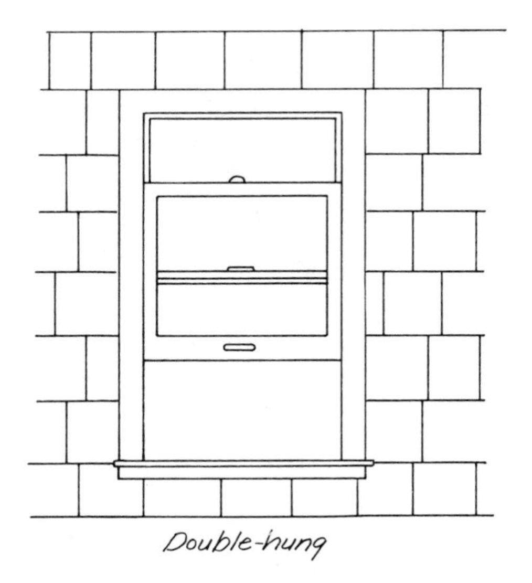

Double-hung

Multi-light Casement

Fixed center with casement sides.

hinges per window, top and bottom. Most often they are hung in pairs and the edge where they meet is rabbeted so that there will be no air leakage. Similar to the casement is a *hopper window,* which is hinged at the bottom and swings inward at the top. It allows for ventilation but keeps rain out. An *awning-type window* is hinged at the top and swings outward from the bottom. A *transom* is a small hinged window (hopper or awning style) that is placed above a larger fixed window or over a door. It provides for extra light and ventilation. Another window style is the *sliding window.* This moves horizontally along a set of tracks or guides built into the sill and the head jamb. A *fixed window* is stationary and allows for light but no ventilation. If it is a large window, it should be either plate or tempered glass for safety reasons. Tempered glass seconds can often be bought directly from the manufacturer at considerable savings. Tempered glass cannot be cut and is exceedingly strong once in place. Like auto glass, if it is broken when being installed, it will break into small, round, dull pieces.

Anatomy of a Window

In general, windows have two main parts: the *sash,* which holds the glass (often inaccurately referred to as the frame), and the true *frame,* which fits into the wall itself and will hold the sash. All window frames have three parts: the top, or *head;* the sides, or *jambs;* and the bottom, or *sill.* The sill is made or placed with a downward sloping bevel that allows water to run off. A *drip groove* cut on the underside will keep rainwater off the wall. The sill projects beyond the outside sheathing of a building, while the jambs and head are cut to meet flush with the exterior sheathing

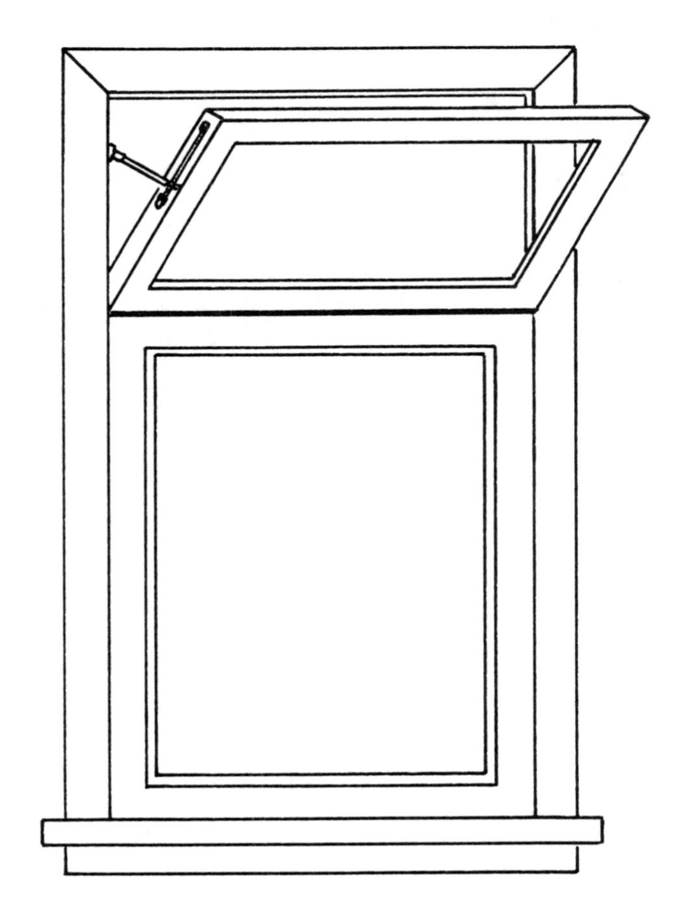

Transom Top

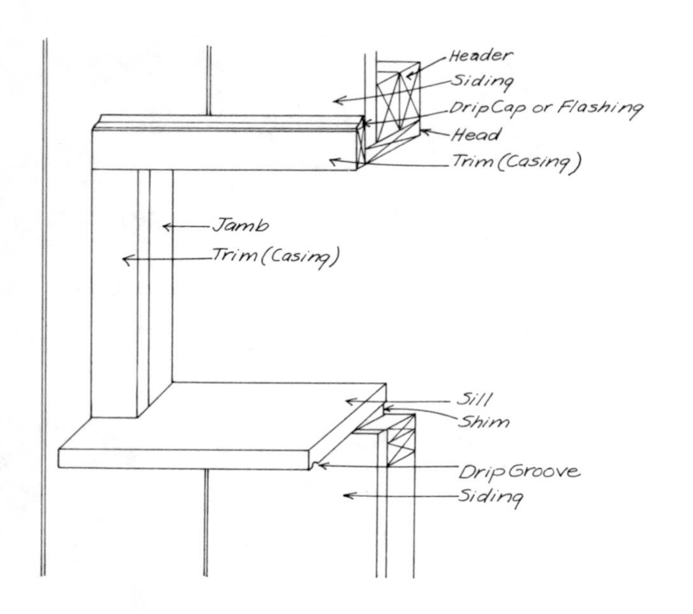

and the inside paneling. This allows for inside and outside trim or casing to be used to finish the joint between the wall and the frame. The entire window must be carefully constructed and sealed to protect against rain seepage. A *drip cap* or piece of curved molding can be placed over the head casing, or flashing may be fitted over the casing, shedding water shingle-style away from this joint. It doesn't hurt to put a bead of Dap or caulking compound along the top edge of the head casing, either.

When a sash is divided and has more than one piece of glass in it, the smaller pieces are called *lights* (as in "a six-light window") and the bars that separate them called *muntins.* Wood-framed windows of all types are available new or used in single- and multi-light. I prefer a conscious use of both, which is rather unconventional. At first I wanted only multi-light windows in my house because they have a more charming and cozy effect. However, I later found that I missed the drama that comes from watching a rainstorm through a large single-paned window.

Putting a window in a house is like turning a camera to the outside world; make sure you are looking at something you want to see. Think of it as framing a picture. If you are building in a city, privacy is an important consideration when deciding how to get adequate light. The use of skylights and the use of windows placed high on the wall are two possible solutions. A long narrow row of windows that goes across a whole wall is called a *clerestory*—it might be useful. It is often used in conjunction with two shed roofs of different heights.

One advantage of being able to cut glass is that you can fit odd-shaped windows into areas that normally couldn't be glass. Often such windows can be

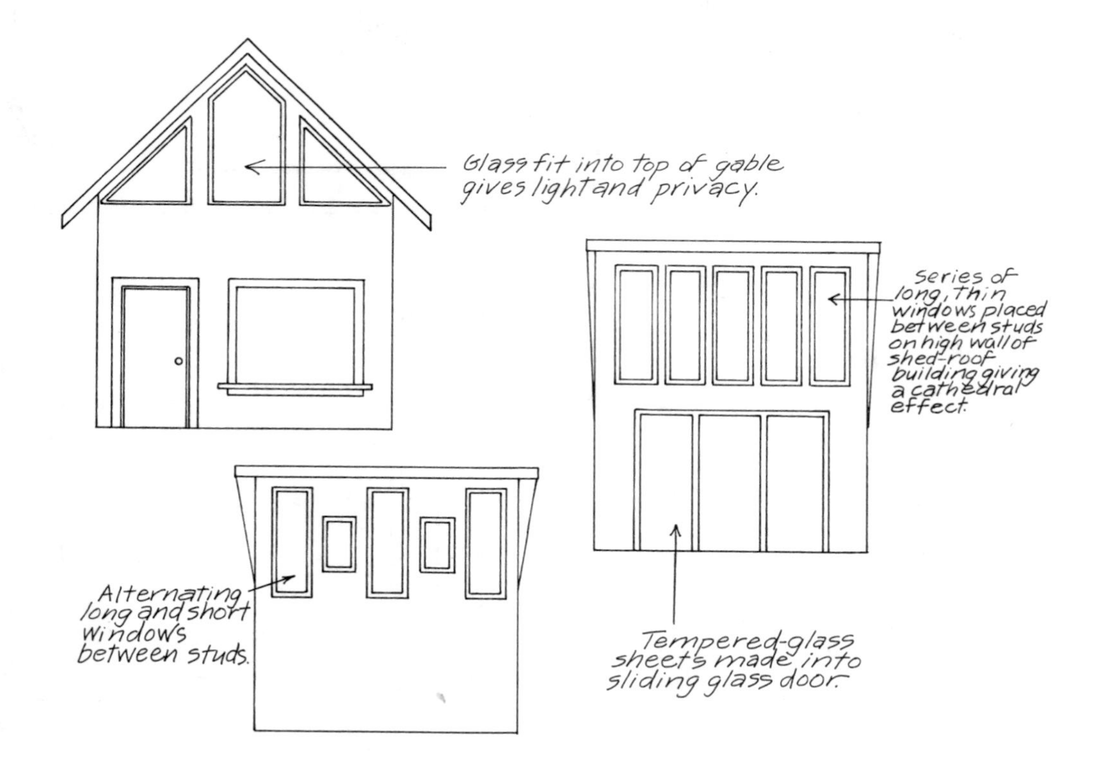

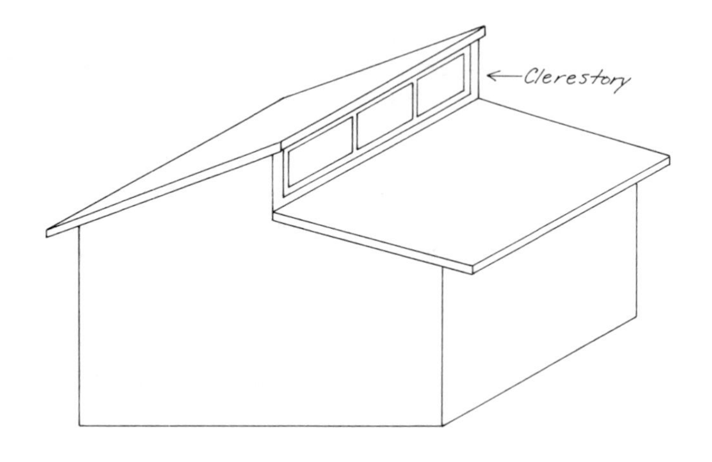

put right between existing studs. To make these windows, I cut pieces of glass to the sizes and shapes I wanted. These windows were unconventional in that they had no wood sash.

To position them, I cut pieces of wood to function as the jambs and head and ripped each of these lengthwise. The first half of each piece was nailed to the stud. A bead of caulking was put on the edge and the glass placed against it. Another line of caulking was put on the outside of the glass, followed by the second part of the jamb and head to create a kind of sandwich (the glass in between the pieces of wood and the caulking forming a gasket-like seal). This is not the professional or conventional method, but it has functioned well for me and is very flexible for fixed, out-of-the-way, and odd-shaped windows. If the glass should break, one must pry off the outside casing and the outside half of the jamb to replace the glass, but this shouldn't be any more time-consuming than the usual method of removing the stops and taking the glass out of a wood sash.

Trim or Casing

Even a small plain window can be made special by putting a little extra time into the trim boards. This can be done on the interior as well as exterior walls. The purpose of the trim boards is to keep rain out and to finish the area where the siding meets the window. Illustrated examples show how a few extra minutes with coping saw, saber saw, or rasp can add originality to a room. Natural redwood against Sheetrock or white pine next to a darker wood accentuates the contrast.

Lest this all seem in excess, let me say that a soft chair by a favorite window can be a very special place to view the world from. Likewise, bay windows with window seats offer a meditative corner for work or dreaming. My parents' house had a curtain that closed the bay window area so that you were cut off from the rest of "reality." Although I haven't duplicated that feature yet, I can't imagine why, as nothing was more pleasurable than bringing a book to that secret spot.

Cutting Glass: Or Bring In the Band-Aids!

If you are like me, the idea of cutting glass might at first bring a shiver of apprehension. So first a word of encouragement. Cutting glass is not difficult; it takes time and practice, but you will get to actually *like* the sound of glass cracking. It is an exciting process, and if you are careful, there is no reason to cut yourself. Cutting glass is basically a two-step procedure. First the top surface of the glass is scored (a light scratch) with the glass cutter. Next the glass is broken along that scored line.

In order to begin you will need a glass-cutting tool (which can be purchased for about a dollar in most hardware stores), scrap single-strength (SS) glass from old windows or a local glass store, a jar with some kerosene in the bottom to keep your cutter in, a flat surface to cut on, some padding for the cutting surface (plywood, pressboard, or an empty record cover will do), and a whisk broom to keep your surface clean of glass slivers.

Thoroughly clean your piece of glass, as dust will impair the action of the cutting wheel. Mark a straight line with a ruler and a fine felt-tip pen or a special glass marking pencil. Place this line perpendicular to your body. To start, hold the cutter between your index and middle finger with your thumb pushing from behind—similar to the way you would hold a pencil. An alternative method, and the one I like best, is to make a fist around the handle of the cutter and let the wheel extend out from the bottom. Hold the cutter at the far edge of this line and pull the cutter toward you, pressing firmly downward. It is not necessary to bear down with all your weight. The cutter should make a continuous scratching noise. DO NOT BACK UP OR ATTEMPT TO GO OVER ANY PART OF THE SCORE as it will damage your cutter and will not correct a mistake. Practice just scoring the glass as many times as you need until you can keep a consistent, continuous pressure. Dip your cutter into the kerosene between each cut; this will lubricate and cool down the cutting wheel.

Now clean your table and take a fresh piece of glass. Score it as described. Pick it up and place your thumbs on either side of the score, grasping the glass from underneath with your fingers, which are curled into a loose fist. From the bottom, press upward and outward (your little fingers will move toward each other) with a quick sharp motion. The glass should break smoothly down the score. What is actually happening is that you have created a run that will continue down the scored line, widening and deepening until the glass breaks. With practice, you will be able to control that run. Sometimes it is necessary to help the run get started by tapping the underside of the scored line with the metal ball at the other end of your glass cutter. Using a definite sharp motion, you will see the score deepen as you move the tapping down the score line. Be careful that the glass doesn't fall and break.

Another method of breaking the glass and one that is necessary for long pieces is to first score and then place the glass so that the "discard" portion extends about an inch over the side of a table. Place one hand firmly on the portion of glass resting on the table, and with the other hand grab the extended part next to the score line or at the far edge of the glass. Press sharply downward. In order to break small strips of

(Photo by Carol Newhouse)

glass, it is necessary to use pliers to grasp the edge of the part you wish to remove. Then use the same rotating motion you would use if you were using your hands to break the glass.

It is best to attempt to break the glass immediately after the score is made, as the stress patterns are freshest at this time and you will get the cleanest, easiest break. Sometimes it is necessary to remove a small remaining nub. This is best done with a pair of glass pliers or ordinary pliers that the temper has been taken out of. To do this, heat the pliers with a torch until red-hot, dip in cold water, cool and dry, and presto: grozing pliers. *Grozing* is the method of chewing away the ragged pieces of glass that are sometimes left.

Restoring Used Windows

Interesting used windows can often be purchased at a considerable savings. The important thing to consider in the price of used windows is how easily they can be restored and how much time this will take. Always check for rot in the wood sash. Take a pocketknife and poke it into the wood. Check out the joints on the top and bottom. Pay special attention to the bottom, since it takes the most weathering. If the wood is soft or decayed, probably dry rot has started and you would be better off finding different windows. If the wood is sound but the joints are loose, it is possible to drill, dowel, and reglue. If all the joints are loose, it is better to pass. A solid coat of paint will make repainting easier, but flaky paint must be wire-brushed, sanded, and repainted. It does take extra time. It is important that all window sashes have an adequate cover of exterior enamel or other protective coating, something like Thompson's water sealer. Otherwise, fir or pine sashes will begin to rot in a few years. Check the glazing. Most windows will have a beveled wood molding around the glass on one side (this faces inward). Facing the exterior or weather will be a border of glazing compound, such as Dap 33 or putty. The glazing compound seals the glass to the wood sash and keeps water out. It also helps hold the glass in place. If the glazing is dry and flaking off, it must all be removed and replaced. If you can find windows with glazing intact, it will save you a lot of time. If not, or if it is necessary to remove the glazing to replace a broken pane, reglazing a window is discussed next.

Glazing a Window

The first step is to remove all the old putty. If it is hard, try using a screwdriver and hammer to break it loose. Be careful of your glass. Hard compound can sometimes be softened with a soldering iron. Once the glazing compound is removed, you will see little metal triangles, or *glazier's points,* which hold the glass in place. Remove these with pliers and take out any broken panes. Clean out the recess in the sash until it is completely free of any old compound. A putty knife is useful here. Now paint the exposed wood with linseed oil. This will prevent the dry wood from absorbing the oil out of the new glazing compound and thereby weakening it.

Next apply a thin layer of compound along this recess, and set the new glass in it. This is called *back-puttying* and forms a tight seal between the glass and wood. Take your glazier's points and carefully press one end of the triangle into the wood with a screwdriver or the edge of your putty knife. Leave enough of the triangle projecting to hold the glass firmly in place. An average-size window needs two to three points on each side.

The next step is to cover the points and to make your exterior seal by applying glazing compound around the edge of the glass. Read the specific directions on the can of compound carefully. If it is a cold day, almost all compounds need to be warmed for easiest use. Using a good clean flexible putty knife is essential and not having one makes the job almost impossible. Roll the compound into strips and pat it into place. Now, with a sure firm stroke and your clean knife, press the putty against the wood and glass, forming a sort of triangle of compound. Try to do this in one deft motion, since going over it several times usually just botches it up. With a little practice, you'll be sailing right along. The key is in holding your putty knife at just the right angle (you'll have to experiment at this) and keeping a firm, continuous pressure from top to bottom (or side to side). If you make a mistake and have to redo the line of putty,

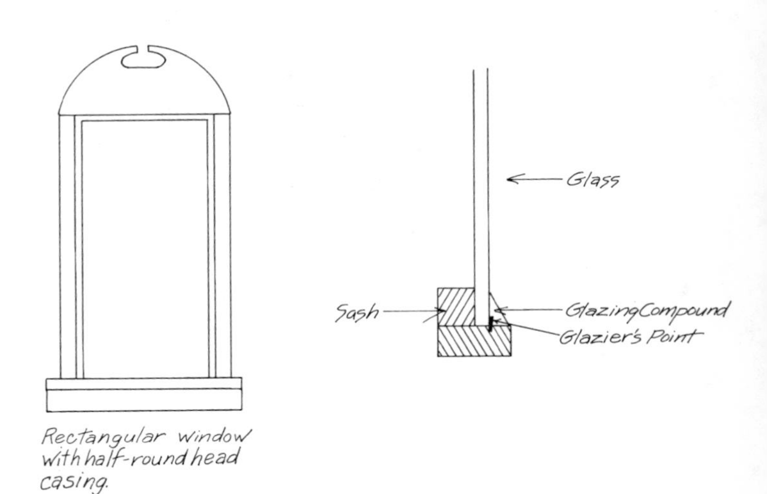

Rectangular window with half-round head casing.

(Photo by Janet Cole)

it's best to remove all of it and start over. Trying to "patch" seems to intensify problems! Don't be discouraged if it takes a while to get the right touch for this job—it is definitely a knack that will come to you all of a sudden, but it does take practice. Using a really good quality glazing compound helps immensely. Don't use old compound that has lost its elasticity and life. When you're done, check directions on the can to see how long until the compound is dry. The final step is to paint your final coat onto the sash, covering the glazing compound. A good job should last many, many years.

You may also find an old window with a wood bead or molding holding the glass in place on the exterior as well as the interior. This molding is called a *glass bead* or a *putty bead.* Small brads hold each bead in place. If you have to remove the bead to replace a pane of glass, work from the *inside* of the window. Pry the bead up (beginning at the center and working toward the edges of each piece) using a tiny Wonder Bar, a putty knife, or a chisel. The bead is fairly delicate but it has some play in it. You can use pliers to pull the brads loose as you work. When you are ready to put in the new piece of glass, make a little seating of putty all along the edges where the pane will fit. Put the glass in place and then replace the beads. Put the shorter ones in place first. Longer pieces may be bent slightly at the center and eased into place. When all the beads are in place, tack them in with brads or tiny finishing nails.

Why Start with Something Easy?

It seems as if all this happened a long, long time ago because Jules and I are now putting the finishing trim on the exterior of our house. This is the story of an owner-built home, a house built to code in one of the strictest, code-wise, counties in California. It was built in our era of stiffening restrictions on building, in a time of codes and regulations. But take heart, you can wade through it all and build yourself a fine home, as we have. You will learn how to deal with officials who govern everything from the height of the walls to the type of septic system you use. You will learn about yourself, also. By the time you finish, your tale of changes, surprises, and frustrations will be as varied as ours!

It began in November 1976. After a long search, we had finally found a perfect piece of land. It afforded about 3½ acres of growing space, included a secluded building site among the oaks, and was far enough away from town, yet not totally isolated. For six months we had looked at existing houses on land, but found nothing: either the price was outrageous or the house or land was not suitable. Finally we decided to build our own house on this great piece of property. Here was a task we relished as a dream come true.

The first decision after acquiring the land was what type of home we should build. We considered domes, a fast, less expensive, and county-code-approved way to set up housekeeping. While adobe block sounded good, adobe mud is not available in our area. Our final decision to work with an experienced architect and design a wood frame house was based on the fact that our location necessitated that we build to code. A detailed set of blueprints would pass inspection quickly and help us with construction nuances. A more conventional house would have a better resale value, this being a lesser yet necessary consideration. We put a lot of thought into the design of the house. No wasted space was a must. After talking together for some time, we discussed our ideas and priorities with an architect and came up with the two-story, wood-sided house that is our home today.

The blueprints that passed the building inspection department included 400 square feet more than we built. Since a deletion is much easier to pass than an addition, we have had no problems with that. This is one example of how you can use a basic design blueprint, submit it to pass your building inspection, then add on or modify after the final inspection clears. We found that things like solar collectors, fireplaces and wood stoves as sole sources of heat, outdoor privies, and pole supports or vigas (roof beams) are all very difficult to pass. One solution is to install the "minimum" requirements, pass the inspection, then tear out the "required" equipment (such as electric heaters), sell it, and install your own designs anyway.

(Photo by Rebecca Lawson)

Before we submitted the house blueprints, we had to deal with the environmental health department ("the desk opposite the zoning department and adjacent to the building department"). This bureau governs septic permits and well bacteria tests. Passing this department is a prerequisite to getting a building permit. The total process of septic and well installation, then approval by the environmental health department, took us a frustrating two months. Experienced builders can cut this step to two weeks. The time difference involves your credibility as well as your knowledge of county bureaucracy and codes. Our credibility was almost zero considering our sex and youth (both women and aged twenty-one and twenty-five). These officials simply did not take us seriously! Many times we were given half-correct advice or partial answers because people didn't think we knew what we wanted or could understand their explanations. We learned to ask more than one person about the same problem to be sure that we were getting the right answers.

(Photo by Rebecca Lawson)

Our first purchases were tools. A friend started us off by giving us an extensive tool list, which we followed pretty closely. These beloved and basic implements carried us through the whole building process. Some tools (like the nail puller and pry bar) have been used more than others (like the hammer and nails), but all were necessary at some point in the work. To the list of tools recommended to us as "the most important and often used" we have added others that we found to be desirable to have on hand. Our tool list reads like this:

Tools each person should have:

- tool apron (varied styles, so try them on and then choose)
- hammer (a 20-ounce framing hammer or 16-ounce straight-claw)
- tape measure (Stanley's has a wide blade that we liked—choose one with indication marks for stud placing)
- pencils (a must)
- hard hat (not a necessity—but not a bad idea!)
- combination square

Tools to share:

- power saw (circular saw with extra blades; 7¼″ is a good size)
- saber saw (not absolutely necessary, but helpful)
- drill
- handsaw (for general use, we chose an eight-point Diston)
- nail puller
- pry-bar
- levels (one shorter, 16″ to 22″, and one longer, 26″ to 34″)
- chalk line with plumb bob (this combined tool is not as commonly used as two separate tools, but we liked ours—you'll need a vial of chalk, too, for refilling the line)

10-pound sledgehammer
framing square
nail sets
miter box and backsaw

At first you may feel intimidated by your tools—or clumsy with them. When we first began, I had no experience with carpentry tools. Jules had taken a carpentry course at school and built some small structures for friends, so she was familiar with most of them. I was afraid of our power tools—the Skilsaw and the large drill—and had to force myself to use those tools until I became comfortable with them. Simpler tools like the combination square begin to feel "natural" after a little practice using them. Sometimes situations occur where a certain tool is an obvious choice, but you don't think of using it. For example, moving walls into place calls for that 10-pound sledgehammer. Leveling walls can be aided by a fence puller/stretcher. Learn to think about each action before you attempt it: How will the tool work? What can I expect? Why? Using your tools properly can save you energy.

Beyond our basic tool acquisitions, there were other tools that we rented or borrowed. These are the more expensive and/or larger tools like the ditch trencher we needed to bury our water, telephone, and electric lines underground. The county specifies the depth at which these lines must be covered, and we had 270′ to bury! We rented a ½″ right-angle drive drill when wiring our house—we installed our own electric wiring and used this drill to make holes through the wall studs. We also rented scaffolding, a must since our house is two stories high. Ladders worked for some things, but were not sufficient when it came to lifting heavy sheets of plywood up to the second level. Installing windows is also easier if you have adequate scaffolding. Be sure that you plan your work around the rental so that you don't waste time and money keeping the rented equipment longer than needed. In general, it's a good idea to plan your day's work and set goals for yourself. Things won't always work out exactly on schedule, but this discipline will help you to feel less overwhelmed by the whole project.

Another important thing we learned was the necessity of knowing the language of building. Without knowing the jargon, you can't communicate with building code officials, other carpenters, co-workers—or anyone who might help you if you only knew the right way to put the question! There are many ways to learn: you can take a beginning building course, hang around a building site, read carpentry books, and talk with more experienced friends. We bought *The Uniform Building Code* (abbreviated version) and several general construction books. There are limitations to reading books and asking questions, as you can well imagine. Many times a book would take some bit of information for granted, or use a term that was undefined; this is both confusing and intimidating. Asking questions can be just as confusing, as each person asked seems to give a different explanation or description of a technique. We were surprised to find that most people we questioned about problems in building or about things we didn't understand were helpful, although somewhat amazed to find that we were serious. Yet once this was established, they were also very friendly. One thing to take into account when asking questions over the phone (or even in person) is that the person you're asking doesn't know all of the details of your project and may not know how to fit the general information into your specific question. This is where we had trouble. An answer may seem complete when you ask it, but once you are actually working you see how much more you might need to know to do the job completely and correctly. Many times a person with experience takes it for granted that you know certain facts or information—which you don't—and leaves those extra parts out of the explanation. This can be frustrating, to put it mildly. We learned to look carefully at other building sites and situations, to ask questions and look again, and to try to compare what we saw with what we were doing. In this way we discovered alternatives and solutions to problems that we just couldn't find answered in books, or even get answered by people we asked.

But on with the process! The foundation was surveyed, built, and poured for us. It had to be a solid concrete wall (continuous foundation), which involved building wooden forms and drilling into our unusually hard sandstone site. The expense and difficulties of doing it ourselves persuaded us to contract this one step out. We hired a woman contractor who gave us the best bid for the job and was very helpful. The crew included her cousin, her husband, and two other workers. Meanwhile, we contacted four local lumber companies for bids on our major lumber needs and deliveries. Although the main concern was price, we also considered their reputation for fast delivery as well as their consistency in wood quality, choosing the company we felt would give us good lumber at fair prices. We learned that it's a good idea to be present for any deliveries made and to check the entire load to make sure that you aren't getting cracked or marred wood, and that your order is complete. We had to send back an entire delivery of pine decking because we were planning to use the lumber exposed and what arrived was stamped with indelible black ink. We went to the lumberyard and hand-picked unstamped pieces. It's good to know, too, that a lumber company will give you a bid that lasts for only a specified time (sixty days for ours). That's how fast prices spiral. Also, lumber that is bought and stored on the site prematurely will tend to warp or get harder to work with even if it is properly stacked. Try to plan out wood orders realistically to coincide with your work schedule. All finished materials, doorjambs and so on, should be purchased when you are ready to use them; they do not store well.

The moment we had been working toward was the day we could take up our hammers and build! We began by framing the floor, using 2×10 joists placed 16″ on center across the entire foundation. We thought at the time we wouldn't need blocking between the joists because it wasn't required by code. Now that we are in our house, we realize that it would have been better to use it anyway. The blocking would have helped to support and distribute weight better and to minimize play in the floor. Over our joists we nailed sheets of tongue-and-groove plywood, and this completed our floor for now.

The next step was wall framing. Walls go up really fast once you get the hang of it. We learned at this point that trial and error was to be a common element in each new step of building! For instance, the very first wall we built we had to take down and adjust. We didn't account for the corner windows, which need an extra stud for support, so even though we had nailed the bottom plate to the floor with 16d nails, we had to pry the wall loose, pull the nails out, lower the wall, and move all the studs over to make our stud spacing correct. We took extra time throughout the whole building process to do things right, and to work to achieve the overall aesthetic quality we wanted in the house.

A major feature in the design and construction of our house was the use of large rough-sawn beams. These 4×10 beams are exposed on the ceiling of the first floor and give the house a rustic look plus extra strength. To lift these beams into place, we had extra womanpower in the form of three or four friends. With wall framing well learned and these beams placed, some of the tension we'd been feeling began to ease. We were showing ourselves that we *could* do it, and began to feel more confident. As we passed the building department's inspections (spaced at intervals throughout, as when the foundation forms were built, the concrete poured, the floor framed, rough plumbing in, walls framed), we felt better and closer to the end. This project was difficult from beginning to end, as we were carrying the burden of learning as well as tremendous financial and emotional commitments, and were relying upon ourselves in important new ways.

An interesting—and critical—exercise to practice in building is the study of the blueprints. No one ever actually explained to us how to read blueprints. They are

(Photo by Rebecca Lawson)

drawn to scale (for instance, ¼″ to the foot), and with a measuring tape you can figure out what the actual dimensions are to be. There are four elevations: north, south, east, and west. Floor plans and wall and roof framing plans are each handled with separate drawings. We took to reading over our blueprints each evening after work to check out what we'd accomplished and see what came next. We had some trouble reading the blueprints. The architect was told to include as much detail as possible, since this was our first house, but he overlooked that and gave us only the bare necessities. The architect also made a few major mistakes, like designing a stairwell too short for the number of stairs required. We have been told that this happens frequently, since architects are removed from the actual building. We had to ask different people at times to interpret symbols on the blueprints, like electrical symbols (our woman electrician was responsible for this).

Our next step was laying what would double as the first-floor ceiling and the second-story floor. It was tongue-and-groove pine decking in lengths of 8′, 10′, and 14′. We wanted varying lengths so we could stagger the boards to give a sturdier and better-looking floor. We liked the light color of the pine and we saved money by using one layer to double as floor and ceiling. The most important aspect in laying lengths of board, tiles, or shingles is to get your first row straight. Everything else depends on that first row. We used galvanized nails, but even these rust after repeated exposure to moisture, so cover your floor until the second story is enclosed or you will have rust marks on your floor or beams below.

We then framed the second story, a repeat procedure of the first-floor framing, although it went much faster. By now we were getting more optimistic about the project, as we could almost see the end in sight—and then we started on the roof. This roof framing was my favorite part, looking back, but at the time I dreaded it. The main support for the rafters is the roof ridge piece. Our roof ridge consisted of two 2×10's nailed together. The length of your ridge piece depends upon your house length. Our house is 32′ long with a fireplace in the middle, so the ridge piece had to be separated for the flue. This left one ridge beam at 18′ and the other at 10′. Each had to be hoisted atop four 4×4 supports and balanced there until some rafters were nailed in place to anchor it. By the time we were ready to use the lumber for the rafters, the ridge piece was very warped and twisted from exposure, which made it *very* difficult to work with (especially while working up on a ladder!). The rafters had to be cut at correct angles according to the slope of our roof, and that was a hard one. We used a framing square and the step-off method explained in any beginning carpentry book. We then covered the rafters with sheets of plywood, forming the basic roof covering.

We used preframed windows and doors. This is costly, but we felt that we could not learn to frame the windows and doors in time to finish the house in our allotted time. Besides, we were led to believe it was more difficult than it really was. The house siding consists of ⅝″ rough-sawn plywood sheets, 4′×8′ and 4′×9′. These were nailed directly over the wall studs as a siding with 1×3 redwood battens covering the seams and spaced every 16″ (over the studs). This gives the house the look of board siding. We would have liked to use actual board siding, but it was much more expensive and would have taken a lot longer to put on.

At last we got our final inspection. We were so excited to pass that last legal threshold! We could now say, "Oh, it wasn't so hard after all." But of course that wasn't the end. The interior finish work will go on for months. Still, nine months after the first nail was driven, we can begin to appreciate what we have accomplished.

The hardest thing was living in the house while working on it—waking up on the job every day leads to fits of anxiety, but it did give us a better feel for the way we wanted to finish the interior. For instance, we realized how we wanted to arrange the kitchen cabinets, and we realized we needed more ventilation upstairs on one side of the house where no opening windows were drawn in the plans. We liked the simplicity of our structure from the beginning because we knew we couldn't have attempted anything with more complicated features without becoming discouraged.

Since we live in a moderate climate, we can rely mainly on our fireplace for heat and plan to use a solar water heater. We've turned on the heaters only a couple of times during this entire winter.

Generally we are pleased with the way the house looks, although the rough-sawn fir plywood siding is not as beautiful as redwood boards. (It has its own beauty, being one of the simplest, quickest, and most economical ways to side a house.) On one side of the house we used #3 cedar shingles (the cheapest) over exterior plywood as a cosmetic feature and to breakup the monotony of the plywood and battens on a rectangular house. It looks great.

The roof surface consisted of double layers of tar paper over the roof plywood for many months until we got around to laying down the asphalt surface. We chose white roll asphalt roofing because it would reflect sunlight (heat) and is quick to put on.

We realize now that a two-story house entails a lot more lifting of quite heavy materials, such as plywood for siding and roof and roofing materials, which should be considered in planning. Closet space is a boring detail, but it really is nice to have an area for storage! Sheetrock is a quick and inexpensive interior wall covering. We didn't want much of it at first, but since it is required in the stairwell (for fire protection) and in the kitchen and bathroom, we used it in those places. We didn't want it without wood highlights, but it goes up quickly and it also offers some diversity, since you can utilize wallpaper, paint, or textured plaster for a finish. For the bathroom area, where lots of water is splashed around, we used a special Sheetrock called "tile backer." This Sheetrock doesn't mold or crumble as regular Sheetrock would if continually exposed to moisture. Over this we put aromatic cedar closet lining, a special kind of paneling which comes in strips 4′ in length. It takes extra time to install, but it is beautiful and smells wonderful. The rest of the house will be done in wood except for the downstairs floor, which will be done in 12″ terracotta tiles. Interior paneling upstairs in the bedrooms is 1×8 V-rustic pine, normally used as exterior overhang material. It is one of the least expensive panelings and is both sturdy and handsome. There is a lot of flexibility in the types of wood you can use for interior finish, and you can use materials imaginatively to make your house unique. We hunted for redwood trees on our property thick enough to be used as posts for the porch and roof frame, and used them rough. We used 1×6 redwood boards for the floor.

Things are shaping up at the old homestead and we are relieved (and happy!). I would encourage any woman to build something and to consider building her own home. When you construct your own living space, you can know the special deep feeling of satisfaction and security that comes from knowing where to look if there is any problem within the house. It's a great feeling to understand your home!

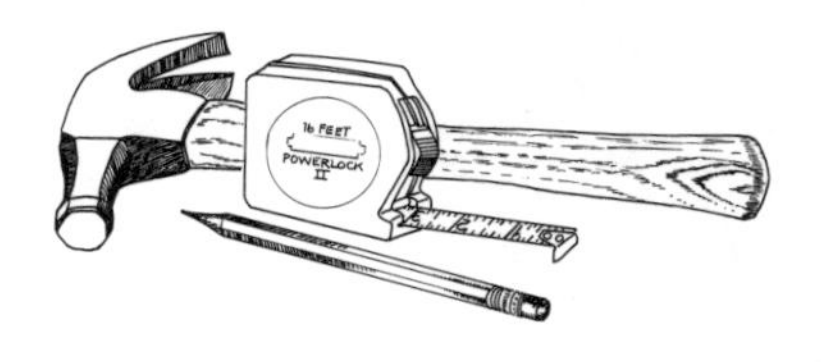

First Steps First

SIMPLE BUILDING LAYOUT

Layout for a small building, a deck, or a room addition can be done with simple tools and can be quite accurate if you work carefully. The purpose of layout is to create on the building site a set of lines that define the structure you want to make. These lines can be positioned so that your building will be perfectly square, rectangular, or polygonal and will be situated exactly as you want it. The layout lines will give you a guide to follow, whether you are going to construct forms for a continuous concrete foundation or plan to use a system of piers and posts for your foundation. If you are going to sink posts as a foundation system for a post and beam building, layout lines can give you an accurate way of locating the positions of these posts.

Basically, this system of layout will give you a set of horizontal strings which define the outside perimeter of your building. The strings are held in place by temporary constructions called *batter boards.* The materials you'll need for layout are simple and can consist of scrap lumber if you have it. You'll need 1×6 (or material roughly this size) for the horizontal ledgers of the batter boards. Each corner of your building will need a set of batter boards, and each of these sets will take one or two pieces of 1×6, about 4′ to 6′ long (more details on this later on). You'll need some 2×4 or 2×2 for the vertical legs of the batter boards. The lengths of these will depend upon your building site (a level site calls for shorter batter boards; a hillside site may call for some extra-tall legs). Scraps of 1×4 are handy for bracing material. You'll need nails for constructing the batter boards—6d or 8d will do; double-headed nails are a good choice as you'll be demolishing your work once the foundation is together.

Most important of all materials for layout is the string that will make up your lines. You can use any string you have, but the best choice is a *nonstretching* type of string. Nylon string is terrible to work with as it stretches easily and will cause you to lose important marks and accuracy. Cotton string is generally better. One of the very best types to use is chalk line string; it is sold in building supply and hardware stores and, as the name implies, is intended as replacement for worn-out chalk line strings. This is a strong, medium-weight string with very little stretch. Sometimes you can find string in a building supply store that is specially made for layout.

Another and final helpful item on the materials list is a handful of small finishing nails; these are used in initial location of lines and are a little more accurate than heavier nails. A couple of felt pens (two colors at least) will help as well.

Tools for the job are also simple. You'll need a hammer, a handsaw or circular saw, a plumb bob, an ax (for sharpening and driving stakes and legs), a level (24″ or 30″), a line level (or a good long straightedge), a 50′ tape (if you have one), and a steel square (optional but handy). Layout can be done by one person, but at many stages it is definitely helpful to be working with a partner.

Of all of the stages of building, layout can seem like the most time-consuming and the least rewarding. Everything you do should be as accurate and careful as possible, as the entire building will rest on this work. If you rush and decide to ignore that ¼″ mistake, you may find it miltiplying itself to frustrating proportions by the time you're trying to fit your roof rafters. My only advice is to take it all slow and easy; it *will* get done—though it seems it never will!

To begin, you need to roughly locate the four corners of your building. If you're adding a porch, deck, or room to an already existing structure, your first line

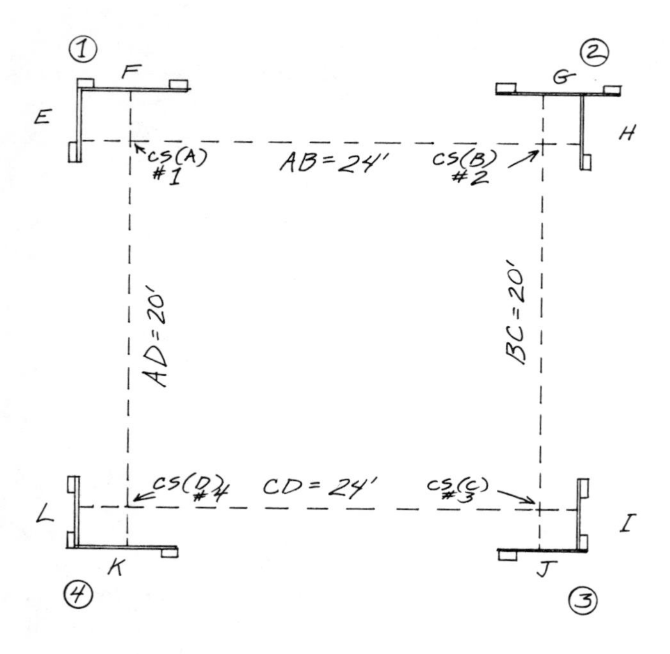

(AB) is already established for you. If you're laying out a polygonal building, you may begin by laying out a basic square (multiple corners/points can then radiate out from this). Let's assume for simplicity of explanation that you are working at making a rectangular structure 24′×20′. You begin by driving a small stake into the ground at point CS#1 (corner stake #1). Drive a small finishing nail into the top (center) of this stake. Now you can measure over 24′ in the direction you want your building to lie and drive a second stake CS#2. Put a finishing nail in the top of this stake and connect your two stakes (points) with a string stretched taut and tied to each nail. If you're adding to an existing structure, you can simply measure 24′ along the building where you want to join your room or porch and mark line AB. A finishing nail driven in at a sharp angle at CS#2 will give you the point to attach your tape to, hold your square against, and so on.

Now you can use your steel/framing square to show you roughly where your next line (BC) will run. Hold the square so that it corners at CS#2. The blade of the square can be held along the line AB and the tongue will then point out in the right (literally) direction for line BC. If you're working with a partner, one of you can hold the steel square in place and take the end of the tape; the second person can walk out with the tape to a point 20′ away and drive in a third stake, CS#3. Put a finishing nail in this stake and stretch string from CS#2 to CS#3.

Now you can check to make sure that the first corner you've established is square (90°). The method used to check the squareness of corners in laying out a building is called the *3–4–5* or *6–8–10 method.* It employs the principle of geometry that in any right triangle the sum of the squares of the two shorter sides will equal the square of the longer side. A triangle with sides of 3′, 4′, and 5′ is a right triangle ($3^2 + 4^2 = 5^2$) and any triangle with multiples of 3–4–5 will likewise be a right triangle (6–8–10 or 12–16–20, for example).

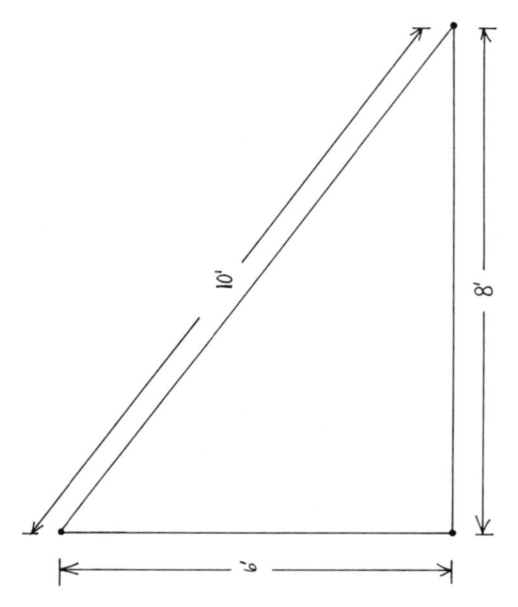

Go back to CS#2. Measure from that point out 3′ on line AB and make a mark on the string itself. Use the same color felt pen (or ballpoint) for all marks at this stage of layout; later, if you have to correct your lines in some way, you can use a second color and not confuse yourself over which mark is which. This advice comes from experience! With two or three changes of points as we adjusted lines, had problems with strings stretching, and so on, my partner and I found our tempers a little frayed by "It's the fatter mark to the far side of the medium-sized mark. . . ." Using different colors to mark at different stages helped!

Now measure out 4′ from point CS#2 along line BC and make a mark. Stretch a tape between these two marks and you should come up with a measurement of 5′. Stretching this tape sounds easier than it is. You have to be very careful not to pull on the strings or change their positions in any way if you want an accurate reading. At this stage you don't have to be a perfectionist, though—that comes a little later. But you can practice! If you find that your measurement reads less than 5′, you have placed CS#3 too far in and you should move it out until the measurement reads 5′ and the corner at CS#2 looks square. If you're reading more than 5′, you've placed CS#3 too far out and should move it in. In general, using a larger multiple than 3–4–5 will give you a more accurate reading of a larger building. In this case, I'd go to at least 6–8–10 for later, more perfect readings. For now you need a more or less rough location of your four corners, so the 3–4–5 will do.

If you are working alone on layout, you can still measure out and mark your 3′ and your 4′ points; but to stretch a tape between them, you'll have to drive a stake at one of these points with a finishing nail to attach your tape to. The work will go more slowly because, if your 5′ is off, you'll have to move the strings and stakes, recheck, and so on with each step a separate one.

Once you've established that your corner at CS#2 is fairly square, you can go on with locating your final corner CS#4. Use your steel square at point CS#3 to show you roughly where to run line CD. Measure out 24′ and drive stake CS#4. Use the 3–4–5 method to square corner CS#3. Now you can run string from CS#4 to CS#1—your final line, 20′ long. Check the square of corner CS#4 and then check the square at corner CS#1.

(Photo by Carol Newhouse)

With all four corners placed, there is another way to check that your proposed building is laid out square. If all corners of a rectangle or square are 90° (square), the diagonals will be equal. A line from point A to point C should be equal to a line from B to D.

The next step is to construct batter boards—one set per corner. There are two ways to build batter boards: *straight*—a horizontal ledger held up by two vertical stakes or legs; or *right-angle*—two horizontal ledgers meeting at a right angle and supported by three vertical stakes or legs. The right-angle batter board is greatly preferable as it is stronger; the two pieces tend to brace one another so that the tension of a layout line pulled taut won't pull them over, even slightly. Batter boards should be constructed outside of your actual building—at least 3′ or 4′ beyond each corner—so that they can be left undisturbed by excavation, form building, or pier setting. They should be built level to one another—the horizontal ledgers of each set of batter boards should be level to the ledgers of all the other sets. This is important for making accurate measurements of your lines. If you try to measure from point B to point C, for example, and one point is a few feet higher than the other, you are actually measuring a *diagonal* between the two points, not the real distance corner to corner. Consider the slope or grade of your building site and begin with your highest corner. This way, all other batter boards can be built up to be level with this set. If you start with a low corner, it may be impossible to construct the batter boards in a high corner to be level to this one.

The ledgers of the batter boards should be higher than the top of your proposed foundation wall; if your walls will be 12″ high, make your batter boards at least 18″ or 20″. Cut the legs or stakes long enough so that you can drive them into the ground and sharpen the end to be driven in. I like to drive two legs in first, then tack the horizontal ledger in place (use your level here), drive the third leg in (your steel square can show you where to place it), and attach the second ledger (level this one to the first). If you want to be extra careful, add bracing to the batter board until it is quite rigid. If you are working on a sloping site and have to build any set of batter boards over 3′ tall, it should be braced as a matter of course. Likewise, if you're working in very sandy or otherwise unstable soil, bracing is necessary.

In building your second set of batter boards level to this first set, you can use either a line level, a water

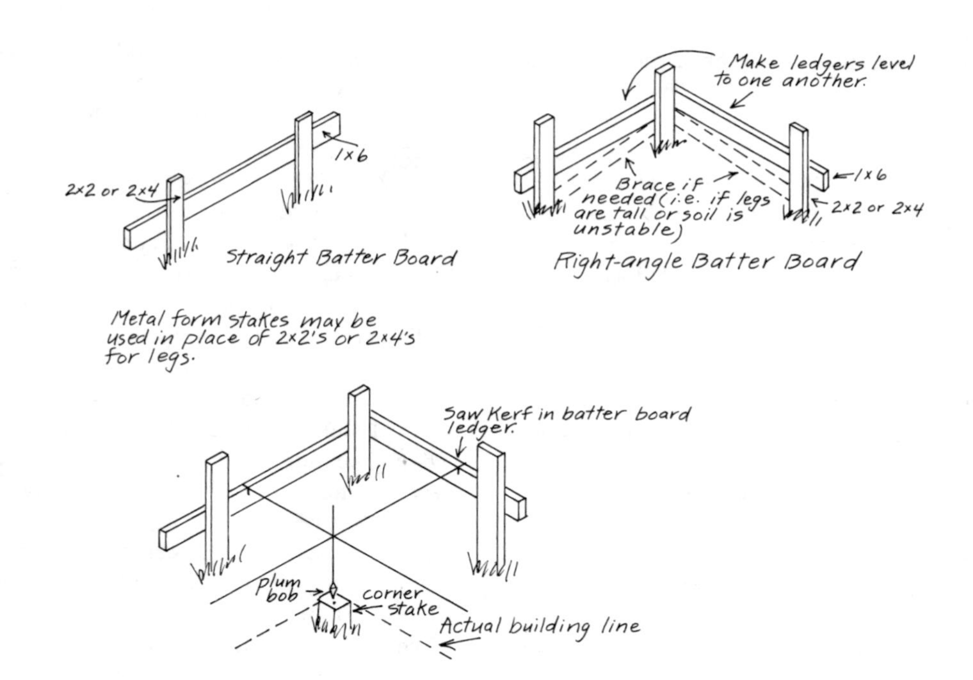

level, or a straightedge (with your 24–30″ level). Dale McCormick explains how to construct and use a water level in *Against the Grain* (see "Resources"). I use a line level, which can give a pretty accurate reading if the string is stretched really tight and you are careful. Another alternative I've read about is to drive a series of stakes from one set of batter boards to the next, leveling each stake to the one before it. This method seems risky to me as any error(s) could be compounded over the distance. You might want to try it to see how it works, though. Still another method is to run a continuous batter board around the entire perimeter of the building. This could be useful if you're laying out a polygonal and/or very complicated building without the help of a transit (a special leveling and measuring device). Whichever method you choose, go ahead and construct all four sets of batter boards. When these are done, you are ready to place your actual layout strings. These strings will define the exact outside line of your building, so they must be placed carefully and accurately. At this point you can remove your "rough" lines running between your corner stakes. Leave the stakes in place, as you'll be using them as a reference one more time.

Rather than tie or nail your layout lines in place, try using a saw kerf (cut down about ¼″) in the batter board where you want to secure each layout line. The kerf is cut into the top edge of the ledger board. This kerf will be perfectly accurate and will stay in place, unlike a nail, which might get bent over or pulled out by accident. You can remove your lines and replace them as you need to without losing any accuracy of placement whatsoever. If the string is wrapped around the batter board and through the kerf once or twice, it will hold very well.

To begin, set the first line of your building by running a string from point E to point H. Use your plumb bob dropped over points CS#1 and CS#2 to show you exactly where to place this string. Cut a ¼″ saw kerf in each batter board ledger and wrap the string in place. Pull it taut as you place it. Now place a string from point G to point J. Use the 3–4–5/6–8–10 method to square your corner. Lines EH and GJ should intersect right over CS#2. With this established, you can go to corner #3. To determine where point I will be, measure 20′ from CS#2, or the intersection of lines EH and GJ. With this point located, make a saw kerf and secure the line at point I; run it over to point L. Again use your 3–4–5/6–8–10 method to make sure corner CS#3 is square (and locate point L accurately). You can now locate point K by measuring 24′ from CS#3. Run line KF, using 3–4–5/6–8–10 to establish it squarely.

Now go back and double-check yourself everywhere. Measure your building lines to make sure they are exactly 24′ or 20′. Recheck each corner to make sure it is perfectly square. For a final check, measure the diagonals from CS#1 to CS#3 and from CS#2 to CS#4. If your diagonals are not exactly equal, something is wrong. It's possible that your batter boards aren't quite level to one another, which will affect the reading of your diagonals. Go carefully through each possibility until you discover and correct the error. Time spent now will save you headaches later. This may be difficult to believe if you've just spent a full day staring at strings, plumb bob tips, and levels—but it's true! Once everything reads perfectly, you've accomplished one of the hardest parts of building. Very few steps will take more patience and yield such tiny "obvious" results.

BUILDING WITH PIERS

A foundation connects a structure to the earth, supporting and stabilizing the building that rests upon it, and protecting vulnerable wood members from contact with moisture and damage from wood-eating insects. Many houses and larger buildings are constructed with continuous concrete foundations, but smaller houses, structures built on sloping ground, and decks and porches may utilize an alternative type of foundation system: piers and posts. This system makes use of a number of concrete piers, which support wooden (or sometimes steel) posts. The posts in turn support girders or beams, which bear the weight of the structure above. While it is not "to code" in some areas to build a dwelling on piers, a pier foundation may be legally utilized for small structures, sheds, porches, decks, stairways, and so on. For any small structure, a pier foundation can provide a safe, permanent, and reasonable alternative to the more conventional continuous foundation if you plan carefully and do your work well.

The pier and post foundation is economical, can

(Photo by Carol Newhouse)

be completed quickly, and demands less skill of the beginning carpenter. Piers are particularly well suited to remote building sites, as the materials are light enough to carry in. Piers may also be used in a house with a continuous foundation; here they are used to support girders that break the span of floor joists, so that the size of the joists (determined by what span they can safely make) may be kept within reason. For instance, a commonly used size of floor joist, the 2×10, can safely span up to 16′2″ (spaced 16″ o.c., Douglas fir or similar). A girder or beam supported by posts and piers may be run at this point to break the span in a house that is, say, 30′ or 32′ wide. The girder is run at a right angle to the direction of the joists. In some instances, piers may be used to directly support the joists of a small structure or a porch (see the sections on "Hexagon Building" and "Yurt Building" for details and photographs of this). Learning how to work with piers and posts is a valuable skill that can be applied to many building projects.

Planning Your Work

As the diagram shows, a pier foundation has four major parts: the footing, the pier, the post, and the girder. This system may be used for the substructure of a deck, a small house, or a porch. Sometimes the footing is not used, and sometimes the girder may be omitted (the joists will then bear directly on the posts). Generally, all four parts of the system will be necessary, and your plans are made around these parts. It's a good idea to draw up plans for your project, whether it is a simple deck or a small house. Draw the plans to scale (say, ½″ = 1′) and use graph paper if you want to be really accurate. Your plans should show all parts of the substructure and how they relate to one another. It should include the size of the pieces—posts, girders, etc.—you want to use for various jobs. You will have to calculate your needs using specific tables or general adaptations based on these tables, depending upon how exact or critical your plans must be. If you are building with a permit and to code, your plans do not have to be elaborate; they can be surprisingly simple if you draw them up clearly and label them.

Plans are usually drawn in two ways: first the *plan,* which shows the structure as though you are looking straight down on it from above; second, the *elevation,* which shows the structure as though seen straight on from one side. Plans should show the spans of girders, joists, and decking if the plans are for a deck or porch; spans of girders, joists, and subfloor if for a house. *Span* is the distance a piece must bridge. The plans should show the spacing of posts, girders, and joists. *Spacing* is the distance from one member to another: for example, the distance between joists or between piers. Spacing is usually expressed *o.c.,* "on center": the measurement from the center of one piece to the center of the next. In most building, spacing follows conventional units which are created to make economical use of materials and to establish uniformity within the structure; spacings for floor joists, studs in stud walls, and roof rafters are commonly 16″ o.c. or 24″ o.c. Tables have been calculated to show the relative strengths of certain sizes of lumber when placed at these spacings. It is usually helpful to plan your work using conventional spacings because you will be work-

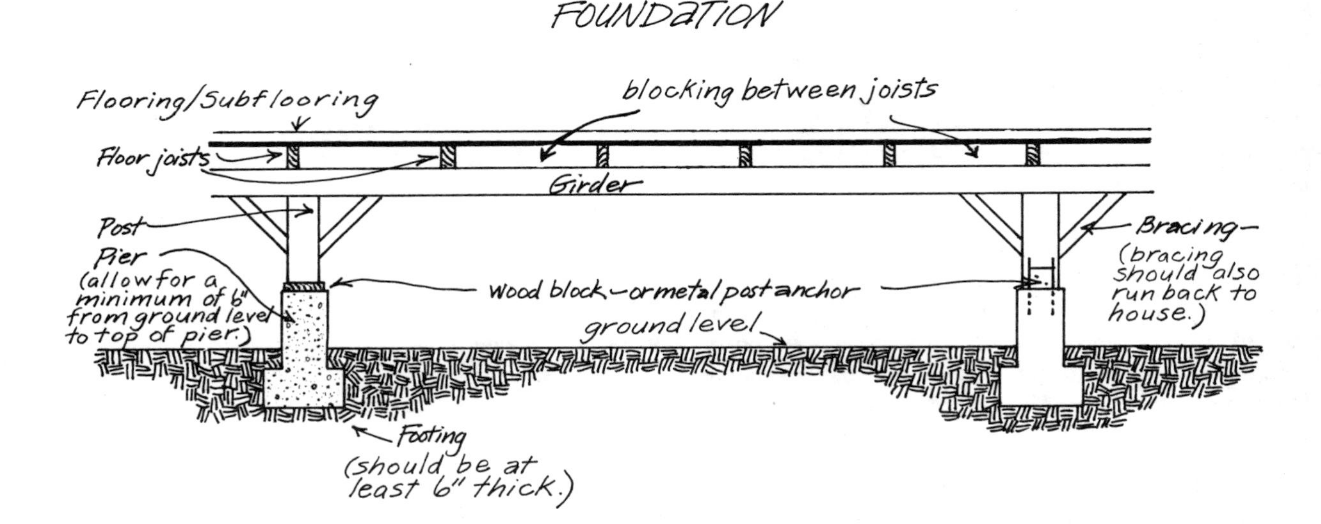

ing with more or less conventional materials and can make use of available knowledge about lumber strength and so on.

If you are planning a deck or porch, one of the best sources for information is the Sunset book *How to Build Decks* (listed in "Resources"). This book has extensive tables of joist spans and spacings, beam (girder) sizes and spacings, post sizes, and so forth. You can look up all of your requirements in terms of how tall your deck is to be and what size lumber to use for each component part. Here is an example of the specific information you might need for planning a deck adapted from these tables:

(All lumber used to be Douglas fir, southern pine, or western larch)
4×6 girders may be used, placed 6′ apart
posts/piers must be located 6′ apart
4×4 posts may be used
joists may be 2×6 material, spaced 24″ o.c. or 32″ o.c.
decking may be 2× material (2×2, 2×4, 2×6) laid flat
maximum deck height is 12′

If you are planning to use piers and posts to support a girder that will break the span of floor joists under your house, you can probably follow general code requirements. Usually the girders are used to support floor loads only: i.e., the girders are not doing any work other than breaking the span of the joists. The building code requires piers/posts placed 5′ o.c. where 4×4 girders are used and piers/posts placed 7′ o.c. where 4×6 girders are used. Girders should be Douglas fir or lumber of similar strength. If your girder is to support extra weight (a bearing partition wall, for example) you will have to go to a heavier girder and/or closer spacing of piers.

If you are drawing plans for a small house, studio, or cabin, your plans will have to take into account the total floor area of the building and the expected loads the foundation will have to carry. Without an architect friend to help you, it's difficult to know how to go about designing your own pier foundation dwelling. Very few carpentry books talk at all about using piers as the primary foundation support (most are geared to conventional continuous concrete foundations, and those that aren't seem to skip over the vital details, assuming, probably, that you'll take your problem to an architect). If you are planning a fairly straightforward, simple structure, you can calculate the approximate weight of your structure and use one of the following references to determine the size of your girders, spacing of your piers, etc.: *From The Ground Up; Wood Structural Design Data; Timber Design and Construction Handbook* (see "Resources").

For a rough idea of what you might find yourself using, one set of USDA plans shows a 24′×32′ house supported on three 4×12 beams (32′ long) with posts and piers placed 8′ o.c. The Uniform Building Code calls for girders carrying floor loads—breaking the span of joists between perimeter walls of foundation—to be 4×4 for spans up to 5′ and 4×6 for spans up to 7′.

There is a standard formula for calculating the weight loads of a structure according to the size (floor space and number of stories) of the structure. This will give you a rough idea of what your house will weigh and what live weight (you, your things, and your friends) the structure will support. The formula appears in *Modern Carpentry* (see "Resources") and a couple of other carpentry books in more or less detail. However, the following should offer enough guidance to help you calculate the weight of a small house or studio.

First, figure out the square-foot area of your structure. This is simply length × width = sq. ft. area (all measurements in feet). Now multiply that figure × 10 (the "dead load" of the first floor: i.e., the weight of the structure itself). Next multiply the sq. ft. area of your structure × 40 (the "live load" of the first floor: i.e., your friends, things). Next calculate the sq. ft. area of your roof. To do this, you have to measure and calculate for the entire roof surface; it will *not* be the same as your floor space unless you have a perfectly flat roof. If you have a gable roof, you'll need to know the length of your rafter × the width of the house for each side of the roof. If your roof is a shed, you'll need to multiply the length of a rafter × the width of the house. If you're planning a polygonal building, you'll have to figure the sq. ft. area of each roof section and add them all together. Once you have the sq. ft. area of your roof, multiply this number first × 10 (the dead load of your roof, figured for the heaviest roofing—wood shingles). Then multiply the same number (sq. ft. area of roof) × 30 (live load on your roof—calculated for wind, snow, rain). If your plans include a second floor, you'll have to go back to your sq. ft. area of the house again and multiply first × 20 (dead load) and then × 40 (live load). Each partition wall is figured in at 20 × sq. ft. of floor area. I'll assume that you don't plan plastered ceilings, but if you do, look at the *Modern Carpentry* formula (page 116) or look in *The Uniform Building Code* for weight allowances. To look at all of this more graphically:

Approximate weight of the structure =

sq. ft. area of structure × 10 (10# per sq. ft. dead load, first floor)

+ sq. ft. area of structure × 40 (40# per sq. ft. live load, first floor)
+ sq. ft. area of roof × 10 (10# per sq. ft. dead load)
+ sq. ft. area of roof × 30 (30# per sq. ft. live load)
optional + sq. ft. area of structure × 20 (20# per sq. ft. dead load, second floor)
optional + sq. ft. area of structure × 40 (40# per sq. ft. live load, second floor)
optional + sq. ft. area of partition walls × 20

Once you know the approximate weight of the house you want to build and can figure out what size girders to use and how the piers/posts must be spaced, you can draw out a plan for your work. The accompanying "Foundation Plan" shows one very simple version that might be like yours. The size of posts you use will depend upon the size and weight of the structure you are planning. For a small, light structure, such as a low deck, you can use 4×4 posts. A 4×6 is more desirable but more expensive. You may want to use 4×6 posts for your corners and centers and "fill in" with 4×4's. For larger, heavier structures, you should increase the size of your posts, using 4×8's or, as the plan shows, poles of a good size. One way, of course, to avoid using really large posts, girders, and even floor joists, is to space things closer. You should figure out which alternative is more economical for your particular building: spacing larger pieces farther apart, or putting smaller pieces closer. You'll also want to think in terms of lifting and handling your girders, posts, and joists. If you are going to have to use two or three pieces for one girder, joints should be planned over posts and strengthened with metal connectors if possible. Don't forget that you can use built-up girders. These are made by splicing two pieces together to make one piece of the required size. For instance, you can make a 4×12 by splicing two 2×12's together. A built-up girder is not quite as strong as a single, solid piece of the same dimensions, but may be a necessary choice under certain conditions. Large-dimension timbers are sometimes hard to find, and if you don't have the time to special-order them, or the cost would be too high because they aren't a stock item, you may substitute a built-up piece. If you're carrying materials into a remote building site, it may be easier to carry in smaller pieces and splice them together on the site.

Once you've decided how to space your girders and posts and piers, you can lay out your building and begin work on the foundation. There are a number of alternatives to consider for your piers: using precast piers, casting your own piers, using concrete blocks, and even substituting embedded poles for piers. The "Hexagon Building" article discusses using concrete block piers and Susun's "Incredible Woman-built House" talks about embedded poles as a foundation. The FHA pamphlet *Pole House Construction* (see "Resources") has some very good, usable information about soil types, setting poles in the ground, and so forth. For the remainder of this section, I'll share some information and experiences about precast piers and casting your own.

Using Precast Piers

Commercially available piers are precast concrete

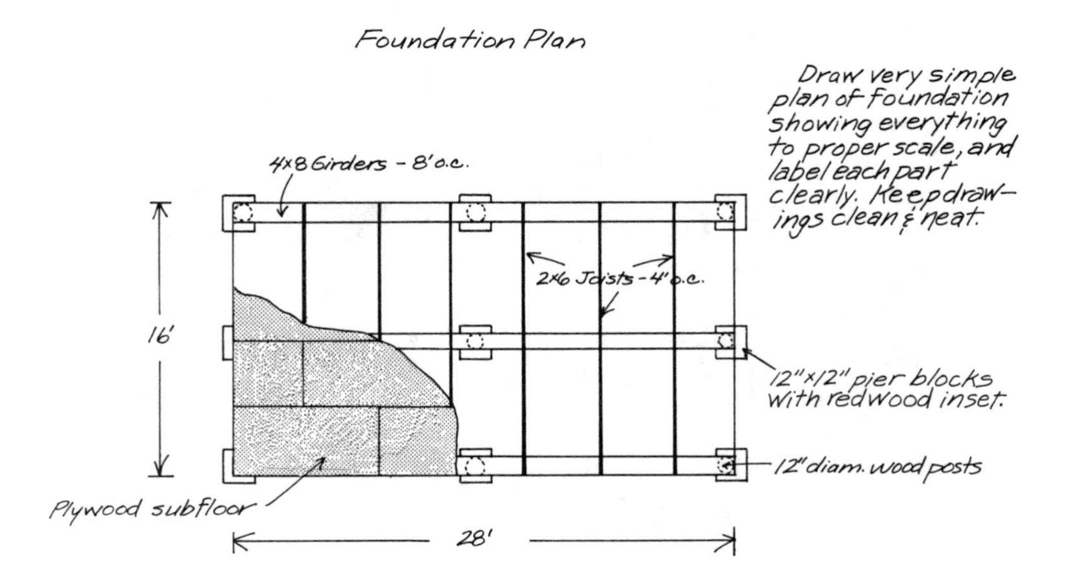

blocks, tapered from about 10″ square at the base to 8″ square at the top, and usually about 8″ high. A piece of good-grade, decay-resistant wood is embedded in the top of each block. This piece provides a nailing surface so that you can toenail a post firmly to the pier. Precast piers are cheap, readily available, and easy to work with. One person can lift a pier, so that working with and transporting them is not too difficult. Piers of this size are fine for supports under small structures, porches, and decks, and to support girders that break the spans of joists under the house. In buying precast piers, try to choose those that have the wood blocks set squarely in the center. This will make lining up the piers easier, though it is not absolutely critical. Avoid buying any piers with defects in the wood blocks, such as knots or splits, as these can make secure nailing of your posts difficult or impossible. Most building supply yards or lumber stores will think you are a bit eccentric to insist on choosing your piers, but insisting upon your right to do so will make your work much easier.

Precast piers may be set directly in prepared spots or on poured concrete footings. Both procedures are detailed below. Using the piers without footings is cheaper and somewhat faster, but the footings provide a much more stable base for your structure. If you want a stronger and/or more massive pier, or want to custom-build piers to your site and needs, you will have to cast your own. This will involve quite a bit more work and some more money for materials, but

will give an even sturdier base for your structure. A reasonable alternative would be to cast your four corner piers and use precasts for the remainder.

The procedure for using precast piers without footings is as follows. With your building lines in place, you can locate the position of your first pier. Begin with a corner. Place a pier or a piece of wood of approximately the same size as your pier beneath the intersecting lines of your corner. Drop the plumb bob on the outside edge of this intersection; this will locate the outside edge of the post that will support this corner. Your pier will sit roughly under this point. With this established visually, you can dig out an area about 18″ square, removing any sod and digging down to solid ground. Try not to excavate too much, as you want the top of your pier to sit well above ground level. Try, too, to make your work fairly level even in this rough stage. Once your hole is roughed out, drop your plumb bob again to check that you are in the right location. Then take a hand trowel or any suitable small tool and level your surface. This is surprisingly exacting work that takes a highly meditative (or comatose!) state of mind. Some spots seem to level out quickly and easily, but others defy all of your attempts. Use a torpedo level to check your progress. Always *remove* earth to establish level; never add earth back, dear as the temptation may grow. Your pier must rest on solid ground to receive and distribute the weight of your structure, and earth added back creates an unstable condition. When you've established a level surface by scraping and removing what's needed, put your pier in place and check its level by placing your torpedo level on the wood block; check all directions, including diagonal. You may need to tip the pier out to scrape dirt from under one corner or edge; again, do *not* add dirt to make the pier level! At this point, you should use your plumb bob to line up the pier exactly as you want it. I like to mark the wood blocks with lines showing precisely how the post will sit: for example, a 4×4 post will sit centered on the pier block, and you must allow for the space all around. This is something you must consider and adjust for in advance if your building is to turn out exactly right. When you have the pier perfectly level and exactly in place, fill back in around it and tamp the earth gently to compact it and help hold the pier in this position. You are ready to go on to another corner.

The procedure for using precast piers with footings begins the same: locate your approximate pier position using your plumb bob. Dig down about 8″, continuing until you hit stable soil, or until you are deep enough that your footing will be well below the local frost line. A footing about 24″ square and 6″ to 12″ thick should be sufficient. If you are short on cash for materials, you can use some gravel in the footing hole to stretch your concrete (see Helen Garvey's *I Built Myself a House,* listed in "Resources"). When you are ready to pour your footings and set your piers, soak the piers beforehand to allow the concrete and pier to bond to one another. The book *How to Build Decks* (see "Resources") suggests setting the piers in a bucket of water and positioning them five to ten minutes after the footing is poured. Helen Garvey, in a personal letter, described her process as follows: "I did not let the footing cure before putting in the pier. . . . I wet the pier (so it wouldn't drain water from the concrete too fast and dry it unevenly) then set it in, putting concrete over the base of the pier to hold it in place. And I got the pier level while all that was wet." However deep you make your footing hole, remember to adjust the thickness of your footing so that you have 6″ to 8″ clearance from ground level to the top of your pier. Be sure that your pier is set in level, and be sure it is set exactly according to your building lines. Read through the following section for more information on footings and working with concrete.

Casting Your Own Piers

The first step is to dig or "rough out" the holes that will provide footings for your piers. If the soil you are working with is fairly stable and not really loose and crumbly, you simply have to dig out a hole with sides as straight up and down as possible. Perfection isn't necessary, but don't be overzealous in your digging. The hole you make will be filled with concrete, and you should never throw dirt back in to make do. In this instance, careful work definitely equals economy. If you are working with soil that crumbles and caves in easily, you may have to make a simple form box to contain your footing concrete. Dig down to stable soil and construct a box of appropriate size and height (read through the information on pier form box construction, below, and adapt as needed). This form box can be held in position with stakes, and your pier form positioned on top; you will have to improvise to fit your situation and needs. In most instances, you can dig your footing holes and they will be sufficient. Always dig down to stable soil for your footings even though it may make your footing larger than "minimum requirements." If you live in an area of deep frosts, your footings should be deep enough below ground level to be protected from the movement of freezing and thawing ground. You can check with the local building department or other builders to see what this depth should be. In heavy-frost areas, it may be as much as 3′ to 4′.

To determine the position of your footing hole, use the building lines you've laid out and a plumb bob. Begin with a corner pier. The intersection of your two lines in the corner shows you where your corner post should sit. The two sides of the post must line up exactly with your building lines. This means that you must situate your footing, pier, and wooden block or post anchor with this in mind. Looking at the situation illustrated ("Foundation," near the beginning of this section), you can see how the post anchor/wood block exactly locates the post that sits on it; the pier below extends beyond the edges of the post all around; the footing below that extends beyond the edges of the pier. By dropping your plumb bob from different points, you can see clearly where your rough footing hole needs to be dug. The size of your footing will depend upon the size of the structure or stairway or deck that you are building. For a small house or cabin, you may want to use a 4×6 post, a 10″×10″ pier to support it, and a 24″×24″ footing below. The thickness of your footing should be at least 6″ to 8″, and more if you have to compensate for deep frosts. The base of your pier will be slightly below ground level and the pier itself should extend at least 6″ to 8″ above ground level to protect your post from contact with earth and moisture. Your posts will extend up and be cut off to a height level with one another, so that your piers may be cast independently (i.e., not level to one another).

With your footing hole dug, you are ready to construct a *form box* for your pier. A form is a temporary "mold" that holds wet concrete until it hardens or "sets up" in the shape, size, and position desired. You can either build one form box and use it in turn for each pier, or you can build a set of pier forms and pour all of your piers at one time. The form box must be strong enough to hold the pressure of the outward-pushing wet concrete without being tipped out of level or forced out of position. It is held in place by a series of braces and stakes, which likewise must be strong enough and placed correctly to resist this pressure. When the concrete has set up, the braces and stakes are removed and the form box itself stripped away. You can build a square or a tapered box: the square box is a little easier to build; the tapered box will give you a pier that sheds water better if exposed to the weather. I will describe how to build the square or rectangular box; for details on making a tapered box, see *Practical House Carpentry,* listed in "Resources."

Your form box will be open-ended, four sides with neither top nor bottom. If you are building on fairly level ground, your form can be pretty much square or rectangular. If you're building on a hillside, you may want to make a box that fits to the contours of the slope. Form material is usually 1×, and you can use any square-edged (not tongue-and-groove or ship-lap material) 1× you have. Plywood is easy to work with (especially if you have power tools) and makes a smooth, strong box. Plywood is especially well suited to building forms that must be shaped to the contour of the ground. If you are working with narrow boards (1×4 or 1×6), you may consider covering horizontal joints with cleats or battens on the outside of the box. If your box is built tightly, you may not even need these cleats; chances are that the concrete will not ooze through the joints. The most important thing to remember in building a form box is that you will have to take it apart later. Everything should be done with this in mind. Because concrete bonds readily with metal, any nail head or point exposed *inside* your form box will make dismantling the form harder. Plan all of your nailing so that nails are not exposed inside the box. Using double-headed nails will make your dismantling work much easier, too.

In cutting the material that will make up your form box, remember that your finished pier will equal the *inner* dimension of the box: a fact important but easily overlooked. Assuming that you want a pier 8″ high and 10″ square, for example, you must cut two 10″×8″ pieces and two 11½″×8″ pieces; the longer pieces will overlap the other two, so that the box is perfectly square.

The next step is securing the form box in place over your footing hole. You may, of course, choose to pour all of your footings first and then construct the form boxes over the partly set concrete in the footings. It's a little easier to pour both footings and forms at the same time, as you'll have all of your concrete working tools there and can do all of the mixing, etc. at once. The footing and pier will be a solid piece, too. The form box must be positioned so that the lower edge is slightly below ground level in the footing hole. It may appear that, when you pour in your wet concrete, the concrete is just going to flow up out of all of that open space at the foot of the form box, but this is not what happens. The concrete will fill the footing hole and squeeze out a little as you fill the form box, but it is thick enough to hold its shape and not come oozing or pouring out. There are many ways to secure a form box in place, and you can use your ingenuity with the materials you have available and the peculiar needs of your building site and forms. One way to do the job is as follows:

2×2 ledgers are nailed onto the form box on parallel sides. Position the form box exactly where you want it by dropping your plumb bob from your building lines. (Remember that you are lining up to a post anchor or a wooden block—you may want to skip ahead at this point and fasten the anchor in place to make this step easier.) Secure the form box in place by driving in form stakes or wooden stakes at the ends of the ledgers well beyond your footing hole. Level your form box and securely nail the ledgers to the stakes. The form box now hangs suspended in place on its ledgers. For extra strength, two more ledgers may be added perpendicular to and just below the original two; secure these with stakes, also. If you are casting a fairly large pier, you may need to do additional bracing to make sure that the pressure exerted by the wet concrete doesn't tip the box out of alignment. If your form boxes are going to be subjected to a building inspector's approval, make sure that they are strongly built and well braced. (Watching a building inspector use our precisely located and painstakingly leveled form boxes to steady himself along a steep hillside convinced me that we hadn't overbuilt them!)

There are a few alternatives to building form boxes. Some concrete companies sell ready-made, heavy cardboard cylinders for use as pier forms (one company calls these Sonotubes). These may be cut to size, used, removed, and discarded. I've read of people using asphalt roofing paper (rolled into cylinders and secured with wire) or clay pipe for forms. Anything that is stiff and strong enough to contain the wet concrete and that won't bond to it may be used. Probably building wooden form boxes is the most common choice because the forms are easily built with readily available materials and may be braced to take a lot of pressure from the concrete.

With your form box in place, you are ready to position a post anchor—if you're using one, and if you haven't already skipped ahead and done this step. You may just want to wait until you are pouring the concrete and set the anchor directly in place in the wet concrete. Setting the anchor in place ahead of time allows you to have everything perfectly in place; you won't have to be dropping a plumb bob to line up the post anchor, or checking the level of the anchor, while dealing with wet concrete and all the mess it creates. Preplacing the anchor also allows you to take down your building lines—a plus that you'll appreciate when wielding shovels and guiding wheelbarrows

around the forms. A simple way to set the post anchor is to cut a piece of 2× material just long enough to sit across the top of the pier box. Tack the post anchor to this piece, using double-headed nails. Set the anchor and ledger on the form box, drop your plumb bob to line up the anchor exactly where you want it, and tack the 2× firmly in place. After you've poured the pier and the concrete has set up, the 2× may be easily removed.

It's a good idea to use rebar, or reinforcing steel rods, in your footings and piers. Concrete has a lot of compressive strength (it can take a lot of weight pushing directly down) but much less tensile strength (it has less resistance to twisting or bending under pressure). Proper use of rebar can increase the tensile strength of your pier and footing, giving your structure a much stronger base. It is especially important to use vertical rebar if your piers and footings are tall. There is more information on using rebar in the section on "Continuous Concrete Foundation." If you are building to code, with a permit, the building inspector will want to see your form boxes with rebar in place before you pour. If you are building without a permit, you can choose to place your rebar directly into the wet concrete when pouring. Generally a couple of pieces of vertical rebar will strengthen a pier and footing. If your piers are large, you may want to place as many as four vertical pieces in each one, and even add a square of horizontal pieces in the footing. Rebar should always be placed so that it is completely embedded in the concrete; don't place it so that an end touches the ground or protrudes beyond the concrete as the end will rust and the reinforcing strength of the rebar will be lost.

There is one final, optional step in constructing your form boxes, and that is oiling the forms. Coating the inside of the form boxes with oil will prevent any tight bonding of wood to concrete, and make it very easy to strip (remove) the forms later. You can use cheap motor oil for this coating. If you're using metal post anchors or rebar, it is critical that you don't splash oil on these pieces as they, too, will be prevented from bonding with the concrete. I've removed wooden forms that have *not* been oiled and found that they come off pretty easily, so it's not really a necessary step.

Working with Concrete

Dry concrete is a mixture of cement, sand, and gravel. These ingredients are mixed together in proportions that are "standard" for certain jobs. A common mix is one part cement to two parts sand to three parts gravel; this is expressed as a *1–2–3 mix.* Water is added to the dry concrete just before the concrete is put into footing holes or forms. The water and cement in the mix react chemically and a process begins called *setting up;* the concrete begins to stiffen or "harden"—actually it is bonding together. Concrete is usually mixed, ordered, and spoken of in terms of cubic feet and cubic yards. To determine how much concrete you'll need for your footings and piers, you have to figure out the total volume of each footing and/or pier, add all of this together, and add in some for waste. It is easiest if you start by converting the measurements of a footing to feet and fractions of feet. For example, 6″ will become ½ ft. or 0.50 ft. You can then multiply width (of the footing) × length × depth to get the total volume, which is now in cubic feet. To convert cubic feet to cubic yards, divide this figure by 27 (1 cubic yard = 27 cubic feet). Once you've figured out approximately how much concrete you'll need, you can decide how to buy or order it.

Large amounts of concrete (1 to 5 cubic yards or more) may be ordered ready-mixed (also called *transit mix*) from a concrete company. A truck will deliver the wet concrete to your site; the concrete flows down a chute from the truck into your forms. Smaller amounts of concrete are usually mixed on the site; you can buy all of the dry ingredients separately and mix them yourself, or you can buy dry *ready-mix,* which is all premixed—you just add water and stir. If you want to mix your own concrete, read through the sections in *How to Build Decks* (see "Resources") on estimating quantities of materials and on mixing concrete. It tells you exactly how to buy your sand, cement, and gravel and exactly how to mix and add the right amounts of water. This is one of the clearest and simplest explanations of this work I've seen.

I've always used dry ready-mix for footings and piers. You can handle the sacks easily and mix small amounts on the site. You don't need a mixer and can carry your materials and tools into a remote or difficult site. One sack of dry ready-mix will give you about ⅔ cubic foot of concrete. It is a little more expensive than buying and mixing the ingredients separately, but for relatively small amounts of concrete the price difference may be minimal.

For your pour (the actual mixing and shoveling of wet concrete into your footings and form boxes) you'll need: a wheelbarrow, shovels, a water source, a plunger (a piece of 2×2 or 2×4 about 4′ to 5′ long), and some hand trowels. The wheelbarrow is used for mixing the concrete; the shovels, for mixing and moving it around. The plunger is worked up and down in the freshly poured concrete to force out air pockets. The trowels are smoothing tools, used to finish the tops of the piers. If you're working with one or two partners, you'll want to distribute the work involved: one person can mix and shovel the concrete; a second can work with the plunger, and so on. Be careful when mixing the concrete to avoid inhaling the cement dust—an irritant that's not especially healthy for your lungs. Use a respirator or at least tie a bandanna across your nose and mouth for protection. Once the cement in the concrete gets wet, this danger is over. The same precaution should be followed when handling the empty sacks later.

It's important to pour all or most of the concrete for each pier/footing at the same time, otherwise you may find yourself with sections that have begun to set up and won't bond well with concrete added later. If you're doing a succession of footings and piers, try to fill up one entirely before going on to the next. If you're using ready-mix, follow the directions on the sack exactly—the amount of water always seems too little, but as you mix and fold it in, it will "stretch." Adding too much water will wash out your mix, diluting the cement too much; this will make it difficult for your concrete to set up properly. As you fill up a footing/form box, use your plunger to settle the concrete into all corners and to eliminate air pockets. It's better to overfill the form box than underfill it. As it fills the form box, the concrete will ooze up a little from the lower edges of the box. Don't worry about this now. You may want to divide the work so that one person goes along mixing the concrete and filling up the footings and forms while her partner comes

(Photo by Dian)

along behind, using the hand trowel to smooth excess concrete away from the top of the pier and making an even finished surface. If you're setting in wooden blocks or have decided to put your post anchors in during the pour, this is the time to do it. Try to keep the top and edges of post anchors or wooden blocks as clean as possible. After you've done two or three piers, go back and clean up the concrete that has oozed up out of the footings. Press it down or scrape it back so that the entire form box is fairly clean and accessible. You should be able to see the lower edge all around; this will make it easy to remove the box once the concrete has set up.

When all of your footings and piers are done, go back and double-check everything to make sure that any newly set post anchors or blocks line up properly with your building strings and are level. Check to see that all form boxes are free of concrete around the lower edges. Make certain that all tops of piers are smooth and post anchors are clean, and so forth. Wash down all of your tools and your wheelbarrow as soon as you're done, taking care that wash water doesn't run down into your footing holes and dilute your concrete.

Depending upon the size and thickness of your footings and piers, the concrete should take three days to a week to set up enough so that you can remove the forms. There are some concrete additives that will make the concrete set up faster, but usually forms should be left on at least three to five days. The concrete should be kept damp all during its curing time, especially if the weather is hot or dry. You can give the forms and footings a light spraying with water once or twice a day. Wrapping them in wet burlap will help if it's really hot. If the weather is cold and there's danger of frost, the concrete should be protected with canvas, straw, or layers of burlap. Removing your form boxes at the end of the curing period is simple work: a Wonder Bar or small pry bar can be used. All of your work is rewarded now with a set of smooth, sturdy piers ready to accept your posts and the next stages of construction.

CONTINUOUS CONCRETE FOUNDATION

A continuous foundation running under the exterior perimeter of a structure will support and evenly distribute the weight of the structure. Commonly, the foundation of a house is made of reinforced poured concrete. It may be constructed instead with concrete blocks, bricks, rock, and so on; older houses were sometimes built on "skids" (large wooden timbers or rough-cut logs). In areas of mild weather, low foundation walls are built of concrete and a "crawl space" of 18″ or so is provided under the house. This crawl space allows for ventilation and lets you have access to floor joists and plumbing. If you've ever had to work in one of these spaces, you can see why the word *crawl* is used. (Not mentioned are the protruding nails, the cobwebs, dirt, and assorted unpleasant debris accumulated—under and through which you must crawl!) In cold-weather areas, many houses are built with full basements. Another alternative foundation system is the slab-on-grade (see Dale McCormick's *Against the Grain,* listed in "Resources" section, for specifics on this). A continuous foundation is a reasonable project to tackle if you have some carpentry experience and want to build your own house, add a room to an existing house, or replace sections of your foundation that are in poor condition. There are a lot of details and suggestions in the following pages—don't be intimidated by their sheer bulk! Like most carpentry skills, form building and working with concrete are much easier to share in an actual work situation; they are difficult and tedious to explain, but the doing will probably be easier going than reading about them.

A continuous foundation consists of two basic, integrated parts: the *footing* and the *foundation wall.* Both are made of concrete, which is poured in a wet (plastic)

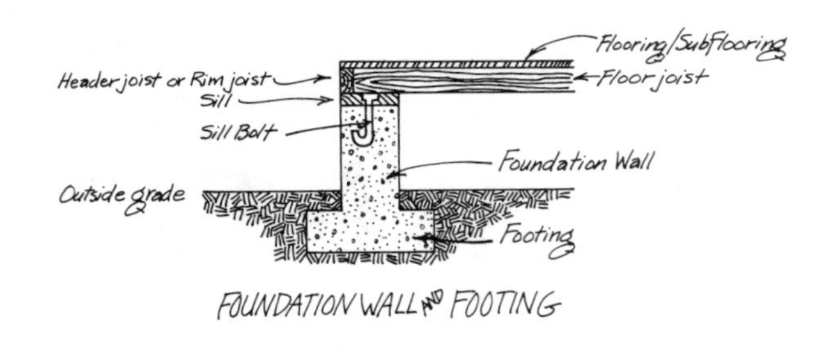

FOUNDATION WALL AND FOOTING

state into forms meant to contain it until it sets up (hardens) enough to hold its own shape. The footing provides a broad base running under the entire foundation wall; the wall, in turn, provides a base and anchoring point for the house. The foundation wall acts as a protective barrier for the wooden members of the house: it holds them elevated above the ground so that the wood is not contacting earth, which, with its moisture and insects, can quickly rot wood in direct contact with it. Along the top of the foundation wall runs a piece called the *mudsill* or *sill.* This is a piece of redwood or other decay-resistant wood, usually a 2×6, which is anchored to the foundation with special bolts embedded in the concrete. This sill forms the point of connection between house and foundation. Foundation work on your house will take you from layout through digging trenches for your footings and walls up to the point of placing and securing the mudsill. Beyond that, you will begin framing up your house and see your ideas actually taking shape.

To give you an overview before the onslaught of details, foundation work consists of:

Clearing and leveling of the building site as needed. For a good work situation, all debris should be removed from the area, and vegetation that might obstruct your work must be removed. Try to leave as many trees and plantings untouched as possible; you'll appreciate them later, even though they may cause you a little extra work now. If you're building in the city, make sure that your plans are properly "set back" (that you are building the required number of feet in on your lot). A building site that is sloping may be leveled off by power machinery, or you may choose to build your house to fit the site.

Layout. This involves locating and setting up a series of building lines (strings) to show you exactly how the house lies. It will give you guidelines for digging trenches and locating the form walls that will contain and shape your foundation.

Excavation of trenches for footings.

Construction of form walls. This includes bracing, leveling, and otherwise securing these walls.

Inspection of forms and reinforcing bar by the building department if you are building to code and with a permit.

Pour. This is the actual pouring of the wet concrete into the footing trenches and forms.

Curing. A period of time must elapse before the form walls are removed from the hardened concrete.

Stripping (removal) of forms.

Backfill of gravel and earth around the footing and lower sections of foundation walls.

The footing and the lower part of the foundation wall are dug into the ground to protect them from the movement of the earth when it freezes and thaws. In cold areas, the base of the footing may be 3′ or 4′ down. In warmer areas, the base of the footing is usually set into the ground a minimum of 18″ or 12″. The depth of your footing will depend upon your geographical area, and local building codes will specify the depth considered minimum (safe) for a stable foundation. Whether or not you are building to code, you should take these minimum depths into consideration. If your footings are too shallow, the movement of the earth might crack or shift your foundation. Besides considering the local frost line (depth to which the ground freezes in winter), you should plan to dig footing trenches down to solid soil. You will want your house to rest on a stable base, so any loose fill or pockets of loose earth should be dug out.

The thickness and width of a footing depend upon the size of the structure and the stability of the soil it rests upon. In general, a footing should be a minimum of 6″ thick and 12″ wide for a one-story structure. For a two-story structure, the footing must be a minimum of 8″ thick and 16″ wide. A general rule of thumb is that a footing should be computed according to the size of the foundation wall it supports: the width of the footing should equal twice the thickness of the foundation wall; the thickness of the footing should be equal to the thickness of the foundation wall; the depth of the footing should be 12″ to 18″ minimum, or well below the frost line.

To exactly locate the position of your footing trenches, you can use the building lines you established during layout. You have to consider the thickness of your foundation walls. Generally, a one-story structure has at least 6″-thick foundation walls; a two-story structure has 8″-thick foundation walls. The building lines you placed during layout show you the *exterior perimeter* of your foundation wall. By measuring in 6″ or 8″ (or whatever) on your batter boards and running a second set of strings, you can locate the *interior perimeter* of your foundation walls. Now you can drop your plumb bob at points every 4′ or so along both sets of strings and, depending upon how wide your footing will be, establish where to dig your trenches. For example, if you are planning a 6″-wide foundation wall with a 12″-wide footing, the footing will extend 3″ beyond the wall (your strings) on each side. Locating and digging trenches for your footings is not precision work, but you'll save time, energy, and concrete if you are fairly accurate. If you are planning to have a backhoe (excavating machine) come in to dig your trenches, you'll have to mark the trench perimeter lines with powdered chalk, powdered lime, or some other temporary system. You may want to do this for your own digging so that you can take down the building lines while you work. An alternative to hand-digging your trenches that is not quite as expensive as hiring a backhoe operator is to rent a trench-digging machine, which you can run yourself.

If you are digging your trenches by hand, try to make the sides of the trenches as straight up and down as possible to conserve concrete. Try to disturb the soil beyond your actual trench as little as possible. A square-point shovel is a good tool for this work, although you may have to break up the ground first with a mattock or pickax. All soil should be removed from the trenches and carried *away* from the area. Don't be tempted to heap it up in the center of the building or too close by. It will get in your way later on and is easier to deal with at this stage.

If your house or structure is wider than 16′, you'll probably want to plan the use of a girder or set of girders positioned so that they will break the span of your floor joists. This will allow you to use floor joists of reasonable and common sizes, rather than trying to plan for the enormous material that could span 20 or 30′. Any girder which you plan to use must be exactly located at this early stage because you will want the ends of the girder to be "keyed in" to your foundation wall. You'll want the top edge of the girder to sit exactly flush with the top surface of your mudsill, so you'll have to calculate out how deep the key or slot for the girder must be made. A common-sized girder for this job would be a 4×6; this piece, set on edge, can span about 7′. With a 7′ safe span, the 4×6 girder will need a supporting post (with pier) every 7′ or closer. Each end of the girder will need a slot in the foundation wall 4″ wide (this allows a small air space on each side of the girder) and 4″ deep (this takes into account the actual 5½″ dimension of the girder minus the 1½″ thickness of the sill plate you are matching to). The slot should go at least 4″ into the foundation wall to give the girder end an adequate "seat." You should locate the exact position of each girder and build little wooden boxes into your foundation wall to make the slots or keys for girder ends.

If you are building on a sloping site, you may have to construct a *stepped-off* foundation. Here the footing and foundation wall are made in sections conforming to the slope of the site. The sections are dug out at different heights and connected by short vertical and horizontal walls (the steps). For details on doing this, see *How to Build a Wood-Frame House* (see "Resources" section).

If the soil you are building on is very loose or sandy, you may have to construct form walls for your footings. You can follow the procedure discussed further on, adapting it to your needs. Usually footings are simply poured into the trenches; the sides of the trenches create a natural form wall.

The next step in making this foundation is to construct your form walls. Like the form boxes discussed in the preceding section, these are simply temporary wooden molds that will hold the concrete while it hardens and then will be removed. Concrete is a mixture of cement, sand, gravel (aggregate), and water, poured in a liquid state. It must be contained until it sets up or hardens—a process called hydration. Form walls are built of wood and must be constructed to contain the concrete exactly in the shape desired. In its liquid state, concrete exerts tremendous pressure, so that the form walls containing it must be very well braced and strongly built. Material used in constructing forms includes the lumber for the walls themselves, material for bracing the walls, and metal fasteners or tie wires used in holding the walls in proper alignment. Material for the walls may be 1×, 2×, or plywood. This material may be reused later in the house, so try to plan it into your building. If you use 1× or 2×, try to use widths that will give you as few horizontal seams as possible. Your choice will be made partly according to the height of your walls (foundation walls should be a minimum of 6″ above grade, but can vary greatly according to your site requirements, the crawl space you want, the design of the house, and so forth), and partly by what you can integrate into the building as useful material: 1× can be reused for sheathing, 2× for subflooring or structural material. Wood used for forms will be slightly damaged, usually containing nail holes, some discoloration from the concrete, and so on, so you will probably want to use it as "unexposed" material later on. Both 1× or 2× material should be square-edged; don't use tongue-and-groove or shiplap-type material, as you'll have problems with concrete oozing into the joints and this material will also be more difficult to strip down (remove from the finished concrete wall). The lengths you choose should again be determined by future uses of the material.

Your form walls will have to be constructed in sections that will be spliced together for a continuous wall. Plywood (½″ or, preferably, ¾″) is an excellent material for form wood as it is smooth, strong, and can be easily cut to size with few, if any, horizontal joints. It is ideal for building difficult forms: for example, those that must conform to sloping, uneven areas. The only criticism I have to offer is that it is hard to reuse cut-up pieces of plywood. If you can reuse it, it's a pleasure to work with. Whether you choose

(Photo by Jeanne Tetrault)

1× or 2× material or plywood, try to buy a fairly good grade of wood with as few defects (knots, splits, etc.) as is economical. Avoid material that is badly warped or cupped; these problems will make your form building much more difficult than need be and make it hard to create a really level, plumb wall.

There are numerous ways to build and secure form walls, and you'll probably find yourself evolving your own preferences and methods. The following is intended to provide basic information and suggestions. For other ideas and explanations, see *Against the Grain* and some of the general carpentry books listed in the "Resources" section. What is most important in building form walls is to make them strong, plumb and level, exactly located, and clean. As metal bonds with concrete, all nails should be placed so that heads and points do not protrude *inside* the forms. Rough edges will leave impressions on the concrete, and inaccurate placings will make your walls uneven. Because you will later be dismantling all of your work, it's a good idea to use double-headed nails, which are easy to pull out and will make stripping the forms a simple job. As you work, try to avoid kicking dirt and/or debris into your trenches—all of this has to be removed before you can pour, and it's a good idea to keep everything clean as you go.

Form walls are constructed in parallel pairs, with the distance between the two members exactly equal to the thickness of your foundation wall. There are two methods of building forms to the right height: one is to build them slightly higher than you want your foundation wall to be; the second is to build them exactly to the desired height. I can see little advantage to the first method, as you must then run a ledger or marker around inside the form walls to show you where the concrete should be poured and stopped. If you build the walls to the exact height, you can simply fill them to the top and then easily float (smooth) or finish off the surface. The components of form building are:

Walls These are built of 1× or 2× material or plywood and are held in place by braces. If you choose to pour your footing and foundation wall as one unit, the walls are held more or less suspended in the footing trenches by their braces. A second method is to pour the footing first and then build the foundation forms anchored (temporarily) to the footing. This is explained in *Against the Grain.* The advantage of pouring the footing and wall as one unit is that you can do all the pouring at once, which is important if you are having concrete delivered ready-mixed by a truck. The form wall is built in sections that are spliced together. It should have a continuously level top edge, and the sides should be plumb.

Studs or Stiffbacks Vertical 2×4's are nailed to the wall every 2′ or so. If you are using 1× or 2× material, these studs will act as vertical splices holding your wall together.

Walers These are optional horizontal stiffeners, usually added to high (4′ or more) form walls. They are lengths of 2×2 or 2×4 material placed every 2′ up the wall and running continuously around the entire wall. Good straight material should be used for walers.

Bracing Bracing is added to hold the form wall upright in a plumb position, and to help resist the outward pressure of the freshly poured concrete. How intensively you have to brace your wall depends upon its height; higher walls obviously need more bracing. Normally you can brace the wall every 4′ or so, making use of the stud/stiffback at each location. A triangle is one of the strongest bracing configurations. Using the stud as one leg of the triangle, you can add a piece of 2×2 or 2×4 at the bottom, at a right angle to the stud (see photographs). This piece may be nailed securely in place. The third and longest leg of the triangle may be made with a piece of 1×4 angle-cut at each end to fit the situation. A longer bracing system (a triangle with long legs) will be stronger than a shorter one, so if you are bracing a tall form wall, you may want braces that run out 4′ or 5′. Before nailing the 1×4 in place, check to see that your wall is plumb at this point.

The bracing system will help to hold the wall level as well as plumb, but as you work you may also need to set temporary stakes in place to hold the wall. Metal form stakes are excellent for this purpose. They may be rented at most tool and equipment rental places for pennies a day. They are metal stakes with holes drilled for placing nails through them. They are light, strong, and easy to work with. Metal form stakes come in several lengths, so you can choose those most suitable for your particular project. After you nail your 1×4 piece in place, drive a form stake (wooden or metal) at the point where the 1×4 and the lower leg of the triangle are joined, and tack the brace securely to this stake. This should hold your wall rigid.

If you rent metal form stakes for your work, be sure to rent a form stake puller at this same time; as the name implies, this is a tool especially made for pulling out the stakes. You can pull them out by hand, but it's not easy. Also try to pick out stakes that aren't "mushroomed" at the top (the edges not splayed out). Mushroomed tops may chip and send pieces of metal flying when you drive the stake in. You can file off the splayed edges if you want, but to save time, choose stakes with clean, solid heads to begin with. Another thing to watch for is the bent stake; this will be hard to work with and is best avoided.

Joints, where one section of form wall joins another, are a weak point in the wall and must be carefully braced. If they aren't, the wet concrete may push the form wall out at these points, causing the wall to zigzag or bow out. This is especially critical if your walls are high. Another problem area is a corner; this should be extra well braced, too, as it receives considerable pressure from the concrete.

If you are going to be doing a lot of form building, you may want to invest in sets of metal bracers/aligners which take the place of wooden bracing systems. These are manufactured and sold by companies like Burke (see "Resources").

Ties and Spreaders Many devices and systems have been developed for securing parallel form walls in place opposite one another. As the pressure of the wet concrete tends to push the walls outward, devices are made to tie or hold the walls together. Spreaders are used to keep the walls aligned an accurate, exact distance apart. The old system for building form walls made use of tie wires and wooden spreaders. The spreaders are pieces of 2×4 cut to the exact thickness of the form wall. They are placed between the walls every few feet. At or near each spreader, a piece of tie wire (this may be purchased in small coils or rolls at most lumber stores) is threaded through the wall. The tie wire passes through the first wall, goes through the second wall and is wrapped around the stud or a waler, is passed through again, and threaded back through the first wall. The two ends of the wire can then be twisted together. By twisting tightly, you draw the two walls toward one another and wedge the spreader in place. The tie wires are left in place when you pour in the concrete; the wooden spreaders must be pulled out just before the pour.

Newer systems of ties and spreaders employ metal spreaders (which may be left in place) and special metal form ties or rods. The form ties or rods take the place of the tie wires and may even act as spreaders. They are easily put in place. Some form ties are slipped between the horizontal joints in the form walls (if you're using 1× or 2×); they are held in place by wedges, cones, or other devices that are fitted in place outside the form wall and hold the tie tightly in place. After the concrete has been poured and allowed to harden, the wedges or cones are removed and the ends of the ties snapped or twisted off to leave a clean surface. Form ties must be bought to size: for example, for an 8″- or a 6″-wide wall. Form rods are simply ½″ steel rods threaded at each end. You have to drill to place them. They are held in place with washers and nuts and act to spread and tie the wall simultaneously. Again, the protruding ends of the metal ties are broken or cut off once the concrete has hardened.

Metal form ties, spreaders, and rods can be purchased from some building supply and lumber stores, but may be a "specialty" item you'll have to search for. Try the Yellow Pages of your phone book and look for a concrete company that might sell concrete forming accessories. You can usually buy ties and other items by the piece, rather than having to buy a full, unopened box; they are not expensive and can make your work go much faster.

Rebar Rebar or reinforcing rod is placed in the forms as you work because it must usually be placed at certain levels within the forms and tie wires or form ties may be in the way if you wait until the end to put

(Photo by Jeanne Tetrault)

the rebar in. How much rebar you use and where you place it depends upon the size of your walls, the structure you're building, and like considerations. Here again, you may want to (or have to) check out building code requirements. A typical requirement would be: "Two #4 (½″) or #5 (⅝″) reinforcing rods running horizontally every 18″ the length of the footings and walls." Rebar is purchased from the lumber company; it may be cut to length with a hacksaw or with a metal-cutting blade in your jigsaw or reciprocating saw. Don't buy rusty rebar—the concrete won't bond properly with it—and don't leave rebar lying around outside where it can become rusty.

Rebar is usually placed in sets of two rods at each level, and these are set a few inches apart. The first pair should be suspended off the ground in the footing trench. You can use tie wire to hold the rebar, or buy special little concrete blocks to hold this first set. These blocks are about 2″ to 3″ square, with two wires protruding from the top. Locally, they're called "adobes." They are simply set in place along the bottom of the trench; the rebar is laid across them, and the wires twisted around the rebar to hold it. Where you have to "join" pieces of rebar to make a continuous piece, you simply overlap a foot or two. The next (higher-up) sets of rebar may be hung in place with tie wire or rested on form ties or form rods.

Vents, Pipes, Etc. If you're going to place metal ventilator screens in your walls, need sections for pipes to pass through, and so on, you can construct little wooden boxes or tubes, which are placed in the form walls where you need them. Later these boxes may be knocked out, leaving the hole through the poured concrete wall.

Sill or Anchor Bolts To attach the wooden structure of your house to the foundation wall, you'll be using special metal anchor or sill bolts. These must be embedded into the wet concrete with their threaded end protruding enough so that the sill (usually 2×6) may be fastened in place. Half-inch sill bolts are a standard size; these must be long enough that at least 7″ can be embedded in the concrete. They are usually spaced every 6′, with one placed within 12″ of every corner. Placing the sill bolts carefully is important so that when you come to attach the sill plate, your connections will work; they should not be embedded at an angle, or too deep or too shallow. You can place the sill bolts directly into the wet concrete, but an easier, more accurate method is to place them with simple ledgers when you build your form walls. A piece of 2×4 can be cut long enough to span from one wall to its neighbor, resting on the top edge or surface of the walls. Drill a hole for the sill bolt and slip it into the 2×4, with its nut and washer to hold it loosely in place. Using these ledgers, you can place all of your sill bolts where you want them and be assured that they will be perfectly in position. The 2×4's can be tacked in place. After the pour and setting up, the 2×4's can be removed and the sill bolts will be protruding to exactly the right height.

When your form wall is built, your rebar in place, and so on, you are ready to estimate and order the concrete you'll need, or buy materials if you plan to mix your own. As mentioned earlier, concrete is a mixture of materials. The ratio of that mixture is critical as it will affect the workability of the mix, the way it sets up or hardens, the strength of the finished wall, and so on. A mix ratio may be expressed as: 1:2:4, which means 1 part cement, 2 parts sand, 4 parts gravel. This is a common mix for foundation walls. For footings, the mix is usually 1:3.75:5. If you plan to mix your own concrete, be sure to explain exactly what you need to the yard people where you buy your materials. The size of the gravel will affect the properties of your finished concrete. All of the materials you use should be clean if the mix is to bond properly. You will probably want to rent a small mixer to get your concrete thoroughly and properly mixed. If you are ordering your concrete as ready-mix from a concrete company, you'll have to figure out your needs in cubic yards (which is how concrete is sold). A cubic yard equals 27 cubic feet. You can estimate what you need by first converting all of your measurements to feet and fractions of feet—for instance, 8″ should appear as ⅔ foot. Take the total length of your footing and multiply it times the thickness and times the width. Divide this figure by 27 and you have the cubic yards of concrete needed to fill your footing. Now do the same for your foundation wall:

$$\text{cubic yards of concrete needed} = \frac{\text{width} \times \text{length} \times \text{thickness}}{27}$$

You should add about 10 percent overall for waste and for variations in your actual needs, so that what you will order will be: total amount needed for your footing + total amount needed for your foundation wall + 10 percent. If you are going to run piers and girders to support your floor joists midspan (more on this in "Building with Piers" and "Posts and Girders" sections), be sure to add in some for your footings and piers.

Most concrete trucks carry a maximum of 5 cubic yards of concrete. You can usually order fractional loads (¼, ½, etc.) at slight extra cost. You may be asked what "slump" you want. A slump test is one used to measure the consistency of the concrete mix—the concrete is put into a special cone, stirred, and then the cone is removed. The slump refers to the relative consistency of the particular mix, which is determined by measuring its outward spread under these conditions. Unless you decide to become a concrete expert, you should be content to answer that you just want the right slump for the job—in this instance, for the footing and form walls of a house. Also, if your building site is difficult to reach, you should know that most concrete trucks can be outfitted with a special pump and a long, flexible hose to reach hard-of-access areas. The hose is valuable to know about if you are working on an existing house, replacing sections of foundation, or even adding a room around in back. There is a charge for the use of the pump and hose, but it may be necessary and is not exorbitant. Normally the concrete is poured down a chute into the forms. Special systems of chutes may have to be rigged up to reach all areas of your forms.

There are a couple of things you may want to do the day before the pour. Some people like to oil the inner surfaces of the forms to make them easy to strip (remove) later. Concrete will not bond to oiled surfaces, so this may be very helpful. You can use any really cheap motor oil. Be very cautious, though, when applying the oil, because if you splash it on your rebar or form ties, *these* won't bond properly with the concrete. Another suggestion is to soak the forms themselves about twelve hours before the pour so that the wood won't absorb a lot of water from the concrete (and change its ratio and ability to set up), and so that the forms themselves won't swell and distort.

Another thing to do before the pour is to organize your work crew and decide how you want to distribute and handle the work. A pour takes place very quickly and there are (it seems) hundreds of things to do all at once. You are working under the literal pressure of the concrete pouring out of the chute, and the time limits of having to take care of certain jobs, such as finishing the surface of the concrete, before the concrete begins to set up. Depending upon the size of the foundation you are pouring, you'll probably want to work with at least three and up to five or six people. The jobs will be:

Handling the chute itself. This means guiding it into the forms, signaling the truck driver when to let the load come down and when to stop, and so on. It's pretty heavy work, so you'll probably want to take turns at it. If you're using a hose, this is even heavier work (it helps to carry the hose braced over your shoulder). With a hose, you may need one person as an intermediary between the woman handling the hose and the truck.

Consolidating the concrete. As the concrete pours into the forms, it will be full of air bubbles and pockets. *Consolidating* means working the concrete to smooth it out and to eliminate these air pockets. This can be done by hand, using a wooden stick, which you plunge up and down into the concrete. A good long piece of doweling works well because it is smooth and easy to handle, but it should be fairly thick. The woman or women doing

(Photo by Jeanne Tetrault)

this job will have to move along pretty quickly to reach all areas of a large foundation. There is a tool for this job, the mechanical vibrator, which you might want to consider renting, but unless you really know how to handle and judge it, you risk overvibrating the mix and disturbing its balance. I've been part of the pour crew for a couple of good-sized jobs and we've gotten along perfectly well plunging and consolidating by hand.

Moving the concrete. You may need one person with a shovel, whose work will be to push the concrete along in the forms, spreading it beyond the reach of the chute. The chute is heavy and awkward to handle, so a hand shovel comes along well as a backup. If you're pushing and shoveling during the pour, you'll have a special appreciation for well-built, sturdily braced forms.

Finishing the surface. The top surface of the concrete will have to be smoothed and finished if you are to have a good surface to attach your sill to. The first step when a section of wall has been filled with concrete is *screeding*—using a piece of 1× or 2× lumber to push off the excess concrete. You hold the board upright and firmly against the top edges of the form wall and use it as a kind of scraper. The board is called a *strike-off board;* the work is simple and will be obvious once you're on the job. After screeding, a process called *floating* must be done. Here you use a metal or wooden float to smooth the top surface further. A float is simply a flat surface with a handle on top. Floating must be done quickly, while the concrete is still plastic. Don't overdo it, or you'll pull too much water to the surface. Again, the job is not too difficult if you keep in mind that what you want is a nice smooth surface. A steel float will give the very smoothest finish.

Troweling. This is a final step to give the absolute smooth, dense finish. The tool is a specialized one: a metal trowel with offset handle.

In planning how to handle the work, remember that each person can cover more than one job; the chute handler can later be the one to do the screeding or troweling. You'll probably find yourself changing some of your plans as the actual pour takes place and you see that you need an extra person at one job and fewer at another. It helps to have things fairly organized, though. Another important matter is clothing. Wear your least favorite rags and a pair of old boots—it's not unusual to find yourself more or less wading in and bathed in concrete! Work gloves are a good idea, as concrete tends to dry out your hands and can even cause burns.

Final details once the pour is under way include

(Photo by Jeanne Tetrault)

As we learn more and more about designing houses that make use of natural solar energy, conventional foundation and flooring systems will be altered. It may well be that most houses will be built with massive slab (concrete or concrete over crushed rock) floors that can act as solar collectors. These slabs will have to be placed in rooms with large areas of glass that allow sunlight to strike the slab floor during the day; at night, the glass window areas must be covered and the slabs will release heat stored up during the day. We may find ourselves designing and building houses with slabs in certain areas and conventional foundations and wood floors in other areas. If you build with a continuous foundation now, you can easily pour a slab within the perimeter or within specified areas later. See the "Resources" section for books about solar energy designing, or check your library or bookstore for recent publications.

making sure that any metal form stakes that have been driven in to hold walls temporarily and are now becoming embedded in concrete are *removed.* Those holding your bracing are far enough away from the concrete to be safe. Don't forget to pull out any wooden spreaders. When the concrete is all poured in, one person should go around with a shovel clearing away any excess concrete from the bottom parts of the form walls; this will make it easier to strip the wood later. You should be able to see the lower edges of the wood. As you finish with your tools, stack them in one place so that you can wash them down as soon as the work is finished. It's much easier to wash off wet concrete than to chip away dry concrete! Form stakes, likewise, should be washed off when you're through with them.

Forms must be left in place for four to seven days, unless you've used a special quick-set additive and can take them off in a couple of days. The forms and concrete must be kept damp throughout this period, which means spraying them down with a fine mist once or twice a day. If the weather is really hot, you may want to cover the walls with wet burlap or special paper (available from some concrete accessory companies) to keep the concrete damp. Letting the concrete set up too fast will result in a weak foundation. The final step in your foundation work is to strip the forms once the concrete has cured well. This means undoing all of your work, and a Wonder Bar and pry bar will be the tools of choice. All of the wood that you plan to salvage should be denailed at this point and thoroughly scrubbed down. A wire brush helps to loosen the concrete and won't damage the wood too much if you're careful. If you've used form ties, this is the time to break off protruding wedges, or cones. Once the form wood has been removed, you should give the concrete walls a light spraying once a day for two or three days. Again, the weather will have an influence upon how much effort you have to put out. Concrete doesn't reach its full strength for about twenty-eight days, but if you've waited patiently for your walls to cure for five or six days, you can safely move on to setting your sill plates and beginning your framing. This work, following the slow and tedious labors of layout and form building, will actually feel like flying!

Star Building

Every house at Cabbage Lane is a schoolhouse, a house of learning, a place constructed by students—women learning how to build, inspired by their dreams, needs, and one another. Every bent nail, pounded flat, is part of some woman's journey to knowledge of self-sufficiency. That drafty window was probably the first that woman ever hung, and a little gap was minor compared to the feat accomplished. Women come, learn, continue to come; we build, repair, learning from the mistakes of sisters before us, elaborating dreams and creating new ones. Layer upon layer, each structure is built with womanpower and woman spirit; each house is begun by some, helped by many, finished by others, repaired by still more. . . .

The five-sided star cabin stands on a small hill overlooking the garden. It marks the entrance to the canyon, and the view from this cabin encompasses both the canyon, stretching back into secrecy, and the hills that roll down to the road, to the small town, to the freeway connecting up with the world beyond. A link between these two realities, the cabin was built as a place where a woman might rest; it is a safe space, a quiet and comforting haven where one can nurture dreams and gather strength.

The star cabin is so called because of the five-pointed star that forms the ceiling of the cabin and supports the roof. This star is constructed of poles—trees felled in the nearby forest. Stripped of their bark, the poles are a soft golden brown. Their bold strength and clean lines create a sense of safety and openness. The pole framing of the roof forms an expansive star stretching over the cabin's inner space. When you lie on the bed inside, the convex star draws you up and into her vortex, the queenpin. Lightness and ease are in the movement, yet the physical structure is heavy and solid. The work to raise this ceiling and roof came from many women, all just learning to build. It was a challenge that taught us that we could reach beyond the limits of what we'd known and done—that we could imagine, design, and build a structure as graceful as it is sound.

The star cabin was built primarily with hand tools. There was no electricity on the land where it was built, so most cutting was done by hand (many women learned for the first time how to use a handsaw). A nearby neighbor's power-equipped workshop allowed us to do some cutting (of plywood sheets and compound angles) with power tools. There was not much money available for this building, so we improvised as much as possible, using native poles and scavenging for recycled materials. The hand tools we used were a basic collection of saws, hammers, squares, and levels. We learned never to underestimate the value of a *sharp* saw, and we came to appreciate the versatility of simple hand tools.

The foundation for the star consists of 8″×8″ blocks of creosote-treated wood set on top of pads of poured concrete. After laying out the basic shape with line level, strings, and batter boards, we dug holes for each of these foundation supports. The holes were filled with wet concrete. Into the concrete we placed pieces of vertical rebar (one per hole, centered, and left extending out about 6″). After pouring the concrete, we kept the pads covered with wet burlap, allowing them to dry slowly, protected from the hot sun. The next step was to drill a hole into the bottom (center) of each wooden block. Each block was then fit over a piece of rebar, thus resting on and secured to a concrete pad. With these in place, 2×6 floor joists were stretched out from the center block like spokes of a wheel to each corner. Also, 2×6's were laid from corner to corner, forming the outer perimeter of the cabin. Additional pieces of 2×6 were nailed in between the "spokes" as blocking. Across the joists we stapled tar paper and cardboard to form a layer of insulation for the floor. Above this came the floor boards, 2×6 tongue-and-groove, salvaged from a local sawmill and sanded smooth.

In framing the walls, we used traditional stud framing, 2×4's spaced 16″ on center, with top and bottom plates. The trickiest part of the wall framing was in the five corners where the walls met. These corner studs had to be cut at an angle so that they would meet up flush with each other. The angle cut was figured out beforehand, and the cutting was done on a neighbor's table saw (ripping these studs on an angle by hand would have been nearly—or virtually—impossible to do). All of the wood for the floor framing, the floor, and the wall framing was milled lumber (used). When we came to the roof, we decided to use poles from the forest. We could have them for the labor of cutting and stripping, and their beauty and strength would last a long time.

The construction of the roof was the most difficult and interesting part of the building. First in the construction of this roof was the raising of the queenpin, a five-sided block of wood that forms the center of the star. Our idea was a roof that would support itself through the interlocking pressure of rafter poles centering in, fastened into, the queenpin. This would enable us to have an open interior space, with no distraction of upright support poles. The concept of raising this queenpin high in the air, and then attaching the roof poles, seemed outrageous in the beginning, but the idea of raising it on a tall tripod that could later be removed seemed a reasonable inspiration. The tripod would be a temporary support structure holding up the queenpin at the desired height. One end of each roof pole would be fastened to the queenpin; the other end would extend over the wall plate, forming the overhang of the roof. Each pole would have to be notched carefully to fit snugly onto the wall plate and be nailed in place. When the roof framing was finished, the tripod would be removed and the queenpin and roof poles would form a whole sitting somewhat like a Chinese hat atop the five-sided cabin.

"This is never going to work—we're just guessing."

"No, no. It's more than guessing. Look, it's got to work. We'll nail three legs to the sides of the queenpin and raise her up there. Then we'll bolt the roof poles to the pin and to the top of the wall plate. Presto!"

"Then we detach the legs and the whole thing stays up there? Do you really believe that?"

"Where can it go?"

"Down on our heads! Or it will push the walls out—"

"It can't come down—the poles can't fall into the center. It would have to push out at the walls. . . ."

"But Nelly went to look at the hexagon. They used a cable—we could run one through the five roof poles and that would draw them together. They couldn't split out at the walls."

"Well, part of me believes in it—and a whole lot of me doesn't!"

"There's only one way to find out—let's do it."

So we worked on. We cut the queenpin, a piece 16″ long and 5½″ wide, on a neighbor's table saw. This transformed a four-sided block of wood into a five-sided pin for us. Next we had to figure out the length of the tripod legs, according to the desired height and pitch of the roof. Figuring that the queenpin should extend 8″ above the tripod legs, we subtracted 8″ from the desired height of the ceiling. This figure equaled the measurement of a straight line from mid-queenpin to floor. This straight line meeting the floor formed a right angle. A tripod leg going from queenpin to floor completed a right triangle. The length of a tripod leg was thus equal to the length of the hypotenuse *(c)* of the right triangle: $c^2 = a^2 + b^2$. Having acquired this measurement, we then cut three 2×6 boards the proper length, allowing extra for angle cuts where the foot of each leg would meet the floor. These three legs were nailed to the queenpin. We left the heads of the nails protruding so that they could be easily removed when we were ready to dismantle the temporary support structure.

One evening near dusk, we raised the queenpin, center of the roof to be. Balancing the tripod legs, lining them up so that the homemade plumb bob hanging from the center of the queenpin was dead center to the cabin floor, we secured our tripod by nailing blocks of wood around the legs. Just as we'd imagined, the queenpin was now suspended high above our heads. We gathered together, arms around each other, staring up at our work. Tomorrow we would try our first roof pole. We were actually doing it! There was not an experienced carpenter among us—we were learning to build by building! We were using hand tools, some homemade like our plumb bob (a V door hinge hung from string). Suddenly the possibility of our own success caused us to waver. . . .

"Well, she's up there. I still wonder what happens when we take the legs down. . . ."

We grinned at one another, uncertain, glad for the cover of the nearly dark night.

"Guess we'll just have to find out."

The daring was rushing our blood.

That night was full of dreams about rafter poles, angles, formulas. Our next task was to figure out how to cut the poles so that they would meet flush with the queenpin. There were no books that we knew of that addressed this sort of problem, and no one we knew who had built a structure like this one. The women who built the hexagon had faced similar problems, but our solutions for our five-sided structure would have to be specific. We knew that we would just have to go on figuring things out as we came to them. Because the pole (rafter) formed the pitch of the roof, we knew the cut that joined it to the queenpin would have to be angled. Before raising the queenpin, we had nailed to each of its five sides a block of wood that would

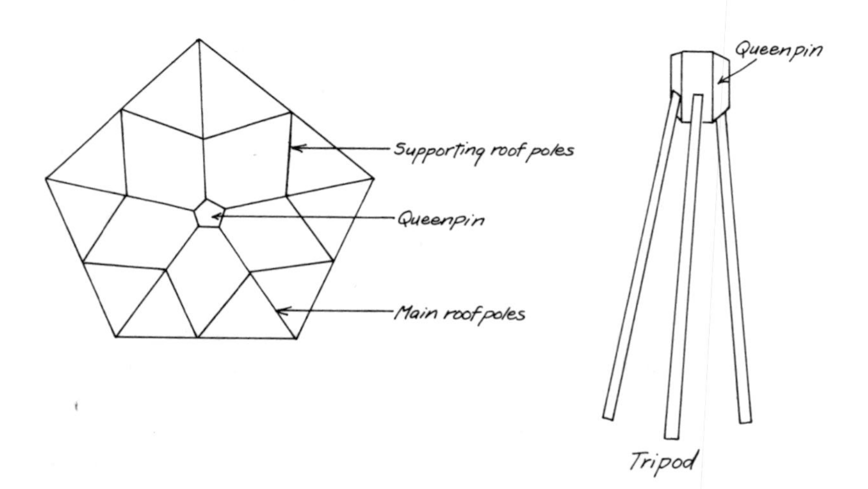

act as a ledger to support the rafter pole before it was fastened to the pin. These ledgers would be removed when the rafter poles were secured in place. To determine the angle of the cut, we stretched a piece of string taut from the top of the support block/ledger to the inside edge of the top of the wall plate. This string simulated the *bottom* side of a rafter pole actually in place. We then held a square with one edge on the string, the right-angle corner touching the queenpin, and the other edge extending above the string. We measured up the square's extended edge 3½″ (the average diameter of the poles). At this point we held a straightedge parallel to the

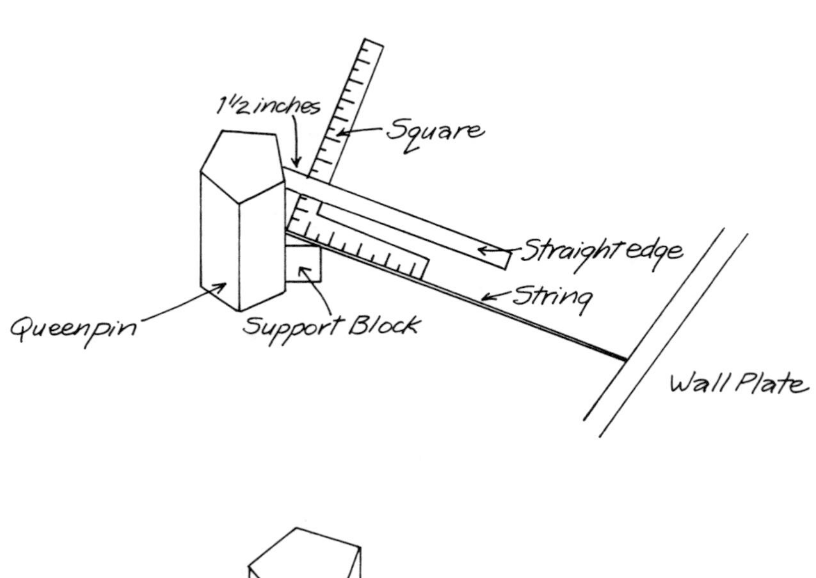

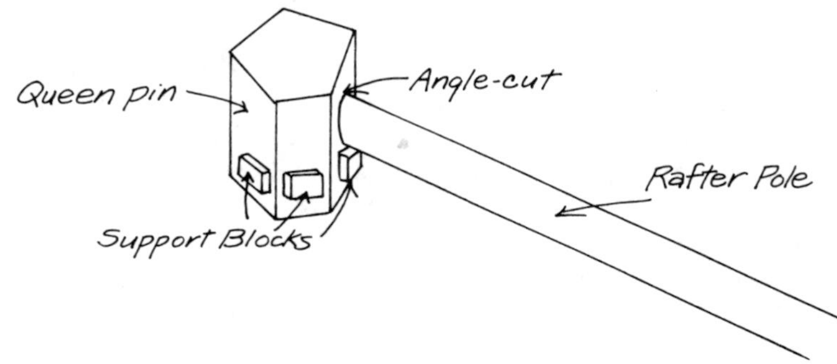

(Photo by La Rosa Ponderosa)

string and measured from the point to the queenpin. It came to 1½". This measurement was the additional length that the pole had to be on its *top* side in order to meet the queenpin. We knew the angle cut this measurement gave us would have to be approximate, as all of our poles were not exactly the same size. But it gave us a place to begin. Using this information, we then built a miter box specifically constructed to cut our poles at that desired angle. Our miter box did not have sides of equal length; instead, one side was made 1½" shorter than the other. When a pole was placed in the box and the unequal edges were used as a saw guide, the resultant cut gave us the desired angle.

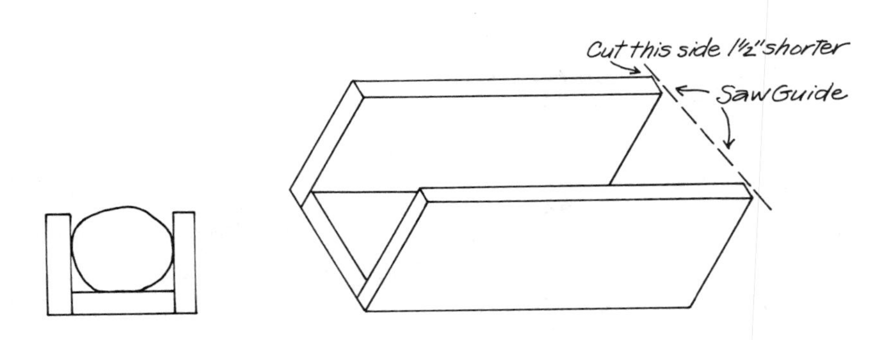

Poles are not always perfectly straight: they have bumps and knots, variations in diameter, and often slight curves and crooks. When a pole is being placed horizontally or on an angle, any curve should extend to the side, not up or down. In using the pole this way, you take advantage of its greatest strength and flattest surface. As we cut each pole, we first determined how it bowed or curved and how it should be placed as a rafter. When we placed it in the miter box to cut, we made sure that it was lying correctly. We cut an angle on each end of each pole; one end would butt into the queenpin, and the other would create the overhang of the roof. We made our overhang cut so that the angle was parallel to the angle resting against the queenpin.

Next we positioned a pole with one end on the support block attached to the queenpin and the other end extending over the wall plate. The angle cut was not sitting flush against the queenpin because we still had to cut the notch to seat the pole snugly on top of the wall plate. With the pole in position, we marked for this notch by eye, allowing leeway for the pole to fall down into place when the notch was made. We chiseled away a bit at a time on this first pole, wanting to be sure that we didn't overcut. That first pole got raised and lowered more times than any of us cared to count! As we came closer to the best fit on the wall plate, we had to return to the queenpin end and do some rasping. Back and forth we went, from queenpin to wall plate, rasping and chiseling until the pole fit as snugly as we could make it. With the pole in its final position, we marked the queenpin end for a hole that we would hand-drill through. A heavy 6" lag screw would be used here to attach the pole to the pin.

Once more we took the pole down and drilled this hole as planned. Then we measured 26" from the queenpin end (shorter side of the angle cut) and hand-drilled another hole sideways through the pole. We would later pass a cable through the holes on the five poles and secure the cable ends together. This cable would provide an additional security to the roof structure. A piece of metal tubing was placed inside each cable hole to protect the wood from the tension and wear it would be subjected to. When this pole was ready, we raised it up for the last time. We used

(Photo by La Rosa Ponderosa)

the hole in the queenpin end as a guide to drill down into the pin, then used a ratchet to turn the lag screw down in place. The notched end of the pole was nailed into place on the wall plate. One pole secure! The roof was on its way!

That first pole was the hardest. From there we moved with growing confidence through the other four poles, each fit getting a little better, trusting our eyes and our hands and our tools a little more. Finally all of the poles were up and secured in place. The tripod still stood, supporting the queenpin. Skepticism and optimism about removing it raged. But first the cable was to be put in. The cable had been purchased, a long, oily, heavy metal snake, coiled around itself. We unwound it and passed it through each hole in the rafter poles. One end was looped through a turnbuckle and then bolted to itself three times, using U-bolts. We would need to draw the cable as tight as possible before securing the two ends together. To facilitate this, we hooked one end of a come-along to the already looped end of the cable, and the other end of the come-along to a vise grip attached to the open end of the cable. We began closing the come-along until the cable seemed tight enough. Then another woman looped the loose cable end through the turnbuckle and began bolting the two thicknesses of cable together.

It was a late afternoon; new women were working with us. They had come for a few days. We were laughing and joking at so many of our heads and hands being up there, near the queenpin, holding on to the come-along, rafters, ladders, and the socket wrench used to tighten the bolts. We were singing, too:

Women who sing have strong faces,
Women who sing have strong faces.

Women who build have strong bodies,
Women who build have strong bodies. . . .

The job was almost complete. Forest, one of the new women, was tightening down the last bolt. "Don't tighten it too much or it will break," one of us advised. Forest looked over with a frown. She didn't say anything, but her face reflected years of messages we have all received: women can't build, aren't strong enough to handle heavy timbers, aren't meant to use tools. . . . We all felt a little jolt of response, our own message: we have to work extra hard to do the job. . . . Forest tightened with conviction. "This roof is going to stay up!" and crack! she broke the bolt in half. Shock leaped from her eyes. "Why it was all a bunch of lies! I can tighten this bolt enough to break it!" We all laughed.

Women who build have strong bodies,
Women who build don't know their own strength. . . .

The broken bolt was replaced after a quick trip to town. The come-along was removed, and the turnbuckle twisted a few more times to tighten itself. It was done. The roof framing was finished, the cable in place. It was hard to believe. Still the tripod remained. Should we take it down now?

"I'm scared."

"I'm *positive* it'll be all right. Those rafters aren't going to go anywhere."

Three hoots carried through the air, the call for dinner.

"Let's do it later."

So we left the site slowly, looking back, taking time to let it sink in—that we had built this structure, that we could do it.

After dinner another of our new helpers, Ruby, went back and knocked out the tripod legs. All remained steady and stable. As the night sky was losing its last light, some pink clouds still hovering overhead, some of us walking back along the canyon to our sleeping spots saw the silhouette of the star cabin in the evening light. The long legs supporting the queenpin were gone. Now the rafters were her strong and noble supports. Her crutch was gone. She stood on her own. We knew her strength and ours were one. We hooted and howled, gave thanks to the forest that surrounded for giving the trees that would give us this shelter. That night was full of dreams of floating through the sky, free and unsupported.

Now the five main roof poles were up. Would we do additional poles from the center of each wall to the center of the rafters, forming the star shape? The idea was beautiful. We could all see it with our mind's eye. But could we start a whole new series of angle cuts, each one demanding special attention because of the varying shapes of the poles . . . and the difficulties of matching pole to pole at the rafter ends . . . ?

"No, it's too much work. I can't take any more angles!"

"But it needs the support."

"Not that much, one per side will do."

"Then we wouldn't have the star shape."

"It sure would be beautiful."

"This cabin is going to be here a long time. Lots of women will live in it. Let's make it as beautiful as possible."

And so we began adding the star poles. Again the first one was a tedious process of estimated sawing, chiseling, and rasping. One angle had to be cut through the bottom of the pole, where it met the wall plate, and the other angle cut on the side of the pole, where it met the rafter pole. We just got that first pole up when all work came to a dead halt. Nelly, to whom the cabin was originally to belong, moved away from the canyon. The trauma of her departure and the threat of coming rains stopped the work. We turned our attention to other tasks: gathering and cutting winter firewood, finishing work on the main house, making money to see us through the winter. All winter the star cabin stood, complete only in its framing. The open walls and roof called for more work. But the dreamer behind this building had left, and another dream was yet to come.

Spring and summer brought another tumble of changes to the canyon. New women came to spend time. Some stayed, realizing they wanted to live here. Other women who had been here for a while moved on. It was a time of change, transition, and there wasn't the focus to start work again on the star cabin. But in all the speculation about winter living, the cabin was included. We joked about how we *just* had to throw up the roof and walls to have another living space. Then late summer brought with it the need to make firm decisions about winter living. Five women, two previous residents of the canyon and three new women, made a commitment to live together on the land through winter. In order to make that possible we would finish the star cabin.

So, once again, the crew of women gathered at the work site. Except for one woman, it was a totally new crew. Once again, most of the women had very little carpentry experience. Bookshelves and bed frames were the major past accomplishments. The new women began examining the pole construction with interest.

"Well, how exactly do you cut these angles?"

"Not exactly at all—you just start working on them."

"There must be an easier way."

"Whatever you can think of would help."

"Maybe we don't have to do two poles from each wall. One would probably be enough support."

"But then we wouldn't have the star. . . ."

And so, of course, the whole conversation was repeated about the value of the star. New faces, new voices, many of the same words, and in the end, the same decision. This woman house was to have a star on its ceiling!

One star pole was in place. Another had one end cut to fit the wall plate. We began with this one, working on its other end. We made all of these cuts by eye, gaining leeway by moving the pole slightly on both the wall plate and the rafter pole. After working for a few hours, two of the women dropped their tools in disgust and walked away.

"This is crazy! I can't figure it out!"

"I'm going to work in the garden."

"Listen, I know it's hard at first, but you can do it. It just takes time. We did it last year. You can, too. Take a break and try it later."

The next morning the women returned to the site, carrying new tools of determination and patience. The work moved slowly, but one by one the poles went up. Gradually the star of so many imaginations began to appear in material form. It was still an open star, leading up to a full sky.

These women worked on, and many more came to help. The crisp, fall days were a constant prod to move along with the work. After the roof framing was complete, we began nailing on the ceiling boards. In order to have the pole framing and the star still visible from the inside of the cabin, we nailed the ceiling boards on top of the framing. The boards inside the star were all going in one direction, and the rest of the ceiling boards in another. This accentuated the star shape.

We decided to use fiberglass insulation sandwiched between the ceiling boards and the outside roof, so we needed to build a frame that would hold up the outside roof and provide a gap sufficient for the insulation to lie in. We built this frame out of old 2×4's, building one section of the roof at a time. Each section was the shape of a triangle, with additional pieces inside the triangle for roof support. We nailed the framing to the poles and top plates.

With the framing complete, we cut sheets of ½″ plywood for the roof itself. This cutting process involved passing the plywood up and down a few times to get a good fit. Only when the plywood was ready to nail down did we cut and put in place the fiberglass insulation. We tried to work with and be exposed to the fiber glass as little as possible. With the fiberglass in place, we nailed down the plywood boards. On top of the plywood we used 90-pound roll roofing as our final layer, cut to the shapes needed and nailed in place.

While this roof work was being done by some, other women were hinging windows and working on the wall insulation and siding. For insulation, a layer of single-sided builder's foil was stapled over the inside of the wall framing, shiny side facing in. Another layer was stapled over the outside of the wall framing, shiny side facing out. While stapling up the foil, we were careful to seal all seams with masking tape so an airtight space would result. Then rough-surfaced 1×6 cedar boards were nailed directly over the foil (to the framing). These boards were used on both the inside and the outside wall.

One very chilly, windy day, it was clearly time to move into the cabin. The cabin was not completely finished—more windows needed to be hinged, more outside boards nailed up, a door hung. But the weather was not waiting, and the woman for whom the cabin was to be a winter's home needed her space. So we stopped "building" and "made do." We nailed a board diagonally across the foil on the one unfinished outside wall so that the wind would not tear off the foil. We installed the remaining windows "temporarily," holding them in place with nails. That first door was a heavy rug hung against the early winter chill. Later a wood door was hinged and hung in place.

Susan moved in and slept the first night in the star cabin. Many more women have lived there since, and many more to come. It is a woman's house, whole, holy,

(Photo by Carol Newhouse)

solid. It gives shelter and freedom. Lying on the bed in the cabin, looking up, you are carried to the center of the star, the queenpin. From there it is free flying for any woman's dreams:

> Women building dreams have strong faces,
> Women building dreams have strong bodies . . .

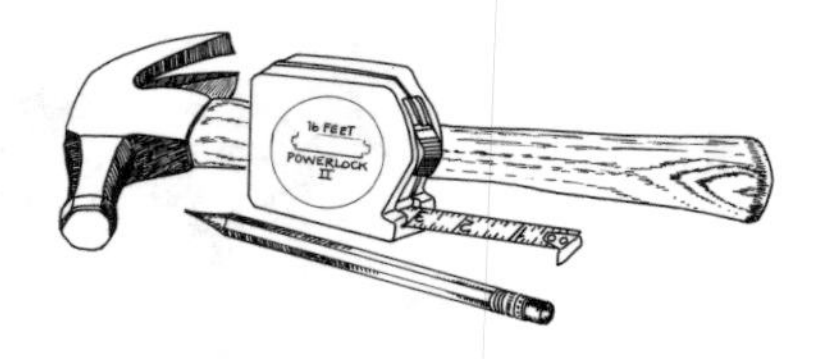

From Your Foundation Up

POSTS AND GIRDERS, SILLS AND FLOOR JOISTS

Once you've cast your piers, poured and cured your foundation, or embedded poles, you are ready to begin framing up your house. For clarity, this work should be discussed separately for three possible foundation systems: the house on piers, the house on a continuous foundation, and the house on embedded poles. There are, of course, other variations that will need special solutions. The following is intended to give you some general ideas on procedure as well as some specifics to follow or adapt from.

The House on Piers

The first step is to decide what your minimum crawl space will be. Crawl space, as mentioned earlier, is the space between the ground level under the house and the girders and floor joists that support everything. You should plan to leave at least 6″ to 8″ so that the house won't be sitting right on or too close to the ground, but you'll probably be happier later if you leave yourself a minimum of 12″ to 18″ crawl space at your lowest points. This space will allow you access to the underpinnings of your house later, if you need to work on pipes, floor joists, and so forth. On the other hand, setting the house too high in the air will create problems of stability (the house will tend to rock on its "stilts" and will have to be particularly well braced) and insulation (cold air flowing under your floor will make the house very hard to heat). Choose a crawl space height that gives you a reasonable work space but isn't too extravagant.

Once you've decided upon a minimum crawl space, go to the corner of your house situated at the highest point of the site, assuming you're building on sloping ground; if not, you can begin at whichever corner you like. Cut a post that will give you the crawl space you want, taking into consideration the height of the pier already in place. It's very important in cutting posts that you make square cuts at both ends. Otherwise, the post and/or the girder resting on the post will be able to rock and you'll create an unstable situation with attending problems. Of course, you can always add some shims later (and may have to in some instances), but it helps to start with square-cut ends.

Place this first post on its pier and check it with your level to make sure it sits plumb. If it doesn't, recheck your cut. Is it square? If the cut is square, check the level of your pier block—it may have gotten slightly out. If the pier is at fault, you'll have to use shims, driving them between the pier and post to level the base the post sits on and allow the post to stand plumb. When the post is plumb, fasten it to the pier, either by toenailing it to the wood block on the pier (two 8d galvanized nails per side will do; larger nails will tend to split the post or block), or by attaching the post to the post anchor, using lag screws or bolts, with nails set through the small holes on the anchor.

Now you have a choice of procedures. You can use a line level and go to one of your other corner piers. Stand a length of post material on the pier there and check to see that you're holding it more or less plumb. Then stretch the line level string taut from the top of your first post to establish the point at which you should cut this second post to make it level to the first. If you've bought your post material in long lengths and don't want to balance one of these monsters in place while you run lines and check level, just substitute a piece of any straight scrap material. You can mark it and then take your measurement for the post. You can do this work yourself by using temporary braces to hold posts and tacking the line level string in place, but it's much easier to work with a partner. Following this procedure, you can set posts at all four corners (or more) of your building. Make sure each post is secured in a plumb position as you go. Setting each post plumb is really critical because the whole weight of your house or structure will bear down on the posts; a post can take considerable weight or pressure if it's standing straight (plumb), but if it's tipped or twisted, it could give way under strain later. With all corner posts set, you can run a string or straightedge from one to the next and use this to mark each intermediate post exactly right. Or you can set one or two intermediate posts along the sides where the girders will run, set the girders, and come back later to "fill in" all of the intermediate posts. Intermediate posts along the sides or edges of the building perpendicular to the girders—parallel to the joists—should be placed after all of the girders are set and the joists have been added; this way, they can be tucked up under the joists to help support the edge joists under the nonbearing walls.

Before going on to setting girders, I should mention an alternative to using your line level to set posts. An even more accurate way to do this work is to use a straightedge and level between posts. A good straight length of 1×6 is perfect, if you have one that is long enough and not bowed. Running a straightedge from corner to corner will be difficult unless you're working on a very small structure. You may need to go from your first corner post to an intermediate post and work your way around the building that way. In either case, you can set the 1×6 on edge and run it from your post to the next feasible pier; set your level on top of the 1×6 and use it to establish a level height for your second post. Don't forget when you hold up a post or piece of scrap material to get a measurement to hold it plumb; otherwise your measurement will

(Photo by Carol Newhouse)

be inaccurate and this will throw off all of your work. As you go along, check and recheck to make sure that all posts are lining up with their tops level to one another. Any that are slightly off can be dealt with later by driving or placing shims under the girder, but any that are badly out of level should be replaced.

The next step is to measure and cut your girders. These are usually run along the longer dimension of the house: in a 32′×20′ structure, say, along the 32′ edges); the floor joists run across the shorter dimension, perpendicular to and resting on the girders. If you have girder material of the right dimensions, you can measure and cut lengths that are reasonable to lift and carry. Remember to center all joints over a post. Make all of your cuts square so that you can make neat, strong joints. If you have to make a built-up girder (for example, you can make a 4×8 by splicing together two 2×8's), cut the individual pieces to length before splicing them together. Splice the pieces carefully, using your combination square to make sure that edges are flush. The usual procedure is to stagger nails in a zigzag pattern, anywhere from 10″ to 32″ o.c., with two nails set parallel to each other at each end. You should use 16d or 20d nails and drive them in at an angle so that they don't break through the other side of the material. The built-up girder should be nailed from both sides for maximum strength.

Before you place your girders, cut some short (1′ or so) lengths of 1× material to use as *scabs* on your posts. (A scab is a piece of lumber spliced or tacked on to cover a joint, create a ledge, strengthen an area, and so on.) They should be tacked in place on the outside edge of each post or on every other post; they should be positioned so that they stick up 6″ or 8″ above the post. These scabs will be very helpful when you lift a girder into place; they'll keep it from slipping off the posts, allowing you to relax as you go about your work. Before you place any girder, take a careful look at it and determine how it bows. Remember that you always use a piece set *on edge* for maximum strength. Each piece of lumber will usually have a slight bow or curve and you'll want to set each girder so that the crown, or high point, of the bow is *up;* later the weight of the building will push down on this high point, more or less straightening it out. Make a pencil or chalk mark on the crown edge of each girder in advance, so that when you go to pick it up you can lift it and place it easily. Be careful lifting and moving girders if they are large and heavy: you can hurt your back all too easily if you're not thinking about what you're doing. If possible, try to schedule your work so that you're doing this lifting early in the day, when your muscles are fresh and not fatigued.

As you place each girder, double-check yourself to make sure everything lines up and is level. Each post-girder connection should be tight and neat; add shims where needed. The girders should be toenailed to each post they rest on, but until all of your girders are placed it's a good idea just to tack a few nails partway in. These will hold the girders while you make a last check to make sure that the girders are level to one another and all the posts are plumb. When you're satisfied that everything is right, nail the girders to the posts, toenailing them with 8d or 10d nails. You may want to use metal connectors to strengthen these girder-post joints. There are various styles and weights, ranging from the simple tie strap (a flat metal plate that could be nailed over the joint) to more elaborately formed post caps, which come in specific sizes for different-sized posts and girders. T and L straps might also be used. Metal connectors may be nailed and/or bolted in place. As you secure the girders, you can remove the 1× scabs that helped hold the girders in position.

At this point you'll want to add some sort of *cross bracing* to your posts to keep them rigid. This bracing

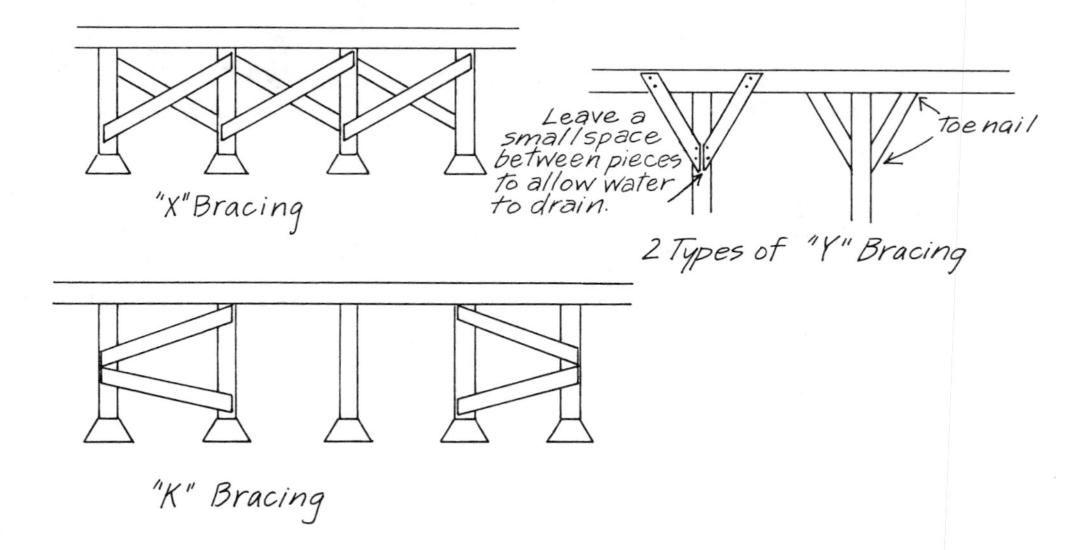

should run from one post to the next and is usually applied in an X, K, or Y pattern (see illustration). If you decide on the X pattern, it is easily done with two pieces of 1×4 or 1×6 nailed to opposite faces of the two posts being joined. Two nails (8d) at each end of each brace will hold the bracing securely. The K bracing may be used on smaller structures, where you don't really need bracing on every post. It may be constructed with 1× material and used on corner posts only. If you plan to put sheathing and/or siding over this lower part of your structure later, be sure that your bracing is nailed to the inner faces of posts so it won't be in the way. You may want to incorporate your bracing in ways that will let it double as nailers for your sheathing or siding later. The Y style of bracing may be done with 1× material nailed to the faces of the posts, or it may be made with 2×4 material that is set in, as shown. The set-in 2×4 bracing is quite strong and has a neat appearance. Finishing up the bracing of your posts is the final step in this stage of your framing work. Next will be the cutting and setting of the floor joists, so you can skip ahead to this section/work.

The House on a Continuous Foundation

The *sill* or *mudsill* is a continuous piece of decay-resistant 2×6 or 2×8 that is placed on and bolted to the foundation wall. It forms the point of connection between the foundation and the wood framing of the house. Thus, when your foundation is completed, the next step is to cut and place your sill. Although the sill functions as a continuous piece, it is made up of a number of pieces that follow the perimeter of the building, meeting at each corner in a simple butt joint. It's a good idea in planning out and ordering your sill material to work with as few pieces as possible; this means choosing the longest lengths you can get and minimizing joints. Sill material is usually redwood or cypress but may be other material pressure-treated with wood preservative to make it last. It should be clear material, free of knotholes or other defects, and should come in good straight pieces. Sill material is usually fairly expensive, so look carefully at the pieces you're buying and reject any that seem warped or bowed as these will cause you headaches at the least, and if they're really bad, they may be unusable. Try to plan for and buy lengths of material that will give you a minimum of waste. If you're going into or calling a lumberyard to order sill material, be sure that you specify "mudsill" or explain what you're planning to use the material for; "sill" is a term sometimes used for *window*sill material, too, and you may find yourself in a confused conversation!

Sill sealer is a waterproof material that comes in rolls; it is sandwiched between the sill and the foundation as an extra moisture barrier and a protection for the sill. In some areas of the country, metal *termite shields* are used between the sill and foundation. The termite shield is a wide piece of metal flashing placed over the top surface of the foundation before the sill is put on. Termite shields were intended to prevent the movement of these wood-eating insects from the soil up into the wooden parts of the house. They have proved to be less than 100 percent effective, though, and are being used less and less frequently. The "modern" solution to the termite problem in affected areas is to treat the soil around the foundation walls with toxic chemicals. Another solution is to keep the area under and around the house clean and free of debris, and to check periodically to make sure that there is no ground-to-wood contact happening and that framing members remain clean and dry.

If you don't use sill sealer between your sill and foundation, you may want to treat the bottom of each sill piece with creosote or a good wood preservative just before placing it. You may also want to make a "bed" for the sill to set on; the old-fashioned *grout bed* was made up of 50 percent cement and 50 percent fine sand, mixed with enough water to give a fairly loose consistency; this grout was troweled in place just before the sill pieces were set on the foundation—hence the term *mudsill.* If you use sill sealer, the sealer will compress down to make a bed for the sill.

To begin the actual placing of your sill, measure and cut the lengths that you need. You'll have to drill holes in each piece of sill material that correspond exactly to the positions of the sill bolts protruding from the foundation wall. You can take careful measurements to get these locations, but there is a shortcut method that can be very accurate if you're careful. First, remove the nuts and washers from the sill bolts. Take some chalk line chalk or crayon and rub it over the tip of each bolt. For the next step, you'll need a work partner. Each of you will take an end of the length of sill material you want to mark for drilling. The sill should be positioned so that it sits in from the outside edge of the foundation wall a distance equal to the thickness of the exterior sheathing you plan to use. For example, if you're using 1× (nominal) sheathing, back your sill in ¾″ from the outside edge of the foundation wall. Balance the length of sill material in place over the protruding sill bolts and give it a couple of taps with your hammer; when you lift the piece away, each sill bolt will have left a crayon or chalk mark which you can use to center your drill. If you want to be extra careful when doing this work, place a piece of scrap wood over the sill material before hitting it with your hammer. Holes should be drilled slightly larger in diameter than the sill bolts; this will give you a little allowance for warp in the sill material, slight variations in the bolt placings, and so on.

Just before placing each length of sill, put your sill sealer (or termite shield, or grout bed) in place. Use a piece of scrap wood to protect the sill if you have to tap it down in place with a hammer or small sledge. Put all sill pieces in place before you add washers and nuts and tighten down the sill bolts. As you secure the sill, make sure that it is level all around. You can use thin wooden shims or mix some grout for problem areas. When the sill is fastened down all around, go back to your corners and toenail adjoining pieces together.

If you're using girders to break the span of your floor joists, they can be set in place now. The ends of the girders should fit into the slots or keys make earlier in the foundation wall. They may be supported at mid-span with posts and piers, or may be divided into lengths, which should meet over a post or pier. The top edge of each girder should be level with the top of the sill. Remember to set girders according to their bow, crown side up. Before placing a girder that fits into the slot in the foundation wall, cut a small piece of roll roofing and place it so that it will act as a break or moisture barrier between the girder and the concrete. To locate and measure for posts needed to support girders mid-span, use a straightedge or line level as described in the "House on Piers" section, or place the girder and then use it in locating your posts and piers. Metal connectors may be used to strengthen joints where girders and posts connect or where two pieces of a "continuous" girder butt together.

The House on Embedded Poles

The FHA booklet *Pole House Construction* (see "Resources") is a good source if you're using embedded poles for your foundation system. Most simply, the framing of this part of your house can be done by spiking or bolting pairs of 2× material to the poles to act as girders. The two pieces making up each girder are attached opposite one another (with the pole sandwiched between) and should be carefully placed so that they are level with each other. If you are using large enough poles, you may be able to cut a small notch in each side to act as a "seat" for each girder piece. You can use a straightedge and level or a line level to establish parallel sets of girders at exact level heights to each other. A second way to frame up this part of your house if you're using embedded poles as a foundation system is to cut all poles so that their tops are exactly level to one another and then add girders that rest on these "posts." The procedure for this work would be fairly similar to that described in the "House on Piers" section, as you've simply substituted an embedded pole for each post and pier.

(Photo by Janet Cole)

Floor Joists: Layout and Framing

The floor joists carry the load of the house and transfer it to the foundation via the mudsill or the girders. When you begin to set your floor joists in place, your house will begin to take shape in a special way; no longer just a set of perimeters of parallel lines, it will be laced together with a clear skeletal structure, becoming more distinctly defined. Floor joists are usually 2× material set on edge, evenly spaced, and making a span that is calculated by their size, o.c. spacing, and the weight of the building. The chart in the "Building Code Tables" (see end of book) shows some safe spans for common sizes of floor joists; you can also refer to technical books of architectural design. For most small to moderate-sized buildings, the table should provide the information you'll need. The most common o.c. spacing for floor joists is 16″; this spacing will allow you safely to use medium-sized material (2×8, 2×10, or 2×12) and will be properly spaced for common subfloor material.

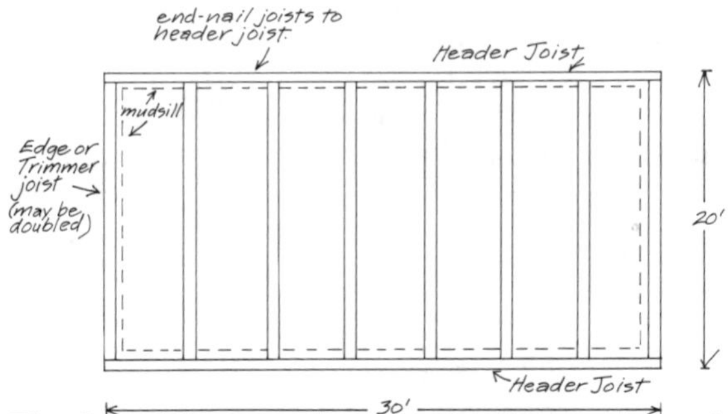

The floor joists usually span the shorter dimension of the building: for example, in a 20′×30′ building, the joists would span the 20′ dimension. Unless the building is quite small, the joists will need an intermediate support of a girder supported on posts and piers. A single joist may make the complete span, or two joists may meet and be spliced or joined together over the intermediate girder. In this case, the two joists may overlap one another and be nailed together, or they may be cut to butt into one another and be connected together with a metal tie strap. Butting the two joists end to end will allow you to keep consistent o.c. spacing.

If you are building on a continuous foundation, your floor joists will rest on and be attached to the mudsill. A piece called the *header joist* is run perpendicular to the joists; it sits on the sill at each end of the house and all joists butt into it. For example, in the 20′×30′ building, the header joists would run along the 30′ dimensions. These would be made with material the same size as the joists, set on edge and toenailed to the sill with 10d or 12d nails 16″ o.c. The two joists running along the 20′ edges of the building are called *end joists,* or sometimes *trimmer* or *stringer joists.* It is common to double these joists (i.e., splice two together for each end of the building). These are end joists toenailed into the sill with 10d or 12d nails 16″ o.c. The usual procedure is to set the header and trimmer joists in place and to mark the o.c. locations of the other joists directly on the header joists.

Each joist may then be cut and put in place. Check each one carefully for even a slight bow and place it crown up. Toenail each joist to the sill with two 10d or three 8d nails, and then end-nail it to the header joists with three 16d (or 20d) nails each end. If your house will have interior partition walls, you should place double joists under any partition walls that run parallel to the joists. Areas that will take extra weight (for example, under the bathtub) should likewise have double joists. If you plan to use a rock wall inside, joists should be doubled or otherwise "beefed up."

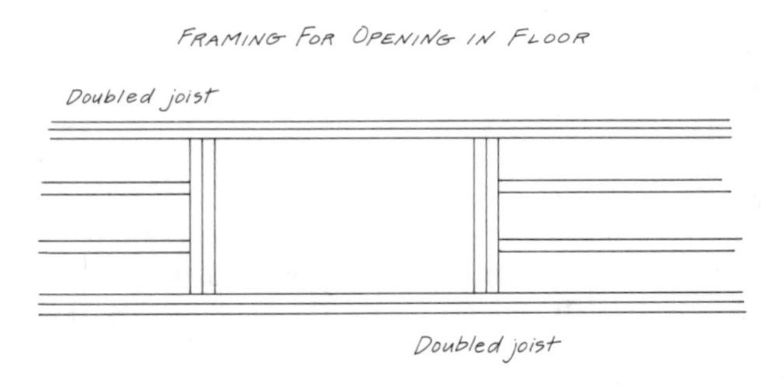

Framing in for a floor opening requires some extra material, too. Chapter 11 in *Against the Grain* (see "Resources") shows exact details for this work. Basically, you'll want to run double joists along each side of an opening and doubled joists, called *headers,* at each end. Any joist whose normal span is interrupted by an opening is, of course, cut short and must be nailed to the doubled header joists. Whenever you double up your joists, they should be nailed together with 16d or 20d nails placed about 16″ o.c. Follow the process described earlier (under "The House on Piers") for making a built-up girder.

If you're building on a post and pier or an embedded-pole foundation, your floor joists may be attached in one of two ways. They may be placed on top of the girders, with a header joist used over each girder as it was used over the mudsill (explained above). Or the joists may butt into the girders and be supported by metal joist hangers or a ledger strip nailed to the girder. This strip should be made of 2× material. The width will be determined by the size of your girder and your joists: it should be just wide enough to allow the joist to sit on top of it, with the top of the joist level with the top of the girder. The ledger should be securely nailed to the girder; set nails in a zigzag pattern and face-nail at a slight angle. Setting the joists on top of the girders is probably the stronger and more economical system.

Bridging or blocking is placed between floor joists to keep them from twisting and to help transfer weight loads more evenly over the floor surface. A line of bridging or blocking is usually placed every 8′ or so.

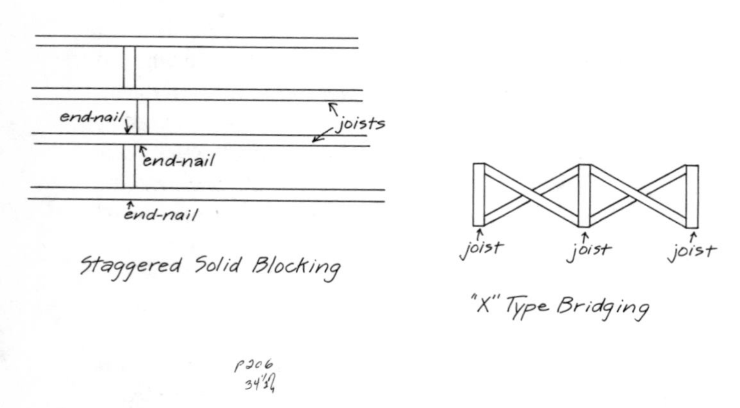

Bridging may be made of 1×2 material cut to fit in an X pattern (two pieces) between two joists. Each end of each piece of bridging is toenailed to the joists with two 8d nails, but the lower ends of the bridging are usually left unnailed until the subfloor has been put on. Prefabricated metal bridging is also available for use between joists; this serves the same purpose as the 1× material and is quickly installed, but is a more expensive option. Solid blocking is a third alternative; this is made with material of the same size as your joist material and is nailed in place between joists. Lines of solid blocking should occur every 8′ or so. This type of blocking may be end-nailed (two 16d nails per end, each piece) if you stagger the pieces as shown.

Framing for exterior doors should be considered at this stage in building as you may need to cut into your joists to allow for the pitch of the doorsill later. Whether or not you are building to code or following exact plans, you'll probably want to design and build an exterior door that will be weatherproof and watertight. To accomplish this, you'll want a slightly pitched or beveled doorsill that can shed water away from your doorway. Usually the doorsill is set in so that its upper face or level is flush with the finished floor of the room; the joint of floor and sill is then covered with a thin oak threshold that is both cosmetic and practical, as it hides the joint and also protects it. A standard sill pitch is 2 in 12: that is, the sill is placed at an angle that allows a drop (or rise) of 2″ for every 12″ of run. You may elect to use this standard or make one up for yourself that seems reasonable and easy to deal with; whatever your choice, you would do well to make some provisions now for each door to your house. If you aren't sure at this stage exactly where you want your doors, or don't feel ready to take on this work yet, you *can* do it later. It will just be a little more bothersome, as you'll have to cut through subfloor, too.

If you feel ready to take it on at this stage, here are some basics. Figure the length of space you'll need for the door: this should equal the outside width of the doorframe. Then figure out the width of the space you'll need, which should equal the thickness of your framed wall (probably 3½″) plus the thickness of your planned interior wall (probably ¾″). The depth of the space you need is next: this should equal the thickness of the doorsill material plus the pitch of the sill. Don't forget to subtract the thickness of your subfloor and your finished floor from this total depth. With all of your figuring done, you can measure and mark the space for your door. Then cut down and chisel out whatever you have to from your header joist or edge joist. Cut back intermediate joists (any that butt into the header or edge joist in this area) slightly if necessary. The actual amount you remove from each joist may be quite small, but it will help your work later if you've planned ahead with some of these tinier details.

SUBFLOORS

A subfloor is nailed over the floor joists and provides a work platform for the framing of walls; it later serves as a base for the finished flooring of the house. The subfloor may be 1× material, usually no wider than 1×8, or plywood (4′×8′ sheets in thicknesses of ½″, ¾″, or more, depending upon the spacing of joists and the intended finished floor). There is tongue-and-groove plywood subflooring (in 4′×8′ sheets), which is slightly more expensive but makes a tight floor because the edges of the sheets interlock. If joists are spaced widely apart, or the house or structure will be bearing an unusual amount of weight, 2× material may be used for the subfloor. In some instances you may want to make a single floor with subfloor and finished floor one and the same. In these cases, a tongue-and-groove-edged plywood would make a good choice. You'll have to add extra blocking so that all edges of each sheet of plywood will rest on solid blocking or on joists.

Usually 1× subflooring is laid diagonally; 1×4, 1×6, and 1×8 in Douglas fir, yellow pine, or southern cypress are commonly used. Face-nail 1×4 and 1×6 boards with two 8d nails into each joist; 1×8 boards require three 8d nails per joist. The ends of each individual board should be placed over joists, so that all joints in the floor occur over joists. If the subfloor is applied diagonally, the finished floor may be either parallel to the joists or—more commonly—perpendicular to them. Oak flooring strips, parquet blocks, and flooring planks are commonly laid over 1× diagonal subflooring.

Plywood is a much more widely used choice for subflooring nowadays. It is relatively inexpensive, strong, and readily available and can be applied very quickly. A special grade of plywood called *plyscore* is made especially for subflooring. If your joists are 16″ o.c., you can use ½″ plywood, or ⅝″ for a stronger floor surface. For wider-spaced joists, go to a heavier plywood, ¾″ or more. Plywood sheets should be placed so that the longer (8′) dimension is perpendicular to the joists and planned so that successive courses of flooring break over different joists. The pattern of your sheets will depend upon the size of your floor, but generally you should try to stagger your sheets/joints as much as possible. For a large-surface floor, you can use the basketweave pattern. Two sheets are laid side by side with their 8′ dimension perpendicular to the joists, then a sheet is laid parallel to the joists, followed by two more laid like the first two. This pattern is alternated every time you start a new row.

Sheets of plywood should never be placed tightly together; regardless of the "pattern" of your flooring, always leave about ⅛″ between the edges (8′ dimension) of sheets and 1/16″ between the ends (4′ dimension). This spacing, or something roughly equivalent, will allow the plywood to change slightly with changes in air moisture.

Plywood may be nailed to joists with 8d common nails (for ½″- to ¾″-thick sheets) or with 7d ring-shank or threaded nails (same sheets). Thicker plywood requires 10d commons or 8d ring-shanks. Nails are spaced 6″ o.c. along all edges and 10″ to 12″ o.c. along intermediate joists. It's a good idea to lay out your plywood sheets before doing your final nailing so that you can make sure that everything lines up properly. You can tack each sheet in place with a couple of nails driven partway in; then if you have problems lining things up, the nails can be pulled and the sheets can be easily adjusted.

Most types of finished flooring can be applied over a plywood subfloor. If you are using different kinds of flooring in different parts of your house, you may have to add an underlayment of plywood or fiberboard in rooms where thinner floor coverings (linoleum or vinyl tiles, for example) are to be used. Putting down the finished floor is one of the later steps in house building. The subfloor provides a platform that can take the abuse of wall framing or sudden rainstorms before the roof goes on. Seeing your partly finished house with an inch or so of water over the entire floor surface will make you appreciate the rough-finish sequence in floor making! Of course, you can drill holes to let the water out (as we did), but you'll be glad that it's just the plywood subfloor that took the drenching.

Finishing up the last nailing of the subfloor is a wonderful stage in house building: now there's a place to stroll around on, imagining the view from a specially planned window, or daydreaming about a cozy fire

(Photo by Dian)

in the wood stove some winter afternoon. For once, all the agonizing precision over foundation measurements and all of those 16d's driven into joists and blocking will seem worth it. It becomes much easier to imagine the walls flying into place and a roof overhead—an actual house taking shape, solid and believable.

ROUGH-FRAMING THE WALLS

Conventional stud wall framing is an easy and efficient method of construction which makes use of materials that are commonly available, relatively inexpensive, and not difficult to handle. There are two types or styles of stud wall framing: platform framing and balloon framing. Any single-story house and many two- or three-story houses are constructed with *platform framing.* The walls of the first story are built and raised upon the platform created by subflooring laid over the floor joists. If there is to be a second story, ceiling joists are supported by the first-story walls, and when flooring is attached to them, a second platform is created. Walls for the second story are then built and raised on this platform. The walls for different stories of the house are thus constructed and raised as more or less independent units. In *balloon framing,* the vertical members (studs) of the walls run all the way up two stories. Ceiling joists for the second floor are supported by ledgers let into the studs at appropriate heights. With this system of framing, the house has a common wall for two stories. Because most of the framing for houses, room additions or partitions employs the platform framing method, this section will cover that style.

The anatomy of a stud wall is simple. There are horizontal members at the top and bottom of each wall called *plates;* between the plates are regularly spaced vertical members called *studs.* The framing material is usually 2×4, in a #1 or #2 grade of a wood such as Douglas fir, yellow pine, or hemlock. In most areas, precut studs are available. These are 92¼″ 2×4's in a special stud grade (between #2 construction grade and #3 standard). If you're framing up conventional 8′ walls, precut studs can be a great time-saver and slightly more economical than buying 8′ 2×4's to use.

In special cases, 2×6 framing material may be used for walls instead of the regular 2×4 material. For example, if insulation exceeding the 1½″ thickness of a regular 2×4 wall is needed or wanted, if an extra-strength wall is desired, or if space for large plumbing pipes is needed, a 2×6 wall may be called for. This will raise the cost of the framing material significantly, though, so the substitution is not a common one.

Besides the top and bottom plates and the studs, the wall may have special framing for doors and windows, some type of bracing, fire blocking, and special insets where interior partition walls connect to exterior walls. All of these components are discussed below. A distinction should be made at this point between the term *wall,* which refers to the exterior walls of a building, and *partition wall* or *partition,* which refers to interior walls. Another distinction is made between *bearing* walls or partitions, which support the weight of a ceiling or roof, and *nonbearing* walls or partitions, which support only their own weight.

The walls of the house take the weight of the roof and transfer it down to the floor and foundation system. Depending upon the size and design of the house, certain walls will bear more or less weight. In a gable-roof house, for example, some of the walls will take the direct weight of the rafters (roof); other walls take only their own weight. If you disturb the transfer of weight by placing a door or a window in a wall, you have to make allowances for redistributing the weight and pressures. The usual spacing of studs is 16″ or 24″ o.c.—16″ o.c. for most construction; 24″ o.c. for small, single-story structures and for nonbearing partition walls. If you put a 2′-wide or a 3′-wide door into a wall, you have to remove some of the regular studding. To redistribute the weight that the regular studs would have taken, a horizontal piece called a *header* is run across the opening. The header is supported at either end by a *trimmer stud,* which is attached to a regular stud (see illustrations). If the opening is for a window, a second horizontal piece called the *rough sill* is placed along the lower edge of the intended window (more on all of this to come). Above the header, and below the rough sill in the case of a window, short studs are placed that continue the regular 16″ or 24″ o.c. spacing of the studs. These short studs provide nailing surfaces for interior wall coverings and for exterior sheathing.

Bracing adds rigidity and strength to a wall. It is usually placed at each exterior corner of a house and/or at the end of each wall. If your house is large, you'll want bracing every 25′ as well. Bracing runs from the top plate to the bottom plate at about a 45° angle. If the "run" of a piece of bracing is interrupted by a door or window opening, the bracing may be constructed in a K form. *Let-in* bracing is 1×4 or 1×6 material that is set into notches cut into the studs (see photograph). This bracing allows for a smooth surface on the faces of studs (necessary when you come to nail on sheathing and paneling). A second type of bracing is placed between the studs and is made of the same size material as the studs. A series of small

(Photo by Carol Newhouse)

pieces must be angle-cut at each end and fit and nailed into place, forming a "continuous" diagonal brace. The very fastest and cheapest bracing is metal bracing—metal straps with predrilled holes for nailing. It is set on the inside of the wall (running diagonally like the let-in bracing) and nailed to each stud it crosses. The interior wall covering is later placed over this bracing. There is an alternative to either style of bracing on exterior walls: you can use plywood sheathing and eliminate the need for bracing altogether. The building code in most areas will allow you to use 5⁄16″ plywood sheathing and no bracing. The plywood sheets, if of proper thickness and properly nailed, will add the rigidity of conventional bracing.

Fire blocking or *fire stops* are horizontal pieces of 2×4 (or 2×6, if your walls are framed this way) nailed in place between studs at a height about halfway up the wall. In the case of a fire, this line of 2×4's will help prevent a strong updraft of air that could spread the fire rapidly behind the walls. This extra blocking can be incorporated into the later nailing pattern for interior wall coverings if you think and plan ahead. It can also be used to correct bowed studs and to add rigidity to the walls.

Assuming that you're planning out your walls without already made blueprints, you can now think through details of each wall and make up a materials list. It would be a good idea to skip ahead and read through details on header size, framing for windows, and so on, before tackling this. In planning out your walls, try to think in terms of manageable sections. You'll be doing your rough framing on the floor platform and then lifting walls or sections of walls up in place, so the size of each wall or section should be reasonable. Two women of average strength can lift a 16′ wall section without too much straining, so you may wish to plan your wall sections around that. If you have plenty of womanpower on wall-lifting day, you can build in bigger sections; if you're building alone, you may want to keep your sections 10′ or less.

To calculate how much framing material you'll need for a given wall, figure out:

The length of your bottom plate
The length of your top plate (doubled)
Number of studs
Number and size of headers, trimmers, etc., for each window or door
Bracing (if needed) and fire blocking

Make a list for each wall. If your roof is a gable, remember to figure in studs for the gable ends. When you have a list of all material for all walls, add in a dozen or more extra studs—windows and doors always seem to take more material than you imagine they could, and you're bound to run into a poor piece of lumber or make a bad cut now and then. If you're far from the lumber store, add in some extra header material for the same reasons.

Most of your framing will be done with 16d nails (common, ungalvanized). The most economical way to buy these is in 50-pound boxes. How many pounds you'll need will depend upon considerations such as the number of doors and windows you're framing in for and linear feet of walls. Buy good nails; you'll save yourself endless frustration and pounds of bent nails. I can't give much of a guideline in terms of how many boxes of 16's you might need; we used about 25 pounds framing the walls of a 16′×24′ cabin (lots of windows, two doors). It's much better to overestimate your nail needs—you can always use them later. You'll also want to add some 10′ or 12′ lengths of 1×6 to your materials list for temporary bracing when you lift your walls up. Later you can no doubt find ways to recycle these pieces.

Framing up a wall is a lot more fun than writing or reading about it. There's something very pleasurable

Header Spans for Window and Door Framing *

Material size:	Can span up to:
4×4	4′
4×6	6′
4×8	8′
4×10	10′
4×12	12′
4×14 (#1 grade)	16′

* Material to be Douglas fir #2 grade or lumber of similar strength. Material to be used *on edge.* This table is intended for general reference; for more exact specifications, consult with your local building department.

If you are framing a door opening for a nonbearing partition wall (interior), a single 2×4 may be used as a header.

in this "rough" work—the clean, uncomplicated joining of pieces, the sudden amazing form of the structure emerging (a wall where there was only a platform; a window to look through), the satisfaction of seeing your day's work standing. . . . The tools you'll need are basic: your hammer (a 16-ounce or, better, a 20-ounce), tape, combination square, saw (power or handsaw), a cat's-paw (in case you make a mistake), nail nippers (likewise), sledge (for slamming the walls into place), a level, chalk line, and pencils. Have your patience in hand, too—this work may be simple, but it takes *absolute accuracy.* You don't want to find yourself trying to fit a window into a space that is 2″ too small, or wondering how the door could possibly go *there.* I can't overemphasize the importance of checking and rechecking yourself as you work!

Framing Up a Simple Wall

Your floor platform is the place to build your walls. Clear and sweep it clean first so that you have plenty of work space. If you're working with a partner or partners, divide up the walls and space so that you have plenty of room. Have one or two areas designated for cutting. Keep your material stacked and orderly. This is especially important if your framing stretches over a number of days; material left in a careless pile can become badly warped and miserable to work with. Begin with your simpler walls or wall sections so that you can get a feeling for the work before it gets complicated.

Every carpenter probably has a slightly different procedure for framing up walls. Here's one way:

First measure and cut your bottom plate and one top plate (the second top plate is added later). Lay your two plates out parallel to one another and approximately a stud's length apart. Now is a good time to mention that even though studs are supposed to be machine-perfect and true, a few sneak through now and then with imperfect ends. A quick check with your combination square of each and every stud you use will ensure that you catch any problem studs *before* you nail them in place. Now nail the two end studs in place; two 16d nails should be driven through the plate into each end of each stud. Use your combination square to make sure that plates and studs are joined squarely. I usually find myself kneeling on the plates and/or studs to hold them in place while I nail. (This is hard on the knees but helps to steady the work!) If you're doing a lot of rough framing, you may want to invest in a pair of protective knee pads. Carpenter's overalls have cushioned knees that help a little, too. With your end studs nailed in place, you can measure and mark for your other studs. You'll want to mark both the top and the bottom plates so that you can line up each stud quickly. Be very careful to be consistent about the end you begin your measuring and marking from. Most tape measures have special markers for conventional o.c. spacing. I use a 16′ Stanley Powerlock II, which has a tiny arrow for 16″ o.c.; with this tape and regular centering, marking for wall studding is easy and fast.

Remember that you want your studs 16″ *center to center.* If you make a mark at each center point, it would be hard (time-consuming) lining up the stud correctly. Here's a quick method. Hook your tape over one end of your bottom plate. Pull it out past the 16" o.c. spacing. Back up ¾" from 16" and make a little *V* to mark that point. Now put an *X* to the right of your mark; this will show you to which side of the mark you're going to want to place your stud. The ¾" backup allows you to center the 1½"-thick stud properly. Now go on by pulling your tape out farther to the 32" o.c. and repeat the same procedure: back up ¾" from that, make your mark, and make an *X* on the right side for your stud. Continue along until all of your stud placings are marked. Now take your combination square and use it to extend each *V* mark down the inner face of the bottom plate (or sole plate). Now you'll have a square line to place the stud along, plus your *X* to show you which side of the line to go to. Begin at the same end of the wall you chose to start out at and mark your top plate the same way.

It's important to keep your wits about you when you do this measuring and marking; if you daydream and put an *X* on the wrong side of a mark, you'll end up with a peculiarly tilting stud in your wall. This can be corrected, of course, but it saves work if you don't make the mistake to begin with! It's also extremely important that you always begin your measuring/marking from the same end of the wall. If you're making your wall in sections, take this into account. Not only should you begin at the same end, but you should make certain that your o.c. markings are made as though the wall were one continous piece. Where the wall is broken and joins, you'll have two studs together and only one of these will be properly o.c. That one is properly in line and should be exactly 16" center to center with the next *regular* stud in your wall.

When all of your stud placings are marked on the bottom and top plates, you can nail the studs in place: two 16's per stud end, driven through the top plate or bottom plate into the stud end. Again, use your lines to make sure that each stud is sitting squarely.

If your wall or wall section is part of a corner of the house, you may want to use it to make a special double corner that will help when you come to placing Sheetrock or interior paneling. If you simply joined two walls with their normal corners, you'd find yourself with a problem when it came to nailing on interior wall coverings. One of the two adjoining walls would not provide a nailing surface for the wall covering. This is easily seen if you take two short pieces of 2×4 and pretend they're the end studs of adjoining walls. The walls join in a simple butt joint: the edge of the end stud for wall #1 meets the face of the end stud for wall #2. If you were to nail Sheetrock to wall #2, you could use the 1½" edge of the end stud as a nailing surface. Wall #1 offers no nailing surface at all because the end stud is hidden in the corner. To provide a nailing surface, you have to make a double corner of some sort. One way to do this is to nail several short lengths of 2×4 to the end stud of wall #1 while you are rough-framing it, and then add another full stud, which is nailed to these *spacers.* The end result is a "sandwich" 4½" thick. This will give you a full inch of nailing surface for your interior wall once the corner is nailed together. This may sound complicated, but it's actually very simple once you've worked with the pieces or seen it done.

Each exterior corner of your house should have a double corner framed in before the walls are raised, to save much time and trouble later. There are several other ways to make these double corners, so you might like to look over examples in other carpentry books, or look at the rough framing on other houses under construction.

A second consideration in wall joinery is the place where an interior partition butts into an exterior wall. This should be planned for in advance, too. Again, there are several options for framing. Basically, you want to create nailers for your interior wall coverings (Sheetrock, wood panelling, or other), and you need to create a nailer for the partition itself so that it can be attached to the exterior wall. One way to accomplish this is to use three full studs at the point where the partition butts into the exterior wall. These are positioned and nailed into place, using techniques similar to those shown in the sketch of a double corner-stud wall.

Bracing may be added to the wall at this point or later on, when the wall is standing. If you're using 1×4 or 1×6 let-in bracing, it's easier to place this while the wall is still down. The bracing goes on the exterior face of the wall, which is the side facing up now. Ideally, this bracing should run at about a 45° angle from the top plate in the corner of the wall down to the bottom plate. For the simple (no windows, no doors)

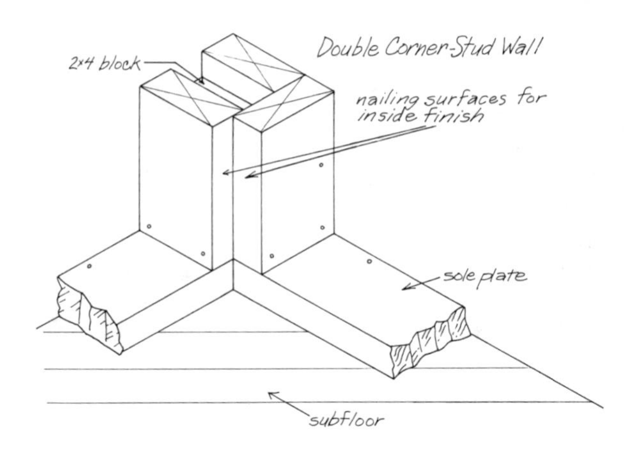

(Photo by Carol Newhouse)

wall section, you can place this bracing easily. The length of 1× material can actually be laid in place and each stud marked where it must be chiseled or routed out to receive the bracing. The bracing should be angle-cut at the top and bottom to fit against the wall properly. Cut out each stud as needed and then place the length of bracing and nail it (two 8d nails per stud).

If you use bracing between the studs rather than let-in bracing, it should be made of material the same size as the studs (2×4 or 2×6). You'll have to cut a series of small pieces, which are slipped into place and nailed between the studs, forming a "continuous" brace. Each of these pieces must be angle-cut at both ends. To get the right angles and lengths for these pieces, you can tack a guide strip (a scrap piece of 1×2 or whatever you have) onto the wall where you plan to run the bracing. This piece should run from the top plate in the corner of the wall to the bottom plate at about a 45° angle. Lay a length of 2×4 along this guide strip and mark the pieces you'll need. Alternatively, you can use your bevel and tape to get the angles and measurements for the pieces you'll need. The top and bottom pieces of each line of bracing will have to have a double angle cut at one end in order to fit into the corner of the wall and against the last stud. When you are ready to nail your bracing in, use two 8d nails for each end of each piece of bracing. These are toenailed in, and it helps to start your nails *before* you slip a piece into place (nailing between the studs is an awkward affair).

Fire blocking may also be put in place while the wall is down. You may want to wait until all of your walls are completed and up, though, and make use of scrap pieces of 2×4. Whenever you decide to add the fire blocking, you can follow a simple procedure. In measuring for a piece of blocking, make your measurement at the bottom of the wall rather than at the midpoint where the blocking will be nailed in place. If any of your studs are bowed, measuring at the midway point will give you a slightly inaccurate reading. Measuring at the bottom plate will give you the true reading, since the studs are nailed exactly in place along the plates. Most of your fire blocking will be cut to uniform lengths that fit between the studs, but wherever you have a window or door or the joint between two wall sections, you'll have to cut blocking pieces to special lengths. You may choose to run all of the blocking at the same height, and in this case you'll be able to end-nail some of your pieces and will have to toenail others.

It helps if you cut two scrap pieces of 2×4 to the

height you've chosen for your blocking; these can be stood in place as temporary ledgers while you nail a piece of blocking. Using these aids will also enable you to nail the blocking at a uniform height without having to mark and square each stud. Alternatively, you may choose to nail every other piece of blocking 1½" higher than its neighboring piece, which will allow you to end-nail every piece. Here, too, you might want to make use of scrap 2×4's cut to the needed heights to help steady the pieces you're nailing.

With your wall or wall section completed, you're ready to lift the wall up into place. First, you may want to measure in 3½" (the thickness of your wall) from the outside edge of your floor platform and snap a chalk line. This line will show you exactly where to line up the inner edge of your wall (bottom plate); it's an option, because you can also line your wall up flush to the outside edge of the floor platform. Push the wall over so that it's lying with its bottom plate a few inches from this chalk line, or 5" to 6" from the edge of the platform.

At this point, you may want to tack a piece of 1×6 to the wall to serve as a temporary brace when the wall goes up. This piece should be tacked with one nail to the last stud on a corner wall, or to an intermediary stud. Place it so that it can swing out from the wall at about a 45° angle when the wall is standing; the single nail will allow this brace to pivot freely. If you put the brace on the end stud at a corner, you'll be able to tack it to the outside of the floor platform; if you place the brace on an intermediary stud, you'll want to tack a short piece of 2×4 to the subfloor to nail the lower edge of the brace to when the wall is up. If you have plenty of extra hands available to help when the wall goes up, this brace can be tacked on later, when the wall is standing.

To lift the wall, start at the top plate end. If you're working with one partner, position yourselves at opposite ends of the wall. One of you should have a hammer and nails ready in her tool belt or apron. If you have a bigger crew, spread yourselves out along the wall. Lift the wall from the top plate end and walk toward the bottom plate, lifting as you go. The wall will thus be "walked" up bit by bit. When the wall is standing vertical, it can be kicked or sledged over into place. Do this very gently, a little at a time, so that you don't lose control of it. When you have the wall pretty much in place—lined up to the outside edge of the platform or the chalk line—swing your brace into place. You can use a level to get the wall close to plumb (perfection isn't necessary at this stage), or just sight it as close as you can. Nail the brace securely with three nails, triangle-fashion, top and bottom. This should hold the wall until you're ready to move it

(Photo by Janet Cole)

later, when you square up the whole building, wall to wall. If you're worried that one brace won't be enough, add another, but you'd be surprised how well one will do. Because the wall is skeletal at this stage, wind can blow through freely with very little resistance; so the wall can stand with pretty meager bracing through some fair winds.

If at any point in your wall framing you find yourself confronted with a standing wall that has a stud obviously nailed wrong (it will be "leaning" in place), you can remedy the mistake without taking the wall down. Usually the tilting stud means you've nailed to the wrong side of your line, and a quick check with your tape will reveal the error. You can hit the stud loose with your sledge—a few gentle but firm taps right near the plate will usually knock the stud loose—and can then pull it free and cut off the tips of the protruding nails, using nail nippers. If you find the wall is moving when you're sledging, try standing on the bottom plate to steady it. Place the stud correctly and toenail it in place.

Framing the Wall with Windows

A wall with a window or windows will take slightly more time to frame up, as there are more cuts and complications. Still, the work is simple and clean, and once you've framed in one window and seen how each piece fits in, the next one will go much faster. Each window will have to have a rough opening framed in, which allows space for the window itself plus the frame surrounding it plus some space for squaring up and leveling the window and frame. If you are working with blueprints and/or with prehung or premanufactured window units, you should have ready access to your rough-opening dimensions. Blueprints should give you rough openings and the height of your rough sill. If you have no blueprint but are using a prehung window unit, measure it and add ½″ to the height and to the length for your rough-opening dimensions. The height of your rough sill in this case will be determined by the placing of your window and how far up you want the lower edge to be from the floor. If you're going to make your own window frames and need to figure out your rough openings, you'll have to know:

The size of the window, or combined size of a pair, if, for example, you're using a pair of casement windows together. In measuring the height of a window, take the measurement from the top of the sash to the lower end of the bevel at the bottom of the sash. In measuring the width of a pair, set them together and take the full and exact measurement.

The size of the jamb material. Usually nominal 1× (actually ¾″) material is used for window jambs, but you may be using recycled materials and have some true 1×.

The size of the sill and angle you'll be placing it at. You can buy special 2× sill material designed to be installed at about a 15° angle, but you might want to use other 2× material or even 1× tilted at any angle you feel will shed water properly.

With this information assembled, you can figure out your rough opening. To get the *width,* add 2″ to the actual width of your window (or pair). This will allow for two ¾″ side jambs plus ¼″ on each side of the frame for leveling. If you use actual 1″ jamb material, add another ½″ to this width. To get the *height* for your rough opening, add 4¼″ to the height of the window—more or less, depending upon the sill material you're using, the thickness of the jamb material, and the angle you're placing the sill on. The 4¼″ allows for a ¾″ (actual) head jamb, with conventional 2× sill material at about a 15° angle, and a ¼″ to ½″ play above the head jamb for leveling.

Next figure out the height you want to place the window at. You may have planned this already, but now that you're actually walking around on the floor platform you might decide to move a window or change locations of one with another. With this decided, you can set about your framing. I like to precut all pieces for a rough opening and then assemble them all at once. Make a list of the pieces you need and the mea-

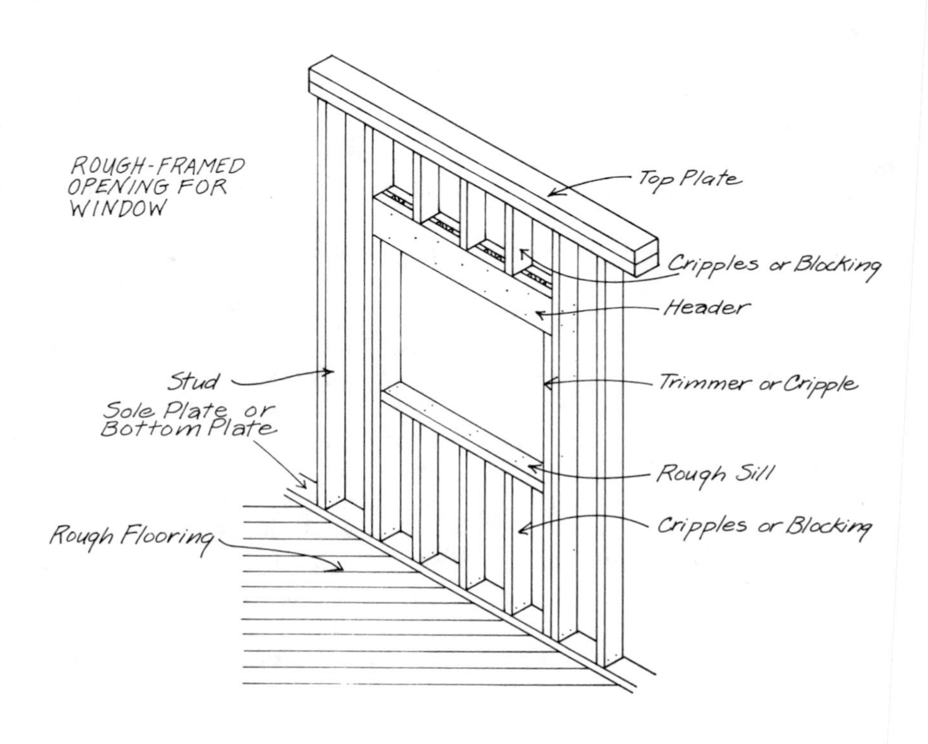

surements for each one, and as you cut a piece mark it with pencil or crayon. You'll need:

A **full-length stud** to each side of your window opening. You may want to use one of your regular o.c. studs for one side. The other side will be an "extra" stud in the wall. It's important that you continue the regular o.c. spacing of studs throughout a wall—when you come to sheathing or paneling later, you won't have to stop to add in extra nailers.

A **trimmer stud** for each side of the window. These studs will support the header. The length of each trimmer equals the rough height of the window (as figured above) plus the distance from the floor to the top of the rough sill. For example, if the rough height of your window is 34½", and you plan the rough opening of the window to begin at a height of 36" from the floor, your trimmer studs should each be cut 70½" long.

A **header.** The length of the header should be equal to the rough width of the window, as figured above, plus 3". This allows for 1½" each side—the portion of the header resting on the trimmer studs. You may use a solid piece for the header or build up a header by splicing two pieces together. In the case of spliced headers, you'll have to add a piece of ½" plywood or particleboard to make the header equal in thickness to the studs in the wall. This will give you a flush nailing surface for later sheathing or interior siding. The plywood piece may be sandwiched in between the two header pieces or added after to one face. In making the built-up header, it's important to splice the pieces very carefully with all edges square to one another. Use your combination square to check as you nail. The header can be spliced with 16d nails driven in at opposing angles in a staggered pattern, with pairs of parallel nails at each end.

Rough sill. This piece makes a base for the actual finished sill to set on. It is usually made with one 2×4 cut to a length equal to the rough width opening of the window. If the window is wider than 4', a doubled rough sill is used.

O.c. stud fillers above and below the rough opening. These short pieces of studding—also called cripple studs or cripples—are used to continue the o.c. spacing of the studs above and below the window opening, so they should be cut to lengths needed to fit. They can be added later.

With all of the pieces for your rough window opening cut and ready, you can cut the top and bottom plates for your wall section. Nail your two end studs in place and mark your 16" (or 24") o.c. locations as explained earlier. Remember that if this wall section joins another, you have to make allowances for your centering so that the extra stud doesn't throw things off. Add all regular studs until you get to your window location. Measure and mark for the studs on either side of your opening: the distance between the two studs will equal the length of the header. Before placing the studs, nail a trimmer stud to each. Put the studs in position and nail them as usual, two 16d nails per end. Then slip your header into place. If there's room, nail the header with two 16's driven through the studs on either side and into the ends of the header. If you can't manage this because adjoining studs are too close, toenail the header to the studs. The next piece to put in place is the rough sill; this can be toenailed, two nails per end. Then nail the short studs below the rough sill and above the header (again, toenail with two nails per end). Now double-check all of your work. Is your rough opening correct? Are all of your o.c. marks on the bottom and top plates occupied by either a full or a short stud? You can finish up the wall by adding bracing, if needed, and fire blocking and double corners. Lift the wall or wall section into place and secure as outlined earlier.

Rough-framing the Door Opening

The door opening is framed very similarly to the window opening, except that there is no rough sill. A header is used over the opening, supported on either side by a trimmer stud, which is nailed to a full-length stud. The usual formula for determining the rough opening for an exterior door is simply to add 2½" to the actual height of the door and 2½" or more to the width. The amount you add to either height or width depends upon the material you're using for jambs and the type of sill you use. Jamb material may be 5/4 thick or ¾"; manufactured door jamb material is made of interlocking sections ribbed on the back to minimize warping and is 5/4 thick. You should allow at least ¼" on each side of the door for swing and ¼" to ½" for shimming (to plumb). Remember in figuring the rough opening for your door and cutting trimmer studs that the section of bottom plate under the door will be *cut out* after the wall is secured in

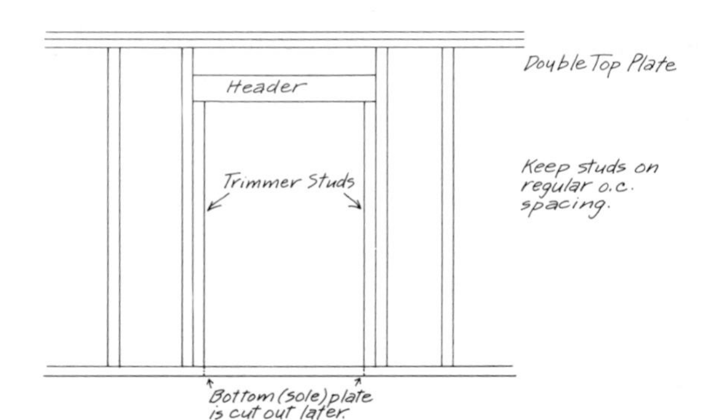

place. This gives an extra 1½" to be considered in calculating the rough door opening and length of trimmers. Cut and assemble header, trimmer studs, and adjoining full studs. This will complete your door framing and you can go on to finish up the wall as usual.

Framing In Gable Ends

This part of wall framing is not actually done until later, when the roof rafters are in place. The normal 16" o.c. stud pattern is then continued between the top plate of the wall and the rafters located directly above the wall. The studs may be angle-cut to fit under the rafters, but a better solution is to notch each stud to fit into place. In this case, each stud must be custom-cut to its position. A notch should be made, 1½" deep (or the thickness of the rafters) and angle-cut at the bottom so that it fits snugly in place. This procedure takes more time than simply angle-cutting the studs to fit up under the rafters, but is a way of providing a nailing surface for interior walls.

Framing Partitions

Interior walls are rough-framed exactly as you've done the exterior walls, except that you may space studs 24" o.c. for nonbearing partitions, and doorways in these nonbearing partitions will need simple 2×4 headers. Some partition walls may require 2×6 framing, especially in areas where large plumbing pipes have to pass between the walls. Closets may be framed in very lightly, with 2×3 material in many instances. Partitions are usually framed in as time permits; bearing partitions should be framed at the same time that exterior walls are done, but nonbearing partitions can be done later if you want to get the roof on first.

Squaring and Securing the Walls

If you've been very careful in all of your work, squaring up the building and nailing the walls securely in place goes fairly smoothly. This work follows the construction and raising of the walls, which are now held in place by temporary braces. Each corner of the building must be checked to make sure that it is perfectly square: i.e., that the walls are running out from one another at 90° angles. This is very important because the success (and ease) of your roof framing will rest upon the squareness of your top plates and the building in general. If you are ¼" off in one corner, this can create problems for your rafters. As usual, there are many ways to approach this work and every carpentry book you read will probably suggest a slightly different method. Here are some of the basics for you to follow until you evolve your own tried and true method.

Each wall should be plumbed up as you work and will be joined to the wall next to it in two ways. First, the corners of the walls will be spliced (nailed together); second, the double top plate will be put on in an overlapping fashion. The walls will be nailed to the floor substructure. Later, the wall sheathing and siding and the roof will add to the overall cohesion of the building.

You can begin with any wall. First, check to see that it is standing plumb and that it is closely aligned with the outer edge of the floor platform. Do the same for the adjoining wall. Now use the 6–8–10 method described in the "Layout" section earlier in this book to determine whether the corner formed where these two walls meet is square. Adjust the walls as needed by loosening the braces that hold them and sledging them into place bit by bit. If you're working on a small building, you can do all four corners as accurately as possible and then measure the diagonals of the building to double-check yourself. On a larger building, you can create smaller sections to check this way by running chalk line "divisions" if necessary. When you're satisfied that the walls are all plumb and square, nail your corners together with 16d or 20d nails. The second, or double, top plate can be added as you go; take care to measure and cut so that all joints are overlapped and so that the double plate serves to secure the walls together. The nailing pattern for adding the second top plate should be staggered 16's face-nailed about 16" o.c. If you set the nails in at an angle, they'll hold better and be less likely to come through the lower of the two plates. Give your corners a second quick check (6–8–10) and then nail the walls down to the floor by driving 16d or 20d nails through the bottom plate roughly 16" o.c. Place the nails so that they pass through the subflooring into a joist or header joist, not just into subflooring. At this point, unless you've left bracing or fire blocking until the end, you're finished framing your walls and can contemplate closing them in.

A Hand-built House Is a Special Kind of Home

Winter came early the year I built my house. Whoever heard of pouring rain all through August in southern Oregon? This was a true test of my endurance. Every day I'd get out there, and as soon as I'd start hammering, the drizzle would turn into a downpour. I spent those weeks in frustration, attempting to fight the weather and losing. I tried to do other things. Everything was a dismal failure. Canning jars would break, lids wouldn't seal. It got so bad that I'd even go drown my sorrows at the Wonder "Blur," a local tavern.

The rain also gave me a lot of time to think, a dangerous activity, for often my head would fill with doubts. Were these bad omens?—the heavy rains, the awful rowdy neighbors who just moved in. . . .

If only no: lights, dogs, neighbor noise, cars . . .

There'd be just: mountains, moon, silhouettes, creek sounds.

But even if the woods of southern Oregon weren't paradise, there I was in the midst of a project and determined to see it through. Builder's blues would come and go, but the house would keep on growing.

It would never have occurred to me five years ago that I could build a house. Living in the suburbs of New York City, one didn't think of such things. Besides, the biggest thing I'd ever built was a spice rack. For anyone living in or near a city, with little or no building experience, this idea could sound completely off the wall. Furthermore, money and land are scarcely available to many women (I was very fortunate to have a sister who owned land she was willing to share). Yet even if you hate to saw, are allergic to poison oak, and wish you didn't have to get up so early in the morning, you can still build a house. All you have to do is follow the example of the Chinese on the Long March and take it one step at a time.

Getting Started

Construction follows a certain indisputable logic. In most cases you start at ground level and go up (treehouses excluded). But the process of building a house involves a whole stage that precedes actual construction, generally known as planning. Deciding where, how, and what to build and collecting resources and tools can demand as much time and energy as the actual building.

Recalling my own experience, this was the hardest part to get through. There is real danger of getting hung up at this stage because the decisions can seem endless. If you find this happening, take heart—it's just the chronic problem of getting started. Once you get over this inertia and move on to construction, it's all up from there.

(Photo by Marsha Emerman)

Here's what goes on during planning. You have to reckon with the following: choosing a site; drawing plans; the county building department (if you choose to work within the law); tools and materials; budget; and resources like books, advice, and finding people to help. These all require mental energy, which can be a lot more exhausting than physical energy, especially when it involves what feel like earth-shattering decisions (sure, you might shatter a little earth when you start digging for your foundation, but after that it all gets simpler).

Choosing a Site

I spent about two weeks tramping around on the land, exploring the possibilities. I thought most about: privacy, access, and light. Privacy meant distance from other houses and the road. Access meant how easy it would be to get tools and materials to the building site, and to go to and from the house in winter. It also would affect the installation of electricity and a water system later on. Light meant how the building would fit into the terrain, how dense the surrounding forest, how high or low on a hillside. Weighing these factors, I found access and light demanded some compromise of privacy. Of the sites under consideration, the one that had easiest access and best light was on a hillside, making it more visible to the road and to a neighboring house. Also, being visible would make compliance with building codes more crucial.

I went back and forth many times over the relative merits of two sites. One, which was nicknamed "the Island," was more secluded (I'd have to build a footbridge over which to haul in materials) and closer to the creek (I'd hear the creek at night, but the sandy soil might not support the foundation). It was also a spot I had come to know and love, having pitched a tent and lived there for four months, two summers back. By comparison, the hill site felt unfamiliar but had special features. Light filtered through the trees in an appealing way and birds could be heard singing. There was a clearing just large enough to accommodate a 16′×16′ structure and surround it in a ring of slender trees that at night formed a perfect arena for the stars. Beside the clearing stood a small yew, a tree sacred to the American Indian and used for making the wooden bow. After much deliberation, I chose the hill site and to that piece of ground committed my future fantasies.

Drawing Plans

People often ask if I designed the house myself. Although vaguely modeled on some similar structures in the neighboring countryside, the design was just the simplest way of fulfilling certain basic criteria. I wanted a house that would be large enough to feel comfortable for one person, but small enough to be inexpensive to build and easy to heat. The 16′×16′ dimensions seemed appropriate. Why square instead of rectangular? Well, someone suggested that square would feel more like a circle instead of a box, and it made sense.

I wanted a sleeping area separate from the living and cooking area, and a loft could satisfy that, provided it had sufficient headroom. I was never crazy about lofts where you bump your head sitting up in bed, and as I needed only 5′2″ headspace it didn't seem so hard to construct a stand-up loft.

Given the 16′×16′ downstairs dimensions, a loft of 8′×16′, with half the downstairs floor space, was adequate. Since standard stud wall construction uses 2×4's which, together with the floor and ceiling plates, total 8′, that would be the height of the walls. If the walls were to be 8′ high, the loft floor could be built over that, as a separate unit. That would keep the 8′ ceiling height downstairs, which felt more spacious.

An alternative would have been to build an interior loft once the basic structure was complete, a method common to post and beam construction. I chose stud wall construction because, with my limited experience, it felt simpler to do everything in order, from the floor up. Somehow it was impossible to imagine building the roof before the walls, although I'd read about doing it that way.

Having determined dimensions and a basic shape, my other concerns were to allow for lots of light (so I could feel more like a plant in a greenhouse) and to keep the space as open as possible by avoiding center poles or posts and by use of a ladder rather than a staircase to the loft.

Before drawing the actual plans, I decided to alleviate the fear of reprisal from an irate county if I dared to construct a home that had not been given the "Good Housebuilding Seal of Approval." So it was off to the building department to find out how to become official.

Building Codes

We've all heard that laws are on the books for our protection. That contention begins to look suspect when one finds oneself in the dubious position of applying for a permit to construct an outhouse or to remodel a room in an old house. Nevertheless, in most areas codes exist and we have to deal with them in one way or another. At least one positive contribution of building codes in Josephine County is that they have helped to create coalitions of old-time farmers and more recent back-to-the-land folks who are equally opposed to all the hassle and interference.

So, faced with the question of compliance with codes, one has to consider several things: Is it worth the trouble if they find your unapproved structure? How much more expensive will it be to comply? Can the county building code personnel be helpful? My decision was a compromise. Instinctively, I was disinclined to comply. It sounded about as much fun as going in to pay a slew of parking tickets. But because my house was distantly visible from the road and other approved constructions would be going on nearby, it seemed inescapable. So I applied for a studio permit. The distinction between studio and house is that a studio does not require a septic system, as does a house, because one does not officially "live" in it. A studio can be any size, can have electricity and plumbing. In order to construct a "nonresidential" studio, it is helpful if there is an existing house on the same property. Otherwise, code requirements for a studio are the same as for any other structure. The codes stipulate such things as diameter and depth of concrete piers and footings for the foundation; lumber grade; dimension and span required for the subfloor, floor, walls, and roof, as determined by the size of the structure.

My first visit to the building department was intimidating. They told me I needed to present a foundation plan, a side-view section, and a floor plan detailing window and door placement. They further instructed that three inspections would be done: the completed foundation and subfloor; the framed skeleton of walls and roof rafters, and the finished structure, including fire safety. Being forced to present my fantasy on paper actually turned out to be helpful. The plans I drew were rough and did not detail all areas, but they provided some guidelines to follow. It's amusing to look at these plans now. The first drawing looks like a cover for a school term paper, complete with teacher's corrections. The foundation and subfloor plan was more specific. It called for nine concrete piers and footings on 8′ centers. The concrete would be poured into holes dug 12″ deep and 18″ in diameter (for Oregon frost line, that's deep enough). Some kind of metal strap connecting the concrete to the wooden posts would be set into the wet concrete. Then 6×6 posts would be bolted onto the piers.

My actual foundation is constructed just as I drew it for them, with one exception. The building department will in all cases tell you to use the most expensive items, prefabricated metal joist hangers and post caps and Select-grade lumber. How rigid the department is in enforcing these requirements will vary. I tried to meet their standards as best I could afford, with some fudging here and there. For example, they red-penciled onto my plans little items like "steel post bases cast into concrete (Simpson or similar)." These cute little items cost $15 apiece. Economizing, I got a tip on using railroad crossties that could be found along the tracks for free. These were 1′ steel straps that took only a welding torch to put holes in and served the purpose fine. I rarely used "Select Structural #1 grade lumber," at many times the price of other grades and not always better, but if I got an occasional #1 board I always made sure to put it in a place where the #1 stamp would be sure to catch the building inspector's roving eye.

In the side-view section there is considerable detail, but it is far from the complete picture. In drawing it I followed examples supplied by the department, as well as consulting outside help. As a final requirement, I had to present a floor plan; this was a complete hype. At that time I had no idea where I wanted the windows and doors to go, so I just stuck them all around each wall on the drawing. In retrospect, dealing with the building department was not entirely bad. It helped me to clarify some things and to get more organized. It didn't actually alter my plans much. There was just one instance—how to support the loft—where code compliance made me spend a lot of time and energy on a problem and use a solution I might not now have chosen.

Submitting the house to inspection was the least desirable part of dealing with the building department. Although Larry, the friendly building inspector, came out only twice and approved everything without qualification, his presence on my land and in my home was not by invitation. If the department would stick to offering guidelines, advice, and information to people building their own homes, and enforce its restrictions on building contractors or landlords who don't maintain their dwellings in decent living condition, it would be providing a public service.

Tools and Materials

Once you've chosen the size of the structure, tools and materials determine its cost, ease of construction, and appearance. Plans in hand, you enter the vast marketplace of building supplies and begin shopping around for bargains, for laborsaving devices, and for the new wardrobe of beautifully grained woods, uniquely framed glasses, and so forth.

If you enjoy shopping, this can be fun, but also frustrating. Budget limitations play a big part. So does the amount of time you have to spend. It's possible to build something beautiful for very little money with imaginative ideas, love, and labor. Gathering natural materials or finding old buildings to tear down in exchange for lumber are two constructive suggestions. Ordinarily, the more time you have to collect materials, the less money you'll need to spend. If you have time to read the building column of the local newspaper, listen to the "bargain roundup" on the radio, check signs posted at the country store, and stop off at every yard sale, flea market, and auction you see, you'll find lots of cheap building materials. Chances are, these won't fit exactly into your plans, so it'll mean making adjustments. Even when buying new material, it's better to shop around. Lumber prices go up and, rarely, down. One day 2×4's will be cheaper at one store, the next day at another. Small mills may have some good deals.

In the case of building a house, consider the following issues, as yet pending decision: New Versus Used Lumber, Borrowing Versus Buying Tools, and Power Versus Hand Tools.

New Versus Used Lumber I wound up using mostly new lumber, owing to a time restriction. My goal was to have the basic structure completed in four months and be living in it by winter. At that pace, obtaining used lumber was too time-consuming. I did a bit of this, and found that in tearing down a building you have to be very careful not to rip the boards, and will spend a long time pulling nails.

Used lumber has some real advantages. It's cheaper, stronger (they used to make lumber full dimension—a 2×4 was a 2×4, instead of a 1¾×3¾), and more aesthetic (no ugly stamps that say "Utility grade"). And it's being recycled. When one lives in an area where poor logging practice has destroyed much of the forest and threatens to continue, the ecological advantage of used lumber becomes clear.

When buying new lumber, I found that most of what is sold, except perhaps for the most expensive grade, is of very poor quality. Most lumberyards are loath to let you pick through the piles for the boards you want, but whenever possible I did. Sometimes you'll go through ten boards to find one usable one, without a lot of knots, cracks, or improperly milled edges. Judging lumber is a skill acquired by necessity, or else you pay a lot of money for a lot of junk.

People in lumberyards can be helpful or hostile, particularly to a woman. Building supply places were usually helpful, since the salespeople are used to dealing with customers. Large mills will occasionally sell to the public but prefer not to. They don't like to be bothered with small orders, let alone with a woman who pulls up in her pickup and begins to load her own lumber.

It's important to use the best quality and strongest type of lumber available for all the parts of the building that will support weight (floors, walls, rafters, roof). I used fir for all the structural lumber and the exterior siding, and cedar for the interior paneling and ceilings.

Milled lumber is invariably square (rectangular). So is my house, and this squareness got tedious. Square lumber is easier to work with than poles, because the pieces are relatively uniform. This simplifies matters when it comes to laying a floor or a roof. Preferring their roundness, texture, and grain, I had resolved to use poles for roof rafters, where they'd be most visible. Later, I had to abandon this plan when building progress was too slow to allow for finding the poles, stripping them, waiting for them to season, and getting them in place. So there are just two poles inside my house, under the beam that supports the loft. But I found other ways of incorporating round shapes—a pole ladder and natural edges on tables and counters.

Borrowing Versus Buying Tools It's much better to have your own tools, for practical and psychological reasons. Running around trying to locate the right tool when you need it will waste a lot of time. Forgetting to return tools right away or breaking them won't make the lender very happy, either.

When I bought my first tools, I felt like a *real* carpenter (being brought up like most girls, I never did play around with tools much, so I guess it was like having a bunch of new toys). I bought all the basic items: an 8-point crosscut saw, a 16-ounce hammer, a 16′ measuring tape, a combination square, a wrecking bar, a chisel, and carpentry pencils. Later, I got a few other things like a utility knife, chalk line, and nail set. I had use of a level, a brace and bit, and a longer framing hammer on extended loan. Clad in my shorts and canvas apron, tools in hand, I was ready to work.

Power Versus Hand Tools Nowadays people rarely rely on hand tools alone. After all, if an electric can opener does the job faster than a manual one, how much more so the difference between a power saw and a handsaw. Yet one can still find people who swear they can saw through a board faster by hand. Such zealous characters are far outnumbered by people who think anyone who prefers using hand tools is nuts (we're in the machine age, and just because Grants Pass has a caveman for a symbol doesn't mean you have to pretend you're living in the paleolithic)—why do more work than you have to? Well, several reasons. I didn't install electricity until the house was complete and I was living in it. A power hookup could take several weeks, and I didn't want to wait until then to begin building. It would also mean getting another permit.

Since I owned no power tools, there was no need for on-site power. My brother-in-law was generous in allowing me to use power tools at his building site when I really needed to. In a few cases, I'd truck my lumber over and cut it there. More often, I'd stay at home and hand-saw (it might take more elbow grease but it was faster than transporting the lumber back and forth).

In the course of using hand tools, I got to like them. It feels good to make a nice smooth cut with a handsaw. The saw becomes an extension of the arm, its rhythm becomes your rhythm. And hand tools are blessedly quiet, compared with power tools.

Unless someone is a "power freak" or an anti-power purist, I'd suggest that a combination of hand and power tools, whenever each seemed most useful, would be the way to go. A small circular saw would probably be the most versatile power tool to own.

Budget

My structure cost about $1,300 to build. That is an approximation, based on records I kept of all my purchases until the final stages. When I add on later expenses, I figure on having somewhere around $1,800 in the building.

For a house of its size, the investment seemed reasonable, certainly a lot better than spending the same amount in rent. The house will probably be around for a long time, whether I live in it or not, so it seems like a deserving project. Besides, it gave me a chance to learn new skills and create something lasting.

My expenses broke down like this:

Foundation	$ 70	Electricity	$200
Lumber	800	Water system	100
Insulation	100	Stove and pipes	150
Roofing	100		
Glass	60		
Tools	40		
Miscellaneous	100		
Building permit	30		
	$1,300		

Since each builder's budget will be entirely different, this is intended only as a sample. One could spend a lot more or less, depending on available resources and priorities. Sometimes what seems like a large expense will be worth it in the long run. For example, plywood is very useful in adding strength to a building (bracing isn't always sufficient). Although it's expensive, I used plywood sheathing on the back wall, an area 16′ wide by 16′ high, beneath the exterior siding. This served to eliminate the wobble from the loft. Fiberglass insulation can be another big expense, but will save on wood chopping or utility bills later on. My suggestion is to decide what's important and then economize where it won't affect your comfort.

Resources

Superwoman might build a house single-handed, but the rest of us need help. Part of the building process is learning how to find available resources and use them. If the project is to be a collective one, this becomes the job of many people pooling ideas, skills, and energies. If it is primarily an individual project, as mine was, a lot depends on knowing how to seek help when needed.

Books and articles can be useful. Two sources I'd recommend are *Illustrated Housebuilding* by Graham Blackburn and *I Built Myself a House* by Helen Garvey (see "Resources"). The second book was encouraging and helped to demystify the building process. It was less useful as a how-to on building than as an example that building need not be the exclusive realm of experienced carpenters. Since I didn't do post and beam construction, a lot of the techniques described weren't relevant to my structure. *Illustrated Housebuilding* was well detailed and easy to understand. All books, however, are most understandable in conjunction with experience. I remember when Paula and I began framing the first wall. We had *Illustrated Housebuilding* open to the chapter on framing and kept running back and forth to the book to look at each picture as we worked. It didn't prevent us from doing a whole bunch of things backwards.

Another thing books won't do is tell you to take a break when you're working too hard, or remove a splinter, or come by to admire your progress, or bring you a cold lemonade. Friends, however, will do these things, which can make all the difference between a good day and a lousy day.

Friends are probably the most valuable resource. I was lucky to have help from a lot of different people, some of whom had already built houses. When I visited them, I found myself staring at whatever part of the house I happened to be working on at the time. Suddenly their floor would take on a new fascination. "Hey, Marsha, what's wrong, you're not eating your dinner and you keep staring at the floor," they'd say.

Dozens of people actually showed up to help. But just counting on them to show up didn't always work, so I tried to prearrange to have someone come whenever there was a job that needed two people. I did a bunch of labor swaps with other people who were building. Paula was building a house at the same time, and Marilyn was adding a room onto hers. The days we worked together were among the highest, since there was real camaraderie and sharing of decisions. It was best working with someone on an equal footing, in terms of experience and knowledge, rather than directing or being directed by people who knew less or more about building.

At first, when I lacked confidence in my own skills, I was often confused by seeking help and getting conflicting opinions on how to do something. Gradually it became clear that there were always several approaches, methods, or solutions. Eventually I was able to say, sure, it can be done that way, but this will work, too.

So, having chosen a site, drawn plans, had them approved by the county building department, bought tools and materials, read books, consulted friends, I stopped and thought, "This is ridiculous—what am I doing all of this for when I could rent a nice clean apartment in the city with running water and lights all hooked up for just a few hundred bucks a month?" I was getting cold feet. Then I remembered what they said about Rome and that if I didn't get started soon my feet were going to get a lot colder. It was time to put my nose to the grindstone and get it polished.

To Work

June 4 Clear site.
June 7 Square corners.
June 8–9 Dig holes.
June 12 Unload cement, build forms.
June 13 Haul stones, find metal straps to anchor posts to cement.
June 14 Pour foundation. It takes twenty-seven 90-pound bags of Redimix (builder's bisquick). Unmolded, they look strange. They look like birthday cakes! Hope it's all still square.

That's a composite version of the first ten days of building from my journal. (I'll continue to refer to my journal for a general sense of the progress of the building, adding some specific information on how things were done.)

There are many ways to do a foundation, depending on how big the structure, how solid the ground, and whether you want it to stand for the next twenty or two hundred years. It's cheapest to hand-mix concrete from cement, sand, and gravel, but Redimix is convenient and contains no harmful preservatives. If your structure is accessible by road, you can hire a cement truck and pay by the yard for the operators to mix and pour the concrete. Or you can rent or borrow an electric cement mixer. I used the old shovel and wheelbarrow routine and, with two of us, it was done in one long, arduous day.

June 15–18 Let cement sit, covered with wet burlap bags (you're supposed to wait at least a week). Lumber trip to Selma and Merlin. At site with transit, measure heights required of posts to bring it to level. Cut posts and set in place (a transit isn't necessary but makes it simpler).

June 19 Fussing with posts, trying to get square, setting strings wrong. Confused and concerned over all the adjustments needed. The two purchased metal post anchors (couldn't find enough of the railroad ties) are for 6×6 posts and I'm using 4×6. The post anchors were set in different directions into the cement, and they're not perfectly square. I'm 3″ out of square.

June 21 Get help unraveling problems. Reset strings and get posts as close as possible to square. Bolt in posts and put blocks and shims where needed (shims are a miracle device that can make everything okay, just when it seems hopelessly screwed up). Set the three 4×8 girders on the posts. Two out of three are badly bowed.

June 22–23 Treat posts with creosote. Hold a beam sit-in. It takes six people sitting/standing on those girders to get them down so they can be bolted on. (You might check your lumber out at the yard to avoid this, but it can still happen. Under sufficient pressure, however, a bow can be straightened.)

June 24 Nail together sills and joists into a box. Set the rest of the joists (2×6's that support the floor) in place. Marilyn arrives at 4 P.M. and we nail all the joists and blocks into place by 9 P.M. Good working rhythm, high on seeing so much done.

June 25 "Herr Redtag" (nickname for the building inspector) comes today and approves the foundation and subfloor.

June 26 Begin laying tongue-and-groove flooring (2×6).

I did my floor diagonally. It's stronger that way and looks nicer, though not intended as finished flooring. This job starts out easy and gets harder toward the center of the floor. If the diagonal is slightly off, it keeps getting more off with each board. The solution is a wood-splitting maul, a scrap board so you won't smash up the tongues, and lots of banging to drive the boards together. It's best to have two people on the long boards.

July 5 The floor's finally nearing completion (hand-sawing off all the overhanging ends is slow work). Night of the sixth, with four boards to go, I do a midnight moonlight dance on my new floor to celebrate. Imagine how it feels to be on your own private stage, in the midst of the woods, no one to watch but the stars, your body delighting in the night air.

July 12–15 Gathering windows, planning where to put them—so hard to imagine in the abstract. Moved table and benches onto the floor for a writing place.

Window-shopping used to mean strolling along Fifth Avenue in New York with Ma, admiring expensive clothes. Now it's shopping for windows, big ones, multipaned ones, ones that don't have purple frames with six layers of paint to strip.

I overdid the windows. It seemed outrageous to close anything off, when it was all so wonderful out there. I was forgetting that you need wall space inside a house, and that too many windows weaken the walls. Consequently, I had to omit some of the windows later, after they'd been framed in. By placing windows high up on the walls, especially in the area planned to be the kitchen, there was space left under them for a sink, counter, and so on. Their high placement also let in more light. I framed in a front door, but had to go back to frame in a door onto the back porch.

In retrospect, my approach to window placement was silly: I literally ran around on my floor holding up windows in different places to see how they looked—they looked pretty funny in midair. However silly the method, the results were just fine.

July 16 Begin framing. Paula and I build the first part of a wall.

July 17 Awakened by thunder and rain at 5 A.M., breakfast at Hungry Wolf (forgot to mention that I'd moved onto the floor and was sleeping out there every night). Begin work again and at 5 P.M. finish framing the wall, and three of us lift it into place. First wall is up!

July 18 Work alone all day, finish most of the north wall.

Framing is enjoyable work, and one of few jobs that is just as easy to work on alone. The walls can be built entirely on the floor and then lifted. It's exciting because, as each wall goes up, the building takes shape. People said that framing would go fast after the early stages, and it's true.

July 19 Finish building wall number two and raise it up.

July 20 Work alone again till 9 P.M. Finish west wall. Good, exhausted, proud feeling.

July 21 West wall goes up. Its corner doesn't meet the north wall. Leave it for now and continue building south wall.

July 26 Finish and lift south wall.

Remember that corner where west and north refused to meet? Well, that caused a lot of grief. A good 4″ separated them. Tried pounding, pulling, pushing, all to no avail. Then, magically, some friends showed up from Keto's garage and suggested using a car jack to hook on one wall and crank it over. It worked!

Got the old slow-down blues. Oppressive heat, a whole week goes by peeling poles, drilling, and doweling, all in preparation for the "BIG BEAM" (a 4×16 piece of lumber, the dimension required by code to support a 16′ span holding up the loft). This beam had to be special-ordered from a mill and was an unwieldy piece to work with. But it allowed for keeping the downstairs space open and uninterrupted by vertical poles or posts.

July 31 The BIG BEAM arrives, and four of us lift it—a moment of terrified suspense until it sets down comfortably onto the dowels. It looks oddly out of proportion with the slender framing lumber.

August 1–8 Heaviest rains for August in Oregon history!

August 2 Do floor joists of loft, same method as downstairs.

August 3 Begin tongue-and-groove on loft floor. Boards won't drive. It begins to pour. Feeling frustrated and homeless.

August 8 Finish loft floor. Full moon.

August 11 Tear out a section of wall to frame in a back door. Do bracing of downstairs walls with diagonal pieces of 2×4 (like pulling out stitches in sewing, it's a pain to go backwards, but not so bad once it's done).

August 13–16 Rain all weekend. Canning peaches.

August 18 Elbow-grease farewell gig at the county fair.

August 16–25 Framing loft walls. Same procedure as downstairs.

August 26–29 Get a complex explanation of the use of the framing square to figure rise and run of rafters. "Theoretically" you cut one and the rest should be the same. (The theory is thrown off if your building isn't perfectly square.)

There are ways to do rafters without using a framing square. You just have to do some climbing around on top of the walls taking measurements, so as to make the needed adjustments in the angle and depth of the rafter notches.

With the rafters notched, the house appears as a completed skeleton. Got to get a roof on before it rains again. This calls for some concessions. Sure, plywood looks ugly from the inside, but it'll be covered with all that lovely, sparkly silver insulation until I get around to doing an interior ceiling.

(Photo by Marsha Emerman)

August 30–September 1 Put up plywood on both roofs.

Plywood is remarkably fast to work with, compared to individual boards. It's a drag to cut with a handsaw, however. Shop and reject plywood (grades of plywood with various defects) can be gotten a lot cheaper than the better stuff, which can run as high as $12 per 4′×8′ sheet. Mine cost $3.00 a sheet at the plywood mill. Not all cheap plywood will be worth buying. Sometimes the glue doesn't hold and the layers may start peeling apart.

September 2 Tar paper and roofing. Good old green rolled roofing to make the house blend in with the trees, so when those helicopters go over they won't see it.

How come the guy in the building supply store told me there's no difference between the two kinds of roofing cement? One of them is thick like black licorice toffee (get that kind); the other is runny and gooey and drips all over everything. Be sure to cover all exposed nailheads and seams with the stuff.

September 3 The roofs are done! (It'll sure feel good to stay dry when it rains.)

September 6–12 Odds and ends, felt papering, trip to purchase 1×8 fir siding, more wall bracing.

September 12 Begin siding, cover half a wall—beautiful! Now the house really has a shape and light patterns.

My siding was done vertically in the board-over-board fashion. That means that I left 4″ to 5″ between every two boards and then covered that space with a third 1×8 board. This gives a similar effect to board-and-batten, but saves having to rip wider boards into narrow battens. The siding lumber came in mixed lengths, so in some cases the boards could run all the way from the bottom to the top of the loft, which helped tie the whole building together. When friends helped out, we took turns on position; the person below got a stiff neck and eyes full of wood shavings; up on the roof, the other person was getting dizzy staring down.

Putting up the siding took about two weeks. When the siding was finished, major construction was over. By then I'd started a part-time job and house progress was slowing down. Some days slipped by like this:

October 12 Hardly worked on place today, beautiful autumn day. Play banjo awhile, and just be here . . .

During the last week of September and all of October I worked on the final things needed to make the house ready for winter: installing doors and windows, building steps, doing roof flashing, insulation, and the wood heat stove. These final stages merit some description.

Doors and Windows Plastic (4 mil, double-layer) will hold in heat as well as or better than glass, but it won't let you see out as well. There are fancy moldings and sills to use in framing windows, but I did mine simply. I put in my favorites, a set of multipaned french windows and a beveled glass door given by a friend. Some windows I'd collected were too funky to use, so those got covered with plastic.

Steps Steps and rafters still baffle me, so I wouldn't try to instruct anyone else on how to build them. I used half rounds of lumber, milled from our land, and delighted in climbing up and down.

Insulation Fiberglass is expensive, terrible to breathe, and unpleasant to work with, but it will keep the house warm. Builder's foil (foil-backed paper) and Styrofoam are cheaper alternatives. I stapled foil up under the floor, with a network of battens to hold it up, and used 4″-thick fiberglass in the walls and ceiling.

Wood Heat Stove Colette wrote, sitting by the hearth: "The fire is my friend and the work of my hands." In a wood-heated house, fire building becomes a daily ritual and the stove fills a special place as provider of comfort.

Since it can also be a fire hazard if not properly installed, care must be taken. Double-insulated pipe is expensive but the safest. If you use regular stovepipe, you can surround it with asbestos sheeting (sold in hardware stores by the square foot). All pipe and stove must be at least 12″ to 18″ from any wall or burnable surface. I used a flat sheet-metal plate inside and a metal roof jack outside to direct the pipe through. The pipe should go up a little higher than the highest point of the roof. The straighter the pipe is, without bends and elbows, the better the fire will draw and the less likely the pipe will clog with creosote.

Another suggestion is to drill little holes and put three sheet-metal screws where each section of pipe joins the next. This will stabilize the pipe, especially outside when it's windy.

October 25 A landmark day. I finish installing the stove and build my first fire. It's warm! Spend the evening in my house and it feels fantastic!

Moving In

After nearly five months of concentrated effort, it was a relief to be able momentarily to sit down. Modestly furnished with table, benches, chairs, and a bed, the house already felt wondrously like home.

(Photo by Marsha Emerman)

Way back at the start of building, I'd begun to visualize where to put up shelves, hang curtains—all the familiar parts of moving into any new place. Unpacking was to be a gradual process, but a few things wouldn't wait. Up went Carolyn's patchwork curtains and out came Jill's raku teapot, treasured work of friends. Instantly more warmth and comfort.

In December, I took a vacation from my house and went East to see my folks. Old friends and family were excited and asked lots of questions. Showing pictures and talking about it made my little house in Oregon seem like a distant dream.

Returning to Oregon, I began to work on all the many projects remaining. It's amazing that the list of work to be done is always growing. Bit by bit, I added such things as interior paneling (rough-cut cedar), counters, shelves, interior ceilings. Most recently, I built a wonderful pole ladder held together completely by dowels.

Home and Back

Looking through the windows, above my typewriter, I see a million flickering lights, fading into a vast dark space that is either sky or ocean. No, this isn't Oregon, it's the USF law library in San Francisco. Strange to be sitting here this past week taking these memories out of storage. Most incredible is that the information is all solidly there, and only needs to be put into usable form.

Oregon to San Francisco and back. A couple of weeks ago, I was up there visiting. In the space of a week, the weather went from summer to winter and back again. It's like that. A stranger pulls your car from a ditch. People go away and return, like mud puddles in the road.

So soon I'll return to live in my favorite home. It's no exaggeration to say that a house you've built feels like nothing you've ever felt before (story of a love affair between woman and house). Like any other great attachment, it's probably healthy to leave it for a while and return with renewed appreciation and energy.

It's hard to say what were the most valuable things I learned from building: new skills, confidence in finding ways to solve unfamiliar problems, or the satisfaction of working to fill a basic need.

Just as I couldn't imagine building a house five years ago, who knows what will happen in the next five, ten, or fifty years? If there's no major earthquake, my house should still be standing. It's fun to picture myself, at eighty, sitting there on the porch in a rocking chair, strumming my banjo and singing.

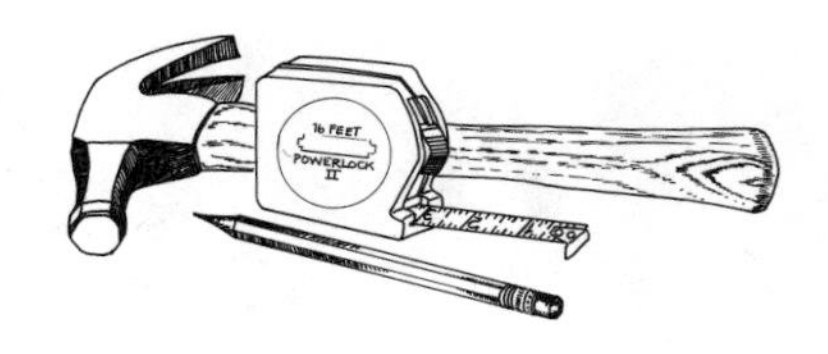

Adding the Roof

FRAMING UP A GABLE ROOF

A *gable roof* is formed when two roof surfaces slope upward to meet at a common peak; at each end of this roof, a triangle called the *gable end* is formed. A simple design, the gable roof can add both charm and space to the smallest structure. It is well suited to later additions—dormers, skylights, and the like—so that framing up a gable roof gives built-in versatility to your house. There are several other types of roofs, each requiring special framing and offering certain features. A *flat roof,* for example, is commonly seen in some areas of the country and virtually unknown in others. It requires a special roofing system of built-up layers of roofing paper and hot tar. The *shed or single-pitch roof* resembles one half of a gable. *Intersecting or multiple shed roofs* are a common feature of modern architecture. The *gambrel roof* is found on the traditional barn; adapted to a house, the gambrel gives extra attic or second-story space and has an advantage in that it can be framed up with fairly small-sized lumber. The *mansard roof* is similar to the gambrel, but rises equally from all four sides (like the gambrel, it has two distinct roof surfaces per side). Of all the possible roofs for your house, the gable is versatile, pleasing to look at and live under, and not too complicated to frame up if you have some basic carpentry skills.

The common gable roof consists of a *ridge piece* or *ridge board* running the length of the building and a series of matched pairs of *rafters,* which are regularly spaced and run from the doubled top plates of the outer house walls up to the ridge piece. Occasionally the gable roof is built without the ridge piece—the rafter pairs meet at the peak of the roof and are fastened together directly. Because the weight and force of the gable roof tends to push the side walls of the building outward, some system of tying the walls together must be fashioned. If your house has a second floor or a loft, the ceiling joists for either will accomplish this work. Lacking either loft or second floor, a gable roof house should be provided with collar beams (also called tie beams) or trusses. A *collar beam* is simply a piece of 1× (or 2×) material that runs across a pair of rafters and is nailed to each, forming a strong triangle that resists the outward pushing of the roof. Collar beams are easy to add to the roof framing. They are usually placed at a position roughly two thirds up the rise (see drawing). *Trusses* are more complex frameworks of wood (or metal) used to help support the roof and to tie it together. They are used on larger buildings; for most small structures, collar beams will suffice. If you'd like more details on truss construction, see Siegele's *Roof Framing* (listed in "Resources").

The language of roof framing includes several key terms that you should become familiar and comfortable with. The *run* equals the distance from the outer edge of the wall's top plate to the center of the ridge piece or center of the roof. The *rise* equals the distance from a line level with the upper edge of the top plates to the tip or peak of the roof. The *span* equals the total distance from the outside of one outer wall to the other; the *rafter span* (as given in rafter tables) equals the distance a given rafter has to span (actually equal to the run). The *slope* of a roof is the incline, expressed as the ratio of the vertical rise to the horizontal run in terms of inches of rise per foot of run. A slope will thus be expressed as "4 in 12" or "6 in 12," which means that for every foot of run the roof rises 4″ or 6″. Another way the incline of a roof is expressed is in terms of the *pitch;* this is the ratio of the vertical rise to the span (twice the run), and it is spoken of as a fraction. For example, if you have a roof with a run of 12′ (span = 24′) and a total rise of 6′, your

(Photo by Dian)

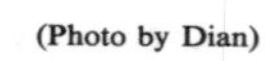

(Photo by Dian)

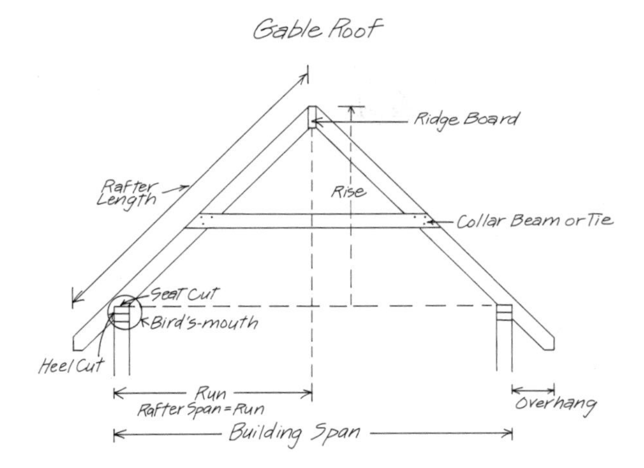

pitch is 6/24 or a "1/4 pitch." It is easy to become confused and think that a roof being spoken of as a "4 in 12" is a 1/3 pitch. *This is not so.* To determine the pitch, you must know the total span and total rise. To go back to our example—a roof with a run of 12′, a span of 24′, and a total rise of 6′—the slope here will be "6 in 12"; for each foot of run, this roof rises 6″. If you are designing a house or other structure yourself, you can draw out various pitches to see what appeals to you in terms of the run/rise ratio. In doing this, consider what kind of roofing you want to use and whether the pitch you are designing is adequate.

After you've decided upon a roof pitch, consult the rafter table (in "Building Code Tables" at end of book) to see what size material will make the span. (Remember the rafter span is the actual distance each rafter will cross. In figuring out the length of rafter material, add on for the overhang—the portion of the rafter that will extend out beyond the wall. The overhang helps shed water away from the sides of building and offers some weather protection to the areas directly below.) You will probably want to space your rafters on conventional 16″ or 24″ centers to maximize use of materials and be able to choose the correct sizes according to strength. In calculating out the number of rafters you'll need, add in a couple extra just in case you make a wrong cut or find a poor piece of lumber in the lot. It's a good idea to consider this at the beginning—and you can always use the extra material, for example, for blocking. If you plan to use a ridge piece, or ridge board, make it one size larger than your rafters in width; for example, if your rafters are 2×6's, go to a 2×8 ridge piece. This will allow for the angle cut at the top of the rafter where it butts into the ridge piece. If at all possible, get your ridge piece in one solid length. If this is out of the question, you can join pieces. Look for good straight material for both rafters and ridge piece. It's possible to improvise a little with your ridge piece (I've used 1× material—and even built with no ridge piece at all), but for your rafters you'll want strong, straight stock. Don't forget to check your rafters and ridge piece for bowing (as with girders—see "From Your Foundation Up") and place them crown side up.

For roof framing, you'll need your regular carpentry tools (combination square, hammer, saw, etc.) plus one specialized tool: a framing square, also known as a steel square. Each rafter must be cut in three ways: a ridge cut at one end, which allows for a snug fit against the ridge piece; a bird's-mouth near the opposite end, which allows the rafter to sit on the top plate; a plumb cut at the end that will overhang. To establish locations and angles for each of these cuts, the framing square is used. The framing square consists of a blade (the longer arm, 24″) and a tongue (the shorter arm, 16″). The face of the framing square is the side with the manufacturer's name on it; the back is the opposite side. How you hold the square as you work is critical, so look at the accompanying drawings as you go along to make sure you are right. In addition to your square, it helps to have a set of staircase fixtures, little brass attachments that can be fixed on the square when you have to make repetitive markings (as in laying out a rafter). The framing square is designed to work in terms of units that correspond to the units of a roof or stair stringer. The roof, remember, is discussed in terms of rise per foot.

If you hold your square with the back facing you (blade in left hand, tongue to the right), you can locate one unit (12″) on the blade. You can then locate the number of inches of rise per unit run on the tongue—say 6″ (for a roof with a run of 12′, span of 24′, total rise of 6′; this is a "6 in 12" roof, or a 1/4 pitch). Place a rafter on a pair of sawhorses; lay it with the top edge (as it will be when in position as part of the roof) opposite you. Hold the framing square against the rafter as shown in the illustration (1). Line AB equals the unit rise (6″ per foot) and line CD equals the unit run (12″). Fixing your brass markers on the square at points A and D will make your work much easier.

Start at the top end of your rafter and hold the square as shown to mark your ridge cut (2). Make a pencil mark at point D and move your square so that the 6″ mark on the tongue (point A in drawing 1) hits point D. Mark that. As you move the framing square, you are marking off units of run that exactly correspond to the proper rise/run ratio of your particular roof. For each foot of run, you'll move the square one unit. So for this roof you should move the square until you have marked off 12 units—or a total run

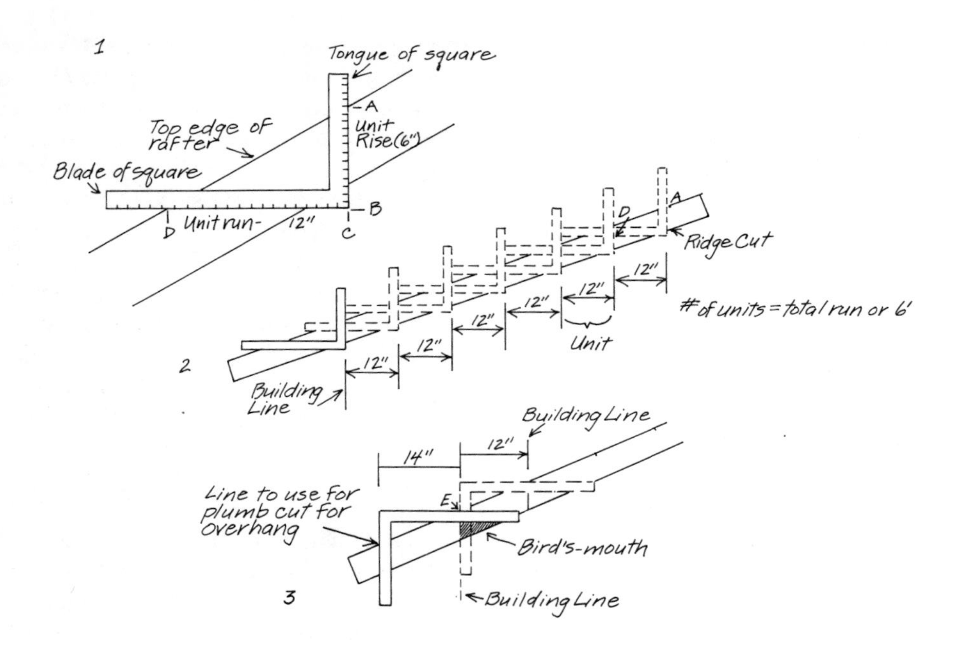

of 12′. To protect the illustrator's sanity, our drawing shows only six units or steps, but this should give you enough of an idea to follow. When you reach the last step, mark the building line as shown. This is the point where the rafter will rest on the top plate (it corresponds to the outer edge of the top plate) and it is used to determine the position of the bird's-mouth.

With the building line marked, hold the square as shown (blade up; tongue down and to the left). Place it carefully so that a 12″ unit hits the rafter at the top of the building line as shown. Draw in the top part (seat cut) of the bird's-mouth (which will follow a small section of the lower edge of the blade); the side part of the bird's-mouth (heel cut) will correspond to that section of the building line. Before you move your square, mark point E. Now move your square as shown to mark the plumb cut for the overhanging end of the rafter. In this example, I've allowed for a 14″ overhang—you can change this to whatever you want. Don't forget that a very long overhang may interfere with opening doors and windows! An overhang that is too short will look stubby and odd. From 10″ to 18″ is a good overhang range, depending upon your building's needs.

There is one more step before you consider making any cuts. If you're using a ridge piece, you have to shorten your rafter to allow for the thickness of that ridge piece. For instance, if you have a 2× ridge piece, each rafter of each pair should be shortened by ¾″ (half of the actual thickness of the 2× stock). Go back to your ridge line and back it up ¾″. This step is left until last to make certain that your original units are counted off correctly. At this point, too, I should mention how to deal with odd units of run. For instance, say that your run had been 12′6″ rather than an even, easy 12′. In this case, you begin by making your ridge line as before. The first unit you mark off, though, should be the odd unit: in this case, 6″ rather than the standard 12″. The line you would adjust is shown as CD on drawing 1. As before, mark point D, and continue on, using regular 12″ units until you've marked your complete run.

Before you make the actual cuts for your rafter, double-check yourself. Count the number of units you've stepped off to make sure they correspond to the actual feet of run. The ridge cut and the plumb cut at the other end should run parallel to each other, and the bird's-mouth should be as shown. When you're satisfied that all is well, go ahead and cut your rafter. The plumb and ridge cuts may be made with a power saw, but the bird's mouth requires a handsaw to finish it. With a power saw, you'll tend to overcut and weaken the rafter. When you're done with this rafter, you should mark and cut a second one. You can use the first rafter as a pattern, tracing it onto the second. Until you're really comfortable with the work, though, it is probably more accurate to step off each rafter in turn.

I should mention here, too, that there are other methods of laying out rafters. The one I've described is the traditional "stepping off" method—the simplest and safest (least open to mistakes). For more information on alternatives and on more complicated roof framing, see Siegele's *Roof Framing,* Dale McCormick's *Against the Grain,* or Wagner's *Modern Carpentry* (listed in "Resources").

Once you have a pair of rafters ready, it's a good idea to set them up in place and make sure they work properly. If you're working on a structure with a second story or loft, this is fairly easy. If you don't have a "stage" to work on, you'll have to make a temporary one. Use material that is sound and make sure that you secure it in place—unattached boards and planks have a dangerous habit of creeping around if they aren't tacked down!

While on the subject of warnings, here's another, personal one: tune in to how you feel and trust your instincts. If you have had a fight with your lover the night before and are still feeling out of sorts, leave the roof work for another day. This applies especially if you're going to be walking around and working on temporary scaffolding and planks. A change in your work schedule won't be nearly as devastating as falling off the roof because you're preoccupied and inattentive! If you're working with a partner and sense that she's having an off day, take a break and talk it over. Maybe you can shift around your work and wait until you can both give extra concentration and absolute care.

Rafter raising is definitely a two- or three-woman job! You *can* do it yourself, but working with someone else will make the day infinitely easier. Every crew will probably develop special techniques and systems for raising rafters. The terrain of the site, the size and weight of your materials, the height of the building, and the number of people working together will all influence how you can best do the work. Basically, what you have to deal with is lifting rafters and ridge pieces up to position and securing them in place. If you haven't already done so, mark off the on-center locations for your rafters along both doubled top plates. Check your first pair of rafters by lifting them up into position with a piece of 2× material (equivalent to your ridge piece) between them. They should fit snugly in place over top plates and butt smoothly together at the peak. If this pair works out, go ahead and cut the rest.

There are two (perhaps more) alternatives for raising the ridge piece. You can raise it up and secure it with temporary bracing. The bracing can be nailed to the top plates at either gable end. The ridge piece should be put in place and made level all along. This is most easily done if your ridge piece is all one; if you're working with sections, you can raise them one at a time as you go along. An alternative is described in *Against the Grain:* the two gable end rafters on one side of the building are attached to the ridge piece and this whole unit is lifted up by three women. Then, while one steadies the ridge piece at the center, the

---

If you frame up a gable roof with an enclosed attic space (as opposed to an open cathedral ceiling with or without a loft) that doesn't have opening windows, be sure to place vents in each gable end of your building. They should be framed in while you're doing the rough work, though they could be added later if you've inherited an already built structure without them. These vents allow for air circulation and will help keep your house dry inside by allowing moisture from the heated living space to pass outside rather than condense on the inside ceilings. A simple rectangular wire screen vent may be purchased at most hardware stores and put into place, or you can buy triangular-shaped gable louvers (ventilated openings with slats, which may be fixed or movable) or make your own. A louver has the advantage of shedding rain; a louver with movable slats may even be closed tight during brief periods of extremely bad weather. In addition to your gable end vents or louvers, you should consider putting some vents just under the overhang of your roof (one at each of the four corners of your building). Depending upon the way you've decided to handle the overhang—leaving it open, or closing it in with soffits—you can place vents in the cornices themselves or in the space between two rafters right above the top plate (an area that is usually filled in with solid blocking). These vents will allow a good flow of air to pass from the eave level of the building up to the gable end vents. They are particularly necessary if you use your attic as a living space but haven't put any opening windows there.

---

other two place and secure the matching rafters at each end. I've always taken the first course—setting up the ridge piece first—but the second sounds like a good idea and worth trying.

Rafters are toenailed to the doubled top plate with either two 10d common nails or with three 8d common nails (this latter nailing is required in the latest Uniform Building Code). If your house has ceiling joists, nail each rafter to the adjacent ceiling joist (four 16d nails). Each rafter should also be secured to the ridge piece (or, if you're building without a ridge board, to its paired rafter). You can end-nail the first of a pair of rafters (two or three 16d nails through the ridge piece and into the end of the rafter) and toenail the second one (three or four 16d nails through the rafter and into the ridge piece). Depending upon the size of the material you are working with, you may have to go to slightly smaller nails or secure the rafters together in another fashion. For instance, if you're working on a very small structure and using 2×4's for rafters, you might want to use smaller nails. If your ridge piece is of 1× material, 16's will split it, so you should use 10's or even 8's, depending upon the size of your rafters.

Once you've established that your rafters are properly cut by the stepping-off method described, you can proceed to cut the rest of the rafters, secure the ridge piece if you haven't already done so, and frame up your roof. If your house doesn't have a loft or second floor (with ceiling joists), you may want to put on some temporary bracing to keep your walls from spreading out while you put up your rafters. Later, collar ties or trusses will take over the work of this temporary bracing and it can be removed. The bracing should be positioned parallel to the rafters and may be as simple as several 1× or 2× boards nailed across from one top plate to the opposite one. These temporary braces may also serve as the basis for a scaffolding or platform for you to walk on while putting up rafters. In setting up this bracing, remember that it should be sturdy but also removable (use double-headed nails, if possible, to make removal easier).

Another step to take before putting up the rest of your rafters is to secure and level your ridge piece. This is best done by putting up the pairs of rafters at either gable end. Nail these in place temporarily (don't drive nails fully in) and check your ridge piece by placing your level on it at several points from one end to the other. If it reads level, go back to each pair of rafters and nail them securely. If not, adjust your rafters until it does. It's a good idea to check each rafter with your framing square as you nail it.

(Photo by Dian)

Another way to frame up a gable roof is to construct trusses: pairs of rafters fastened together complete with ties and/or braces before they are ever lifted into position to form the roof. Using trusses allows you to precut and assemble your roof pieces; it may be a good way to speed up this part of construction if you are working under pressure. For more information on trusses (there are many styles or types to consider), see *Architectural Graphic Standards,* Siegele's *Roof Framing,* or other general carpentry books (see "Resources").

Hold the square in between the rafter and top plate (see photo) and then against the rafter and ridge piece. This will show you whether the rafter is sitting squarely or not. If it is, the framing square will fit snugly in place; if it doesn't, adjust the rafter. To hold rafters squarely in position, nail a temporary diagonal brace (or set of braces) to the underside of the rafters (use scrap 1× material). Later, your roof sheathing will take over this work and the bracing can be removed.

The final step in framing up your roof is to frame in the gable ends. This will finish the walls of your house and add support and stability to the rafters. Using the same size material as for your wall studs, you should continue your on-center spacing as below, framing in for windows or vents as needed. Each stud in a gable end should be square-cut at one end (where it sits on the top plate) and the other end should be given an angle-cut notch that the rafter will fit. The notch should be equal in depth to the thickness of the rafter, and the angle cut at the bottom of the notch and top of the stud should correspond to the angle created by the pitch of your roof. Each stud is toenailed to the top plate and face-nailed or toenailed to the rafter. The framing in a gable end will provide a nailing base for interior and exterior siding, sheathing, or paneling.

Adding collar ties to your rafters will complete this part of your building, though you may elect to wait if you are rushed. The temporary bracing you've run from top plate to opposite top plate can continue to do the work of preventing building spread-out until the ties are placed. The simplest collar tie is a piece of 1×6 or 1×8 angle-cut at each end and face-nailed to each rafter of a pair to form the triangle shown. Depending upon the size of your building, spacing of rafters, and the effect you want, you may choose to put a tie on each pair of rafters or on every other pair. You may want to put on pairs of ties—one on each side or face of every pair of rafters. This is not only more pleasing to look at; it is stronger. You may even wish to go to larger material (4×6, for example) and notch the ends to fit up over the rafters for a really rugged, handsome tie. Whatever your choice of material, the collar ties should be located approximately two thirds of the way up the rafters (i.e., two thirds of the distance up from top plate to ridge piece). Once they are securely nailed in place, you can remove any temporary bracing. The collar ties will add a nice finished look to your rafters and strengthen the building as well.

ROOF SHEATHING

Once your rafters are in place, roof sheathing is added, giving the roof rigidity and furnishing a nailing base for the final roof covering. If your roof is to be covered with asphalt or fiberglass shingles or roll roofing, you will need a solid roof sheathing made up of plywood sheets, 1× boards (either straight-edged or shiplap), 2× decking, or sheets of insulating roof deck. If you plan to cover your roof with wood shingles or handsplit shakes, you can use spaced roof sheathing: 1×4 or 1×6 boards applied with a uniform space between them (this spacing should equal the shake or shingle exposure, which is the amount the shake or shingle extends beyond the overlapping row). Your choice of roof sheathing will depend on several factors. In general, plywood sheathing will go on much faster than individual boards, but the sheets are heavy and awkward to handle. If you plan to have the roof sheathing double as an exposed ceiling inside your house, you may want to use boards rather than plywood for the aesthetics. Natural redwood or cedar boards will make a warm-looking, beautiful ceiling. You may choose regular 2× roof decking (usually pine) if you've framed up your roof with sufficiently strong rafters. This system of roof framing is called plank and beam; it uses large rafters such as 4×6's spaced 6′ or 8′ apart. You may decide to use insulating roof deck, which comes

(Photo by Carol Newhouse)

in 4′×8′ sheets of varying thicknesses, to be applied according to the spacing of your rafters.

Whatever your choice of sheathing material, you will want some system of ladders and scaffolding for your work on the roof. Scaffolding should be strong enough to support the weight of your sheets or boards plus one or two people. Ladders should be in good working condition and set securely in place. Roof work takes a maximum safety consciousness: if you are up on the roof, be careful about how you walk, balance, and handle your tools and materials; if you're down below, keep alert to the possibility of tools and pieces of material that might come sliding down. It's rarely a good idea to work alone on a roof. Having a partner will not only make the work much easier, but if you should slip or fall accidentally, you'll have her help when it might be critical!

Putting on roof sheathing is similar to putting sheathing or siding on your walls: the usual procedure is to start at the lower edges of the roof and work up. If you are working with *1× boards or 2× decking,* try to make use of long lengths as much as you can. This will not only make your work go faster; it will add to the rigidity of the roof. Boards should be placed so that joints meet over rafters. With end-matched boards, of course, you can let them join anywhere, but in most instances you'll be working with common boards that should join over rafters. Try to stagger joints somewhat. The most common board for roof sheathing is the 1×6. Use whatever wood is cheap and plentiful in your area. Boards should be secured with two 8d *galvanized* common nails per rafter if they are 1×6; 1×8 and wider boards require three 8d nails per rafter. At the gable (rake) edges of your roof you can let the boards run wild and come back later to saw them off, or you can set them carefully as you place them. The usual procedure is to let the boards overhang (project) beyond the sidewall covering (siding). How far yours project is a matter of aesthetics and practicality; a really big overhang will require additional support to keep the boards from warping but will give good protection for your walls and will look attractive. You may also choose to cut the sheathing boards so that they are exactly flush with the outer surface of the wall siding. With these few considerations (nailing, positioning, and overhang), your roof sheathing should go on quickly and smoothly.

Plywood sheathing comes in various thicknesses, which must be matched to the spacing of your rafters. There is a "Standard Sheathing" grade of plywood that is most economical to use; it replaces the former "C–D" grade used for this work. The most common thicknesses used under asphalt or fiberglass shingles or under wood shingles are: 5⁄16″ if rafters are 16″ o.c.; 3⁄8″ if rafters are 24″ o.c. If you are using a heavier roofing such as slate or tile, use 1⁄2″ plywood over 16″ o.c. rafters and 5⁄8″ over 24″ o.c. rafters. If you need more specialized information, refer to the "Plywood Details" section of *Architectural Graphic Standards* (see "Resources"). Plywood panels should be placed so that the face grain of each sheet is perpendicular to the rafters. Leave about 1⁄8″ space all around each sheet to allow for expansion. End joints should be centered over rafters. Unsupported edges should be supported in one of several ways: you can put blocking under them and nail them to the blocking; you can use special edge supports called Plyclips; or you can use plywood with tongue-and-groove edges. Sheets up to 1⁄2″ in thickness should be nailed down with 6d nails; sheets up to 1″, with 8d nails. Thicker sheets require 10d nails or 8d ring-shanks. The usual spacing for nails is 6″ o.c. at edges and 12″ o.c. along intermediate rafters. Stagger courses so that end joints aren't consecutive over the same rafter. This may require placing a sheet "vertically" or using a half sheet every other course.

Finishing off the edges of your roof can be a simple or a complicated matter, depending upon the nature of the structure and design you might like. Rafter ends may be left exposed or covered with fascia boards and metal or wooden gutters. The space between the lower edge of the rafter end and the exterior siding of the house may be enclosed with a wooden or aluminum soffit complete with vents and trim, or it may be left open, or "exposed." For more details on finishing this part of your house, see *Architectural Graphic Standards* (pp. 248–49, 257) or Wagner's *Modern Carpentry* (see "Resources").

Insulating roof deck is applied in much the same way as plywood. It comes in 4×8 sheets, which should be set so that ends meet over rafters and edges are supported by blocking. Sheets ½″ thick may be placed over rafters that are 16″ or 24″ o.c. If rafters are more widely spaced, use thicker sheets (up to 3″). Nailing size and pattern are similar to those described for plywood.

Common *roof decking* is applied much the same as 1× boards. This decking is tongue-and-groove 2× material, usually pine or Douglas fir. Joints should occur over rafters unless the material is end-matched. Nailing is done with 16d nails or with spikes—two nails per rafter, face-nailed, and one nail blind-nailed.

When the roof sheathing or decking is completed, you are ready to finish the roof with flashing where needed (around vents and chimneys), metal drip edges and felt paper as required, and the final roof covering.

Moving the "Big House"

Eighty acres of primitive land, it is a wilderness stretching a half mile into the folds of a canyon. Vegetation anchors itself to the steep canyon walls with deep roots. The land is bordered by ancient evergreens, virgin timber. Our fantasies began with our first visits here:

"Why, let's have the main house right here."

"A garden over there."

"We can develop this spring and use gravity to pipe water down to the mouth of the canyon."

"We could put the garden there."

"We'll need a tool shed, a greenhouse, an outhouse, a woodshed."

"How about another cabin here by the creek?"

"I want a round house."

"First of all, we need a road."

This land so clearly used by men, abused, raped by the loggers, by the miners. Its ugly scars are healing now, grown over with soft lilac bushes. No one else wanted to buy this land even at the remarkably low price of $8,000. Were we crazy?

Steep land. Wild land. Good for mountain goats, but for us city folks? Land come to by a group of friends, each of us invigorated by the challenge of homesteading. We were in our early twenties, full of stamina and youthful insecurity. We were determined to make a home here, make it through our first winter without leaving the land.

"They think I won't make it living like this because I'm a woman. I'll show them," pronounced Patti, a strong woman I knew to be a steady worker.

"They think I won't make it on the land 'cause I grew up in suburbia, didn't learn any survival skills. I'll show them!" I said, as determined as my friend.

"They think I won't make it 'cause I'm so young. I'll show them," swore Willie, struggling to appear self-assured in her confusion.

Men were part of the collective and worked the land in the beginning months. We were each trying to clarify our sexual identity and preference.

"I want to try living with women only. After I experience it, I'll tell you whether I like it."

We women loved it. We flourished. We decided to divide the land. Sixty acres were designated women's land. The remaining twenty acres were men's land.

None of us had ever done carpentry before, but that did not halt our fantasies. We planned a 24′×48′ house with rooms to sleep in, to cook in, to sew in. It would be such a beautiful home, this house of our collective dreams. Our dreams took the blows of reality soon enough. Two days under the sweltering summer sun laying out our house, and still the 90° angles were not square.

"Oh, this string will never do! It stretches, and I keep getting different measurements."

"I just read a carpentry book and it says we need a different kind of level, one that would hook onto this string, a line level. I'll have to make a special trip to town."

"This book says we need batter boards."

"But we have no scraps of lumber. And how do you make them? The book makes it look so simple but . . ."

The summer days rolled on. There was a lot of other land work in addition to building. A bulldozer had dug a barely adequate road to the property line, as well as a reservoir a thousand feet up the steep hillside. Patti, Willie, and I hauled white PVC pipe to the reservoir and developed the water system to the garden site, though we still had to haul water to our camping and building area. There was no time now to bury the water line!

Work went slow, slow like the trees grow. But we were city folk and we did not understand that kind of slow. It was late August when we awoke one morning to the sounds of the first rains. We panicked! Still no house begun! Taking a long look at the reality of our situation and our skills, we decided to trim our fantasies.

"How about a small house? We could convert it into a chicken coop later."

"Chipboard is cheap. Let's determine the floor size based on the size of chipboard."

"How about 12′×16′, that would take six pieces of 4′×8′ chipboard."

Money and time were constant considerations. None of us had jobs or the time to work both the land and a job. We all had small stashes of money acquired through previous jobs and gifts from relatives, but no clear prospects of where the next money would come from. With winter close, we also fought against time.

"When do you think the rains will come?"

We wrestled with our fears of the cold.

"How much insulation do you think we need?"

The questions piled up.

"How much firewood will we burn?"

We grabbed for answers to these questions, and our uncertainty produced insecurity.

Our bodies grew stronger by the day. Leg muscles adjusted to steep terrain. Backs, darkened by the sun, grew strong hauling supplies up the side of the mountain. Slowly our bodies became attuned to this new life-style. Our minds cleared after a long day of work-meditation. We were serenaded by the songbird, accompanied by the drumlike beat of the grouse.

"Let's make the house as simple as possible. After all, the rains could start any day now."

"Listen, I looked at old cabins nearby. They're still standing from the mining days and their only foundation is rocks. They've been around fifty years, that's good enough for me."

We dug holes, then went to the creek bed to find the largest, flattest rocks around.

"I think if we dig the hole a little deeper at this corner the rock will sit level."

"Yes, that makes it level in one direction, but it isn't level in the other direction," Patti realized as she stared at the air bubble on the level, which was far off center. "Maybe we could rotate the rock."

"I think we have to forget about this rock altogether and find another." I heard the frustration in my own voice.

Finally the twelve rocks were relatively level and square to each other. Upon each foundation rock we placed an 8″×8″ creosote-treated post and struggled to level these. We did not think to attach the posts to the rocks. I guess we assumed that the weight of the house would stabilize the posts.

Next we headed to town to buy three 4×6 timbers, our beams. These beams were to span the 16′ length of the house and rest on top of the foundation posts. We didn't know that maximum structural support is obtained when a timber is placed on edge (that is, with the shorter dimension horizontal to the surface it rests on). Instead, we placed the 6″ side flush to the foundation posts. Yes, we were learning carpentry by making mistakes! We were unable to find 16′ Douglas fir 4×6's, but were sold 18′ ones for the same price. No matter—we would have a porch!

Each stage of building began with a thorough study of the carpentry books we had. These books were written in what seemed like a foreign language to us, carpentry jargon. We gathered more information by picking the brains of newly made friends, neighbors, and merchants in building supply stores. Carpentry tools were new to most of us. We didn't know how to recognize a good tool or how to take good care of a tool. We didn't know the importance of using the best tool for the job. We bought cheap secondhand tools. Our bodies worked too hard, taking up the slack left by poor tools and by not using the tool best suited to a particular job. There is no electricity in the canyon, so we worked strictly with hand tools.

After we had gathered more information, we placed 2×4 joists on top of and perpendicular to the beams. We later learned that 2×6 joists are more traditionally used; the 2×4's were not considered strong enough for the job we were asking them to do, spanning 6′ from beam to beam. We placed them on 2′ centers, and later we realized we should have placed them closer together. No doubt it would have been so much easier had we known carpentry rules. At least we now knew to place them on edge!

Soon an argument began about whether the joists got nailed down. Our insecurities and uncertainties about carpentry led us into power struggles. We each sounded so confident in our conflicting opinions.

"Let's toenail them down, nail them at an angle."

"No, I'm absolutely *sure* they don't get nailed down."

And though they never were nailed down, as is usually done, the house rested secure. Nor did we block the joists. We must have skimmed past that section in the books.

Now we were ready to lay the subfloor. We scrounged pieces of boards 1″ thick and of varying widths and lengths: old, gray, splintery, and a little warped, but free. Some sawing and hammering and fitting the pieces of the puzzle together, and then, as if by magic, we had a floor! A platform to run on, jump on, leap for joy on! Joy that we, women, city folk, could build, could teach ourselves carpentry.

Next we decided to insulate our floor by placing a layer between the subfloor and the chipboard that would be our finished floor. We chose the least-expensive method, which was not necessarily the best. For our insulation we gathered old newspapers and large cardboard boxes and placed them upon the subfloor. Our vision was limited. We didn't recognize the fire hazard we were creating. Luckily for us, the house never caught fire and we learned that we had made our mistake without losing our home. Also, this method of insulation did not effectively trap air, which we later learned was a primary principle of insulation. Then the six pieces of chipboard were nailed atop the insulation and we were ready to read up on and discuss wall framing.

"Let's use poles. We can gather them from the forest."

"We should use standing dead trees if we can find sound ones. That way we'll not only save money, but also save the lives of some trees. I think it's important to cut as few trees as possible."

So we set out to learn another country skill, felling trees. With inadequate knowledge, it was beginner's luck that the trees all fell where we intended! It was hard to know when our fears were real—that is, when the situation was actually dangerous—since we were unfamiliar with the job.

"Do you think we know enough about felling trees to do it?"

"I'm not sure, but we have to! Let's start with small trees," Willie nervously suggested. "Let's see, the book says to cut a wedge in the side of the tree in the direction we want the tree to fall."

"My mother never said I was going to do this when I grew up." I made a joke to cover my fears.

Climbing up steep hillsides, we searched for straight standing dead trees. Working in pairs, we slowly carved into them with a two-woman saw. Timber! We dragged the trees down the side of the mountain, lifted them into the bed of the truck. Hour after hour, we worked at peeling off the bark with a drawknife, learning that skinning dead trees is a slower process than skinning live trees. Finally we had enough poles. We sorted them. Six-inch-diameter poles would be used for the wall framing; 4″-diameter were set aside for the roof rafters.

"Do you know what this tool is used for?" I asked, hesitantly swinging the 3′-long wooden handle. "It looks like an ax, but the blade goes in the opposite direction."

"It's an adz. We just bought it at the flea market last week. Let me show you how to use it," replied Patti. As she swung the handle, the blade grazed the top surface of a log lying on the ground. "See, it makes a flat surface on a round log," she explained as she chiseled away at the wide end of the pole. "Of course," she added, "I know it would work better if it was sharpened." Tool sharpening was a job that intimidated us, and as a result we often worked with dull tools! We later learned that sharp tools are not only more efficient, but safer as well.

We framed our walls using traditional plate framing even though we used poles rather than 2×4 studs. The poles were "adzed" to a uniform 6″ in diameter so that they would provide a uniform nailing surface for both the inner and outer siding. The poles were spaced 4′ apart. We laid them on the floor perpendicular to two 16′ lengths of 2×6, one at the top and one at the bottom of the poles. We nailed through the 2×6's and into the logs. We used 2×4's to frame the windows and door so that our openings would be square.

It was time to raise the walls. We called for help from neighbors who had raised walls before.

"Some of us should stand alongside the wall and walk it to an upright position."

"I'll secure the bottom plate so it stays in place while you lift the wall."

To our surprise, the first wall was easy enough to lift and position.

"There, it seems to be in place! Let's check to see if it's plumb before we nail it down."

Unfortunately, we didn't know about bracing. When we put the siding up, we realized the wall had racked (or twisted).

The second wall, 12′ long, was raised and positioned adjacent to the 16′ wall. The top plates of the two standing walls were nailed to each other, binding the building together. Once we got the hang of things, the last two walls were easily assembled and raised. A sleeping loft was framed at the high side of the building.

Finally we were ready to build the roof. It proved to be one of the most difficult parts of the building. We used 4″-diameter poles covered by ½″ plywood and 90-pound roll roofing paper. We worked from atop an unwieldy homemade log ladder (another money saver), driving 60d spikes with a sledgehammer. It took us a little while to understand how the logs had to be spaced and squared to the wall plates in order to have a nailing surface for the plywood sheathing. As you can imagine, it was no easy matter to remove the spikes and rearrange the logs! But we were determined to be dry that winter.

As soon as the roof was completed, we moved in. A dry space on the land! Each morning we rolled up our sleeping bags and continued building. We worked on the siding and wall insulation. Windows were hinged. We learned that, since hot air rises, insulating the roof is important. We splurged on 3½″ fiberglass insulation, stapled it to the logs, then covered the insulation with a collage of cloths. We grew anxious as our collective savings dwindled, though the materials for the house cost only $150.

In early October, I was feverish, lying in bed in the loft. The conversation that floated up from below caught my attention.

"This land isn't on your property," announced John Ronke, the man who had sold us the property. He sounded very confident in this stunning pronouncement.

"I just don't believe this! You told us it was on our property when we first chose the building site." I heard the pleading reply.

John refused to take responsibility; instead, he firmly stated that the house would have to be moved! At first I was sure that what I heard was merely a feverish hallucination. A few days later we determined the property lines by sighting from the witness trees (trees that had been marked by surveyors to indicate the corners of the property). It was clear: the house was at least 50′ over the property line! After the initial panic eased, we tried to come to a rational solution.

"Listen, we don't have time to move the house this year anyway. Let's finish it sufficiently to spend this winter in it. Moving it will be next summer's work."

We felt clear, this house was a center; it would become an important part of canyon life, would never be converted to a chicken coop or other minor use.

With no immediate pressure to act on the mistaken boundary, two years passed before we seriously set to the task of moving the house. The two years were filled with periodic suggestions and discussions of how we were to accomplish the overwhelming task.

"First we'll jack the house up with a number of hydraulic jacks. Then we can place logs under it, perpendicular to the direction the house is to move in. We'll have to clear and level a 12′-wide path to roll the house down." It was approximately 100′ to the new building site, located just *our* side of the property line.

"Perhaps we could attach the house to a winch on the front of a four-wheel-drive vehicle."

"I spoke to Sid down at the gas station—he thinks we could move it with several come-alongs secured to nearby trees."

Our schemes were repeatedly stymied as we envisioned losing control of the house, of its slipping down the steep hillside it bordered and crashing into smithereens. It was our third summer on the land when it finally became imperative that we move the house. We had opened the land for women to camp that summer, so we had plenty of womanpower. When we could procrastinate no longer, we chose a safer, less overwhelming method: also one more gentle to the surrounding terrain. We decided that our best solution was to disassemble the house, then reconstruct it at the new building site.

Demolition, the inverse of construction . . . accordingly, we worked from the roof on down. First we cut the roofing paper off with mat knives, cutting inside the roofing nails and thereby salvaging all but the outer perimeter of the paper. We loosened the nails in the plywood roofing with a cat's-paw, then pryed up the 4′×8′ sheets with a crowbar. Detaching the roof rafters, which were secured with 4″ and 6″ nails, proved to be quite a challenge! Next the loft, platform, and siding were disassembled. We released our frustrations as we tore each board down, conjuring up images of the house rebuilt. (*This* time we would correct our original errors!) A few of the poles we had felled and used had considerable rot in them. These we replaced, this time using seasoned poles we had acquired by thinning a dense growth of trees.

Demolition continued, one group of us working at it while other women were laying out the house at its new site. We had learned more over the years, and layout was easier. We decided a concrete foundation would be sturdier and possibly easier to construct than the more primitive rock foundation. We followed the suggestion of another neighbor and created forms for foundation posts by using strips of 90-pound roll roofing paper. The roofing was cut into pieces corresponding to the desired height of each foundation post. We leveled these to one another at this stage of the process. The strips of paper were then rolled and stabilized at the desired circumference with wire (we made a mistake here, making the foundation posts too narrow). We dug holes for foundation pads. Next the cement, gravel, sand, and water were mixed. Finally we filled both the pads and the piers with cement, making sure it didn't dry too fast in the hot summer sun. A foundation was poured for an addition; the new house was to be L-shaped, with a separate space for the kitchen.

We decided that we could take the walls down as complete units and then carry each down the path to the new site. First we removed the windows and door and set them safely aside. Next we removed the outer siding, realizing that the walls would be too heavy to move by hand with the chipboard still attached.

We took a break from our work and returned from the second women's spirituality festival re-energized and eighteen strong, anxious to tackle the task of moving the walls.

"Okay, on the count of three we'll lift the wall. Take care with your back. Lift from the knees. Then we'll walk for the count of twenty. Place the wall down slowly and then we'll rest."

Eighteen women, staggered on either side of the wall, gripping the bottom edge securely, feet planted firmly, peering through window holes at one another, giggling at the strangeness of the task. We were discovering the power in our numbers. The job was far more easily done than we had expected. We repeated the lifting, walking, resting cycle many times until finally the first wall was near the new building site.

"Let's prop the wall against that strong old maple tree."

The next wall was a cinch to move. It was a shorter wall, and therefore lighter. Finally all four walls were propped up near the new building site. Some leaned against the trees, others against the shed.

Next, the ¾″, 90-pound chipboard sheets were pried off the floor and individually carried down the path. The paper insulation was discarded and replaced with builder's foil. The beams were moved individually. We removed the remaining floor, the joists, and the subfloor in three sections. And then reconstruction began! We put the house back together from the ground up. This rebuilding proved to be a never-ending process. There were always improvements to be made. A shelf here, a coatrack there. We were, at last, limited only by women's imaginations! We were discovering that we could teach ourselves what we wanted to know, that common sense was a good adviser, that building need not be mystifying or intimidating. Most exciting, we saw the strength in our growing women's culture.

The "big house." Worked on by woman after woman . . . Willie, Patti, Ellen, Speys, Forest, Fran, Rainbo. Year after year . . . Sharon, Donna, Mary Lois, Tori, Crow, Pelican, Blue. Women learning carpentry by doing it, by guessing, by making mistakes . . . Ruth, Sky, Peggy, Jean, Betty, Suzanne, Lily. The big house became the foundation of life in this canyon. It sits at the opening of the canyon, a place where women come when they want to be open, share their lives with other women.

"Welcome! There's hot tea on the wood cookstove, soup's almost ready."

Women open to change, to hearing each other, to struggling collectively.

"More wood needs to be chopped, water needs to be hauled, there is a *huge* pile of dirty dishes. Everyone's stuff is lying around. I'm *sick* of doing all the work around here!"

The women living in the canyon and the women passing through the canyon, women familiar with circles and women skeptical of circles, form a circle in the center of the rectangular house. The first woman shakes the rattle, accompanied by another woman's steady drumbeat, and tells of her walk through the canyon, of the hawk that circled above her. The next woman screams, releasing her pain, then chants, releasing her joy. We hear a tale of gypsy life sung by a traveling woman. Each woman opens to the circle, sharing her inner space, her dreams and visions. Our shared outer space, the big house, stands as testimony that dreams come true.

Exteriors

SHEATHING AND SIDING

When it comes to covering the rough-framed walls of your house with exterior siding, there are many considerations and choices to be made. Although there are some non-wood siding materials on the market, such as aluminum and vinyl-coated metal, wood remains the most reasonable choice. It is available in a multitude of types for either horizontal or vertical application. It is durable, relatively easy to work with, and attractive. Another plus for wood siding materials is that they can be recycled—even the code-built house in most areas may be sided with used wood if it is in good condition. Wood siding can be broken down into four general categories or types: *horizontal siding,* such as bevel siding and clapboard; *vertical siding* (drop siding such as tongue and groove; board and batten; and others); *plywood panels;* and *wood shingles.* Depending upon your choice of siding and the type of structure you are building, you may have a single wall (siding only) or a double wall (siding applied over sheathing).

The discussion below begins with some of the considerations in choosing a single or double wall and continues through wall preparation and the application of various types of wood siding. Before you make any final choice of siding, take a look at what is being used in your area. The geography and weather of the area you're building in might make a big difference in the suitability of certain siding materials.

Single Versus Double Walls

The rising cost of lumber products combined with a sensitivity to the limitations of wood as a natural re-

source might make you question the wisdom of the traditional double-wall house. The sheathing of a double wall serves several functions, however: it affords extra protection against drafts and moisture; it adds rigidity and strength to the building; it offers some degree of insulation. The double-wall building is much easier to make waterproof, especially in critical areas over doors and windows. If you plan to use plywood panels or wood shingles to cover your house, the single/double-wall issue is solved for you. Plywood panels are designed to be used without wall sheathing; they are meant to be single-wall siding. Wood shingles, on the other hand, are usually applied over wall sheathing; they are inherently double-wall siding. If you decide that you can do without extra weather protection and insulation, are willing to put careful attention to waterproofing, and have provided your house with proper diagonal bracing, you may go ahead with a single-wall siding. Read through the following section for details on putting up a layer of felt or building paper over the wall framing. If you decide to use plywood siding, you should have help. Even the thinner (3/8″) sheets are awkward to handle and the thicker (5/8″) definitely require two or three people.

In general, drop or rustic siding (1× material with specially milled, interlocking edges) will make a stronger wall covering and should be considered for the single-wall house. Horizontal siding will give you more rigidity than vertical siding; this, too, should be considered for the single-wall house. If you're on a tight budget, you might consider a temporary single wall, with later plans for shingles or some other type of siding. Even a very low-cost single wall of particleboard or plain plywood will weather a year or two—you can cover joints with bats or even caulk them. When you have more money, you can finish your house more grandly (and permanently).

First Steps in Siding

Whether you choose a single or a double wall for your house, you should consider an initial wall covering of felt or building paper. If you are using sheathing, this paper is applied *over* board or shiplap wood sheathing; it is not used at all over plywood or fiberboard sheathing, except when wood shingles are used. If you are going to have a single wall (no sheathing), this paper should be applied directly to studs or wall framing. Certain types of exterior plywood siding may be applied directly to studs without this layer of paper. Tightly interlocking edges characterize these types, so if you're using plywood siding check and follow the manufacturer's instructions. Carpenters (and carpentry books) have a running argument over what type of sheathing paper is most suitable. Some recommend using a foil-backed paper that is both water-resistant and vapor-resistant. Others argue that the wood must be allowed to breathe and that an asphalt-saturated felt paper (low vapor resistance) is better. In our area (California coast—with a mild climate), 15-pound (and sometimes 30-pound) asphalt-saturated felt is feasible. Wagner's *Modern Carpentry* recommends the same. If you're halfway up a wall with felt paper and someone comes by and says you're putting on the wrong thing, bear in mind that it's an unresolved controversy!

Whatever your choice, this paper will help to insulate your house (stopping drafts) and will keep your wood framing and interior walls drier. It is usually applied horizontally, beginning with a row or course along the bottom of the wall. It may be stapled in place with heavy staples (staple-gun variety, not the chicken-wire type!) or tacked on with large-head roofing nails. Don't be tempted to blaze along and paper your whole building—a sudden wind could rip all of your work loose and leave you with a mess. Proceed bit by bit, covering a small area and then applying siding over it. The second row or course of sheathing paper should overlap the one below it, by 4″ or so, and side-lap adjoining sections similarly. When you come to a window or door opening, wrap the paper around the 2×4's making up your rough opening and secure it.

The next step to consider is flashing around door and window openings. What you use and how you apply it will depend upon the siding system you're using—and your budget. What you want to consider is how to channel water *away* from these areas so rain or melting snow doesn't drip and leak around your windows and doors. If you're building a single-wall house, this waterproofing may have to wait until after the siding is applied. Then you can fasten a strip of metal flashing over the head (top) casing of the window or door. Use a good grade of caulking under the flashing to create a gasketlike seal. If you can't afford the metal flashing, try a bead of caulking along the joint where the head casing butts against the siding. The first rain should tell you how successful you've been at preventing leaks! You may come up with some other way to flash and/or caulk these areas. Just remember to construct shingle-fashion, so that water will shed over and away from any joints. Don't forget the tendency of water to travel horizontally as well as vertically. A puzzling leak may occur when water is being pulled along by capillary action and drips inside at a most unlikely spot.

If yours is a double-wall house, the window and door flashing should go on after the sheathing is in place and before the siding goes on. Again, there are different ways to approach this flashing, and you may know of a method you've found to work well. A common solution is to nail metal or reinforced, waterproofed paper over each door or window head casing. (Available in some areas, this flashing paper comes in rolls 6″ or 8″ wide and is meant for this kind of work.) The flashing should be fastened to the sheathing and extend just over the top edge of the casing piece so that water will shed away from the joint. Sometimes a wooden drip cap piece is applied first and is then covered with the flashing. When the siding is applied, it will cover most of the flashing, leaving a tiny edge all along the head casing that acts like the lower edge of a shingle, shedding off onto the casing itself. Another material used in some areas for flashing is a polyethylene sheeting. Consult with other carpenters or building supply people to see what's available and commonly used in your area.

Before the widespread manufacturing of plywood, fiberboards, and other alternative materials began, wall sheathing for houses usually consisted of 1× boards—either straight-edged or shiplap—applied diagonally or horizontally. Unless you get a wonderful deal on 1× material or have a lot of recycled wood, you'll probably find it cheapest to go to a plywood or fiberboard sheathing. Plywood is the better choice because it is much stronger; particleboard and fiberboard are, of course, cheaper. If you *are* using *1× material,* the nailing requirements are: material up to 8″ in width should be nailed with two 8d nails per bearing (stud or other framing member); material wider than 8″ requires three 8d nails per bearing. Stagger joints so that they don't succeed one another on the same stud. If you're using plywood, you can use a 5/16″ thickness on walls framed with studs 16″ o.c. and 3/8″ thickness if studs are 24″ o.c. Of course, 5/16″ is a *minimum* thickness and, if you can afford it, 3/8″ would be a better choice. If you want a stronger sheathing that affords a better nailing surface, go to 1/2″ or even 5/8″ thickness.

Plywood sheathing may be applied vertically, horizontally, or in a combination of both. Try to plan your sheathing pattern so that vertical joints don't occur on the same stud in succeeding rows. You may have to add some blocking to your walls so that all edges of each sheet of plywood have support. Nails should be 6d (common or ring-shank) for 1/2″ and less; 8d common or 6d ring-shank for 5/8″ to 3/4″. Spacing of nails should be 6″ o.c. along edges and 12″ o.c. along intermediate supports. Leave a slight space (1/8″ or so) all around each sheet to allow for expansion during damp weather. When you come to a door or window opening, the plywood sheet should come flush with the inside edge of the rough opening all around.

Fiberboard wall sheathing is applied much the same. It comes in panels that are 4′ wide and 8′, 10′, or 12′ long. Thicknesses of 1/2″ and 25/32″ are used. As with the plywood panels, blocking should be added where needed so that all edges of each sheet are supported. For nailing 1/2″ sheets, galvanized roofing nails 1 1/2″ long (with 7/16″ head) are used; for 25/32″ sheets, use a longer (1 3/4″) roofing nail. Nails should be spaced 3″ o.c. at edges and 6″ o.c. along intermediate supports.

Particleboard for wall sheathing is used in a range of thicknesses: 3/8″, 1/2″, 5/8″, 3/4″. For the 3/8″ and 1/2″ sheets, 6d common nails should be used, spaced 6″ o.c. along edges and 12″ o.c. along intermediates. The 5/8″ and 3/4″ sheets require 8d common nails, same spacing.

If you use either fiberboard or particleboard, you might consider using plywood sheets at each corner of your house. These sheets, placed vertically to replace the fiberboard or particleboard at that area, will act to brace your corners and make the building more rigid.

With the sheathing nailed in place, you are ready to install your windows and doors, a step that precedes siding because the siding is usually made to butt up against window and door trim on the outside. For information, see the sections on "Adding a Window" and "Hanging a Door: New Construction." Once doors and windows are in place and all trim is up, there is one final step, besides flashing above your doors and windows, which has already been mentioned. This step is discussed as part of "Corners."

Corners

There are several ways to finish the corners of your house, and some of these must be taken care of before siding is put on. If your house is a simple rectangle or square, you'll be dealing exclusively with "outside" corners; if the house has rooms or porches, extending out from a basic square or rectangle, you'll have "inside" corners as well. To finish an *outside corner,* you may use corner boards. These are usually 1×4's or 1×6's applied vertically; one is nailed to the edge of each wall, and the two corner boards butt together. Less commonly, the corner boards are set flush with the siding and a piece of quarter round is nailed in the resulting groove for a finish corner. Corner boards are usually put on before siding, but sometimes they are applied over the siding and project beyond it. Another way to finish outside corners is to miter the ends of horizontal siding so that they meet in a tight, neat joint. Vertical siding pieces at corners may simply butt against one another. Metal corner pieces are sometimes used, either over wood siding as a finish corner, or under mitered corners or butt-joint corners.

Inside corners may also be finished in a number of ways. A metal corner piece can be nailed in place over the sheathing and the siding placed over it, meeting in a butt joint or series of butt joints in the corner. A strip of wood may be nailed in the inside corner and the siding made to butt into this strip. Metal corner pieces may be used over inside corners to finish them.

Depending upon how you decide to handle the inside and outside corners of the exterior of your house, you either have to work on them before putting up your siding or must wait to add the corner finishes after your siding is up.

Horizontal Wood Siding

From the point of view of aesthetics, horizontal siding will make your house look longer and lower. Clapboards and bevel siding are two commonly seen forms of horizontal siding. *Clapboards* are square-edged boards, usually 1×6 or 1×8, applied so that each board overlaps the board below it, shingle-style. *Bevel siding* consists of boards that are 6″, 8″, or 10″ wide and cut to a slight taper across their width; in place, they will be thinner at the top, thicker at the bottom, which allows them to overlap one another a little more neatly than the clapboards. *Shiplap* and *V-rustic* boards may be used for horizontal siding. Each of these boards is milled to interlock with the board below, and the joint formed helps to shed water away rather than allowing it to be drawn back in behind the siding.

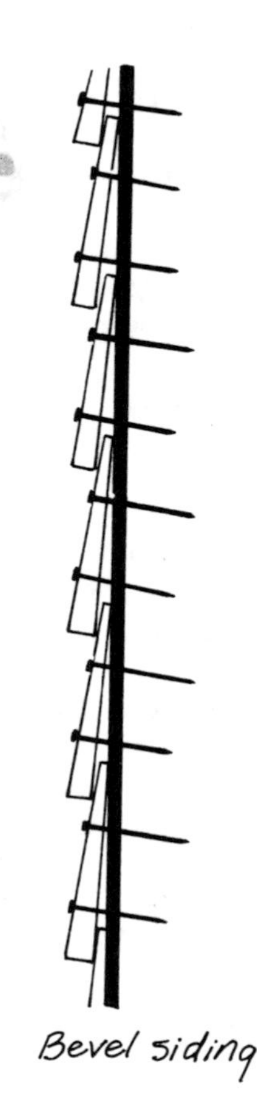

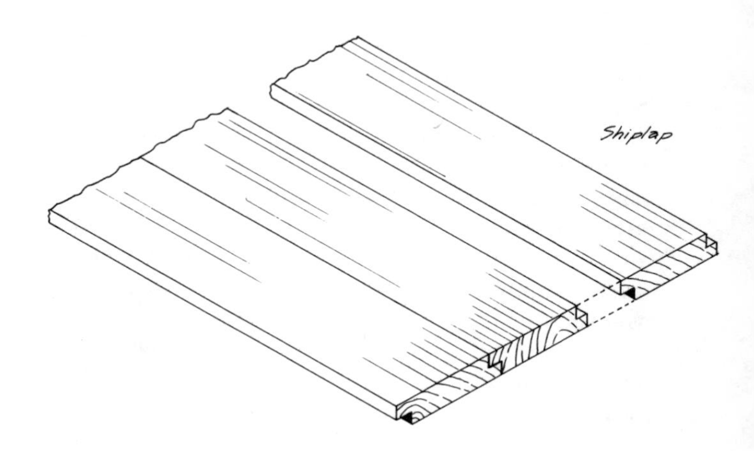

The joint formed by a shiplap is slightly curved; the V-rustic joint is sharper and V-shaped. Tongue-and-groove material is sometimes used for horizontal siding but is better suited to a vertical position.

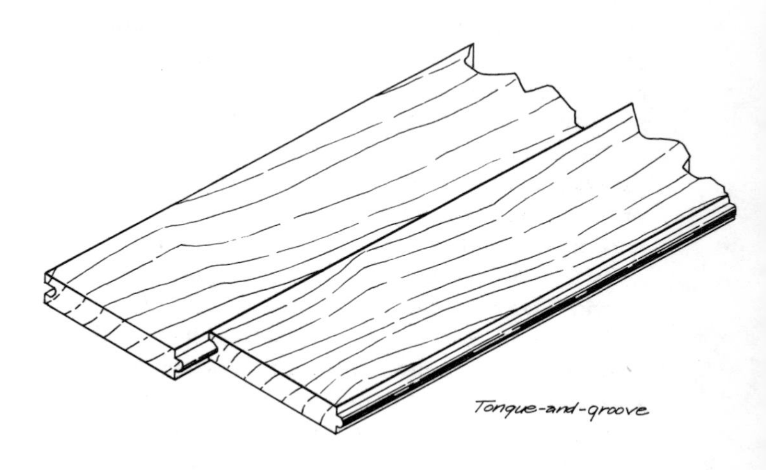

Lumber used for horizontal siding should be fairly dry; green lumber will shrink and leave gaps in your walls. Some siding (cedar, cypress, redwood) can be put on and left to weather without any finish of paint or preservative. Other wood is not as naturally resistant to weathering processes and must be protected with a sealer, paint, stain, or preservative that is periodically renewed (pine and fir fall into this category). The grade of wood that you use will be a matter of economics, aesthetics, and availability. Siding should be of a fairly good grade, since knots, splits, warps, and other defects will make your house less than weatherproof. But if you're working with a limited budget, you may have to put in some extra time to pick and choose through a lower-grade wood that you can get at bargain prices. A unit or two of cheap siding may be divided up, with the better pieces going on the exterior of your house and the poorer pieces being brought inside, where some knots or splits won't be that critical.

In applying horizontal siding of any type, you begin at the bottom of your walls and work up. Try to use boards in long lengths as much as possible. This will lessen your work (in terms of measuring, cutting, and fitting) and will make a more attractive-looking finished job. Try to stagger vertical joints so that they aren't directly over one another. Before you begin, sit down and plan out your walls so that you can make windows and doors look neat; if at all possible, you want to avoid having to notch out pieces of siding over doors and over and under windows. Divide the total height of your wall by the exposed surface of a piece of siding to see how this will work out. If you're left with a few inches, decide in advance how you're going to handle that. Can you change the exposure of your bevel siding or clapboards slightly to compensate? Will you have to rip a piece of shiplap to make a fit? Do the same for your windows and doors. You can shorten this part of your work by making a *story pole:* a piece of 1×4 cut to the exact height of your walls and with rows or courses of siding marked off on it. This can be held up to any window or door and it will show you what you're faced with so that you can make adjustments beforehand.

The first piece of horizontal siding should extend just below the upper edge of your foundation wall, or below the lower edge of your girders if you've built with a post and pier or a pole foundation. If you're using clapboards or bevel siding, you'll have to run a strip of wood around the lower edge of your house to help shim out that first piece of siding so that it won't dip or pitch inward along its lower edge. Siding should be nailed into studs or framing members. It is usually face-nailed with two nails per stud/framing member if the siding is 8″ wide or less; three nails if wider. If you're using a siding you intend to paint later, you might want to back-prime pieces before nailing them up. (This means coating them on the back with a sealer, primer, or water repellant.) This coating will help to keep the paint from peeling off later. Nails for applying horizontal siding should be rust-resistant: use galvanized or even aluminum nails. If you're nailing bevel siding into wood or plywood sheathing, use 7d nails (for nominal 1″ siding); if you're nailing into fiberboard, use 9d nails. Other 1× siding is nailed similarly. The first piece of siding should be nailed on level. Tack one end in place with a single nail. Then use your level to place the board perfectly: set the level against the top *or* bottom edge of the board and move the board up or down until it reads level. Then tack the board in place and proceed to nail it securely. In nailing on the second and succeeding pieces of bevel siding, place two nails per framing member or stud. Nails should be placed so that they miss the upper edge of the piece below, otherwise they may split that piece. (This also allows for expansion and contraction during weather changes.) The 1×6 bevel siding is usually overlapped about 1″; wider bevel siding should overlap 1½″. These overlaps can be slightly adjusted to meet the needs of your building. Drop or V-rustic siding may be face-nailed or blind-nailed, using 7d or 9d nails (depending upon what you're nailing into). If you have problems with nails splitting your boards, try blunting the nails. Be careful, too, not to nail too close to the ends and edges of your boards. Another precaution in putting on siding: make all of your cuts perfectly square. Otherwise your joints will have gaps, which not only look bad but create water traps.

If you are working with really long boards, you may have problems getting them to fit snugly against one another. A warped board can be straightened out with a little extra work and the help of a partner. One person should begin at one end, fitting the board tightly in place and nailing it. The second person can pull the opposite end of the board up or down as needed—she can overcompensate quite a lot and then gradually straighten the board out as you nail it bit by bit. Sometimes a wood block may be used against the lower edge of a reluctant board; you can hit the block pretty hard with your hammer and force a piece into place without damaging it.

A final word on horizontal siding: don't use plain square-edged boards if you want a dry, warm house. This may seem obvious, but perhaps so obvious that it slips someone's mind. Plain boards put on horizontally are virtually impossible to weatherproof. The joints act as open shelves to water—and if you've finished your interior walls, you may not realize that rain is leaking down inside the walls until puddles start forming all around your baseboards! Putting battens over the joints won't help, either: the rain will seep down in back of the battens and run down inside the walls. If you've made this mistake, the best you can do is consider that you've just *sheathed* the walls and add on another whole layer of siding, such as wood shingles or plywood. You might make it through the winter by carefully caulking all the joints, but this isn't really a permanent solution.

If you're going to paint your exterior walls, you may want to sink all of your nail heads and putty over them. Don't do this if you plan to use a stain rather than paint, as the putty will take the stain differently and the holes will show up dramatically rather than obscuring the nails! Another final touch after siding (and prepainting) is to go around and carefully caulk the joints between window and door trim and siding. This caulking will cut down drafts and make it easier to keep your house warm.

Vertical Wood Siding

Vertical siding will make your house look taller but will appear to shorten it length- and widthwise. The most common form of vertical siding is *board-and-batten:* square-edged boards (usually 1×6, 1×8, or 1×10) are applied vertically with battens (usually 1×2) covering the joints between the boards. Rough-sawn boards may be used; these will commonly be in cedar, redwood, or any other species of lumber that can age outdoors without needing preservative or paint. A variation of board-and-batten is *board on board;* here 1× square-edged boards are used vertically but spaced more widely apart (according to the width used). Boards are used as battens, overlapping the two adjoining siding pieces. Still another alternative that is similar to board-and-batten is the *artificial board-and-batten look* created when plywood sheets are applied vertically with 1×2 battens spaced at regular intervals (covering joints between sheets but also placed at "intermediate," and strictly cosmetic, intervals). Specially milled boards such as shiplap and V-rustic may also be used vertically.

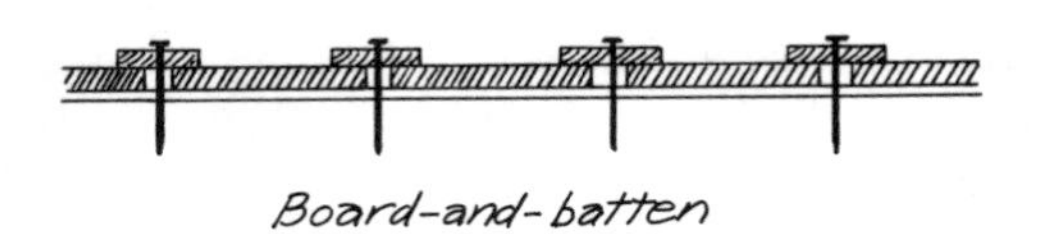

There are a couple of important advantages to board-and-batten vertical siding. This system allows you to use green or wet lumber. If the boards and battens are nailed properly, considerable shrinkage can occur as the boards dry, yet no gaps will show in the wall. A second advantage is that siding boards may be easily

(Photo by Carol Newhouse)

replaced at a later time if they become damaged. The battens can be removed and individual boards pried loose and replaced. For comparison, it is quite difficult to remove a piece of shiplap or bevel siding without damaging adjoining pieces, and very difficult to put a new piece in place without disturbing and/or damaging the pieces around it. Still another advantage (inherent in most vertical siding) is the absence of horizontal joints, which always create some problems for weatherproofing. Vertically applied V-rustic or shiplap siding will make a smooth wall that is both weathertight and attractive. As the procedure for applying most vertical siding is quite similar, it is discussed generally below.

If you aren't nailing to a good solid base of sheathing, you'll have to add some blocking or intermediate nailers to your wall framing before putting up vertical siding. Vertical boards should be nailed about every 4′, so you may nail to your top and bottom plates (if your walls are stud framing), and some intermediate blocking will be needed midway up your walls. Don't forget to apply felt paper as needed, and to set up corners in advance if you plan to use metal pieces under siding, to butt siding up to corner boards, and so on. The siding should extend just below the top edge of the foundation wall or just below the bottom edge of girders so that rain can drip free and not be pulled back in to soak framing members. Begin each wall at a corner. The first board should be carefully placed, checked with a level, and nailed in a plumb position. For most 1× siding, 7d nails are driven through to wood or plywood sheathing and 9d nails into fiberboard. If you're nailing directly into studs and blocking, use 7d or 8d nails. Nails should be placed every 4′ or so, as mentioned earlier. If you're using shiplap or V-rustic, you can either face-nail or blind-nail. The first (corner) board will have to be face-nailed along its corner edge, but the following boards may be blind-nailed through the tongued edge. If you are putting up board-and-batten or board on board, use a spaced block or pair of spacers cut to the width of the space you want between boards. This will allow you to set your boards at a uniform distance from one another in seconds. It will also help you to keep the boards going on evenly and plumb. Don't forget that green/wet boards are going to shrink. You should make allowances for this by placing them slightly closer together; when they shrink, they'll pull apart ⅛″ or more. Ring-shank nails will hold wet lumber more effectively.

When you come to nail on your battens, nail them to one board only. This will allow for shrinkage and expansion/contraction due to normal weather changes and is the procedure for all lumber, green *or* dry. If you nail a batten to both adjoining boards, it may be split when the boards shrink. A second alternative is to nail battens straight through the gap between the boards to the framing or sheathing (one nail centered at each 4′ interval) rather than nailing to the boards.

When your walls are completely sided, corners finished, and so on, caulk any horizontal joints and caulk the joints around windows and doors. Sink and putty over nailheads if you plan to paint your house.

Plywood Siding

Plywood siding comes in sheets 4′ wide by 8′, 9′, and 10′ long. There are various finishes or face appearances, such as texture 1–11 (this has deep grooves cut into the face of each sheet), striated (this has vertical ribbing formed by close, random-width grooves), brushed (the face resembles rough brushed concrete), reversed board-and-batten (grooves are cut at regular intervals to resemble inverted battens), and others (see illustration). Some plywood siding is prefinished with a waterproof coating or preservative and needs no other finish;

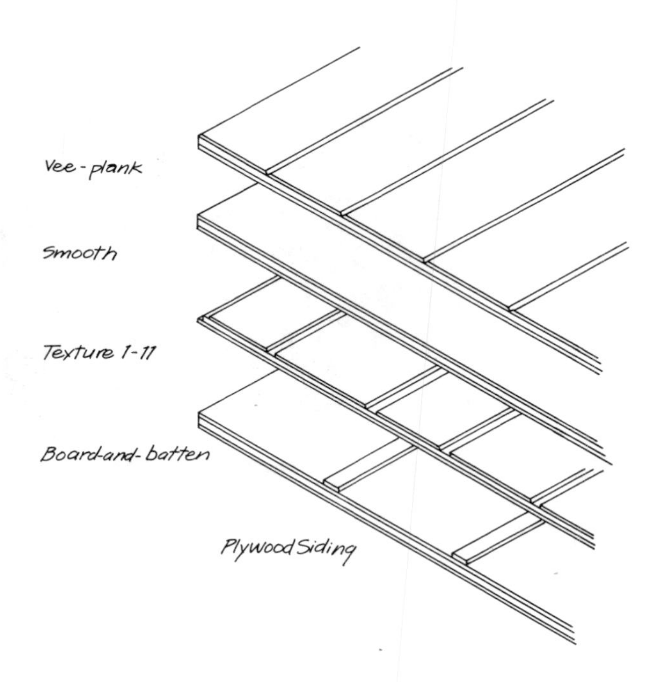

other plywood siding should be painted or stained. Plywood siding is meant to be applied vertically. Side edges are usually shiplapped or otherwise milled to interlock in a weathertight joint. Horizontal joints create a problem unless you can get siding that is end-matched (i.e., top and bottom edges, as well as side edges, are milled to interlock). Or you can buy special Z-shaped flashing meant for these joints. It is set along the top edge of a lower sheet, and the second (higher) sheet set over this flashing. The design of the flashing sheds water out of and away from the horizontal joint. Vertical joints may be caulked or protected with battens if the side edges of your sheets aren't specially milled.

In general, 3/8″-thick plywood siding is applied directly to studs that are 16″ o.c., and 1/2″ or 5/8″ siding to studs 24″ o.c. If used over sheathing, 5/16″ is used over 16″ o.c. studs and 3/8″ plywood over 24″ o.c. studs. In some areas, local codes may specify different requirements that you may have to follow. Nails should be galvanized, or aluminum or other rust-resistant type; 6d nails are used for 3/8″ and 1/2″ panels and 8d nails for 5/8″ and thicker. Nails should be placed about 3/8″ in from the edges of the sheets and spaced 6″ o.c. along edges; intermediate supports can be nailed 12″ o.c. As with other forms of wood siding, joints around windows and doors should be caulked as a final step.

Plywood siding offers an alternative wood siding that is economical, quickly applied, and quite durable. It makes optimum use of our timber resources. With the wide variety of finishes and styles available, it should fit almost any house both aesthetically and practically. Its major disadvantages are the awkwardness and weight of the sheets; these two factors make it difficult for one person to work with this siding. Another more minor disadvantage is that plywood siding adds little insulation to the walls of the house. In most instances, the advantages of plywood siding will outweigh the disadvantages, so that the use of the material will probably continue to grow.

Wood Shingles

There is something special about a wood-shingled house—small or large, simple or fancy, it has a charm that increases as the shingles age. Wood shingles are relatively easy to apply to the exterior walls of your house. They require the simplest tools and some basic attention to details; beyond that, the work is pleasant and your pace can be relaxed. Practically speaking, wood shingles rate among the best choices for exterior siding. They require neither paint nor preservative, weathering into a beautiful, perfectly functional old age. They can be applied to the most complicated walls and produce a tight, weatherproof covering. If a small area needs to be repaired later, old shingles can be broken out and new ones substituted. The only real disadvantages to using wood shingles are cost (they are expensive) and time (they take longer to apply than almost any other form of siding). If you plan to make your house your home for a long time, the investment of both money and labor may well be one you're willing to make—and take pleasure in living with for years and years.

Shingles come in different grades, widths, lengths, and species of wood. Redwood shingles were once common, but as our redwood trees have been recognized as a finite and precious resource, they are becoming rarer. Red cedar shingles are now being used most widely. These shingles are graded as follows:

No. 1 Blue Label 100 percent clear, 100 percent heartwood, 100 percent edge-grain. This is the top grade of shingle, used for roofs and for sidewalls.

No. 2 Red Label over half of each shingle is clear, with some sapwood and flat grain permitted.

No. 3 Black Label a utility grade used for "secondary buildings." Less than half of each shingle is clear.

No. 4 Undercoursing a utility grade that allows many defects. It is used as undercoursing (a first layer that is entirely covered) for double-coursed sidewalls and is used on interior walls. This is a very inexpensive grade of shingles but isn't really suitable for the exterior walls of your house.

No. 1 or No. 2 Rebutted-rejoined this is a special type of shingle that is trimmed square all around and used where you want an exact fitting of shingles against or up to one another. It comes in two grades that correspond in quality to the general grades No. 1 and 2 above.

Within each grade of shingles there are various lengths with special trade names: 16″ shingles are Fivex; 18″ shingles are Perfections; 24″ shingles are Royals. Shingles are sold in bundles, which consist of shingles of equal length but varying widths (from 3″ to 14″). You can also buy shingles in bundles of uniform width; these are called dimension shingles and usually must be special-ordered. In most grades of shingle there are four bundles to a square; shingles are priced by the square: say, $75 a square. How much actual wall coverage you will get from a square of shingles depends upon the length of shingles and the exposure you are using, and also upon whether you are single- or double-coursing your walls. The *exposure* of a shingle equals the portion exposed to the weather (the part not covered by the course or row above it). *Single-coursing* means that you apply one layer of shingles to the wall, with each course or row overlapping the one below it for a weathertight finish. *Double-coursing* means that each course or row consists of a double layer of shingles. This system allows a slightly greater exposure per shingle, but it is doubtful that this makes

(Photo by Carol Newhouse)

up for the cost of doubling each row. The double-coursing style offers a special "look" of strong lines and shadows. For most sidewall shingling, single-coursing is more than sufficient.

The common exposures used for sidewall shingling are:

7½" for 16" shingles
8½" for 18" shingles
11½" for 24" shingles

You can use a lesser exposure if you want, but don't use anything greater. To calculate how many shingles you'll need for your walls, add up the total square-foot area you are covering. Don't forget to subtract for window and door areas. When you have a total square-foot area figure, add about 5 percent to it, to allow for a double-coursed first row, plus some margin for error. Now, using your final square foot area figure and knowing the length of your shingles and the exposure you're using, you can calculate how many shingles you'll need by using the following (adapted from a table published by the Red Cedar Shingle & Handsplit Shake Bureau):

1 square of 16" shingles used with a 7½" exposure will cover about 150 square feet
1 square of 18" shingles used with an 8½" exposure will cover about 154½ square feet
1 square of 24" shingles used with an 11½" exposure will cover about 153 square feet

Dividing your total square-foot area by the appropriate number will give you the number of squares you'll need for the job. It's important to note that the *usual* definition of a "square" of shingles (the amount that will cover 100 square feet) *does not apply to sidewall shingling.* It *does* apply to roof shingles, where a square applied at the proper exposure *will* cover 100 square feet. This can be very confusing, especially because most people are used to talking about shingles in terms of *roofing* squares.

Before you actually begin the work of shingling your walls, you'll have to do some wall preparation and decide how you want to finish your corners. If you are shingling over wood sheathing or over old wood siding, you will want to cover your walls with a layer of building paper before applying the shingles. The Red Cedar Shingle & Handsplit Shake Bureau specifies using "unsaturated building paper between shingles and sheathing." In our area, asphalt-saturated felt is used. In other areas, rosin-sized building paper is used. The best advice probably comes from the people who make the shingles—but you might ask around in your area and see what is commonly used. Whatever your choice, this paper should be applied from the bottom of each wall up, with each row or course overlapping the one below it. If you are putting shingles over a non-wood sheathing (such as stucco or mineral fiber sheets or shingles), put on the layer of building paper and then nail up wood strips to attach the shingles to. These wood strips are added at heights corresponding to the shingle length and exposure you're using.

Once the building paper is on, you should decide how you want the corners of your house to look. One of the quickest (and neatest) ways to finish the outside corners is to use corner boards. These can be nailed in place and the shingles can butt up to them. They may be left natural or painted to correspond to the window and door trim. A second way to handle the outside corners is to use an alternate overlap of shingles for each successive course. In this case, the shingles are made to butt into one another at the corners, and the butt joint is alternated for each row. This method is somewhat slower than using corner boards. You will find that you have to do some custom trimming of these corner shingles to make them fit properly. Look at some shingled houses in your area to see the appearance of corners finished this way and with corner boards. Inside corners may be finished differently, too. The most professional finish is to use a strip of folded metal flashing in each corner and to miter the shingles as they meet over this flashing. A second alternative is to nail a 2×2 wood strip in the corner and butt the shingles up to that strip. A third possibility is simply to butt the corner shingles into one another. Again, you should, if possible, look at several shingled houses to see what kind of finished look appeals to you.

The nails used for shingles are usually galvanized 3d (1½″ long). The best tool for driving these nails is a shingling ax or hatchet (described in the "Tools" section); with its waffled head, it is unlikely to drive the nails too deeply and split the shingles. The shingling ax may also be used to split shingles to widths that you need. Another handy tool for a shingling job is a saber saw (or portable jigsaw). With this, you can cut shingles to size or even shape them with fancy curves to make your house look really special! In lieu of this tool, have a fine handsaw to work with. A combination square, a bevel, and a pencil will be needed if you're shingling a house with a gable (or irregular) roof, since you'll have to cut shingles at particular angles to fit along gable end walls. A nail apron is a necessity for this job. You may want to use a chalk line to mark the wall, or a long strip of wood (more on this soon). Don't forget ladders and/or scaffolding.

If all of the calculating and preparation pre-shingling are rather dry and unappealing, the experience of shingling will be a pleasant surprise. There are few carpentry jobs or house-finishing details that are more enjoyable. Working with simple tools and clean, beautiful materials, you will find your house being transformed bit by bit—and can enjoy being outdoors, working at an even, calm pace. Shingling the walls begins at ground level, and the first row or course of shingles should be doubled and extended just below the upper edge of the foundation wall, or just below the lower edge of girders. Shingles should be applied in random widths and are usually spaced about ¼″ from one another. Use two nails on shingles that are 8″ or smaller in width; three nails on wider shingles. In general, try to use shingles that aren't much wider than 10″ or 12″—these and wider shingles tend to split when you nail them. Nails should be driven about one inch above the butt line of the next (succeeding) course so that, when that next row is applied, the nails will be hidden. For example, if you are working with 16″ shingles and plan a 7½″ exposure, your nails should be driven about 8½″ up from the butt (lower) end of each shingle. Don't smash the nails too hard or they will split or actually drive through the soft wood of the shingle.

When you have the bottom row done all around, go back and double it by applying a second layer of shingles over the first. This layer should be carefully done so that each joint between shingles is covered. As you continue with your walls, keep this in mind: you want to choose the width and placing of your shingles so that all joints are staggered. Whenever you look at a joint between two shingles, you should see the solid face of a shingle beneath. This will take a little practice, but you'll soon find yourself making

(Photo by Lynda Koolish)

choices and decisions almost without thought. When you're ready for the second course of shingles, you may want to measure up and mark the butt line for this row, according to the exposure you are using. You can mark this with a chalk line or you can nail a wood strip in place temporarily. You can even use a level to establish this height, or you can use the marking gauge on your shingling ax to set each shingle at the right exposure. However you choose to do it, try to keep the horizontal lines even and fairly level. When you're done, your house will look the better for it.

There are many possibilities for imaginative application of wood shingles. You can do some walls straight and make others gaily decorative, with the butt ends of your shingles cut like fish scales or triangles. Gable ends of walls may be specially done, or the area right above a door can be made more decorative with specially shaped shingles. An overall design can be created by simply alternating the exposures of your shingles a little—you can do every other shingle with a 6½″ rather than a 7½″ exposure, for instance. As long as you keep the functional needs of your shingled walls in mind (think of how the rain will be shed in each case), you can make your house as fanciful (or simple) as you like!

Atypical A-frame

I never consciously thought about building a house before I saw the land on which it now sits. My mother had purchased the land from a friend at work and invited me to take a weekend trip with her so that she could show it to me. We spent that weekend camping on her ten acres in northern Columbia County in upstate New York, and before we left we went to a local company that displayed precut homes. The only model they had available was an A-frame structure, and I think I would have built an igloo if that's all that had been available at the time. I loved the land from the first moment I saw it and just wanted a house on it as soon as possible. The A-frame manufacturing people said we could have the house by spring. It was precut and ready to assemble (like Lincoln Logs, I thought) and seemed easy enough. There was just one small thing: I didn't think the house seemed long enough. Never having built a house before, I thought I could just add 10′ or 15′ somewhere in the middle and then continue with the original package. . . .

It was an exciting and productive weekend, and I couldn't wait to get home to Jackie and let her in on the big news. I was sure she'd be as happy and excited as I was. Between the two of us, we had just enough money to meet our bills, but I hadn't thought much about the financial end of this scheme yet. Somehow I knew we'd make a way.

Before the winter was over, our deal fell through on the precut house. The manufacturer had sold out and simply left the area. By this time Jackie and I had figured it would be easy to build the extension we wanted, so we now thought: Why not build the whole thing? We were working at JFK Airport at the time and both liked the style of one of the chapels on the airport grounds. It was an A-frame but had a flat roof, which seemed even easier, so we prepared to figure out how to build a similar structure. We set to work with graph paper and balsa wood and made a rough model of the main structure. Figuring ¼″ to 1′, we proceeded to lay the floor plan. My mother contributed a U. S. Government book on *Wood Frame Construction,* and we were ready to begin.

When we began the building of this house, Jackie and I had limited experience with carpentry. I had spent many of my growing years living in a house that was a "Handy-Mom Special," watching my mother (and less frequently my father) repair, remodel, and refinish the home we'd bought. As I grew older, I learned to use a saber saw and swing a hammer. Usually I just followed directions on what was to be done. I was no carpenter, but neither was I mystified by tools and their uses. Jackie and I had shared five or so different apartments, each of which was in need of some repairs or alterations. We were never afraid that we couldn't make a place better than it was! I guess that on a scale of one to ten our carpentry skills rated

(Photo by Dorothy Dobbyn)

about one and a half—but we were willing to try, and we knew that having the tools and knowing how to use them were half the battle. We read a lot of books on the subject of building and began once more to improve our skills by *doing*.

We took another ride up to Columbia County to find a local contractor who would put in the cement foundation. I was twenty-three and Jackie twenty-one years old that winter, and both far from being experienced contractors or builders. The contractor we talked with seemed nice and was available to do the job early in spring, so we agreed on a price and hired him. We told him where we wanted the house to be placed, which direction it should face, and how big it would be. Now we had about two months to figure this whole thing out and try to get some money together. The only savings we had were about $1,500 worth of stocks, which would be just about enough to pay for the foundation work.

Throughout all this early planning stage, neither Jackie nor I ever considered how long it would take to finish this project which had started so suddenly. We never doubted that we could do it, but what we should have asked ourselves was whether or not we needed such a large house! Our final decision was to build the structure to be 30′ wide by 50′ long and 28′ high. We would put in plumbing for three bathrooms and one kitchen. The main floor would consist of a kitchen, dining room, bedroom, large bathroom, and a 25′ by 25′ living room with a cathedral ceiling and a fireplace. The second floor would consist of a bedroom, bathroom, and small deck outside the bedroom (with access through sliding glass doors). We would build a rear deck 30′ by 28′ and a front deck 30′ by 14′. It was rather an ambitious undertaking, and certainly much bigger than we needed. However, that was our plan.

We contacted an architect in New Jersey to give us specifications for the main lumber that we would need. His fee was $75 and we saved at least that amount by using smaller upright beams than we had originally planned to use. There were no restrictive building codes in the county at that time, so there were no delays in that area.

By the first of May the foundation was ready and the first delivery of lumber arrived. We were almost ready to begin. First we needed to get electricity to the land, so we called the local electric company, which installed a utility pole approximately 50′ from where the house would finally stand. We had some tools of our

(Photo by Dorothy Dobbyn)

own and, with the rest borrowed from my mother and aunt, we now had the electricity needed to use them and were ready to go to work on a house that would finally be *habitable* five years later.

I always thought the weather in May was reasonably pleasant, but that month was unreal! When it wasn't raining it was snowing, and the mud and sludge made the conditions even worse. On a vacation from our jobs, we worked the first three weeks of May from early morning until dusk and sometimes fairly late into the night. We slept at a motel during those weeks. Later we would camp on the land and sometimes stayed at our neighbors' house. They were our only neighbors at that time, as the house next door was for sale.

We worked hard and accomplished a lot in those first three weeks. We began by setting the main supports, or Lally columns, into cement footings in the basement. Then we laid the main beams lengthwise. There were four of them, 2×12's butted over the cement-filled columns. Next we laid the sill plate and floor joists, and finally the 4′×8′ plywood for the main floor. We then started to cut and assemble the main A upright supports. These supports were exactly like large trusses. We precut the pieces and bolted the second floor where needed. As this work progressed, we made stacks of huge wooden capital *A*'s. They had holes predrilled for the bolts that would connect them to the floor joists.

After all the A supports were assembled, we contracted a crane operator to come in and raise the individual frames so that we could bolt them in place. It took about eight hours to set all of them in place. We braced one side from end to ground and started leveling, beginning from the braced end. Working late into the night, we had finally straightened and tied together about twenty-one of these A-frame supports. Too exhausted to continue, we went back to the motel for the rest of the night, planning to finish early in the morning. Unfortunately for us there were heavy winds that night, and when we returned the next morning we found that all of the uprights had shifted and the entire job had to be redone. It took about three days to finish this part of our project, and it was a very discouraging experience. I think that this was the first time I began to realize just what we'd gotten ourselves into, and that it was going to take more time and stamina than I'd imagined to build this house. Eventually that job did get done, and we started to nail the plywood to the outside walls.

This was a project in itself! The bottom first course was fairly easy, but considering that there was 24′ of upward-slanted beams left to cover, it kept getting trickier. We nailed 2×4's on the first row of plywood, and I would stand on these ledgers. Jackie and my mother would hand me the plywood while they stood on the main floor, working each sheet up in between the beams so that I could reach it. We continued in this manner until we were halfway up the roof. At this point we had to build a scaffold inside so that the sheets could be lifted up to the heights we were reaching. On the outside, we nailed a series of 2×4's to work as a kind of ladder. With these improvements, we continued lifting and nailing the plywood into place.

Although my parents were separated at the time, my mother insisted that my father come up to help with the construction during the first few weeks. He was willing, although this kind of work was not exactly his choice. I remember him staring at what we had been doing and asking many times what kind of a structure it was going to be. His eye for carpentry wasn't very good, but his heart was generous and we were grateful for his help (after those early weeks, he didn't see the house again until it was practically finished). Once, while I was balancing myself on the 2×4's 20′ up the side of the house, nailing on the outside plywood, my father became concerned that I might lose my footing and fall. He insisted upon tying a rope around my waist for safety. Fortunately I didn't fall, because when I finished the job and

(Photo by Dorothy Dobbyn)

was on the ground I discovered that the knot around my waist was a slipknot! It was good for a laugh—but terrible for a fall!

Our long and short three-week vacation was soon over, and we had to go back to our jobs in the city. We had gotten a good start.

For the rest of the summer, Jackie and I usually went up to work on the house Friday nights, returning to the city late on Sunday nights. My mother often came with us, which had its advantages and disadvantages. She was very helpful and supportive much of the time, but Jackie and I were seldom alone. Also, she was a very early riser and it was pretty hard to try to sleep when she would call out, "Breakfast," gulp down her cup of coffee (which to her was breakfast), and start hammering about two feet from our heads at 5 A.M.! Even with her little idiosyncrasies, we appreciated all the energy and hard work she put in with us that summer. By

October, when it was too cold to continue, we had the rest of the framing completed and plywood put on, and I nailed plastic over the window openings and closed our house up until spring.

The airline I worked for at JFK had a credit union, so I was able to borrow $5,000 to get us started the following spring. By that time, the loan I had previously borrowed was almost paid back. Our goal for that summer was to shingle our enormous roof, put in glass windows and sliding glass doors, and begin framing up the inside ceilings and walls.

Our first step in the spring was to get the shingles. We decided to use three-tab self-sticking asbestos shingles, and estimated we'd need approximately 3,300-square-feet coverage (thirty-three "squares" of shingles). We had a Volkswagen camper at the time and, not knowing how many packages of shingles thirty-three squares would be, drove to Albany and the nearest Montgomery Ward's store to pick them up. We found the shingles we wanted, and ordered them, and the clerk asked where they were to be delivered. We replied that we had a truck and would take them with us. After we drove around to the warehouse and presented our merchandise slip to the clerk, he asked us where the truck was. We pointed to the VW camper. Then he showed us what 33 squares of shingles looked like. We had been planning to put 1¼ tons of roofing material in the van and take it home with us! The shingles were delivered to the house within the next two weeks.

We bought most of our supplies locally, although we sometimes drove to a larger lumber and hardware store about an hour's distance away. Time was valuable, too, and usually better spent in building. Looking back on construction time now, I realize

(Photo by Dorothy Dobbyn)

some money could have been saved. For instance, there is a sawmill in the area that delivers large loads of lumber. We could have bought all rough-cut (unplaned) 2×8's and 2×12's from them at about half the cost we paid for the finished pieces. We didn't know about the sawmill, and, of course, none of the local suppliers we dealt with happened to mention it. One disadvantage of buying from a sawmill is the boards are often a little too long and not squared, but we could have bought a radial-arm saw to make the job easier and still saved money. In fact, the savings in lumber costs would have bought the saw for us. As it was, we bought the saw later anyway and it was a most useful piece of equipment.

After the main shell of the house was completed with plywood and our shingles were going on, we framed the A-shaped front of the house to hold the glass windows and doors we wanted to put in. The front of the house is 95 percent glass. One day an owner of the local lumber company stopped by to see our progress and told us he had two large insulated windows that had been ordered about a year before and that he was stuck with. He said that if we wanted them, he would deliver them and sell them to us for a mere $100 each. They measured approximately 5′×5′ and were ¼″ plate with ½″ sealed air space. We took him up on his offer, but as a consequence had to reframe the entire front of the house to accommodate the different-size-than-we'd-intended windows. We were very pleased with the windows despite this extra work. Unfortunately, about two months later, in mid-fall, the house was very cold. I decided to build a fire, and as the fireplace wasn't installed yet, I built a roaring blaze in a barbecue pit in the living room about three or four feet from one of the glass windows. Electrical tape now holds the bottom left section of that window together, although it wasn't exactly what we'd originally had in mind! I don't build fires near glass anymore.

We had only one instance of male arrogance to deal with. Needing a septic tank, we hired a contractor with a backhoe and had him dig a trench for the pipes from the house to the tank area and a hole to accommodate the septic tank itself. He hit slate about 5′ down, then continued to dig the leach field (the lines allowing dispersal of fluids from the tank). We ordered a tank from a local septic company and they promised to deliver it a day or so later. And so they did: here came a huge 1,000-gallon cement tank on a flatbed truck with a self-contained crane. It was being delivered by the owner of the company, who looked at the hole and proceeded to set the tank in place. The only problem was that the tank stuck out above the ground a good 18″ to 20″ and the pipes coming from the house would have to have been at least a foot above the ground! We told him this would be no good, because, aside from the time and money it would cost us to have these externals filled in, we planned to have a circular driveway that would go right over where the pipes would be. He told us rather curtly to forget the driveway and suggested we back up, turn around, and go down the driveway the other way. We suggested in kind that he get his tank and himself down the driveway and off the land immediately. He got carried away after that and started screaming, so we told him we weren't going to pay him for the tank whether he left it or not. He took his tank and left. We bought a 500-gallon steel tank, put it in place ourselves, set the pipes, and covered it over. It's still working fine.

Although the house finally got more or less finished, it is still changing, and probably always will, at least while I'm living in it. About halfway through the building of the house, seven years ago, Jackie left. At that time we had the incompleted structure appraised, discovering its market worth to be $5,000. We had put about $14,000 into it, however. My first impulse was to move to the West Coast and forget it. I was tired and didn't think I could finish it and wasn't even sure I wanted to. As things worked out, I quit my job a year later and moved into the house, spending the next year doing most of the work myself. Another woman, Marian, eventually joined me. I had just started laying in oak flooring and we finished that and laid a slate floor in the dining area together. Last spring, Marian and I added a loft to the living room, which no longer has a cathedral ceiling, and finally finished another small bedroom this year. Over six years we pretty well finished everything.

When this house was begun, I never realized I had taken upon myself a lifelong project. I'm tired now and I've changed in many ways, and even if I had as much energy as when I started, I would never choose to build this type of structure again. I feel that it drains too much energy just in maintenance and repairs. It was not built with enough attention to utilizing natural resources—for example, not a single window in the house faces south (when we first designed and built the house, concepts like "solar heat" and "solar energy" were not around yet). If I built again, I would plan a house that uses nature and its immediate surroundings. I'd build on the south side of a hill and perhaps have the house dug halfway into the ground. In this way the house could gather and conserve the natural heat of the sun. The A-frame has woodburning stoves and a fireplace; when the fires are burning in cold weather, we have a system connected that furnishes us with free hot water. I'd like to have solar collectors for heating water in the summer and a greenhouse on the south side of the house with access from inside and outside. I view these additions as a kind of self-sufficiency that can minimize consuming of resources and allow some freedom from the system that encourages consumption. Some other important things I've learned in building my house have been to buy only what one really needs, and to buy accessories that will last (for instance, good used restaurant equipment for the kitchen, which will last for as long as needed). I've learned the value of good tools and the importance of building small—a house that will fit you and your needs, without extravagance.

(Photo by Dorothy Dobbyn)

There are some interesting phenomena that occur when you build the house you'll live in. I find that I worry when there are heavy winds: will it be all right? I listen to the sounds of it, and the house is almost alive to me. It's seen my life change; it's seen my relationships change. Sometimes I feel that it owns me. It demands my attention at times when I may not feel like responding but have no choice, as when a water pipe bursts. These crises can make me feel that too much of my energy (and that of others) goes into the house. I think that I would like to be free of it, sell it, move South or West to warmer climates and raise sheep and bees and live in a small, very small, cabin . . . and then I feel the house take over my mind again—I find myself imagining new changes, new additions or projects for it. My mother died last spring and I had her cremated. I've been considering burying her ashes on this land, but wonder what further emotional ties I'll be creating for myself, and what this will mean to my life. This house, my home, continues to weave itself into my future.

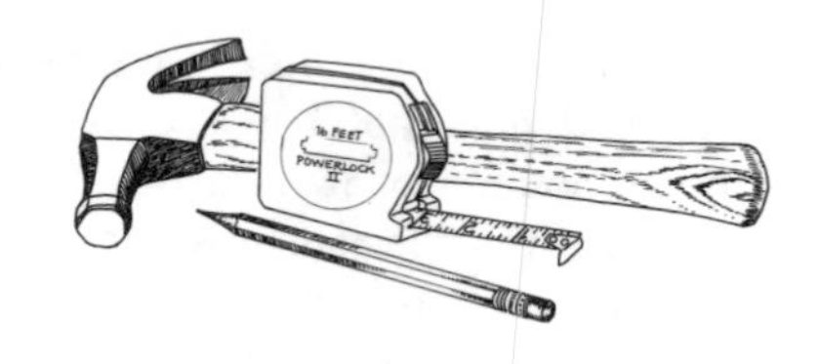

Keeping Out the Rain

ASPHALT AND FIBERGLASS SHINGLES

As the "icing on the cake," the roofing of your house should be fire-resistant, pleasant to look at, and long-lasting, and, most important, it should protect the house from the weather elements. There are two main types of roofing: *multiple-unit* (such as wood shingles, fiberglass and asphalt shingles, and aluminum and galvanized steel sheets) and *membrane* (roll roofing and built-up or "hot tar" roofing). In choosing a new or a replacement roof for your house, you have to consider the cost of the roofing, the slope of your roof, the permanence of the building, and so forth.

Asphalt and fiberglass shingles offer two choices that combine many advantages. These shingles are in a moderate price range, have an excellent life-span if properly applied, are easy to work with, and meet most fire codes because they are highly fire-resistant. They need little to no maintenance and take the weather well. If a small area is damaged, it is fairly easily repaired. Either type of shingle may be used on any roof that has a rise of at least 4" per foot of run. The following discussion is about asphalt shingles, but all procedures and comments apply equally to fiberglass shingles. Fiberglass shingles are rapidly displacing asphalt shingles in the roofing market. They are slightly more expensive but are considered more durable.

Asphalt shingles come in a variety of grades, with different life expectancy depending upon weather and geographical conditions. In general, it's a good idea to choose a good-quality shingle that will wear well and be worth the money and labor you are investing. To estimate how many shingles you need, you have to calculate the entire surface of the roof you want to cover. If your roof has valleys, dormers, and/or a chimney, add extras for these areas. For a shed roof, you can simply multiply the rafter length by the roof width to get your total square feet of roofing needed. A gable roof will mean: rafter length times roof width, then multiply by two. For a gambrel, a hip, or a polygonal roof, you will have to calculate each surface area of the roof and add them all up. Asphalt shingles are sold in squares, which cover 100 square feet of roof surface. They weigh from 150 to 325 pounds per square, with the average being in the 200-pound range. Your roof must be framed with sufficiently strong rafters to take this weight; otherwise, the roof may begin

(Photo by Carol Newhouse)

to sag. You may be able to brace up the rafters to take additional weight.

Asphalt shingles come in various colors; and while it is largely a matter of personal taste, some consideration should be given to the fact that a light-colored roof will reflect the sun's rays, while a dark-colored roof will absorb them, making the house warmer. Also, light colors on a roof will make the structure appear larger; dark colors will make it look lower and somewhat smaller. Asphalt shingles come in two basic types or styles: the single shingle, and the strip shingle. The strip shingle is more commonly used because it takes less labor to apply. One very common type is the three-tab strip shingle. It is 12″ deep by 36″ wide, with two slots or cutouts dividing it into three equal tabs. Each end of the shingle has half of a slot, which becomes whole when butted against the half slot of the adjacent shingle. The slots give the strip shingle the appearance of individual shingles.

Before you're actually ready to shingle, certain preparations must be made. An *underlayment* is placed over the roof sheathing (under the shingles), which protects the sheathing from moisture absorption until the shingles are applied. Underlayment also provides extra protection by preventing wind-driven rain from getting under the shingles. This water could rot or warp the sheathing or cause leaks that would damage interior walls and framing. A commonly used, easy-to-apply material for underlayment is asphalt-saturated felt (readily recognized as "tar paper"), which comes in the form of a flexible rolled sheet, usually 36″ wide, in weights of 7, 15, and 30 pounds (these refer to the weight per 100 square feet). The 15-pound is most commonly used. The felt is simply rolled across the sheathing and stapled down. There should be a 2″ overlap horizontally and 4″ vertically. The staples will hold it down until the shingles are laid, but don't lay shingles over the felt if it rains and the underlayment becomes wet before you get a chance to shingle over it.

The second step is to nail a folded piece of aluminum along the edge of the roof to allow water to drip away from the trim. This piece of aluminum is called a *metal drip edge* and can be cut to the length of the roof as needed. It should be nailed with aluminum nails, which will prevent rust and corrosion, and is applied under the felt at the eaves (bottom edge of roof), but over underlayment up the rake (sloping edge of roof). The metal drip edge also gives a firm base for the shingles.

The first course of shingles is laid upside down so that the spaces between the tabs are filled in when a regular course is laid right side up directly over it. Or you can cut the first shingle so that the regular course will be staggered over the spaces of the first row. The shingles are lined up along the drip edge evenly, and should overhang about ½″ along the rake. Nail shingles with galvanized roofing nails 1″ to 1½″ long, into the shingle about ½″ to ¾″ above each slot. The nailheads are covered by the next course of shingles. This first course of shingles should be doubled; the second layer is added so that all spaces between tabs in the first layer are covered.

A utility knife is used for cutting shingles, and all cuts are made on the reverse side so that the blade isn't dulled by the mineral granules. You can also use a pair of tin snips for cutting—these allow more controlled cuts. The second course (after the first starter course laid right side up) is begun with a shingle that is cut in half so that the slots don't line up with the slots of the first course. The alternation is necessary to avoid a leaky roof. The third course will begin with a full shingle, the fourth with a half shingle, and so on all the way up. The slots of every other course will be aligned, and this is one way to see if the shingles are being laid fairly straight. Another way to be sure is to drop a chalk line over the felt exactly where the top of the shingles of the next course are to be laid. A badly slanting course is not eye-appealing and may cause serious problems when the wind gets at it. One of the best ways to make sure the courses are straight is to get off your ladder and "eye" them from the ground. (Don't forget you're up on a roof or ladder and step back to look at your work!) Continue laying each course, on both sides of the roof, in the manner described, up to the peak of the roof.

In order to cover the peak (or ridge) of the roof, shingles are cut equally into thirds, making each shingle 12″ square. These squares are folded in half to go over the ridge, each one overlapped 6″ for double protection. They are nailed on each side of the ridge; nails are covered by the next square. This is easy to see if you look at any regular asphalt-shingled roof. The squares not only give adequate protection at the peak, causing rain to run off easily onto the shingles already laid, but are a very neat way to finish the roof. If you know from which direction the prevailing winds are coming, the squares should be nailed so they are closed to that direction.

If the roof runs into a vertical wall, flashing should be used for extra protection. The underlayment should run up the wall about 3″. Then install aluminum sheet-metal shingles, which can be purchased in 12″×12″ squares. Bend the shingles at a 90° angle and nail so that half the width is on the roof (over the felt) and half goes up the wall. The regular shingles go over the flashing as close to the wall as possible. Alternate with metal flashing shingles and asphalt shingles covering them to the ridge of the roof. Flashing is installed in the same way around chimneys. The side of the flashing that goes up the chimney can be caulked with roofing cement.

Here are some extra hints. If the material is delivered by truck and the roof is accessible to that vehicle, it is a good idea to unload the truck right next to the roof where the packages of shingles may be placed directly on it to avoid extra lifting. Do not put more than three squares at a time on the roof, as too much weight in one spot may overstrain the rafters. When reroofing, no underlayment is required if new shingles can be laid directly over old roofing; the old roofing serves the same purpose as regular underlayment.

If the roof you are going to work on is very steep, you may feel safer with a rope tied around your waist and attached to the ridge of the roof (no slip knots, please!). While working on a roof last summer, my partner and I found the easiest way to handle the situation was to tie ropes around a ladder laid horizontally along the roof, tie the ropes to 6″ metal spikes nailed into the ridge, and arrange the ropes in pulley fashion so that we could raise and lower it easily to the section being worked on. We simply stood on the side of the ladder instead of the steps of it. It's a good idea to remove tools or any potentially dangerous objects that might be in the path of a possible fall. Look the area over first and get an idea of where the safest place to leap would be in case you do lose your balance. Chances are that will never happen, however, if you exercise a little caution.

And, as a final word, in areas where the wind velocity is very high, it's a good idea to put a dab of roofing cement under each asphalt tab to keep the tabs from flapping in the breeze. Or you can buy self-adhesive shingles, which have a dab of adhesive already in place. Roofing cement is also sometimes needed around flash-

(Photo by Dian)

ing; it can be purchased in cans or in tubes that fit into caulking guns for more precise, less messy application.

Roofing with asphalt shingles isn't difficult or impossible work for one woman. With a little care put into correct installation, your new roof can make your house snug and dry for many years.

WOOD SHINGLES AND SHAKES

A good friend of mine who had been hired to shingle the roof and sides of a very large building once complained to me that it was something like trying to crochet your way around a barn. Although I can empathize with her point of view, I must admit that I find shingling to be very enjoyable, almost meditative work. The tools are simple, the materials often beautiful, and the results satisfying to a soul who desires some order from an otherwise chaotic universe.

Wood shingles are the oldest form of roofing still being used in America today. Formerly they were all split by hand—the thinner pieces being called *shingles* and thicker, longer pieces *shakes*. Today, the terms are somewhat differently employed. Shingles are still thinner, but they are machine-produced and are sawn on both faces (sides), so are relatively smooth-surfaced. Shakes are heavier and thicker and have a highly textured surface. Shakes are still handsplit. There are three styles of shakes produced: *taper-split shakes* (hand-split, with rough-textured faces and one end thicker than the other end); *hand-split, resawn shakes* (hand-split and then machine-sawn on one face); and *straight-split or "barn" shakes* (hand-split with rough faces but a uniform thickness). A good shake or shingle roof will last from twenty to fifty years (or more) if properly applied and kept up. Compare this projected lifespan to a possible five years for roll roofing and a fifteen-to-twenty-year span for asphalt or fiber-glass shingles, and you can see one reason for the popularity of wood shingles. Economically, wood shingles and shakes remain more expensive than other types of roofing (unless you can split your own), so it becomes a matter of balancing aesthetics, longevity, and available funds.

The minimum roof slope recommended for wood shakes is 4 in 12—the roof should rise at least 4″ for every 12″ of run. The minimum slope for wood shingles is 3 in 12—a rise of 3″ for every 12″ of run. Shingles, however, may be used on roofs with lower pitches if the exposure of the shingles is lessened (more on shingle exposures later). As wood shingles and shakes are the heaviest type of roofing commonly used, roof rafters must be strong enough to take the weight (up to 200 pounds per 100 square feet of roof area). If your roof rafters are less than code size for their span (and species of lumber used), you will have to beef them up or reconsider your roofing choice. The

(Photo by Carol Newhouse)

roof sheathing/decking that goes below shingles or shakes may be solid (plywood or boards) or open (spaced 1×4 or 1×6). Usually, in colder climates the boards (sheathing) are laid tightly together and covered with rosin-sized paper. In warmer areas, or for barns and outbuildings, the roof sheathing may be laid with spaces between the boards for a slatted effect. If you are using shakes, you will also need an underlying layer of black felt or builder's paper as a moisture barrier (not used with shingles).

Shingles and shakes are graded according to their quality, and priced accordingly. For example, grade #1 in red cedar shingles is the premium grade, entirely made of heartwood, entirely clear, and all edge-grain. The grade below this one allows for some sapwood and flat grain. The lower grades, known as utility grades, are made from poorer-quality wood and may have knots, sapwood, and other defects. If you are roofing a structure that you hope to see survive many years, investing in a good grade of shingles or shakes may be well worthwhile. If you're roofing a less permanent structure, a lower, cheaper grade of shingles or shakes may be a reasonable bargain.

Shakes and shingles are usually sold by the *square,* which is a unit designed to cover 100 square feet. The shingles or shakes actually come bundled in smaller units that are easily handled. In estimating how many squares you will need for a given roof area, you have to take into account the *exposure* to the weather (inches each shingle extends beyond the row overlapping it) you are using. The standard exposures are: 10″ for 24″ shakes, 5″ for 16″ shingles, 5½″ for 18″ shingles, and 7½″ for 24″ shingles. If you are going to use one of these standard exposures, you can get your number of squares by figuring out the square feet in your roof area, adding some extra for doubled bottom (starter) rows and for valleys, adding some extra for waste, and dividing your total by 100. You may want to buy some lower-grade shingles or shakes for your starter rows if you need to economize; use them as the underlaying.

The nails used in applying shingles or shakes are all galvanized. For shakes, a 6d nail is commonly used. For shingles, nail size varies from 3d (16″ and 18″ shingles, new roof) to 4d (24″ shingles, new roof), with 5d nails used on the second of a double row. If you are putting shingles over an old roof, you will want to go to a longer nail—5d for 16″ and 18″ shingles; 6d for 24″ shingles.

Tools for this work are simple. A tape measure and shingling hatchet are basic. The shingling hatchet has a special sharp blade for splitting shingles to widths you need, and a corrugated nailing head that won't damage the shingles. Some types have an adjustable marking gauge for guiding your exposure. You'll need a nail apron or special shingler's nail stripper (dispenses the nails readily and supposedly cuts time). You may want to use a jigsaw or saber saw for any special cutting jobs. If you are using shakes, you'll need a staple gun and cutting tool for the felt paper. Wear work boots with a good nonslip sole. A chalk line or length of board for setting courses will be needed. Finally, you may have to use some sort of adjustable roof seat or ledger if your roof pitch is steep (check with local rental supply companies, or devise one of your own).

The starting row of your roof is always laid double and should project beyond the lowest roof board by at least 1½″ to form a drip that will cause rainwater to fall clear of the roof. Along the rake or sloping edge of a gable or shed roof, the end shingles should project about ¾″ for the same reason. Joints (where one shingle parallels its neighbor) in the second layer of the starting course (first row) and in each successive row after that must miss those in the preceding layer or row by at least 1½″; otherwise, rain might find its way down and seep through to the roof sheathing and on into the house. In other words, joints in each row should come approximatley in the middle of each shingle in the row below. This is very important and takes constant attention at first. Shingles are spaced between ¼ and ⅜″ apart so that they will not crack when they expand with rain. Usually, shingles and shakes come in bundles of varying widths, and you can use these variations in planning out your proper joint patterns. Shingles may also be purchased in uniform widths for certain special effects.

Lay your starter row, nailing each shingle twice about ¾″ in from the outside edges and about 1½″ above the butt line of the next course (this will be determined by the exposure you plan to use). This placing of the nails will ensure that no nails are exposed. Nails should be driven just flush with the shingle; driving them in too far will cause splitting, and driving them too little will leave projecting heads that will keep the next row from lying flat.

When your first row is done (twice, it's a double), it is time to move on and up. But first, a word about heights. My first day on a roof, I learned what the saying "paralyzed by fear" meant. I stepped off the ladder onto the roof and could not move. I just wanted to cry, I was so afraid, and my fear made me mistrust my step even more. So I had the people I was working with tie a very heavy rope firmly (and I stressed *firmly*) to the top ridge of the roof. I tied the other end around my waist. With this guide, I was able to make it through the first day. The second day, I used the rope only when I was walking around on the roof, and by my third day I felt quite comfortable without it.

Although I did not venture a tap dance, I did manage to look up to see the six beautiful old fir trees surrounding the house, and discover the fact that being high could be just that. I also took a certain joy in the fact that the demands of my seven- and two-year-old children had to be directed elsewhere because Mommy was on the roof. . . . This quiet time alone, among the tops of trees, listening to the wind and the rhythmic tapping of my hammer, is one of my fondest house-building memories.

Now upward. The succeeding courses are set up from the first row a distance that depends on the length of the shingle and the exposure desired. If you measure the distance from the eave to the ridge, you should be able (by finagling a bit) to divide it into a full number of rows, thereby avoiding a short row at the end. Rows are kept straight with the help of a chalk line. Measure up the desired distance, snap your line, then lay the bottom or butt end of each shingle along this line. For greater accuracy, I measured from the bottom of the roof for each line I made, rather than measuring from the row below and risking a possible compound error. Instead of a chalk line, you may prefer to use a long, straight board (1×4, for instance) that is tacked to the roof in the desired place. Shingles are then butted against it, a row completed, and the board untacked and moved to its next position. If you are working with a partner, this method may prove faster.

As you continue your way along your roof, you might meet up with such nuisances as skylights, protruding pipes, and chimneys. You will want to use flashing around these problem areas. Roll roofing or any similar asphalt roofing product may be used here, or you have a choice of various metal flashings. The important thing is to look carefully at the situation and try to work out a system that will channel all rain away, letting it shed off down the roof. The general rule is always to place the lower end of the flashing *over* the row of shingles below it, and be sure that the upper end of the flashing is tucked *under* the row of shingles above it. Use a good plastic-type sealant to seal joints: for instance, if you are flashing a plumbing stack, apply a thick layer of sealant around the base of the stack. Another problem area is in valleys, where two roof lines intersect. Here, too, you will want to use flashing—either roll roofing or metal. It will form a sort of gutter, with the shingles shedding into it along the edges of each roof line. Once again, seal any joints in your flashing with a good sealant or roof cement. More information on preparing valleys before roofing is found in the "Roll Roofing" section following.

Wood shakes are applied in much the same way as shingles. Because of their extra width and length, they may have a greater portion exposed. Because the shakes are not smooth on both faces and wind-driven rain might blow up under, an underlay of felt paper is used. An 18″-wide, 30-pound felt paper is made for this purpose. It is stapled down under each course of shakes as you work. Place it a little higher than the succeeding row of shakes will be so that it won't be visible when your roof is finished. A 36″ strip of paper is used at the eave line.

Assuming you have successfully made your way past the afore-mentioned demons, you might be casting a wary eye toward the top ridge or hip, knowing that something special must be in store. I found that the easiest closure to make is first to nail a flexible metal flashing over the ridge, using asphalt cement or wet patch *under its edges.* Then nail two long boards (called *saddles*), one over each slope of the ridge or hip. Cover nailheads with wet patch or another sealant. An alternative to this is the fancier "Boston Ridge," which substitutes shingles for the saddle boards. The shingles are all cut to the same width and laid with the long edge parallel to the ridge rather than at right angles to it. The shingles are overlapped in the same way as the rest of the roof, but nailed in place so that the exposed side edge is alternated between the two slopes. If these instructions seem confusing, hunt out an already shingled roof with this style of ridge and take a careful look. It is not as complicated as it sounds! Still another alternative is to buy a prefabricated hip-and-ridge unit, which is nailed in place and will do the same job.

Remember that a wood shingled or shaked roof presents a certain fire hazard if you are using a wood heating or cook stove. Make sure that all chimney pipes extend up to code height for safety, and consider use of a spark guard (small-mesh wire) at the top. With these precautions taken, and some minimal care, your new roof should be functional, beautiful, and enduring.

ROLL ROOFING

Roll roofing is another asphalt-based product that fills the need for a low-cost, easily applied, fire-resistant roofing. This type of roofing is mineral-coated asphalt saturated paper; it comes in standard rolls 36″ wide, in a variety of colors (white, blue, green, red, etc.), and in several types. Most commonly used is a roll roofing that is entirely mineral-coated on one side. This roofing is applied in horizontal strips or layers with each succeeding strip overlapping the row beneath it by 2″ or 3″. Each overlapped portion of the roof is sealed with a special asphalt-based lap cement. This

(Photo by Carol Newhouse)

type of roll roofing is suitable for any structure with a roof pitch of 2 in 12 (2″ rise for every 12″ run). The second type of roll roofing is called *19″ Selvage Edge* roofing. This roofing also comes in rolls 36″ wide, but only 17″ of the exposed surface is mineral-coated. Selvage Edge roofing is applied horizontally, too, with an overlap of 19″. The overlap may be sealed with a cold roofing cement (like that used with regular roll roofing), or it may need a special hot asphalt cement. If you are using Selvage Edge roofing, be sure that you follow the manufacturer's directions concerning the type of roofing cement. The advantage of Selvage Edge roofing is that it can be used on roof pitches that are as low as 1 in 12 (1″ rise per 12″ run).

Roll roofing may be applied over old wood shingles, old asphalt shingles, or old roll roofing. In any of these cases, you should go over the roof carefully first. Nail down any protruding nails. Areas that are wrinkled or "blistered" (if the roofing is asphalt shingles or roll roofing) should be slit with a knife and tacked down smoothly. Wood shingles that have buckled or worked loose should be removed and any protruding nails driven back in. If you come across an area of warped shingles or buckled roofing, you may have some roof sheathing boards below that have cupped or warped. These will have to be nailed down, if possible. If they are too twisted to nail back in place, you may have to replace them. If your roof has had areas with bad leaks, you may want to check the roof sheathing in these areas to make sure that it's still sound. Now and then you will come across a structure that has already had two or three layers of roofing (of different or similar types) applied to it. In such a case, your best course of action is probably to strip the roof of its layers—remove all of the old roofing and get the roof back down to a bare wood deck. Replace any boards that are rotted or badly warped. If your roof was originally constructed for wood shingles or shakes and has spaced sheathing (i.e., the sheathing boards have open spaces between them), you'll have to fill in so that you have a solid decking/sheathing for the roll roofing. An alternative to filling in these spaces is to cover the entire roof with plywood sheets. You can use a low grade of plywood for this sheathing.

If you're building a new structure, the roof sheathing should be solid—either plywood or boards that are fitted closely together. Roll roofing may be applied directly to the roof in certain areas, while in others it's a wise precaution to start your roof with a first row or course of underlayment. This underlayment can be a strip of 36″-wide felt or builder's paper. Using an underlayment for the first course of your roll roofing will help to prevent wind-driven rain from getting up under your roofing, and to keep ice and slush from backing up from gutters along eaves. If you live in a mild area or if you are roofing a barn or outbuilding, you may be able to skip the underlayment altogether.

There are two more matters to consider at the planning stages of your roofing project: the first is how to finish the edges along the rakes and eaves of the building so that water will shed off; the second, how to prevent wind from getting under the edges along the rakes and eaves, where it can lift and tear loose sections of roofing. These problems are interrelated. The conventional roofing procedure is to apply roof nosing, or a metal drip edge, all around the roof edges (rakes and eaves) *under* the roofing. This metal flashing channels the water away, making it shed from the roof rather than stopping at the edges and being drawn back along the roof sheathing by capillary action. Placing the roof nosing (flashing) under the roll roofing will help the water shedding situation, but it leaves the edges of the roof vulnerable to wind damage.

An unconventional but somewhat workable alternative is to place the nosing *over* the roofing all the way around the edges. This will protect the edges from the wind and keep your roof intact through some hair-raising storms! If you seal the edge of the flashing (where it lies over the roofing) with wet patch or any good plastic-type roofing compound, most rainwater won't be able to seep under the flashing; it will be drawn back along the roof sheathing. You can't use regular lap cement or tar here, because it softens in the heat and will run and drip on a sunny day, making your house or structure quite unsightly and eliminating any "seal" you might have had. Choose a good-quality roofing compound that won't dry and crack in the cold or melt and ooze in the heat. Unfortunately, this whole method is not foolproof. After a year or two, even the best roofing compound will become somewhat weather-worn and water may begin seeping back along the edges of your roof. You'll have to get up on a ladder and reseal your nosing/roofing joint again. It's a good idea to climb up periodically and check over the whole situation. Hammer down any nails that have lifted and patch any weak-looking points.

An alternative to this setup is to extend the roofing paper 6″ to 12″ beyond the edge of the roof, at rakes and eaves, folding it back up under the overhang and tacking it in place with a 1× ledger or batten nailed along its length (nail through the batten and roofing paper up into your roof sheathing). To manage this kind of "finishing off," you'll need a very warm day so that your roofing paper will be flexible and won't crack when you're folding and nailing. This method will allow water to drip freely from the edges of the roof, and it will keep the edges of the roof battened down (or up, actually) against storm winds. You'll have to plan on a bit of extra roofing all around, but it will be worth this extra expense if your roof stays on and keeps you dry. The main disadvantage of this method is that the roll roofing will tend to age rather quickly along the folded parts and there is very little you can do to repair this damage. With luck, the edges will last as long as your overall roofing does.

Once you've decided how to handle the edges of your roof, you can make up a materials list for yourself. Roll roofing comes in units (rolls) designed to cover approximately 100 square feet per roll. Calculate the square-foot area of your roof. It's much better to overestimate than to underestimate, and as roll roofing is very cheap, you can afford to be a little long on your calculations (rather than finding yourself up on the roof, a quarter roll short). If your roof has valleys or intersections, figure in extra roofing for these. Lay a strip of roll roofing about 18″ wide along the length of the valley, mineral surface down; then lay a second strip over this one, mineral surface up. These strips are nailed in place, and the roofing paper laps over them more or less shingle-style. Try to figure out any and all areas where you might need extra roofing beforehand.

Roll roofing is particularly well suited to polygonal buildings, as it can be cut and shaped and overlapped around the oddest of angles. Here again, you should figure out as nearly as you can how much roofing you'll need—and then buy a little extra.

The total square-foot area of roofing (for any shape of building) divided by 100 will give you the number of rolls you'll need. Usually a lumber store or hardware store will carry one type of roll roofing and that will be your choice, unless you go to another store. Be careful in comparing prices from store to store, because one "$10 Special" might actually be a slightly lighter roll roofing. Most roll roofing weighs 90 pounds per roll, but some companies produce a lighter, cheaper

roll roofing. This lighter variety will be easier to cut, lift, and handle, but probably won't last quite as long. The mineral coating on the lighter types isn't as well bonded and tends to shed off as you're working with the roofing. So my advice is to insist upon (and search for) the 90-pound roofing in a good-quality brand name.

In addition to the paper, you'll need nails, roof nosing or 1× battens (depending upon how you decided to finish the edges), and lap cement (also called roofing cement or roofing tar). You may need special roofing compound for problem areas—joints around stovepipes or chimneys, for example. You *can* use lap cement to seal around these areas, but a good roofing compound works better. (As mentioned earlier, it won't melt and drip—or freeze and crack; lap cement will do both.) Don't be tempted to buy special roofing compound for the whole job, though; lap cement is well suited to its job, which is sealing the overlap areas between sheets of roofing paper. It is inexpensive and you'll need a lot of it—roofing compound for these overlap areas would add an unnecessary expense to your roofing job. The nails to use with roll roofing are special galvanized roofing nails; they have a broad flat head that won't cause tears in the paper. You'll have to choose the length of your roofing nails according to the roof you're doing. A general rule of thumb is to choose a nail that will penetrate about three quarters of its length into the roof sheathing. If you're putting roll roofing over an old roof, get a nail that will be long enough to pass through the old shingles and into the roof sheathing. Lap cement comes in small cans or in gallons; it's much cheaper by the gallon, and will keep well if tightly resealed. How much you need depends largely upon how freely you apply it; usually a gallon will be enough for 2 to 4 rolls (or more). Don't forget to get a roll of felt paper if you want to put an underlayment beneath your first course of roofing.

Tools for this roofing are quite simple. You'll want a hammer, a tape, a chalk line, and a utility knife for cutting the roofing paper. Tin snips work well for cutting, too, especially if you have to make complicated cuts. To spread the lap cement, use a paintbrush or a couple of old wood shingles (these make perfect "paddles"). Two 36″ lengths of 2×4 (or any scrap wood) are handy aids when it comes to laying out the paper and making cuts. Wear clothes that you won't mind seeing slightly tarred by the end of the day—as careful as you might be!

Another "tool" for this job is a nice, warm day: not too hot, but not too mild. Lap cement gets entirely out of control on hot days, and roofing paper is downright nasty on cold days. If you have to work on a hot day, try to keep the lap cement in a cool place until you have to use it, then bring it up on the roof in small quantities (in a can). If you have to work on a cold day, try warming up the roofing paper beforehand. Stand it in the sun or inside a heated room. Otherwise, it will crack and break with handling. Also beware of a windy day. It can send lap cement streaming off your paddle (into your hair and over the roof), and can whip sheets of roofing paper away from you. Another impossible weather situation is a wet one. I've put roll roofing on in the rain (believing that winter had come to stay) and seen my work buckle and pucker all over the roof two days later (when the sun came out). Roll roofing should be put down on a *dry* roof!

When you are actually up on the roof, wear a boot or shoe with a non-slip crepe sole—or work in your socks (bare feet will give you good traction too, but the rough surface of the roofing paper isn't especially comfortable). It's a good precaution to work with a rope secured around your waist (as mentioned in "Asphalt Shingles") if you have some way to tie it at the other end. If you're working on a really steep roof, make yourself scaffolding of some sort. I've used 2×4 ledgers nailed into a roof that was so steep one couldn't stand on it; these gave a series of "steps" that one could safely inch along. As we worked up the roof, my partner and I nailed in the ledgers wherever needed, moving them around as our work progressed. Then we worked our way back down the roof, using the same ledgers, again nailed in as needed. As they were pulled up, a dab of roofing compound was used to seal over the nail holes. If you use a system like this one, be sure to nail your ledgers or scaffolding securely through the roof sheathing into *rafters*. Double-headed 16d nails are perfect for this work.

The first step in your roofing work is to take care of any valleys or intersections, placing two layers of roofing paper, as explained earlier. The second step in roofing is to put on an underlayment if you're using one—and roof nosing if needed. Then you can proceed with the roofing paper itself. The roof is always started from the eaves, and the work progresses on up toward the peak. If, however, you are roofing a gable-roof or gambrel-roof building, leave the last course unfinished for now; go to the other side and work from the eaves on up again. The peak of the roof is finished last. Handling the roofing is easiest if it is in sections of not more than 20′ or so—beyond that, it gets heavy and cumbersome. If you're working with a partner or a couple of people, you'll probably want to have one woman on the ground measuring and cutting the paper, and one or two on the roof, nailing it down and applying lap cement. The rolls are easy to manage on the ground. You can roll out the approximate amount you want and then measure and mark it (make a little nick with your utility knife). Use your chalk line to make a cutting line on the paper. Lay a 2×4 (or other scrap wood) under the line area and use it to bear down on while you cut. Then you can roll up the piece and hand it up to the roof worker(s).

The roofing paper should always be carefully placed before any nails are driven into it. If it's a long piece, unroll it in place along the roof and then give it a few good tugs to straighten it out. Line up edges carefully. The strip should be nailed along both edges and along the top. Don't forget to allow a good overlap if you plan to fold the paper back up under the eaves for a finish. If it takes two (or more) strips to complete your first course, you'll have to smear a layer of lap cement where the second piece overlaps the first. An overlap of about 6″ should be enough, so make your lap cement cover a good 3″ close to the edge of the overlap. If you can, as you do your roofing try to make all overlaps *closed* to the prevailing wind. When the lap cement is all applied, lay the second piece of roofing paper down in place. Stretch it out, line it up carefully, and nail it in place. Again, nail both edges and along the top. The second course of roofing paper will overlap the first by about 2″. While your partner cuts the piece or pieces for the second row, you should make a thin layer of lap cement all along the upper edge of your first row (where the overlap will be). Then place the second course and nail it (along both edges and the top). Each succeeding row will cover the nails of the row below (except the nails that secure the edges). It is important to line your paper up carefully as you go along, thereby preventing dips and buckling. When you've worked your way up to the ridge, leave that last course undone, as mentioned earlier. Go over to the other side of the roof (if yours is a gable or gambrel) and begin again at the eaves. When you've worked your way up this side, you can place the final layer of roofing paper, which will straddle the peak of the roof and overlap on both sides. It should be nailed down along all edges. Now go around the roof and finish all of your edges, either by tacking on roof nosing or by folding the edges up under the eaves and securing with battens. The final step is coating any and all exposed nailheads with a dab of lap cement or roofing compound. You can do this as you go along, but chances are that you'll find yourself stepping or kneeling in these dabs, so it makes a neater job to wait until the end.

If roll roofing is put on carefully and secured well against the wind, it can last up to five or six years. It will help if you don't walk on it unless you absolutely have to. Keeping trees trimmed back away from the roof will add to its life-span, too; branches that rub back and forth against the roof quickly wear away the protective coating of mineral granules and cause early deterioration at these points. Also, keep gutters and downspouts clean to prevent water from backing up in these areas. Periodically you should take a careful look at the roof to make sure everything is tight and that nails aren't popping up anywhere. If you see a section of roofing that's beginning to "lift" in the wind, try to nail it down securely before it tears loose. Use roofing compound over your nails. With some preventive care, your roll roofing can more than repay your investments of labor and money.

Making Magic: Yurt Building

The original idea of building a yurt was formulated when my friend Elizabeth decided she needed a workshop/guesthouse on her property in the beautiful mountains of southern Oregon. I wanted the opportunity to build an entire structure, so the following summer I quit my secure, ulcer-producing job in the city and came to live in the country. There were no indoor toilets, baths were taken in the icy-cold creek, and I faced a new, uncertain life. My goal was to create a small, economical, enjoyable living space that reflected a woman's hand.

Decisions about the structure were determined by the location, purpose, and costs. The building site was an easy decision. Not far, but mostly out of sight of Elizabeth's main house, was a beautiful round opening in the trees on the side of a hill. The space had a large pine tree shading the uphill (north) side of the circle, a madrone tree on the west side, and a view of the mountains to the south. The rest of the circle was second-growth forest. The site suggested that the building shouldn't be rectangular. A softer, more flowing building would lend itself better to the space. A roundish structure seemed the answer. The question then was: what kind of round structure? Several hexagons had been built in the community, and also a yurt.

The yurt had intrigued me for some time. Originally the yurt was a marvelous portable structure conceived by the Mongolian shepherds to stave off the cold winds of the Russian steppes. Ingenious in design, it uses no heavy structural beams or posts for support. Instead, it utilizes a simple method of a tension band around the walls and a compression ring in the center of the roof. The Mongolian yurt is round, with a band of yak hair around the outside of the lightweight, foldable latticework wall. The band holds the bearing force of the roof. The roof is designed of flexible poles. One end of each pole is fitted into a wooden smoke-hole ring, and the other end is attached to the top of the latticework wall. The smoke-hole ring acts as a compression ring to keep the roof from caving in at the middle, and the tension band keeps the roof from flattening the walls to the ground. Blankets are laced to the walls and roof. The appearance is of a plump decorated cake. The entire structure can be dismantled, folded up, and moved about.

Folding the yurt up and moving on to better grazing grounds was not a concern to Elizabeth, so a permanent yurt could be built. Elizabeth liked the yurt because of the large skylight on top. She wanted more windows on the sides than a yurt normally allows (the traditional latticework wall doesn't lend itself to having too many large holes cut into it). I studied existing yurt plans; various groups and individuals in this country have experimented with the yurt form for small and large structures. The plan created by Bill Coperthwaite of the Yurt Institute (see "Resources") seemed the most sound. However, it again did not allow for much window space on the sides of the building, and I felt that the 1×6 wooden wall construction could not tolerate the size and number of windows we wanted. I decided to modify Coperthwaite's plans. In retrospect, I feel that these plans could have been used and the wall structure would still have born the roof weight.

The modification resulted in a twelve-sided building using a tension band and compression ring for the main structural support. The walls of the structure lean outward like those of the traditional yurt. I began by building a hexagonal foundation, then added six more points to it to make a twelve-sided figure. At this point, let me suggest to anyone building a round or hexagonal building a simple way of laying out the foundation without losing one's mind trying to figure where to place foundation blocks. . . . First figure the length of any side of the hexagonal figure. A hexagon is made of six equilateral triangles, so this normally makes the side legs of any one triangle the same as one half the width of the hexagon at its widest point (Example 1).

Now make a rectangle the width of a leg of the triangle (in this case, 5′ × twice the height of the triangle). By now, if you do not have a basic understanding of geometry, it's time to scurry off to the library and get a simple geometry primer book. The rectangle part of the hexagonal figure will look like this (Example 2).

From the center of the rectangle, measure 5′ either left or right (your choice) at a right angle to the long side of the rectangle. For example purposes, I've marked to the left of the figure. From the left corner, either upper or lower, measure another 5′ heading toward the end of your first 5′ mark (Example 3).

Where your two strings or tape measures cross at 5′ will be the final corner of your triangle. Repeat the procedure for the opposite side of the rectangle. Put the center of each foundation or pier block on each corner of the rectangle and at each cross point. Don't forget to put a seventh block in the center (you'll need it later on).

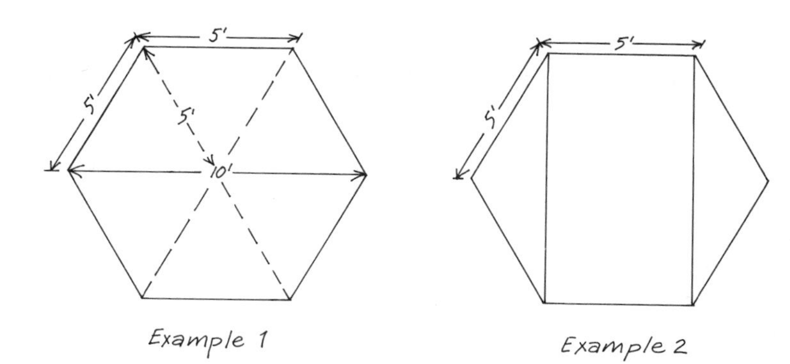

Example 1 Example 2

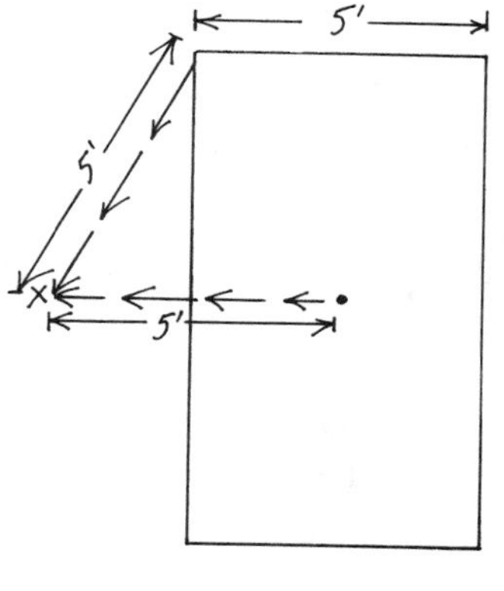

Example 3

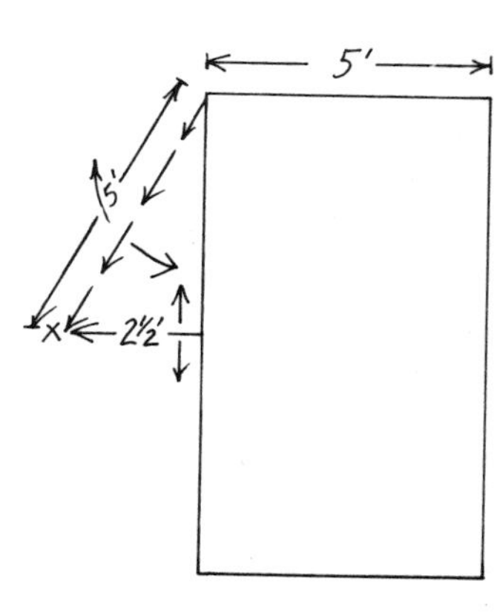

Example 4

(Photo by Ruth Mountaingrove)

If you've managed to get all the way up the side of the mountain with all of your pier blocks, boards, and nails, and you find you have left the geometry primer all the way back home in the city, here's another method that works. It's called "trial and error"—but you can learn something about how to find corners. I will use the same-size hexagon as in the last example. Move a measure of half the distance of the triangle leg (in this example, 2½′) up and down at a right angle to the long side of the rectangle, trying to cross it with a measure of 5′ coming from a corner of the rectangle (Example 4). Where the two cross again will be the last point needed on the equilateral triangle. Also, you can now mark the center of the rectangle and of the other side with either this or the previous method.

Building a yurt will require many quiet moments of pondering over simple geometry. If you like the figuring involved, you'll love doing this building. On Elizabeth's yurt I added six more points to make a twelve-sided foundation system that used thirteen piers. On top of each pier block I set a wooden post. All of these posts were cut level to one another, and the floor joists set directly on top of the posts (the spans were small enough that I felt I could eliminate the use of girders or beams to support the joists). The pattern of the floor framing of the intersecting joists of this yurt was one of the most beautiful parts of the structure—it seemed a shame to have to cover it with the plywood flooring!

The walls of the yurt were framed up with 5½′ lengths of 2×4 nailed to the plywood floor directly over the top of each pier block. These 2×4 studs slanted out from the floor, creating a basketlike appearance (and some problems in angle-cutting and nailing!). The top of each 2×4 was about 2′ beyond the edge of the floor and the bottom of the 2×4 (Example 5). The studs were about 3′ apart at floor level, but because they angled out, they fanned to about 4′ apart at the top. A 2×4 base plate was nailed in between the studs at the floor. After the steel cable that acted as the tension band was placed around the top of the studs and adjusted to the

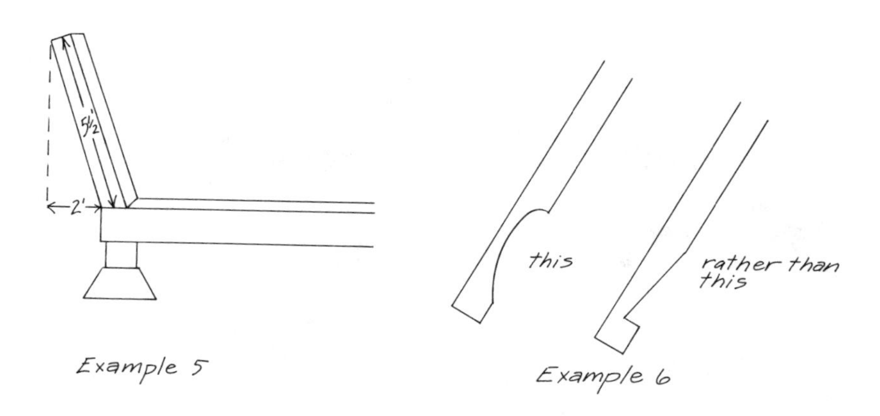

Example 5

Example 6

circumference of the circle, a permanent top plate of 2×4's was nailed to the top of the studs. We tried putting the plate on before doing the cable, but this didn't work very well. It was too rigid. With the top plate pieces removed, the wall had the flexibility needed for placing and tightening the cable.

There are some things to consider in choosing the cable used as the tension band for a yurt or similar structure. We used a rolled-steel cable, plow grade, ⅜" thick. It had a breaking strength of around 2,000 pounds. It is not necessary or even recommended that you use really big cable—it's too hard to work with (just in case you've spied some of that huge, rusty, and free cable the loggers leave behind in the forests . . .). A sod roof will require a heavier cable than a roof covered with cedar shingles or asphalt shingles. Does your climate produce a lot of heavy snowfalls? If so, use a heavier cable. If visualizing weight in terms of numbers (the conventional pounds per square-foot weight of dead and live loads on a roof, and so on) is hard for you, think of something that you know weight-wise. I thought of my VW bus sitting on top of the yurt. That was entertaining as well as educational. Aluminum aircraft cable is lightweight, less strong, and stiffer to use than rolled steel, but it will work. The cable that power companies use as support wires will work—but again, it's stiff to work with. Power companies will often give this material away after they've torn down an old pole. It is not necessary to get galvanized cable, as the cable is protected by the eaves of the roof once the yurt is finished. (The original yurt tension band was a band of woven yak hair, if you care to go that route.)

Putting the cable in place is easier if you have friends to help. We prefigured the circumference of the top of the wall and marked the cable by tying a piece of string around it at that point. When our crew of women pulled up the slack on the cable, we watched for the string marker and knew when to stop tightening. Later, when the weight of the roof was added, we had to tighten the band up a little more. In the photographs, you can see the cable/tension band running around the exterior top edges of the wall studs.

The roof rafters at the wall end had bird's-mouth notches cut in them in reverse manner to what is normally done (Example 6). This was more for appearance than anything else, as the eave rafters were noticeable from the inside—but also because I really didn't know at the time which was the "right" way for them to be cut. The compression ring was the outer steel rim from an old wooden wagon wheel. It was found free in a scrap pile on the back of Elizabeth's land. The iron rim was settled in place at the ends of the 8′-long 2×4 rafters. No fasteners are needed to hold the rim to the rafters. The whole principle of the yurt is that the roof rafters press down against the rim and out against the cable. I must admit that, while working on the roof, I placed a 16d nail in the end of each rafter just below the rim. It was not necessary—it just felt better. Later, these nails were removed.

(Photo by Ruth Mountaingrove)

To cover the skylight in the center of the roof, a large piece of Plexiglas was cut round and used. To keep it from slipping off the metal rim, a piece of rubber hose was measured and cut to the exact circumference of the rim, then slit open and placed around the rim. The weight of the Plexiglas seems to hold it in place—I've seen two yurts in mountain settings (fairly windy) and haven't heard any complaints of the Plexiglas blowing off. In his yurt plans, Coperthwaite makes no mention of having to fasten the skylight cover down.

The construction process was not without its problems. Placing the iron rim caused a major hassle. I discovered that the key to the building was *flexibility.* Keeping the wall studs loose helped in centering the rim. At first the wall studs were secured together with a top plate and the tension band was kept taut. The ring (rim) was mounted on a temporary post nailed to the center of the floor. This didn't work. Everything was too rigid and held immobile. Once the band was loosened and the top plates removed from the studs, we were able to adjust the 8′ roof rafters easily. The temporary pole in the middle wasn't necessary. Instead, the rafters were connected at the rim end with a blocking method common to joist and rafter blocking in standard construction. In essence, a preliminary twelve-sided wooden rim was fashioned. Once this was done, the iron rim could easily be settled into place by hand. The iron rim could withstand the weight of the completed roof.

Before we discovered that we should place blocking between the ends of the rafters, we had problems with the building and roof twisting. The blocking and some temporary diagonal bracing between wall studs and between the rafters stopped this twisting.

In retrospect, I would not have fastened the wall studs to the floor as I did. Instead, I would have run them through the floor to the joists beneath and bolted them to the side of the joist. There are several reasons for this. First, I think that they would be very secure. Second, if the bolt was properly placed, it could be used as a flexible pivot point when centering the rim, and could be neatly tightened up when the proper angle was attained. Then, of course, the stud could be doubly nailed or further bolted to secure it permanently in position. Third, this method would eliminate cutting a critical angle on the wall stud.

Recycled lumber was used in the foundation and floor framing of the yurt. A wonderful old Englishman ran a recycled-lumber business in the nearby town. He tore down old houses and resold the lumber at very reasonable prices. For expediency, new lumber was used in the upper parts of the building. I couldn't always wait for a house to be torn down! Recycled Douglas fir 2×4's of uniform size and strength were used for the wall studs. They were hard and dry and didn't shrink or crack. The outer wall covering was #1 pecky cedar. It was green and shrank considerably. The 2×4 exposed rafters were red cedar. The interior roof was 1×4 redwood fencing boards. Because the redwood showed in the interior, we placed it on an angle (used it diagonally) between the rafters, creating a five-pointed-star effect. Sawn cedar shingles were used on the exterior roof. The door is 5½′ high, rounded at the top, narrower at the bottom, and made of the same pecky cedar as the outside walls. Because the building slants outward, entering through the shorter-than-"normal" door is no problem.

The overall height from the center of the floor to the middle of the rim is 8′. The height of the walls is 5½′. There is a 2′ roof overhang. Large triangular windows are set between the wall studs. The walls tilt out from the floor, making a larger living area than the floor area. The floor is 12′ across and the distance between the top of the walls is 16′. The general appearance of the yurt is of a small, elfin-like dwelling that is harmonious with its surroundings. The smallness of the building is deceptive, for when you enter, the inside of the building seems to open like a flower. Shelves and tables fit back into the walls rather than sticking out into the room. The skylight in the center acts both as a window to the sky and a ventilation opening. The ventilation is achieved simply by sliding the Plexiglas cover slightly off the center of the rim and cracking the door open a bit. The light from the skylight creates an airiness in the room.

The cost of this building was about $700 (in the summer of 1976). Recycled lumber helped reduce the cost. The "store-bought" cedar shingles were expensive but attractive. If a woman had the time, hand-split cedar shakes would be less expensive and lovely to look at. The yurt is double-walled, with double-sided aluminum foil for insulation. If foil insulation is used in your building, remember to leave about 1½″ of air space between foil and walls. Metal foil conducts heat and cold when touching anything and reflects only when not touching. Metal foil insulation works well and is inexpensive.

As this structure was considered a "studio" and not a "house," no building permit was required. (A "house" in some counties is anything with kitchen facilities.) The yurt probably would not have been built if building to code had been required because of the additional costs and inconveniences. Besides the purchase of the permit, we would have had to buy more expensive materials. Also, each phase of the building would have had to wait for approval by the county. In the case of the yurt, our plans and experimenting probably would not have met with approval by the building department!

Four women were primarily involved in the creation of the yurt. Elizabeth had a need for the space and found a beautiful location on which to build. I studied building designs and determined the ways to build, and also worked on the actual construction. Blackberry, who wanted to learn carpentry, helped build the yurt, and Ruth photographed the process. Occasionally a visiting woman would help out. I had been raised part of my life with my grandparents, and my grandfather, a master carpenter from England, sometimes baby-sat me in his shop. My earliest memories were formed pleasantly sitting among the fragrant curls of wood from the plane—curls quite differ-

(Photo by Ruth Mountaingrove)

ent from those most little girls are associated with! I have always done my own carpentry. Blackberry had little previous experience with construction but wanted to learn, and was very strong in math, geometry, and keeping notes.

Figuring degrees and angles was important in this building, as there were no right angles anywhere in the yurt. In fact, the wall studs had three-dimensional angles. The aesthetic pleasure of the yurt made the extra time it took to figure out the angles worth it. Since this was my first building, everything I did took extra time and thought. I have since built a conventional rectangular building and am currently living in it, but I think this will be the last "square" building I will make for myself.

I had a good supply of tools for working on the yurt. There was no electricity at the building site, so all work was either done with hand tools there, or cut below at the barn and hauled up to the site. Although in actual distance the site was not too far from the material unloading point, it was a *steep* hill to climb. I wished for the luxury of a road or donkey to help deliver materials. Each foundation pier block weighed about 45 pounds, but seemed at least 1,000 pounds by the time it got to the site! Blackberry and I devised a two-woman carrying rack to haul some of the materials.

I love the yurt we built. Lessons learned from it were enormously valuable. The

(Photo by Ruth Mountaingrove)

confidence I gained led me to get a general contractor's license and to start my own business building small structures. Every time I go back and look at the yurt, I'm pleased with what I see. I'm not sure that I'd want to live in a yurt permanently—it feels like a beautiful place to go and be for a while and center oneself. I have two friends who lived in a yurt for the winter (in the same area as Elizabeth's). They enjoyed the structure and said they felt strange whenever they entered a square room: it didn't feel as natural as did the round room. They also said, however, that it was hard to find privacy in the yurt—no nooks or crannies to hide in. Now I find myself intrigued with the Yurt Institute's plans for a double-concentric yurt. This building consists of a yurt within a yurt. The living space is around the bracing of the smaller yurt. The smaller yurt's footings provide closet and storage areas in the center of the bigger yurt. With the smaller yurt's living space sticking up above the larger outer yurt, this structure looks like a doughnut with a cupcake on top. It offers some interesting alternatives to the single yurt.

I'm at a loss as to what to say about the complexity or difficulty of building a yurt. Very few "conventional" building methods are used in it, so in a way a beginning builder might fare best with this structure, as she'd have few prior concepts and techniques to get in her way. The principle of the yurt is beautifully simple. It does take a lot of careful pondering. I'm the kind of person who is predisposed to untangling complex problems, so for me the building of the yurt was both fascinating and satisfying.

Interiors

GYPSUM WALLBOARD/ SHEETROCK

Gypsum board has earned a favorable reputation as an inexpensive interior wall covering that is fire-resistant, sound-resistant, and fairly easy to apply. It is usually used as a base for wallpaper, paint, or tile. *Sheetrock* is a trade name for one brand of gypsum board that has become so commonly used that it is now often taken to be a synonym, technically incorrect but generally acceptable. Another term used for gypsum board is *dry wall,* which actually means any material used for interior wall covering that requires little or no water in its application (gypsum board, fiberboard, wood paneling, plywood). Again, the term is so widely used that it is often interchanged. Looking for instructions for working with gypsum board in a carpentry book, you should check all three commonly used terms. Asking for Sheetrock in one lumber store may bring you a rather condescending correction ("You mean gypsum wallboard?") while in another store, that's the term to use. Don't be surprised if you read about gypsum board, go out and buy Sheetrock, and spend your work week putting up dry wall!

The good reputation of gypsum board may become tarnished as we learn more about the dangers of breathing asbestos fibers in a work or home situation. Asbestos is in the headlines these days as a known carcinogenic (cancer-causing) material. At this point, it is the workers in asbestos-product factories who are showing up as the victims of this material. But carpenters and homeowners who put up gypsum board are breathing in the same fibers, which can lodge in the lungs and produce cancer some twenty years later. Perhaps as the hazards of asbestos become more publicly understood, the use of gypsum board will decrease. But until this happens, and until a suitable and safer alternative is found, gypsum board will continue to be used. It is important to know what danger is involved in using this material, and to try to minimize it as best you can. We hope some of the information following will help keep you healthy as well as provide some of the basics about working with gypsum board.

Gypsum board comes in panels or sheets 4′ wide and from 6′ to 16′ long (most commonly used is the 8′ length). It consists of a core of the mineral gypsum mixed with other materials, sandwiched between two layers of special paper. One side of the gypsum board

---

With more and more lawsuits being filed against asbestos-products manufacturers by workers and worker-safety groups, some new joint compounds are beginning to appear alongside the asbestos-containing compounds. These new compounds contain vinyl instead of asbestos (vinyl is a plastic or base for certain plastics). It's quite doubtful that breathing powdered (sanded) plastic is any better for your health than breathing asbestos fibers, so don't be misled into thinking that these new products should be used any less cautiously!

---

sheet is white: this will face into the room; the other side is gray: this will face into the studs or framing of the room. There are special types of gypsum board that come in other colors—a water-resistant variety (for use in bathrooms) in shades of green, for example. There are also vinyl-coated sheets with textured surfaces, available in various colors. *Insulating wallboard* is gypsum board with aluminum foil laminated to one side. Special *fire-resistant wallboard* includes fiberglass reinforcement in the gypsum core. For general use, the standard board is used. This comes in several thicknesses, each with stock uses. Quarter-inch gypsum board is applied over existing wall coverings. For top-story ceilings, ⅜″ is used. Half-inch is the thickness commonly used for walls and ceilings; ⅝″ may be substituted when you want greater noise or fire resistance, or when studs are placed 24″ o.c. and you want a more rigid wall covering. You may also choose to apply a double layer of gypsum board to your walls if you want extra noise resistance in an apartment or "multiple dwelling unit." In most instances, though, you will probably find yourself working with ½″ gypsum board, 4′×8′, in single-layer application.

The edges of this gypsum board may be squared, tapered, or slightly beveled. What you work with will in part be determined by what's available and the "style" commonly used in your area. Usually the squared and beveled edges are most common; the tapered sheets are used where walls are uneven or subject to a lot of changes in humidity. Gypsum board is nailed directly to studs in new construction, to furring strips applied to masonry and concrete walls, or over paneling (through to studs) when old walls are being resurfaced. It may be used either vertically or horizontally. If your room is really large, it may be easier to apply the gypsum board horizontally—you will have fewer joints to cover. It is more common to apply the sheets vertically, especially if the room has standard 8′ walls and sheets will cover floor to ceiling.

Estimating the number of sheets you will need is a simple matter of multiplying the length and height of each wall (or of the ceiling) and dividing that total by the square-foot area of one sheet of gypsum board. This will give you the number of sheets for each area you want to cover. You will have a certain amount of play in this if your room has a lot of windows and doors, but it's a good idea to calculate them as solid, as you will find yourself with some waste and possibly some errors in cuts.

Besides the sheets themselves, you will need tape, joint compound, nails, and possibly corner beads. Gypsum board tape comes in rolls; it is a special white paper, very inexpensive and with no other use that I can think of. Joint compound comes in two forms: a

(Photo by Carol Newhouse)

dry type, which is the cheaper of the two and must be mixed with water for use; and a ready-mix, which is by far the best choice according to every carpenter I've ever talked with (and every book I've read). I've used only the ready-mix, so won't comment on the dry except to say that people who have used both have little praise for it. Ready-mix joint compound usually comes in 5-gallon containers. It will keep for several months if the lid is tight, so you can use up any surplus from one job on a later one. The only estimate of "coverage" I've seen for joint compound is in *The Wall Book* (see "Resources" book list); the suggestion is that 1 gallon of compound will be used for every 200 square feet of wall surface. This is of course very rough. It depends upon the way you apply the compound, how many coats you use, and so on.

The subject of nails used in applying gypsum board is a controversial one. Each book and every carpenter probably have a slightly different suggestion. I've used two basic types: a ring-shank (or annular) for applying the gypsum board to ceilings; and a special "Sheetrock nail," about 5d or 6d, cement-coated, with a concave head that accepts the joint compound (these nails seem very soft and are easy to drive in without damaging the face of the sheet). The Uniform Building Code calls for "No. 13 gauge, 1⅜″ long, 19⁄64″ head for ½″ board nailed to 16″ o.c. studs." *The Wall Book* suggests using 1⅜″ nails with ½″ or ⅝″ board and 1⅛″ nails with ¼″ board (all annular or ring-shank). The size and type of nail you use will depend upon the job you are doing and your own preferences beyond these suggestions.

The final items on the materials list are the corner beads. These are special metal pieces used to cover exterior or outside corners in a room (for example, those in open doorways). They look something like roof nosing, or the metal drip edges used with roofing—but corner beads are perforated. Corner beads come in standard lengths and may be cut with tin snips to fit your needs.

One final thing you may want to consider in choosing your materials is the use of special resilient mountings that may be attached to ceilings or walls for the "ultimate" in noise reduction. They allow the gypsum board to "float," and this breaks the transmission of sound between rooms or floors of a building. They are not difficult to install and not expensive. Ask for the manufacturer's directions and follow these, as each "system" is slightly differently installed. If the instructions you are given seem incomplete or hard to follow, ask your lumber store salesperson if you can borrow the more thorough instructions that are usually provided to the store carrying the product; you can photocopy these, and they may make the job quite a lot easier.

The tools you'll need for putting up gypsum board are probably already in your toolbox, with a couple of exceptions. You'll need a hammer (preferably bell-faced, with a slightly rounded face, and no heavier than a 16- or 20-ounce), a pencil or two, and a tape measure. A straightedge is necessary—you can use a good piece of finished lumber, or a 4′ T square (perfect for this job). You'll need a utility knife and possibly a keyhole saw or a portable (power) jigsaw for cutting the board and making special cutouts, such as for electrical sockets. Special tools you need and may have to purchase are wide joint knives, also called wall scrapers; they look like wide putty knives. The minimum-width joint knife you'll want to work with is a 4″; you will probably also want one 6″ and another 8″ or 10″. If you are going to be doing a lot of this work, you may also want to buy special corner tools; these are joint knives shaped into 90° angles for doing interior and exterior corners (separate tools appropriately shaped). Another handy tool to have is a special metal tray to hold joint compound. It may seem like an extravagance—but a most useful one! You may need screwdrivers for removing electrical plates, a stepladder for reaching the higher parts of walls or the ceiling, a Wonder Bar or pry bar for removing trim around doors and windows, and a Surform for smoothing edges of cuts. You will definitely need sandpaper and a sanding block—or a power sander. The final item on the tool list is a respirator, which you should wear when cutting with your jigsaw and ESPECIALLY when sanding over joints and nailheads.

Gypsum board dents easily, particularly at the edges, so that you must be extra careful in transporting it and/or storing it at your work site. A dented or bent-in edge makes an unpleasant problem when you come to finish the joint that this edge is part of, so it's always better to try to keep the edges perfect until the sheet is up in place. If you are having your materials delivered by the lumber company, make sure you are on hand when they come. You have the right to refuse the delivery if the edges on the sheets are damaged. It's a good idea to have the gypsum board on the job site at least a day or two before you plan to put it up so that it has time to adjust to the humidity of the room. Stack it in a dry place, and stack it flat (not on edge) to minimize chance of damage.

If you are resurfacing an existing wall, you should begin the job by removing all window and door trim, baseboards, moldings, and so on. Any electrical boxes will have to be moved out (adjusted) to sit flush with the new walls. All protruding nails should be removed or hammered flat. If you are working with a bare stud wall and planning to apply the sheets horizontally, you may have to add blocking at the proper height throughout the room so that all edges of the sheets can be nailed down. If you're working on a ceiling, you may want to construct some special T-shaped braces to hold the sheets up while you are nailing them. These should have top pieces that are at least 3′ wide; their height should be calculated so that they fit up snug between the floor and the ceiling (with the thickness of the gypsum board being added a consideration!). The foot of the T should have some sort of protective cushion if you are working on a finished floor. Two of these T braces will make ceiling work a whole lot easier, even if you are working with a partner. Gypsum board is heavy when you're trying to balance on a ladder and nail over your head! Union carpenters who do this kind of work all day have special "headdresses" they wear, so that they can help hold up the work with their heads. I think I prefer the T braces!

If you are doing both walls and ceiling, the ceiling comes first. The gypsum board is applied at right angles to the joists. Nails should be placed about ⅜″ from the edges of the sheet all around. You can set the sheet in place by driving a few strategic nails near each corner, and then go back and fill in. Nails should be driven about 7″ to 8″ apart around all four edges and then along the joists behind each sheet. As you place each nail, drive it just slightly below the surface of the sheet; this "dimpling" will create a depression that you will later fill with joint compound to cover and hide the nailhead. Practice this art very gently until you get a feel for it—if you drive the nail too hard, you'll fracture the surface of the paper. This is forgivable, but makes your work harder and the job less smooth. The adjoining sheet should be butted up to its neighbor, but not too snugly. Leave a slight space for subtle movement.

Walls are done the same way: a sheet started in the corner, tacked in place, and then nailed all around and onto each stud behind. Nails should be spaced 7″ to 8″ apart and dimpled. Remember to place nails about ⅜″ in from all edges. If you have to cut a piece (for example, to fit over or up to a window), make very careful measurements and double-check yourself. To cut the sheet, first make your marks with a light pencil. The sheet should be laid flat and with a good solid backing (on the floor or a table). The marking and cutting is done on the face (white) side of the sheet. Take your utility knife and use a firm, steady pressure to cut along the lines you want. If you have a really steady hand, you can cut freely; otherwise, use your straightedge to keep the knife in line. Next you should stand the sheet up, positioning yourself at the back (gray) side. You can use your knee or one hand to give the sheet a sharp blow just behind

(Photo by Janet Cole)

the line you just scored. The sheet will crack along the line (a little like glass cutting). Take your knife and cut the paper along the gray side, thus completing the separation. You may want to use your jigsaw for cutting gypsum board, but the shower of dust created is not healthy to breathe. Be sure to wear a respirator while cutting, and encourage your partner or co-workers to do the same. Cutting with a utility knife is slower, to be sure, but it creates a lot less dust and does not blow what it creates all over. If you have to make cutouts for electrical outlets and switches, you can use your jigsaw or a keyhole saw. You can also buy a special (and inexpensive) tool for this job—a tiny *dry wall saw,* which has a stiff, protruding blade designed for making these cutouts.

With all sheets nailed in place, you are almost ready to tape the joints between sheets. First, any exterior (outside) corners should be covered with corner beads. These are nailed in place; the nails pass through the gypsum board into the studs behind. This is your final pretaping step. Joint compound is sometimes called *sheetrock mud,* so that the process you're about to begin is *taping and mudding.* Doing this well is an art in itself. It sounds and looks fairly easy, and a beginner can do a reasonable job of it. The art is in making an absolutely invisible joint, which is something you'll learn to appreciate after your first few attempts. The best rule of thumb is to *proceed sparingly*—to apply the thinnest coat of joint compound at each stage, so that you can save yourself sanding later and create a neat, smooth finished joint. The term used to describe filling a joint is *buttering,* and it's an apt one—that's exactly what it feels like.

Begin at the top of your joint (assuming it's a vertical one) and use your joint knife to smooth a layer of compound into the crack between the two sheets. Then take a piece of tape that is generously long for the space (you can hold the whole roll in one hand if you want and just unroll what you need—this is a little harder to manage, especially if you're just learning). Begin again at the top of the joint and, using your knife, press the tape against the joint and into the joint compound with a smooth downward stroke. The pressure you are using should be enough to flatten the tape evenly into place, causing some of the compound to push up through the perforations in the tape. (If you're using tape without perforations, the pressure should be sufficient to make the tape lie down smoothly all over.) When this is done, go back with a very small amount of compound and smooth it over the tape from top to bottom. As you work, you will want to make the edges of the tape blend with the adjoining wall surface as cleanly as you can. This blending out of the edges is called *feathering.* Mud and tape all joints in your wall(s). Each nailhead should also be given a smooth coat of joint compound (no tape).

To tape an inside corner, fold your piece of tape in half lengthwise to make a 90° angle. Then proceed in the same manner: first a layer of joint compound, then smooth the tape into place, and finally recoat the tape with a "skim" coat. For outside corners, you apply the compound directly over the corner bead (no tape). It's quite hard to make a good outside corner, so don't be discouraged if you find it difficult. The corner tool designed for this job makes it a lot easier; it forces the compound into a clean square angle, with fewer rough spots. When you've finished all joints and covered all nailheads, your day's work is done; the

(Photo by Janet Cole)

(Photo by Janet Cole)

compound should be allowed to dry for twenty-four hours. It's a wise habit to clean up your tools as soon as you are done—before the joint compound hardens on them.

When the first coat is dry, give all joints and nail-heads a light sanding. Here again, you should ABSOLUTELY wear a respirator. Most joint compound contains asbestos fibers, which you are releasing into the air you're breathing. If you use a power sander, the fibers and dust are blown all over. Make sure that your partner wears a respirator, too. Sanding is one of the jobs most often given to apprentice carpenters because it takes very little skill and is one of the least pleasant jobs to do. Doing this work day in and day out is a very hazardous occupation! If you have to do it, protect yourself by being disciplined enough to wear a respirator—and a good one at that. Don't forget that your clothes and hair and boots will all be covered with the dust and fibers, and you'll be bringing them home to friends, family, lover. This may seem paranoid, but statistics are showing that many asbestos-related cancers are found not just in asbestos work-

Some Wallboard-related Products and Their General Uses

Insulating Wallboard is gypsum wallboard with one regular paper face and one face (to be used as the backing) of aluminum foil. The foil is laminated on. It acts as a moisture barrier and as a layer of insulation.

Backing Board is used as a base for tile, acoustical tile, and others. Another common use is as a first layer when a double layer of wallboard is applied to the wall ("two-ply construction"). Both faces of the backing board are covered with gray paper, as neither side is intended to be exposed.

Gypsum Sheathing has a core that includes asphalt, and the face coverings (brown paper) are treated to be more or less water-repellant. This sheathing is resistant to weathering, can hold up under damp conditions, and is fire-resistant. It is used as a base for exterior sidings such as wood or asbestos shingles, aluminum siding, and stucco.

Type X Wallboard is a grade of either regular wallboard or backing board that is highly fire-resistant and is meant to be used wherever a fire wall is specified (required by code). Brand names such as Fire Shield and Firecode Sheetrock are used by companies to designate this special type of wallboard.

W/R Sheetrock is the designation for water-resistant panels of gypsum wallboard meant for interior use. These sheets are distinguishable by their green face paper. They are used in high-moisture areas such as bathrooms, kitchens, and utility rooms. W/R Sheetrock is available in special Firecode grade for use where a more fire-resistant wall is needed. If you are applying W/R panels and have to make cuts in a sheet, use a special W/R sealant to cover and seal the newly cut (exposed) edges. W/R panels may be applied with nails or screws, or with an adhesive under some circumstances (when used behind thin ceramic tile, for example).

If you need more details for applying any type of gypsum wallboard or have additional questions about a particular project, the U. S. Gypsum Co. and other manufacturers print small pamphlets and booklets with explicit directions for using all of their products. Usually any lumber or building supply store that carries the products will have a supply of these pamphlets. If not, you can request them from the manufacturer. U. S. Gypsum Co. address: 101 South Wacker Drive, Chicago, Illinois 60606.

ers—their family members are victims, too. Consider keeping a special pair of overalls at work (wear them for the job and leave them there). It's not a bad idea to consider vacuuming your clothes if you've gotten really covered with dust. A vacuum (industrial type) is a very nice tool to have on the job, with the last half hour or so of work devoted to cleaning up your work space and eliminating as much "dust" as possible.

The second coat of joint compound should be applied with a slightly wider joint knife, so that the edges are feathered out farther. Again, be very *sparing* in your application. This coat is allowed to dry (twenty-four hours) and then sanded lightly. A final coat is then applied. All joints and nailheads should have these three coats for best results. For the final coat, you can use your widest joint knife. The last feathering will blend the joint compound against the wall surface. Again, you must let the compound dry thoroughly and then sand over lightly. Don't get lax at this point and forget your respirator. Your walls (and ceiling) are ready now for a dressing of paint. They should be smooth all over, but remember that there is an art involved that may require a lot more practice before your work looks perfect.

Another whole facet of working with gypsum board is repairing it when it has been damaged or broken. If the house shifts over a period of time, structural pieces may move slightly. The nails holding the gypsum board in place may pop out slightly. Extreme changes in humidity may also cause this to happen. Or you may have to repair a hole in the wall or ceiling, large or small. Another common damage is cracking. This usually occurs along taped joints and often happens because the tape has become loose or was never applied properly and had bubbles or blisters of air, which caused drying and cracking over a period of time. For most repair work, you can use a good-quality spackling compound; the better compounds have a smooth texture, adhere well, and sand to a nice clean finish. Tools are simple: a utility knife, hammer, screwdriver, putty knife, wall scraper or joint knife, and possibly a dry wall saw or portable jigsaw. You may need joint tape, too, and even some extra gypsum board.

If your problem is simply nails popping loose, you can drive each one back in and place a new nail (ring-shank) nearby. Dimple all nailheads and apply a thin coat of spackling compound over them. When this is dry, it can be sanded lightly. (Wear your respirator!) The wall will probably have to be repainted if it has damage of any sort, even this minor.

To repair a crack, use your screwdriver to scrape out the crack. If it is over a taped joint, you may have to remove a section of the old tape and replace it with a new piece. Follow directions for taping and mudding as explained earlier, using spackling compound in place of joint compound. You probably can make do with one or two coats rather than three. If your problem is a crack that is not along a taped joint, you can scrape the crack open (try to extend it so that you stop the "movement" of the crack itself), and then cover it with a coat of spackling. Feather the edges as well as possible for a smooth repair. Let the compound dry and then sand. Wearing your respirator for this work is a must, too.

If the damage to your wall is a hole too big to just fill with spackling, or it's a good-size fracture, you will want to cut out that section of gypsum board and then make a "patch" to fit there. The patch should be of the same thickness of material. If you are lucky, you'll be able to nail the patch in place (to studs and blocking), tape the joints, and proceed as though you were finishing up a new wall (use three coats of compound). If there is no good nailing surface or the patch is too small and falls between studs, you can make a patch with flanges. To do this, cut a piece an inch or two larger than the hole you need to fill. Then carefully cut the back (gray) paper and the gypsum filler away so that you are left with an extra edging of white (front) paper and a gypsum patch exactly the size of your hole. Apply a thin coating of joint or spackling compound to the back of the flanged edges and slip the patch into place. The compound-buttered flanges will help hold the patch where you want it. You can then spread a little more compound over the joints to hold the patch, and proceed to tape and mud as described earlier.

Don't be afraid to be inventive when patching your walls! Some carpenters stuff holes with steel wool or even newspaper and apply spackling over that "filler." Others rig up little devices of cardboard and string to fill holes, which are then covered with spackling or joint compound. Often you'll be faced with a situation that just isn't in the books, and your own ingenuity will be the "proper" answer.

ROCK WALL

There is a definite truth to "a dollar saved is a dollar earned." I recently priced interior wood paneling and concluded, after checking out various possibilities, that I did not have the funds to finish my home—at least in conventional ways. The idea of building an interior wall of native rock appealed to me aesthetically and offered an economic alternative: it would be low in cost, utilizing my labor instead of money. Fortunately for me, the wall I considered finishing this way rested

upon a solid, continuous concrete foundation below. If you like the idea of a rock wall, check the foundation below to make sure it can take the considerable weight. You may be able to shore up the floor joists below, or to take advantage of existing concrete foundation walls.

A rock wall begins with the gathering of the rocks. The kind, color, texture, and size are partly a matter of personal preference, and partly a question of what is available for the taking in your area. The area I dwell in abounds in flat slate rocks in hues of browns, reds, slate-gray, and black. The ease of building a rock wall depends upon your selecting rocks that have one fairly flattened side. This side makes a surface that is easily pressed into the mortared wall, preventing the rock from slipping or falling off while the mortar sets. Rocks that refuse to hang on a wall and keep crashing to the floor are harmful to the builder's morale and physical body! The remaining sides of the rock are your choice—material to fantasize with. The more textured a rock is, the more beautiful the addition to your finished wall. Consider, too, the purpose of your wall: if you plan to hang pictures upon its surface, you will need flat surfaces in those areas so that the pictures can hang flat; if your wall is the picture, enjoy all of the sizes and depths and textures available.

Another factor to consider in choosing the proper rocks for your wall is size. It is important to gather larger rocks for the bottom of the rock wall. It is preferable to select rocks with one flattened side to sit upon the floor as a beginning foundation. The flattened surface makes future floor cleaning easier. Gather armloads of small rocks to fill in between larger ones that do not fit together closely. Be sure that you choose rocks relatively light in weight for the bulk of your wall. Large ones may be used if they are not too heavy; otherwise, you will have to content yourself with smaller ones. With your rocks gathered, you have definitely earned your dollar "saved" by now. In the gathering of your rocks, be sure and gather in some of the scenery where the rocks live. . . .

The next step is to go to a lumberyard or hardware store for the few items you will need. If your rock-wall-to-be is an unfinished, bare stud wall, you will have to put up some inexpensive Sheetrock (or whatever you have available) as a backing for your rock wall. You will need some mortar to bond the rocks together. Premixed mortar comes dry in 80-pound sacks. It is easy to work with, requiring only that you add the prescribed water and mix. You may choose the less expensive route of mixing your own mortar (be sure to use extra-fine sand), but I recommend using the premixed. The amount of mortar you will need depends upon the size of the wall you are doing. I built a wall 16′ long by 8′ high and used three sacks. You will also need some 3 percent hydrochloric acid solution, usually available where mortar is sold. It is used to dissolve excess mortar on the face of the finished wall. You may also want to buy some masonry sealer, which is painted over the dry, finished surface to seal and waterproof it. This sealer is not necessary for an interior wall, but it does enable you to clean the wall more easily later: it prevents the mortar from flaking off when the wall is cleaned. You will need a roll of inexpensive felt paper, which is used under certain types of roofing and behind exterior wall sheathing as a moisture barrier. If you are lucky, you'll have some of this already, left over from house construction—or have a friend who has some extra. You will also need some wire mesh to cover the wall (I used

Placed strategically, a rock wall can add a "thermal mass" to your house—an area that will absorb heat from the sun during the day and release it at night. This simple system will add passive solar heating to your house (no set of collector panels, circulating pump, and so on, is used). The wall acts as a collector and the heat radiates directly from the wall into the room at night. To use your rock wall this way, place it on the north wall of a room with windows to the south. Clerestory windows above a set of regular windows will allow you to collect more sunlight during the winter. The heat from the sunlight should collect in the wall and will be stored there throughout the day. At night, windows should be covered with heavy curtains or insulated panels, and the heat radiating from the rock wall will help to keep the room warm.

1″). Tools for your work are simple. You'll need a wheelbarrow and shovel for mixing the mortar, and a rectangular trowel for applying it. You'll also need a staple gun for applying the felt paper and wire mesh to the wall.

Before beginning the actual work of building your wall, you must wash the rocks you'll be using. I began my wall by stapling felt paper over the entire surface, overlapping joints and seams. Next I cut the wire mesh and stapled it over the wall. I placed staples every foot and a half, giving the mortar enough loose wire to sink into, but holding the wire securely enough to give the rocks a surface to adhere to. With this done, you are ready to select the bottom row of rocks you intend to use. Set these down near your work area and mix a batch of mortar. I wore gloves to protect my hands from the lime in the mortar. Pour half a sack of dry mix into your wheelbarrow, scoop a little hollow in the center and add a small amount of water. Mix this in with a shovel, using a push-pull rhythm and making sure that you reach the dry mortar toward the bottom. Add water and mix until your mortar is the consistency of thickened pudding. If you make a mistake and the mixture gets too soupy, add more dry mix. When your mortar is ready, you can use a board or pail to carry small amounts to your wall as you work.

Put enough mortar on your trowel to fill it about a third full. Begin at floor level on your wall and push the mortar into the wired wall with upward-directed strokes. If the mortar sticks, your pressure is correct. If it crumbles off, try pushing it in harder—or reconsider your mix: it may be too wet or too dry. Experiment until the mortar adheres to the wall. When you have 2′ of the wall mortared from the base of the floor up, you are ready to put down the floor base. Put a layer of mortar all along the wall, making it 1″ thick and 1″ deep. This acts as the foundation. Beginning in one corner, take your chosen rock and push it into the mortar, wiggling it from side to side gently so that its entire wall/flat surface sucks into the wall. If it adheres well, add enough rocks in the same manner to complete one row along the floor. Fill in with smaller rocks so that you have a fairly continuous rock surface along the floor, and so that there is no old wall showing. Wipe away the excess mortar and return it to the bucket to use again. Once the base layer of rocks is secure, you can add more rocks in the same manner until you have completed an area about 2′ high. If your base rocks seem pretty firm (i.e., they are not tending to pull away from or slide from the wall), you can continue building your wall. If the first rocks have not set, you should wait for them to become more firmly set before you continue. Mortar will not set entirely firm for about ten hours, but you can work in stages, keeping a careful eye on the stability of the wall as you enlarge it. If you have to stop for an hour or two to let a section firm up, be sure to wash your tools.

When your entire wall is completed and the mortar has had ample time to dry, take a paintbrush and apply the hydrochloric acid solution to the wall. This will dissolve any excess mortar on the face of the rocks. Wear gloves and work cautiously with this solution. If you are using sealer, it can be applied after this step. First sponge off the acid solution—this will clean the wall and enable the sealer to bond more tightly to the wall. Now you can stand back and admire your "earned" beauty. It is wonderfully enhanced by the addition of green plants, and will add a special feeling to your home.

WOOD PANELING

There are few ways of finishing the interior walls of

your house that equal the beauty and warmth of wood paneling. Although solid wood paneling is quite expensive, lower grades of wood may be used, and the "defects" in these grades can create a particularly striking finished wall. For example, if you use a grade of wood that allows a lot of sapwood, the color variations may give your walls multiple tones that break the monotony of the flat wall surfaces. Pecky cedar (cedar mottled with holes and/or damaged areas) and knotty pine make interesting interior walls. Old weathered wood that can't be used outdoors because of splits, knots, and cracks may be used to advantage to finish an interior wall—and in this protected environment, the old wood will last as long as you might want it.

Boards used for interior paneling may be simple straight-edged or shiplapped, tongue-and-groove, V-rustic, and so on. If straight-edged boards are used, you may want to cover the joints between boards with battens. If you have really straight, even boards, you may get by with simply fitting them tightly against one another. If you use green boards, remember that these will shrink quite a bit as they dry and cracks will open between the boards. If you are covering an old wall (one already covered with Sheetrock, for instance), you can paint the wall beforehand; use a color that is close to the color of the wood you're working with, so that if the cracks develop they will be less noticeable.

Paneling may be put on horizontally, vertically, or even diagonally. A diagonal pattern will create quite a bit of waste, so you may want to reconsider whether the aesthetics are worth it. Not only will you have lots of angle-cut end pieces; you'll need many extra nailers in your stud wall or a good deal of infill (pieces added to the wall frame to provide a nailing surface for the wall covering) if you're working with post and beam construction. As the effect of many walls paneled diagonally is a bit overwhelming anyway, you might choose to do just one wall this way.

Horizontal paneling may be nailed directly to the studs, and in most cases you won't have to add any extra nailers to your wall. If your wall is post and beam, of course, you'll have to add infill at regular spacings to serve as nailers. Usually horizontal paneling consists of 1×6, 1×8, or 1×10 boards, which are nailed to the studs with 8d finishing nails. Boards should be nailed at both ends and at center or intermediate points (at least every 3′ or 4′). A general rule of thumb for nailing is to use two nails at each nailing point for boards up to 8″ wide and three nails for wider boards. The nails are face-nailed and should be placed *at least* ½″ in from the edges of the boards to avoid splitting the wood. The nails are set and may be covered with a wood putty stick of appropriate color. Remember to cut boards so that joints fall at the midpoint of a stud or nailer; stagger the joints on adjoining rows or courses so that they don't occur on the same stud. The bottom board for each wall should be level when nailed in place. Don't worry about a crack between the floor and the board: this will be covered with a baseboard later. It's important to begin with a level board if your wall is to have neat horizontal lines when finished.

If you use a very expensive, clear grade of 1×4 or 1×6 tongue-and-groove paneling, you may want to blind-nail these pieces rather than face-nail them. Use a lighter (6d) finishing nail. Nails should be driven at approximately a 45° angle into the base of the tongue and then set slightly below the surface so they won't interfere with the tongue-and-groove joint between boards. Very thin paneling may be applied to a solid wall surface with an adhesive rather than nails. If you're using this type of paneling, follow the manufacturer's directions for installation.

Vertical wood paneling will seem to give more height to a room than the same size (1×6 or 1×8) applied horizontally. You can use one width throughout the room or vary widths in a random or planned fashion. One preparation for vertical paneling involves putting nailers at the top and bottom of the walls (and sometimes at intermediate heights) in the form of furring strips attached to the wall studs. You may be able to nail the vertical paneling directly to the top and bottom plates; in this case, preparation will involve placing nailers at intermediate heights wherever needed (these must be placed between the studs so that the nailing surfaces remain flush). Usually a board should be secured every 3′ to 4′; you should add nailers accordingly. There is quite a bit of flexibility here, so you can place nailers according to the particular wall you're working on. In beginning each wall, make your first board perfectly plumb. Any gaps between corner boards that occur because of this plumbing up may be covered later with a piece of quarter round or other trim if necessary. If you come across a board that is warped and won't fit up tightly against the adjoining board, nail one end in place and then use a block of wood and your hammer or a chisel to bang or force the other end of the board into place. If the board is really difficult, you may need the help of a second person to push the board into place while you nail. As with the horizontal paneling, you can face-nail your boards with 6d or 8d finishing nails (set and cover them if wanted), or you may blind-nail tongue-and-groove or V-rustic boards as described above.

A nice alternative to solid wood paneling is an adaptation of the old-fashioned wainscoting, where the lower two thirds of a wall was covered with wood paneling and the upper third plastered. Following this same idea, you can cover the lower two thirds of your walls with wood (applied vertically or in sheets) and finish the upper third with particleboard or Sheetrock (gypsum wallboard) painted white to approximate the look of plaster. The upper third might also be wallpapered or covered with burlap (in natural or a color) or fabric. This alternative is cheaper than using wood for the entire wall, but, beyond this advantage, it gives the room a lightness and brightness along with the warm tones of natural wood. If you're working with a bare stud wall, you may choose to cover the wall first with Sheetrock, hardboard, particleboard, or even plywood. Remember that Sheetrock will give you some fire resistance that other materials won't; if you're working with a permit, it may be required. If you go to hardboard, plywood, or particleboard you may use strips of lath to cover joints. I've seen some walls beautifully patterned with these strips applied both vertically and horizontally. They may be painted, stained, or left natural, depending upon the other materials you're using and the effect you want.

Finish the upper third of each wall first. If you're painting or wallpapering, extend the paint line or paper below the upper line of your wood paneling. If you're applying burlap to the wall, do the same. To apply burlap, first paint the wall surface a color harmonious with the burlap you're using. After the paint has dried thoroughly, you can apply the burlap to the wall with old-fashioned wallpaper paste. Having a piece of plywood or particleboard to work on helps; you can lay out a piece of burlap and give it a good coating of paste. A broom isn't a bad tool for this job! This is fairly sloppy work, so be prepared for some cleanup. The burlap can be smoothed into place on the wall—and amazingly enough, it will stick! If you have a wood stove or you're using burlap in a kitchen area, you should consider treating the wall surface with a flame retardant. Give the burlap a few days to dry thoroughly before finishing the lower part of your walls.

Finishing the lower walls is simple. Again, be sure to plumb up the beginning board for each wall. Use finishing nails and complete all walls as described above. Be careful to make the top edges of your boards line up. If you have slight variations, leave gaps along the floor (these will be covered by the baseboard). If necessary, run a ledger along the wall (level and tack it in place) and butt all boards up to it. When your boards are all in place, finish the wood part of your wall off by placing a horizontal piece of 1×4 around the entire room. This 1×4 should fit snugly along the top of all your boards; it makes a transition from the vertical lines of the paneling to the flat surface of the wall above. Nail it in place with finishing nails staggered zigzag every 12″ or so and two nails per end.

Another alternative in wood paneling is the wood-veneer plywood sheet. These sheets vary tremendously in quality—and cost. The thickness of the sheet and the wood used for the veneer (the thin surface covering) are most important. The more expensive of these panelings have veneers of exotic woods; the cheaper panels have a veneer of more common wood and may even be vinyl-coated with a "simulated wood grain." Most large lumber companies will stock quite a variety of plywood paneling, which you can look through to find something that suits your taste and budget. The thinner sheets of paneling may be applied over existing wall finishes. If you are working on a new house or a new addition, you'll have to provide a backing for these panels; Sheetrock is a common choice because it offers fire resistance as well as a cheap, solid backing. The thicker plywood paneling—at least ⅜″ thick—may be applied directly to studs. If you are planning to cover old walls that are very uneven, you may have to apply furring strips to the walls to make an even surface for your panels. These strips should be placed every 16″ or so horizontally and added wherever edges of sheets will fall.

This prefinished plywood paneling can be applied with finishing nails or with a special panel adhesive. Follow manufacturer's directions for adhesives. There are nails made to match some paneling. They are finishing nails coated with a coloring that approximates the color of the panel surface so that, when driven in, they are barely noticeable. Size 1¼″ to 1½″ casing or finishing nails are used for this paneling. Space them about 8″ apart and set nails at least ⅜″ from edges to avoid splitting the wood. You may also buy putty sticks in more or less matching colors, which can be used over regular finishing nail heads to hide them. Matching baseboards, corner molding, and trim for around the ceiling are also available; these are color-matched and usually come in a vinyl or other synthetic material. Before you actually begin to panel your room, it's a good idea to stack your materials in the room for forty-eight hours or so to allow them to adjust to its humidity. In applying the panels, try to place nails in darker grooves whenever possible to hide them further. If your panels do not have special interlocking edges (most do), you will have to cover joints with battens later. Begin each wall in a corner; make sure that the sheet is plumb before nailing. If your corner is uneven, you can cover this joint later with trim, or you can scribe the paneling sheet to fit exactly. To cut holes into paneling for electrical outlets, use your jigsaw (saber saw) with a fine blade. To rip sheets to fit or to make horizontal cuts, use a fine blade on your circular saw. When making cuts, remember to place a panel so that the rough side of your cut will appear on the back. In applying each panel to the wall, leave about a ¼″ space at the top and bottom—this space will be concealed with baseboards and trim and leave 1/32″ between sheets. If the room you are paneling is over 8′ high and you are working with 8′ sheets, take time in the beginning to match sheets according to color. Many of the less expensive panels will vary from sheet to sheet, and you want to avoid topping a very dark 8′ sheet with a lighter 2′ section above to complete the wall. This horizontal joint can be covered with a piece of trim; but if there is a big difference in tone between the upper and lower portion of the wall, it will be quite noticeable.

One final idea on the use of plywood paneling is to bring exterior plywood siding indoors. Some of these siding sheets have interesting patterns, colors, and so on; the rough-sawn siding can add a nice rustic finish to the inside wall. Be careful to choose a siding that is not pretreated with a strong-smelling water repellant or wood preservative. These sheets will not lose their unpleasant odor indoors for months (or even longer).

Besides the alternatives for interior walls discussed already, you may consider other wood products not usually relegated to wall coverings. Wood shingles, for example, can make an interesting wall. You can use a really low grade which will be inexpensive, or you can use old, weathered shingles that will no longer perform well outdoors but may be salvaged and reused inside. Strips of aromatic cedar that are usually meant to line closets may be used to cover a small wall or a portion of a wall (they are rather expensive but quite beautiful). Use your imagination to add wood to your home in unconventional as well as conventional ways: the versatility and handsome warmth of this material won't disappoint you!

SOME ALTERNATIVES FOR FLOORS

The finished floor is usually one of the last parts of the house to be completed, primarily to save it from damage during construction. It is also work that can be completed at a more leisurely pace if you're rushing to be in your house before the winter rain or snow. A finished floor is distinguished from the subfloor laid earlier on. The subfloor is usually a rough one made of plywood sheets or 1× boards; the finished floor is a second layer that varies in composition throughout the house. A finished floor for living room, dining room, and bedrooms might be hardwood strip flooring, parquet block, or wall-to-wall carpet; in kitchen, laundry room, and bathroom, the finished floor is usually vinyl or rubber tile, linoleum, or any other covering that is tough, washable, and durable. If you've built a passive solar house with a slab floor, you'll still be facing flooring choices on other levels of your house or in certain areas.

There are many alternatives to the more conventional floorings mentioned above. You can use ceramic tiles, flagstone, even used bricks laid directly on the ground with sand filler between them; you can forget the subfloor/finished floor distinction and use 2× planking for a single-layer floor. You can live with your plywood floor or even 1× boards, single-layer. A floor should be made of any material that is durable, comfortable to walk on, easy to clean, and attractive (to you!). To my mind, the material that meets all of these requirements adequately and one superbly (it is especially pleasing to look at) is wood.

The most beautiful floor that I can think of is one made of hardwood strip flooring such as oak or maple; for this floor, I'd gladly put in extra hours on hands and knees to give it a shine, a polish that glows with all the richness of these woods. But with the astronomical prices of hardwood flooring these days, most of us can't afford such glory! I'm happy to report that there are some alternatives—and handsome ones at that. Softwoods such as pine or fir may be bought in 1× or 2× tongue-and-groove and installed as finished flooring over a subfloor (2× may be used as a single-layer floor). These are much cheaper than hardwood but can be given a protective stain and finish that will make them shine and last. This flooring will be more easily dented, scratched, or water-stained than oak or maple, but if you don't care if your floors stay mirror-smooth, they are an excellent alternative.

One of my favorite floors is made with 2×6 pine decking. This is actually a material intended for roof decking and is fairly cheap, is readily available, and makes a thick, sturdy floor. It is tongue-and-groove, with one face more finished. This face also has an extra little groove or indentation along the length of each board, so that when the decking is in its proper place on the roof, a pleasantly patterned ceiling occurs. Brought down to improvise as flooring, this "good" face would create nightmares, as the groove would gather dirt and dust, tip chair legs, and so on.

When we used this decking as flooring in our house, we sadly turned the better face down. Once the floor was all nailed in place, we stained it with a dark walnut oil stain that gave a surprisingly uniform, deep finish to the floor. After a couple of days, we got down on hands and knees and applied an old-fashioned paste wax to the entire floor. The result? A solid, dark, richly glowing floor that looks and feels underfoot like old colonial-style planking. The wax gives it a durable, water-resistant coating that can be as slippery-shiny as the best hardwood. Our sadness at "losing" the better face of this decking vanished when we surveyed the finished floor! This material is perfectly suited for lofts and second floors—laid over rough-cut exposed ceiling joists, it can make a beautiful grooved ceiling and a fine smooth floor above.

Regular wood strip flooring is usually tongue-and-grooved along its edges and is end-matched (tongue-and-grooved on its ends as well). This means that the pieces interlock securely all around, making a very tight floor with very little movement. If you substitute a different material such as 1×4 tongue-and-groove fir, or the 2×6 pine decking we used, you'll have tight joints along your edges but any butt joints in the floor will be more or less open. If you make any cuts very carefully, perfectly squarely, the pieces can then be butted tightly up to one another.

In laying flooring of any type, you usually place it perpendicular to your floor joists. This is true whether you have a subfloor or are nailing directly to your joists. The exception is, of course, diagonal flooring, which is a common pattern for subflooring and can be used to make an interesting finished floor. If you have a subfloor, consider laying a covering of building paper (kraft paper or felt) over the entire floor surface before placing your finished floor. This will act as a dust and draft seal. If you lay this paper, you may want to snap a chalk line along the center of each joist to help you in nailing later. Flooring should be allowed to sit inside the house for several days before it is applied if at all possible. This will let it adjust to the humidity inside the house.

To begin, lay the first board along the wall, leaving a space of about ½". This allows for later expansion if the air moisture gets high; otherwise, the floor will be forced to buckle when it swells slightly. Set this first piece with its groove toward the wall. For regular strip flooring, use 4d casing nails for ⅜"-thick material and 8d flooring nails for $^{25}/_{32}$" material; for thicker flooring, use 8d or 10d nails. Flooring is nailed every 6" to 8". For most hardwood floors, predrill nail holes to prevent splitting. Face-nail it into the subfloor and/or joist right along the wall, then toenail the tongue edge. To drive your toenails, set them at an angle just above the tongue; don't drive them fully in with your hammer or you'll damage the tongue and make it difficult to slip the next piece up to this one. Use a nail set to finish each nail. This takes extra time but is absolutely necessary if your flooring is to fit together.

Now take your next piece in line and butt it up to the first, nailing as instructed. All butt joints should fall over a joist and each piece should rest on about half the available support. As you work, plan your pieces out so that all butt joints are staggered and don't occur side by side over the same joist. The second row of flooring is set up to the first. If it slips snugly against the first, go ahead and toenail it (through the tongue edge, as described); if it has a bow or bend and won't slide neatly into place, use a wood block to protect the tongue and drive that with your hammer until the piece fits tightly into place. Then toenail it. If you have a really badly bowed piece, you can use a wood chisel to force it in place. Set and toenail the first few feet, then place your chisel at an angle much like that of your nails, blade dug into the joist *under* the tongue of the flooring. Give the chisel a few taps to set it into the joist. Now, by pushing the chisel handle toward the flooring, you can force this piece into position. A large screwdriver may be used to give the same leverage. Once the piece is in place, toenail it at that point. Then move along with your tools and force each section into place, nailing as you go. It's critical to keep all the joints tight and the flooring going on in a straight line; otherwise you'll be in trouble at the other side of your room! Discard any really bad pieces as you go along. When you reach the opposite side of your room, you may have to rip the last pieces to fit. These will have to be face-nailed along the wall; if you set the nails close enough to the wall, they can be covered with the baseboard.

The final step in finishing your floor is to place thresholds over your doorways (these cover the joint between the flooring and the doorsill) and to place baseboards around the bottom edge of the room.

Interior thresholds come in several widths and standard lengths; you should buy one as wide as your door opening and the right length or longer (you can cut it down). These thresholds are usually made of oak and cost from $6.00 to $8.00. To cut a threshold, measure and mark it, with your combination square, and use a fine-tooth handsaw for a smooth, neat cut. Slide the threshold into place and drill holes for your nails (the oak may split if you try to drive nails into it; if it doesn't split, it will most certainly resist the nail, causing it to bend over). You should use 6d or 8d finishing nails and a drill bit just slightly smaller in diameter than the nails. Place two nails at each end and several staggered across the face of the threshold. Using a nail set, set each nail slightly below the surface of the wood.

Baseboards may be purchased as such, with curved upper edges, or they may be simply made with 1×4 or even 1×6 material. They are nailed in place with finishing nails (face-nailed through to the wall), which may be sunk with a nail set and disguised with putty later. Rubber baseboards are often used in kitchen, bathroom, and laundry room areas. An extra piece of molding called a *base shoe* is sometimes added to the baseboard. This is a tiny piece of quarter round (specially milled molding) which is placed along the lower edge of the baseboard and nailed to the floor with small finishing nails. It covers the joint between baseboard and floor and also eliminates a dust-catching corner all around.

Hexagon Building

The hexagon house is a women-built structure in southern Oregon. It is a relatively small house that was intended to be a one-person dwelling. The plans could, however, be enlarged according to your own needs with very few structural changes—a larger hexagon, for example, would require more extensive foundation work and a heavier support system.

The story of the hexagon reaches backwards and forwards, touching the lives of many women. The hexagon was built from May through October 1975, and throughout the six months thirty-one women put energy into making it happen. Since many were visitors who worked for several days to a week, it became a carpentry learning experience for a good number of these women. For several of us, working on the hexagon was an introduction to the trade that now provides our livelihood.

The hexagon was built to be Nelly's home. She had just moved from a partially finished five-sided house (the star cabin), so it felt appropriate to build her a six-sided house—adding one more side as a symbol of the change in her life. There was another connection to the design: my astrological chart consists of two grand trines that cross to make a six-pointed star.

The hexagon is an extra-special building in my life; it was the first complete house I designed and followed through to the finished product. I had dreamed of building a house as far back as childhood and, as a woman raised in this culture, had always felt it would remain a fantasy. Nelly provided me with this opportunity for a reality. The confidence I gained from the experience has helped me build and build ever since. My carpentry fantasies are no longer impossible to me. Seeing more and more women-built structures, I grow more excited: we can learn together, be dreamers and builders as well.

Shannon also began her career as a carpenter by learning carpentry on the hexagon. She was going to go to carpentry school in Seattle when I met her and asked if she wanted to learn carpentry for five months on the hexagon. She's now in the process of building her own home and is supporting herself through carpentry. She and I became partners in carpentry, building several structures.

The hexagon house was built in the country, so no permit was obtained. At the time we didn't really think about this, just built and later discovered we were supposed to have had a permit. The inspectors passed it, though, and tax-appraised it several months after its completion.

As for resources, it was built with mostly recycled lumber that was found or purchased locally, old windows, some new lumber, and sixteen trees (which were thinned from the forest, cut into poles, skinned, and dried). The house, in total, cost $355 in materials. At present-day costs, perhaps you could triple that figure. We had a limited budget, and Nelly put a lot of energy into finding good deals on old lumber and windows so that we could meet our budget.

The uniqueness of this house is in its pole construction, the six sides, the hat, and the ceiling boards forming the six-sided star. The extravagances are perhaps: (1) some lumber waste due to cutting boards at angles rather than building a rectangle or square (boards are sold in 2-foot lengths—for example, 6′, 8′, 10′—and some angular cuts require up to 16″ of lumber wasted; we used most of the scraps, but square or rectangular houses can be planned to have very little waste); (2) building a double roof so that the roof poles on the inside would show and yet the roof could have a 3½″ insulated air space for efficient heating in the winter. Extra lumber was required—2×4's for framing and plywood for sheathing, plus extra roofing.

The trees were cut and skinned and laid in the shade to dry several months before they were to be used. We propped them up parallel to the ground, supported approximately every 4′ to avoid warping and to allow air to touch evenly around the logs. Books we consulted said it should take four to five months in the shade to dry the logs slowly and to avoid checking (splitting along the grain), but we didn't want to wait this long. Several of ours did develop checks, but none were too bad. We skinned the trees with an ax. If you use dead trees, you must use a drawknife. There was much discussion concerning the use of live versus dead trees. We decided to use live trees, cutting them selectively, as a thinning process in the forest.

There are several things I would perhaps do differently now, but I'll explain them more later. Some may just be a matter of taste, for the hexagon seems sufficiently strong. A few other things I'd do differently now because I see them as just plain mistakes and a part of my learning process. For example, the house slants a bit in one direction! In this kind of pole construction, that could have been solved very easily during the building.

A Few Statistics

The floor area is 168 square feet
roof area is 266 square feet
walls are 7′ tall
center top is 8′9″
eave is 1½′

The Foundation

The floor plan was marked with a string, 16′ from one point to the opposite point. Each part of each triangle is marked at 8′. There are seven foundation piers, so the maximum span on each floor support is 8′. The code now requires no more than 7′ spans in Oregon for this type of floor joist.

I decided to begin the leveling of the foundation piers by beginning with the one that would be closest to the ground so that I would make the lowest point of the house at least 12″ above the dirt. Building codes may call for a clearance of up to 18″; check the local code if you are building by permit. Rot and insect damage may cause problems, so, for the foundation to provide a safe surface for your home to rest on, all wood should be a good foot above the dirt unless each board is creosoted or treated with a similar wood preservative.

We used concrete blocks stacked one upon another for our foundation piers. Each foundation was made the same way, but each varied in height because of the sloping land we built on.

We began by digging a level surface for the first block to sit on, first leveling the

(Photo by Dian)

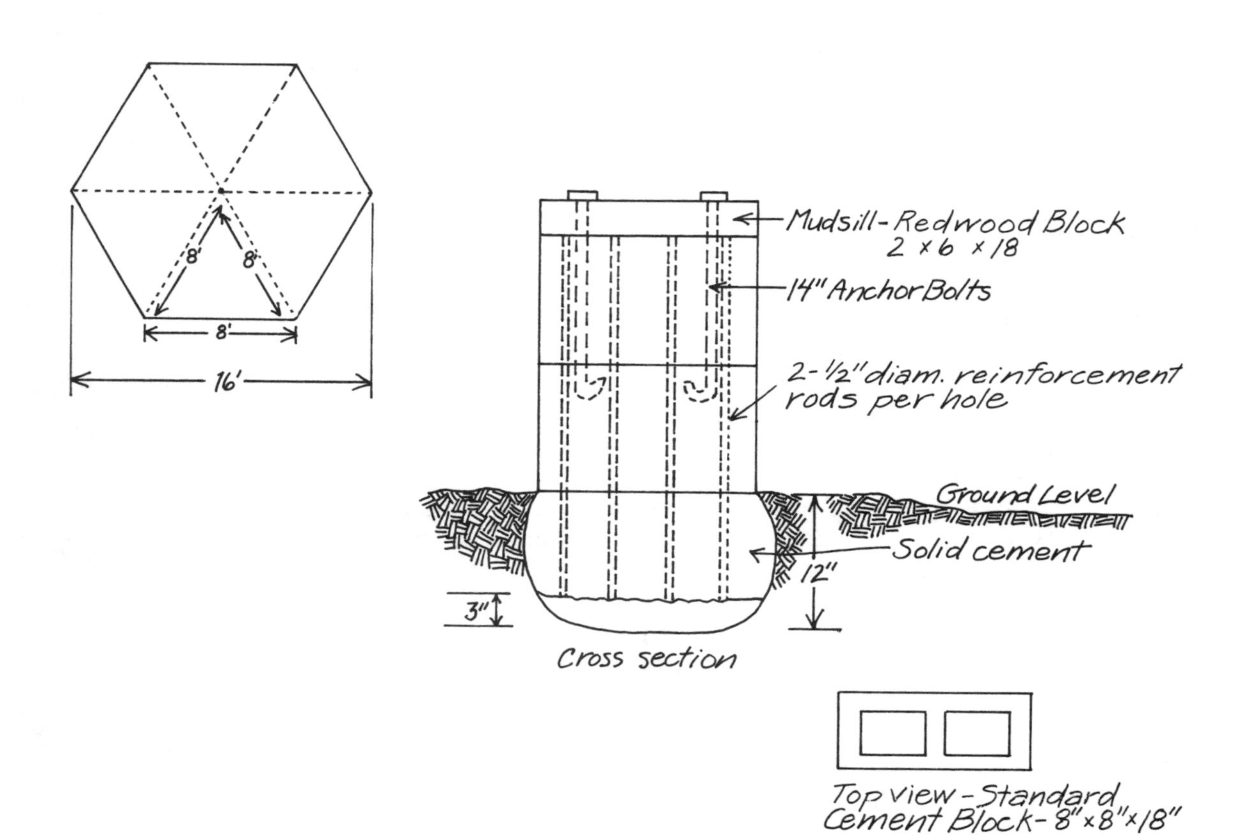

ground on all sides of the block. Then we moved the block and dug a footing hole 1′ deep in the exact location the block had been sitting on. The depth required is determined by the frost line: you must dig 6″ below the frost line. Since in Oregon the frost line is 6″ below the ground, we dug 1′ below the ground level. In areas of heavier snow the footing must be deeper; the code depth in Idaho, for instance, is 2′ deep.

As we dug each center footing hole, we left the edges of the dirt for our block to sit on in its level position. Next we poured a few inches of cement on the bottom of the hole and let it dry. We did this because we didn't want the metal reinforcement rods to touch the dirt and eventually begin to rust. An alternative method is to suspend the rods in the hole so that, when the concrete is poured in, they will have a protective layer of concrete below.

We then placed the cement block level over the hole and cut two reinforcement rods, each a bit shorter than the total height to the top of the highest block. These rods act to connect the entire foundation together when you fill the blocks and footing holes with cement. Before filling the holes with cement, you must make one more part—the wood block (or mudsill), which gives you a nailing surface to connect the house to the foundation. In our area we use redwood, which is resistant to moisture and insect damage. I cut an 18″ piece of 2×6 and next drilled two holes in the board so that the anchor bolts could pass through those holes and into the holes in the cement blocks. I slipped the bolts into place in the mudsill and then, as each foundation block was filled with cement, placed the mudsill board on top, forcing the bolts down into the cement. Some must be tapped down with a hammer. Use of the reinforcement rods and the anchor bolts makes a very soundly held-together foundation pier.

The rest of the foundation piers were leveled to the first by laying a long, straight board on edge from the first one to where the next would be. A level was placed on that board.

Measure the height from the level board to the ground after you have more or less made another flat surface for a block. You must dig down, then, to make the second foundation by an exact multiple of cement blocks. Each of the blocks we used was 7½″ tall and the top plate was 1¾″ tall, so in this case, if the distance for foundation #2 is 30″, you must dig accordingly. Dig here 1¾″ more and level the bottom block, and it should be exactly level with four cement blocks and the top plate.

1¾″	sill
7½″	block
7½″	
7½″	
7½″	
31¾″	total
−30	foundation
1¾″	

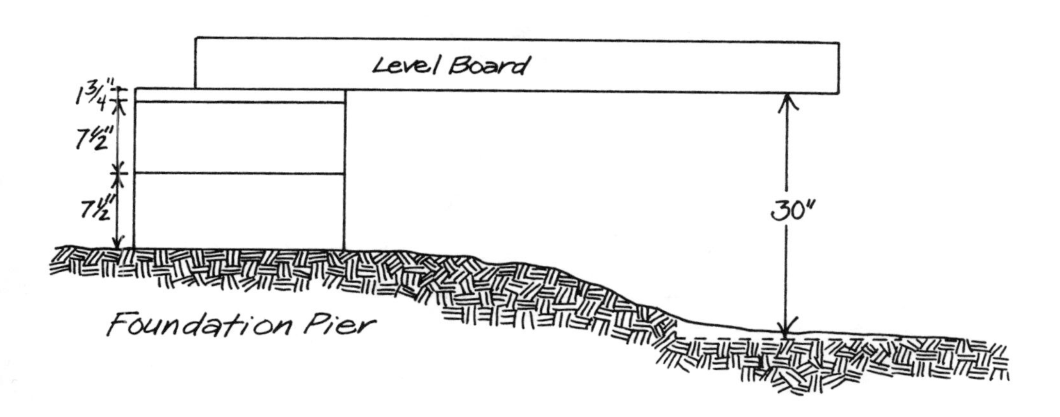

When two of our seven foundations came out short by mistake, it was a lucky 1¾″ short each. We were simply able to put a double mudsill on them.

Each foundation pier in turn is done the same way, and all can be poured at the same time if you have everything ready (and if you have poured a few inches on the bottom of each hole and let it dry previously).

Foundation work is a very tedious process! Although you are making great progress, it feels as though you have done very little and can get discouraging. It is important you spend the time to make it as exact as possible, though, as later stages of the house will be hard to level and frame if the foundation piers are not as level to one another as possible.

Floor

Lumber needed: eighteen 2×6×8′
(Douglas fir, Construction grade)

The Floor Joists Floor joists are the supports that hold up the floor. We learned that they are always put on edge because a board on edge is about five times stronger than one on its side. We used 2×6's, 8′ long. Because of the shape of the structure and the way the floor joists lay, it is difficult to adapt tables of "allowable spans" to know what size lumber to use. I think I would use 2×8's if I were to build another structure this size, or else make the outside frame two 2×6's spliced together.

The ends of each board were cut at angles marked by using a protractor and angle finder. The first six joists (1–6 in the diagram) were cut one end at a 30° angle and the other end at a 60° angle—8′ total length from point to point. Two 30° angles join in the center to form the 60° angle of the triangle. Boards 7–9 were cut 60° angles at both ends and again 8′ from point to point. Boards 10–12 were cut 60° angles at both ends also, but you can see from the diagram why they are approximately 3½″ shorter in length—to make up for the width of the 2×6 coming out from the center.

Nail them all together with 16d nails, placing two nails per end of each piece. We used box nails, which seem to cause less splitting. Also, nail the supports to the foundation mudsills by toenailing (two nails each side).

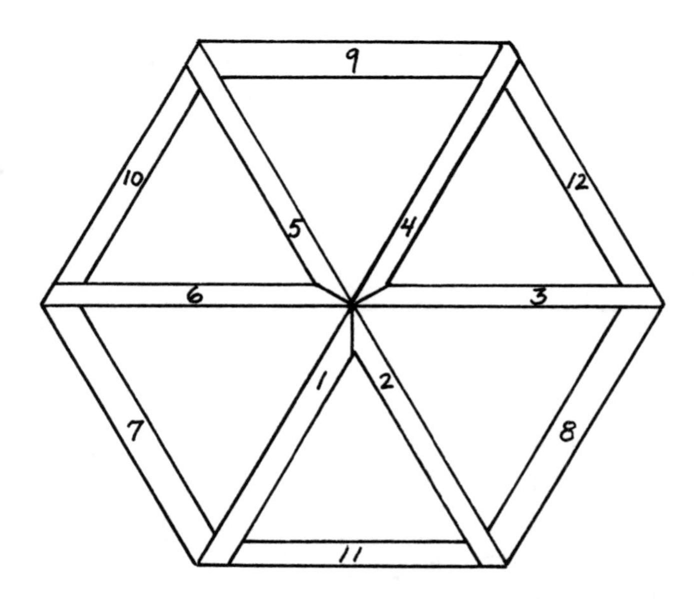

Extra Floor Joists The rest of the floor supports were added in specific places to be used as nailing surface for the plywood floor. We used ¾″ plywood and, since plywood comes in 4′×8′ sheets, we tried to have as little waste as possible.

Put in numbers 1–4 first. You can determine the end angles by laying the board on edge above the existing hexagon and marking it with a pencil from below. Also, it may be helpful to lay one piece of plywood on the structure to mark where the extra supporting joists go. You want these extra joists as plywood supports; place them so that the plywood seam is in the center of the support.

The hexagon was built almost entirely with hand tools. Cutting all those 60° angles and then the additional floor support angles by hand in the hot summer sun was quite a challenge! It took Gypsy and me ten hours each to complete the floor supports, insulate, and cut and nail down the floor.

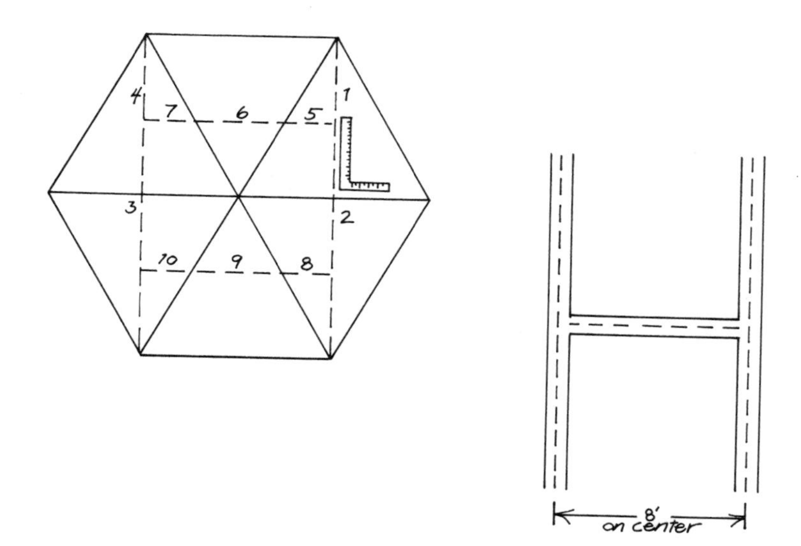

Insulation We used fiberglass insulation under the floor, and again the angles made it a bit more difficult. You have to cut the insulation angles as best you can and staple-gun the insulation to the top of the joists. Remember always to put the silver side of the insulation facing toward the living space. Under the floor, the silver side would be up. One thing I would strongly recommend about insulating a floor this way (exposed below) is to staple tar paper to the bottom of the joists, covering the entire underside. Animals tend to get into the fiberglass if it's exposed. Dogs rub against you after they lie under the house and you itch for days; birds use it for their nests, and on and on. It is *not healthy* for your animals' skin or your own! It is a good idea to keep a separate set of clothes to use solely for insulating days—long sleeves and long pants! Wear a respirator to protect your lungs. Don't forget that fibers clinging to your clothes and in your hair will be carried home.

Plywood Flooring It took six sheets of plywood to complete the floor. Two had to be cut lengthwise, removing 1′ to make an 8′×3′ sheet. Two had to be cut exactly diagonally. The diagonals would be affected if the hexagon measurements were not close to exact, so we laid the full sheet on the triangle and then marked it from below to be sure it was accurate before cutting it.

The plywood sheets were nailed to the floor joists; for this work I suggest using cement-coated plywood nails (3d) or ring-shanks for a tighter hold, as with time and use the floor can loosen a bit. Another suggestion is using 2×6's or 2×8's for flooring instead of plywood (mainly for longer life and less bounce), but our budget was minimal, so we used ¾″ plywood. You could also use a heavier (1″) plywood. If you have the money, consider using a regular 2×6 or 2×8 tongue-and-groove flooring. When these boards shrink slightly, there will be no spaces between your floorboards. Generally, plywood is used only as a subfloor and covered by 1″-thick wood, particleboard, tile, or some other floor covering.

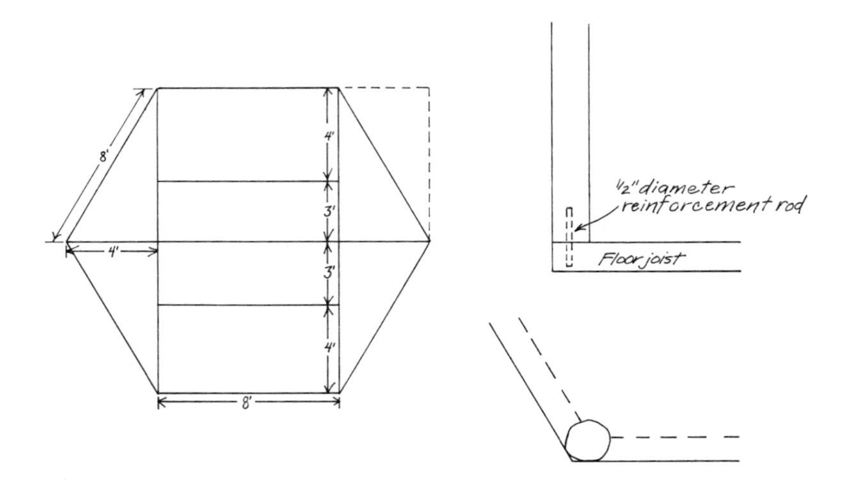

The Corner Posts

From here on, the house basically changes to modified post and beam construction. The corner posts were connected to the floor by drilling a 5″ hole into the floor, making sure the hole was drilled in the center of a floor joist, and drilling a 5″ hole up into the bottom of the corner post. Drilling the floor hole first and then measuring for the proper position of the corner post seemed most accurate.

We situated the corner poles to be plumb with the outside floor surface so that they could later become a nailing surface for the outside walls. The extra parts of the logs now show in the inside of the house.

We used standard metal reinforcement rods (which sell in 20′ lengths, so you need to take a hacksaw shopping with you to be able to put them in your truck or car) to make these connections. We hacksawed them 10″ long: one for each of the six corner posts. These posts are 8″ diameter and 6′7″ tall. After sitting the corner pole on the rod, you need to nail one or two temporary supports from the pole to the floor to hold the corner post plumb and at right angles to the floor.

After all six corner posts are well supported, you are ready for your beams (beam poles, as I call them), which would be called top plates in stud wall construction. Here's where my slanted walls come in. I didn't realize the importance of bracing the corner poles firmly, so I didn't brace them sufficiently. Later the whole thing turned, or racked, a bit, which became clearly visible only after two walls were completed. I wanted to tear them off and force the poles back with rope, but Nelly decided it wasn't too obvious and we didn't correct our error.

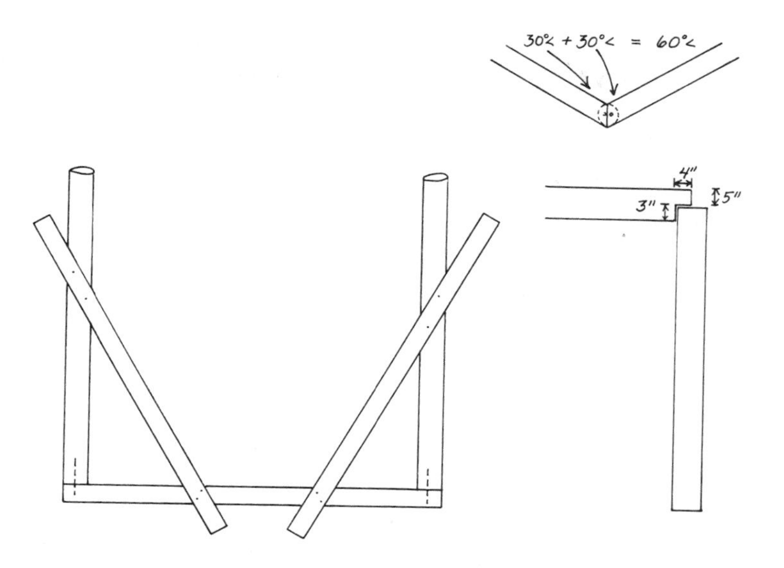

The Beam Poles

The purpose of the beam poles is to connect the corner poles together and to provide the surface to connect the roof rafter poles (beam poles are also called *purlins*). The six beam poles are 8′ long and 8″ diameter. Remember that poles shrink a bit when they dry, so if you want 8″ diameter poles and you cut live trees, they must be at least 9″—and that's *after* the bark is removed. Count on 1½″ to 2″ of bark, which could mean 11″-diameter trees when first cut.

Since the corner poles are 8″ diameter, and you must rest two beam poles on top of each corner pole, you have 4″ of space to lay each beam pole on top of the corner post. We chiseled the ends of the beam poles in 4″ and up 3″ so they would have a flat surface to rest on, and left 5″ above the corner pole.

From the top view, the beam poles also had to be at 60° angles to allow for the hexagon formation.

The beam poles were made permanent with 10″ reinforcement rods—twelve of them (one on each end of each pole). We drilled from the top with a ratchet brace and ½″ bit down through the beam pole at least far enough to make a small hole in the corner post for marking purposes. Then we removed the beam pole and drilled another 5″ down into the corner post. Then we completed by pounding the 10″ reinforcement rod down through the hole to finish the connection at each end. You may need to hammer hard or drill a bit farther than 10″ total, as some sawdust remains in the hole and fills some space. Believe me, it's hard work drilling downward while balancing on a ladder 8′ in the air! It's difficult to get good leverage. It took a total of seven hours' work to make these twelve connections, and you need more than one person working so that a beam pole won't fall while you drill.

Now we are ready for the roof poles, or rafters, but first comes the construction of the hexagon hat.

The Hat

This was by far the hardest part of the house to figure out and mostly it was done by trial and many errors! The thing that made the hat so appealing to me was the idea of being able to support the entire roof of a building or room without having

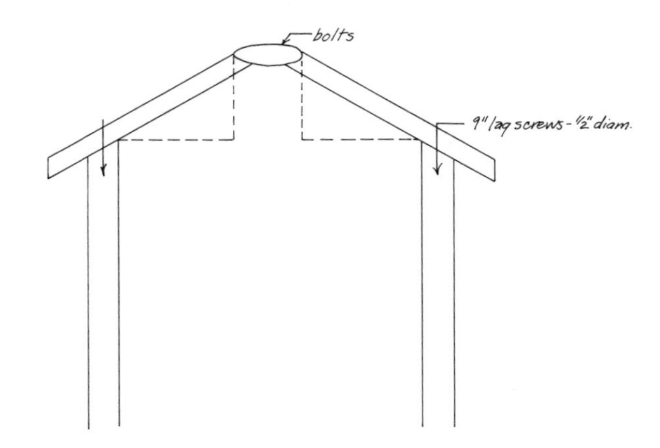

to have a center pole or inside wall to help hold up the roof. The house can be one big room. I first read about the hat on page 131 of the book *Shelter* (Shelter Publications, P.O. Box 279, Bolinas, Calif. 94924; $10), but it was described for an octagon and I couldn't understand it in terms of compound angles, so I decided just to try, adjusting things as I worked.

I used power saws to speed up the hat process (mainly because I didn't know what I was doing and I knew it would take a long time). It took me eight working hours to build it. Then I moved it to the hexagon. It weighed about 80 pounds and was definitely too awkward and heavy for one person to carry very far.

The purpose of this hat was to connect twelve 6″-diameter by 11′-long poles together at the top center of the house, with the other end of each pole bolted in place around the top of the outside wall, thus holding the roof up by centrifugal force. The center of the house can't sink when the roof poles are the longest side of the triangle and are strongly connected at both ends. Each triangle also has an opposing triangle, so it can't move downward.

I'll try to explain the hexagon hat process as best I can, but I'm sure you will have to experiment some, also.

Materials needed: 18 bolts—⅜″×6″ long
(to hold the hat together and to bolt the logs into it)
6 bolts—⅜″×3½″ long
6 bolts—⅜″×4½″ long
(to hold hat together)
washers on each end—60 washers, but buy a few dozen extra—I'll explain later!
1 sheet of ¾″ plywood, 4′×8′
5 2×4's, 6′ long

Using 2×4's on edge for the triangles meant I had 3½″ of space (width-wise) to slip the logs in and I wanted to slip them in 6″, so I began the process by building six triangles (or A shapes). The triangles had to be 60° triangles, so again six triangles would fit together to form a smaller hexagon—only this one slants up in the middle, forming the base for the slope of the roof.

The pieces for the six triangles were cut and each one nailed together to make six separate triangles that would lie on the ground beside each other to form the 36″-diameter hexagon—18″ radius. One way to avoid problems when you later bevel the triangles is to nail the triangle 2×4's together only on the top half of the triangle, as the lower half will be sliced or beveled off later and you could end up cutting

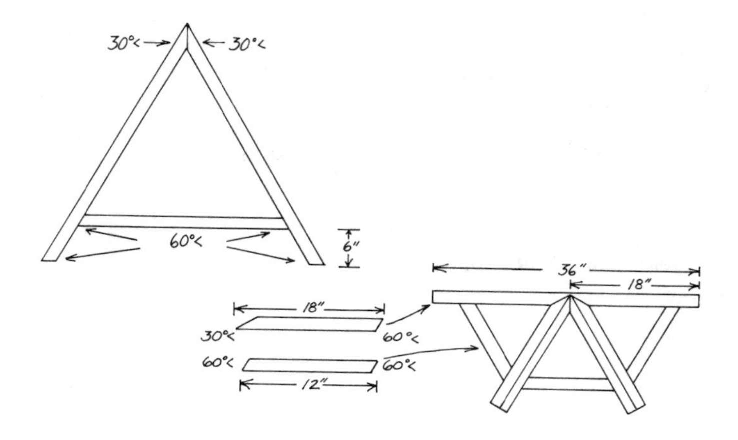

into nails. When you have finished beveling and it all fits together, you can add extra nails.

The triangles were constructed with the 2×4's on edge. The next part is where the compound angles come in, and I have yet to understand the proper formula. Here's how I experimented. The hat had to have the same rise or slope as the roof. The walls were 7′ high, the center was to be 8′9″ tall, and the center of the room was 8′ from the wall. I decided to work on the ground to begin with, as it was clearly easier! So I cut a round log 1′9″ (the difference between the wall and the center height of this building).

Now I could begin making the hat, or cap, for my hexagon roof. I put my 1′9″ log in the center of my working space. From the base of the log I measured out 8′ and marked the point. I temporarily placed a board from the 8′ mark to the top of the 1′9″ log in the center. This gave me the profile of my roof. Next I placed the apex of one of the six triangles I had made over the center of my 1′9″ log, with the base next to the temporary board. I marked where the base of the triangle met the board. From that mark I used a level to determine exactly where the bottom of my triangle hat would come on the center log. When my level was straight, I then marked where it hit my center pole. My mark was 6″ down from the top of the 1′9″ pole. I had the 6″ center pole that provided the profile for building my hat triangle.

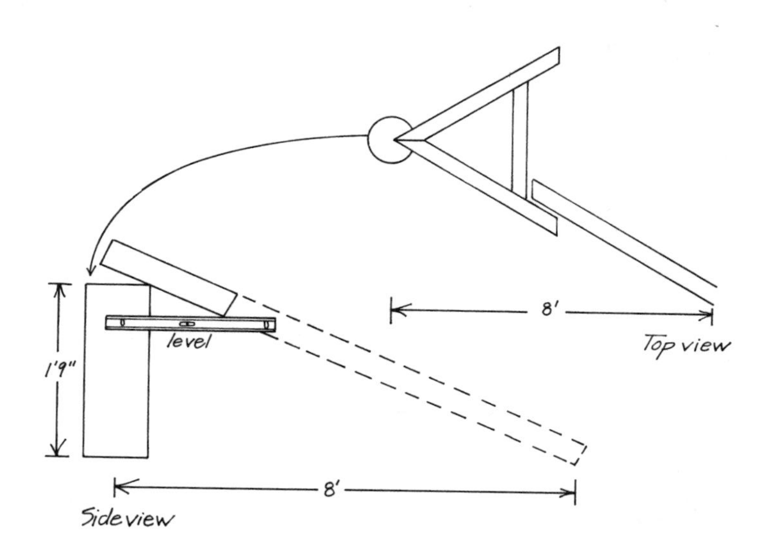

Now you can sit the points of all six of the triangles on top of the 6″ log and resting on the floor, and begin beveling the sides of each so they again fit together tightly like a hat. Bevel (or in other words remove the dark part on the diagram) between the triangles from the end all the way up to the point. I gradually removed this portion by sawing with my Skilsaw set a little steeper each time until they fit together flush enough to bolt them permanently in that position—the points all joining above the 6″ log and the ends all touching the floor. This center log is only temporary. It is used to create the right slope; when the hat is bolted together it will retain that slope with the center pole removed.

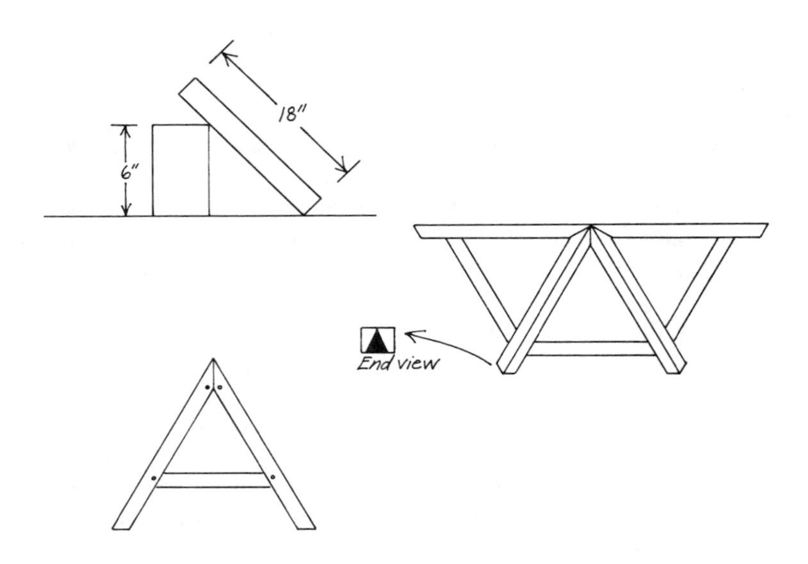

Now, bolting the hat together:

There are six ⅜″×3½″ bolts with a washer on each side—one on each of the six outside connections. There are also six ⅜″×4½″ bolts. These should be 1″ longer because drilling a ⅜″ hole inside these triangles is a bit tricky and has to be done at a slant, so I used longer bolts. In some cases I used extra washers when the slant wasn't as steep and the bolt threads ran out. Otherwise, you need to buy bolt lengths according to each hole.

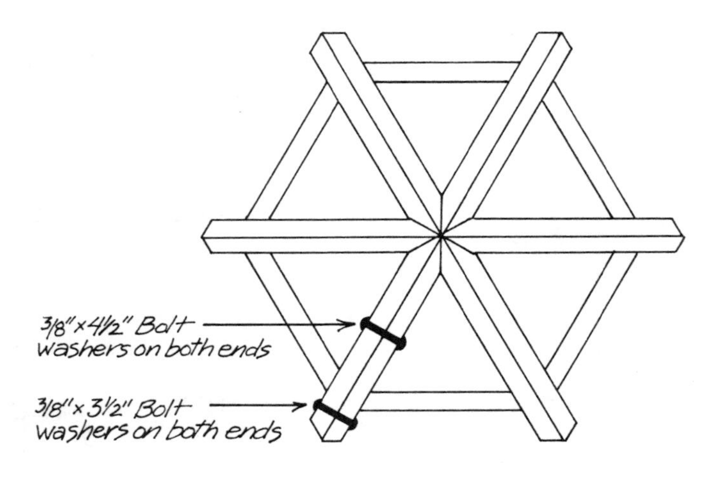

The final work on the hat was to nail ¾″-plywood triangles to the top and bottom of the hat, thus creating the space for the logs to bolt through and also to strengthen the hat in its permanent position. The six outside, or top, plywood triangles are 18″ on all sides. Begin with that. The bottom triangle must be cut smaller because beveling the lower side of the triangles created a smaller triangle in width but not in length. After I nailed the top plywood triangles on, I turned the hat upside down, sat it in a large bucket, and measured each of the bottom (not top) triangles. I knew they would be 18″ long—I just had to figure out the width. I tried to make them as accurately fitting as possible because they show inside the house as part of the finished product.

After the last six triangles are nailed in place, I drilled the final six holes to bolt the hat one last time—through the plywood on both sides. Now the hat was ready to bolt in the twelve poles that support the roof, as shown.

If your roof is to be steeper, the hat can be figured the same way, just using a higher temporary center pole. We made the roof pitch a rise of 3″ per 1′, as that gave us enough pitch for rain runoff, but used minimum building materials. The steeper your roof, the more materials required. Any less pitch would have created water-shedding problems.

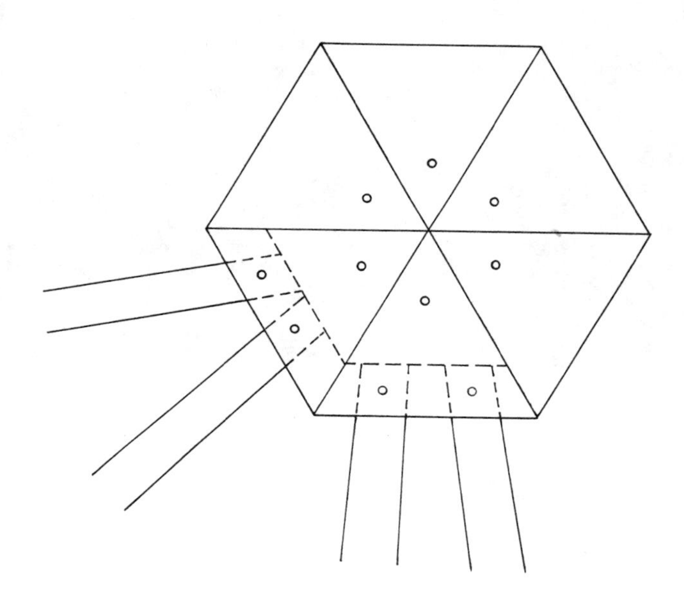

The Hat Raising and Rafter Poles

Since the hat weighed 80 pounds, since it needed to be strongly supported while the roof rafter poles were bolted, and since we needed a place to stand while we drilled holes from above, we cut a center pole 8′5″ tall to rest the hat on. The hat's height made up the difference of the center, making a total of 8′9″ high. Then Canyon took on the project of building scaffolding all around the center pole. She built it of recycled 2×4's, which later were used for the second part of the roof but during the hat raising provided a ladder, a strong support for the center pole, a way to rope the hat in a level position, and a surface to stand on while bolting the

(Photo by Dian)

poles into the hat. The scaffolding was a time and energy saver in connecting the total structure—roof supports to wall supports.

Many of us took part in the hat raising. The hat was lifted and set on top of the center pole and tied as level as we could determine. We began bolting the logs into the hat, alternating between one side and the opposite. Here's where we used the twelve 3⁄8″×6″ long bolts—again, washers on each side.

We chose to put a 1½′ eave all the way around the house, so the rafter poles were cut 1½′ longer than the outside walls to create space for the eave.

Each of the twelve rafter poles had to have one end sawed and chiseled so that 6″ would slide inside the hat structure: that meant 6 inches by 3½″ thick (the thickness of a 2×4 on edge). Since the rafter poles were 6″ diameter, we cut each to slide into the hat.

Each triangle of the hat houses two rafter poles. When drilling the holes into the hat, we tried to space them around the hat as symmetrically as possible, as they are a visible part of the finished product. Also, we tried to space the rafter poles evenly around the beam poles, one as close to each corner as possible. These rafter poles create the six triangles to support the roof above.

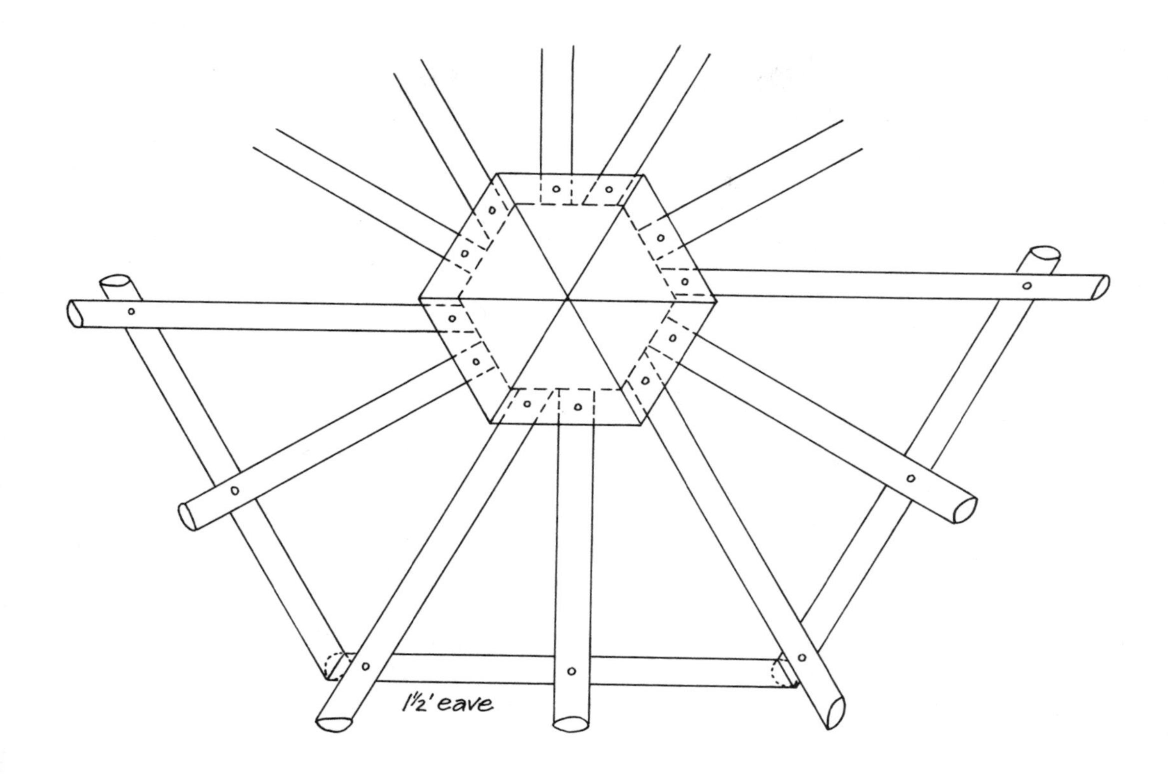

It took Gypsy and me eight hours working together to bolt the logs into the hat and securely bolt them to the beam poles. We countersunk each of the twelve ⅜″×9″ lag screws. Countersinking a bolt or screw means sinking it below the surface of the lumber, leaving the top surface smooth (as a nailing surface, in our case, for a lower false ceiling). When countersinking, you need a ratchet set with a short extension to tighten the bolts down. First we drilled a hole about 1″ deep and 1″ diameter

(Photo by Dian)

(the diameter has to be large enough that the ratchet socket can fit down inside). Then we drilled about 6″ farther with a ⅜″ bit. Again, be sure to use a washer on the top. Washers tend to disperse the weight, which helps prevent cracking at any individual hole.

Once all twenty-four bolts were tightened, we tore down the scaffolding and removed the center pole. I was nervous, so while I was sawing I was preparing to run. Amazingly enough, the theory was correct, and the hat settled downward only a tiny bit!

The Roof

The roof has four layers above the rafter poles.

False Ceiling The first layer creates the star formation on a false ceiling. If your hexagon is totally geometrically accurate, each ceiling board would have 30° angles at each end, but ours wasn't (remember poles aren't always shaped alike), so it was more accurate to mark each board individually. The rafter poles were our nailing surface, and we used recycled 1×8 weathered lumber and 8d box nails to secure them—three nails each end.

We chose to begin the star formation by measuring one third of the distance of the rafter poles and laying a board at those points just to mark the first board. We marked a line parallel to the pole and on center so each board could share a nailing surface.

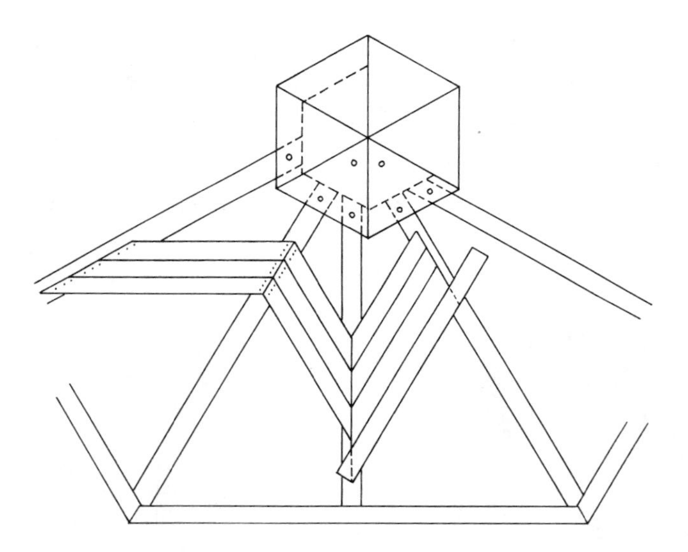

We created one six-sided star all the way around the roof, then followed that pattern until the roof was covered. We left a 10″ square hole to frame in the stovepipe (we used metalbestos, a very safe insulated pipe). The hole, of course, can vary depending on the stovepipe size for your wood-burning stove. Because of the high fire risk of wood-burning stoves, I always recommend installing stovepipe and roof jacks that are up to code. Check your local building code, or talk to someone knowledgeable in a hardware or building supply store.

Shannon, Nelly, Susan, Ginger, and I worked forty-nine hours total among us to complete the star. We worked out an assembly-line system: one person on the roof marking, one nailing, and several handing boards up and down and sawing all the angles.

Spacers and Insulation After the star was created, we began building the spacers to hold up the actual roof and to create another 3½" air space for the fiberglass insulation. We used recycled 2×4's. If you use recycled lumber, be sure to measure the thickness of each piece to be sure that they are the same. Older 2×4's were made 2"×4". Gradually they have become 1½"×3½", but the name hasn't changed. Just years ago a standard stud 2×4 was 8' long—now it is sold 7'9" long and still called "standard."

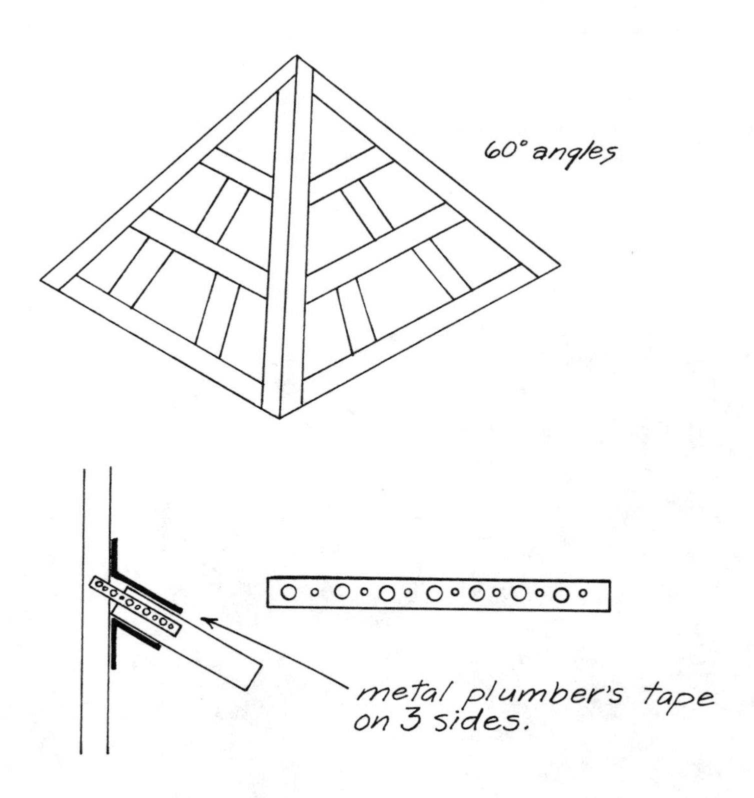

We built six more triangles. The triangles had to extend from the top point of the roof to the end of the eave. These triangles required extra boards (the location of these are more clearly described in the "Plywood Roofing" section) about every 16" to provide proper strength and nailing surfaces for the plywood roof above. Be sure the 2×4's are on edge. The angle cuts for the spacers need not be exact as long as they can be connected tightly. Some places we didn't cut the angles at all and connected the boards together by nailing metal plumber's tape to each side and top of the 2×4's. This was a time-saver and was sufficiently strong for the purpose. If you try this technique, be sure to nail the tape with at least two nails on each end of the tape.

We used tar paper between the spacers, and it served several purposes. The recycled 1×8 ceiling lumber had some cracks, a few warps, and several knots missing. The silver side of the roof insulation could be seen from the inside, and we thought that would distract from the feeling of the wood ceiling. So we stapled the dark tar paper over a lot of the seams, knots, and cracks. No silver shone through the cracks or knotholes. We also stapled it along the end parts above the eave to insect- and bird's-nest-proof the roof as an extra precaution.

Again insulation was cut and laid in between the spacers. This time we placed the silver side down, so that it faced the living space.

Plywood Roofing It took ten sheets of 4′×8′×½″ plywood to complete the roof. We laid each sheet on the roof and marked it with a chalk line. Be sure that you locate the extra 2×4's parallel and square within the triangles so that they provide two nailing surfaces for your plywood. One full sheet of plywood, cut at angles, fits on each triangle. The rest had to be measured and pieced together to fit the remaining nailing surfaces. It was easier to make the seams exact by laying each piece of plywood on the roof and marking it, then cutting it down on the floor. The plywood was the hardest material to saw by hand. You need several people to hold the wood so that one half doesn't bend down and bind on the saw. It took us twenty-two hours total to cut and nail on the hexagon plywood roof. It is worth taking extra time to make the seams as level with each other as possible. Later on, while you are putting on the roofing paper, it is easy to force a tear in the paper if one plywood piece is slightly higher or lower than the adjoining other.

(Photo by Dian)

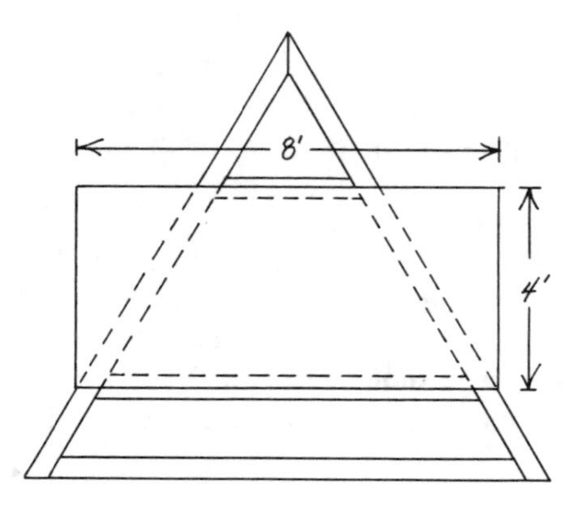

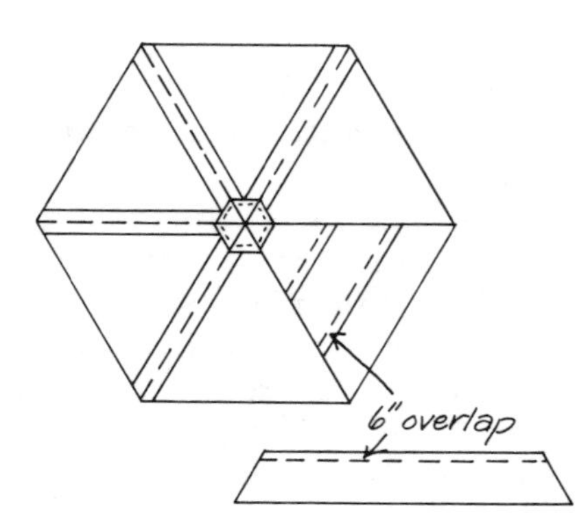

Roofing The materials needed were:

1 roll felt or tar paper
4 rolls (100 sq. ft./roll) asphalt roofing paper
2 gallons lap cement
¾″ roofing nails

We began this process by stapling strips of felt paper approximately 2′ wide along each seam of the triangle from the top point of the roof down to the end of the eave and also along all the seams of the plywood. We did this for added leakage prevention. It is often recommended or required by code that the entire roof be covered with felt or tar paper first, but we did only the seams. The hexagon hasn't leaked in two years now, so it appears to have been adequate. Then we roofed each of the six triangles of the roof separately with asphalt roofing paper, starting with the bottom and overlapping each piece approximately 6″. Be sure to start from the bottom so the water drains downward onto the next section. I've made the mistake of starting at the top and had to tear that piece off and begin again. We put about a 3″ strip of lap cement (roofing tar) under the bottom of the first piece of roofing

(Photo by Dian)

paper (which had been cut to measure from the longest point), then nailed along the bottom approximately every 6″. The tar seals the nail holes. Next we put tar along the top edge of the first piece; the bottom of the second piece is laid over this seam of tar and nailed down. We continued that way to the point of the triangle. Since each layer overlaps the seam of tar below, you don't see the tar at all when it is finished. After finishing all the triangles, we did the final roofing by cutting six strips of roofing paper approximately 18″ wide and slightly shorter than the length of the top point of the roof to the end of the eave. They were tarred and nailed along the seams of the six triangles. Then one last circle, approximately 18″ diameter, was tarred and nailed over the top point of the roof, overlapping these other 18″ wide strips. This created the necessary overlapping of all the various segments, forcing the water to flow downward from one segment over the next in a shingle effect. We used roof nosing (L-shaped metal flashing) to complete the border around the hexagon roof, holding the roofing paper and making an attractive "finish."

The Hexagon Walls

The walls were framed to house the various locations of windows and door, to create nailing surfaces for the siding, and to stop any possible racking. The modified post and beam hexagon construction itself provided enough strength to hold the weight of the roof. Thus it was only necessary to use this 2×4 framing as filler (or infill) in the walls where headers, nailing surfaces, and cross braces were needed.

We used double-sided aluminum foil insulation stapled on both sides of the studs to insulate the walls. Considering the size of the hexagon, we felt fiberglass insulation was necessary only under the floor and above the ceiling (this seems sufficient for Oregon winters). We used 1×6 recycled redwood shiplap on both outside and inside walls. The contrast of the light-colored poles and the weathered lumber on the walls and the ceiling creates an interesting contrast inside the house—part of the hexagon's uniqueness to me.

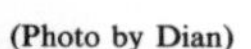

(Photo by Dian)

Many woman-hours (285½!) went into the building of the hexagon to bring it to the point of being a ready living space. That energy felt overwhelming to me—I still feel the strength of all the women who created it each time I am there. My memories of the process include throwing my hammer accidentally through the biggest window of the house, falling off my ladder, carpentry mistakes I learned from, and muscles growing sore, then strong. My favorite memories are of connections made with the other women I worked with, feeling the strength of our house building together. It's an experience I'm glad I was a part of.

Moving Beyond: Earning a Living

UNION WORK, UNION WAGES

When I joined the Carpenters Union in June 1974 I owed money, wore shoes with holes in them, and had just returned to find myself a stranger in a town I'd once known. I was excited and hopeful at the prospect of "going big-time," of working on long-term projects, with large crews, of no longer living hand-to-mouth from small jobs—hard-won, soon finished, and always on the outside of the industry that builds most of America. I hoped for some security: a higher wage, medical and unemployment benefits, and steadier work than I'd had doing small jobs on my own. Most of all, I wanted training. I looked forward to learning from craftsmen who'd had a sense of the trade since they were knee-high, who had carpentry in their bones.

When I left the union a year and a half later, I had paid my debts, gotten a first-hand look at what we used to call "mainstream America," and developed a healthy anger at a system that discourages learning and encourages, instead, divisiveness, exploitation, and tedium at all levels. And I was angry at many of the men who participated in that system.

Strong and active physically, I felt dead in spirit. I had met and worked with men who had carpentry in their bones; I'd paid off my debts; I'd been able to afford a dentist for the first time in years. I knew what it was like to work on a fifteen-story apartment building in downtown Oakland with hundreds of other workers—carpenters, electricians, laborers, glaziers, Sheetrockers, tapers, and ironworkers. But the cost had been high. I was tired of being treated like a "freak," tired of feeling bored a lot of the time and alien always. Though more skilled at the mechanics of carpentry—hammering nails and using a circular saw—I knew little more about how to plan and execute a project than I did the day I joined the union. Aside from the satisfaction that came from having tackled something hard and succeeding at it, and from having tasted a foreign culture, the rewards had been mostly financial. I returned to working on my own and with other women.

I was lucky to have alternatives to union work. I had learned beginning skills from small-time, individual carpenters in a rural county in California, and I was living in a part of the country—the San Francisco Bay Area—where there are dozens of women working in the trades, by far the majority of them outside the union. These women have become my partners and my friends. I had no dependents: I could exist through lean times without worrying about other mouths to feed. And I was white, born of middle-class parents, educated, taught from childhood that I could do in life whatever I chose.

Now, two and a half years after my leaving the union, a group of us have formed a women's construction company, Seven Sisters Construction, Inc., in Berkeley. Three of us are in the process of applying for our contractors' licenses. As I work and learn more carpentry, I make a better living each year—though still below what I earned in the union, even at apprentice scale. I am as challenged in my work as I choose to be at any given moment, and use my head regularly. A day does not pass in which I do not learn something new about the trade. I work in a climate that aims to be one of friendship and support; on stormy days, we try to work things out.

Journal Entry, July 1974

We have joined the carpenters union. Three of us—Priscilla, Connie, and I—three of the members of our women's crew. There are thirty-six women apprentices in the Bay Area, we learned today, in all unions—plumbers, Sheetrockers, electricians. . . .

We are proud!

And yet—a caution lies upon our shoulders. We women must band together, and use our tiny membership to good ends.

The tale of entering? Roundabout.

There was no problem with union officialdom. In San Francisco, in Richmond, in Hayward, male officials and female secretaries in one-story, new-looking, unimaginative brick buildings gave us the hours we must come, the names to call. . . . Four of us sat in a meeting room full of young men and listened to an old-timer rattle off our benefits: seventy-five cents an hour vacation pay, eighty cents pension, sixty cents medical. . . . We were given "letters of subscription," our "hunting licenses"—the union's agreement to accept us if we found jobs for ourselves between June and mid-July, the "hunting season" for union hopefuls. The assemblage was pathetic—no communication among us, kind sticking to kind: black to black, women to women, and white males isolated each in his own shroud.

Given our letters, we looked for jobs. The union had treated us just like anyone else; now we had to find contractors who would agree to hire us for at least three months. We sought out the office trailers on every building site we saw, from Hayward to Marin County. We saw more naked calendar blondes with silicone breasts than *Playboy* offers in a month.

"You wouldn't want to work here. The foreman's language is worse than a longshoreman's."

"Why aren't you married?"

"He'll probably hire you, he likes girls."

We showed pictures of the two-story, octagonal goat barn we and other women had built the year before in Mendocino County.

"Sure your older brothers didn't help you?"

And a response, the tenth "no" in a day, one afternoon in Hayward, that reduced me to howling and threats later riding home:

"Honey, I'm sure you can do the work. But this is a three-story apartment house we're building here. I can't have my men falling off the second- or third-story walls because they're all thinking to themselves, 'I'd like to screw that broad!' " (That was the argument women got when they first tried to get jobs in offices as secretaries, my mother told me, when I described the scene to her later.)

My father writes in the midst of it all: "I admire greatly your courage here. You must not let it make you bitter, however. . . ."

At last, jobs. Connie and Priscilla were hired one afternoon in Marin County, by a husky, broad-shouldered young contractor . . . "Let me see your hands."

"Don't let them touch your hands," Connie said afterward. Instead, Connie showed her muscle. "We're strong. We have our own tools. We built this barn." That evening, the call: Connie and Priscilla were to come to work next day. As for me, I was to continue looking for myself, relieved no longer to be hunting with carpenters more experienced than I, more confident now, glad to make statements for my own self, not to hang back, extricating myself from claims which did not apply to me: I had few tools, no calluses.

At the bottom of a dead-end street one day I saw a young man, almost a boy, standing alone by an unfinished house. Tired after the eighth interview of the day (the seventy-fifth since I had begun looking for union work), I was turning my truck around to go home, surprised to see him, certain the job was too small to need another worker, discouraged, afraid. I told myself: soon as you start failing to ask, that's where the possibilities end; stopped up the street, went back and asked. And that's the job I got.

In the four years since I joined the union, some changes have come about that make it easier for a woman to get a job. Unemployment is less than it was when I applied; and now a person wanting to join the union may sign up at the union hall and obtain a hunting license anytime, instead of having to wait for the short "open season" to come around once a year. When I joined, an employer was required to hire a new apprentice for at least three months; now a contractor need not promise a job of any certain length. The union has removed its upper age limit (thirty-one) for apprentices, thereby making union jobs available to women who work at other jobs or raise children and then turn to carpentry.

The most publicized change came in January 1976, when the California Apprenticeship Council, under pressure from women's groups, held a public hearing to consider an amendment to the California Plan that would be the first step in establishing an affirmative action program for women seeking apprenticeships in the trades. Two months before the hearing was to be held in Los Angeles, a group of tradeswomen, both union and nonunion, organized to support the amendment. Calling ourselves Women United for Apprenticeship, we worked with other women's groups, writing press releases, putting up posters, contacting union groups and Apprenticeship Council members, and urging women throughout California to appear at the hearing and testify in favor of the amendment. Twelve of us formed a Bay Area Women's Caucus of the United Brotherhood of Carpenters and Joiners of America.

On Friday, January 30, our group joined hundreds of labor representatives and women's supporters in a day-long verbal battle, listening to men who protested there were no women to fill the jobs, said that they were "doing all they could" to get women into the unions, or asserted that, unlike Russian women, perhaps, American women just plain lack the physical strength to do the work required in the skilled craft jobs. Woman after woman stood up and refuted them. We reminded the council that women had slipped easily into men's shoes when workers were needed during World War II, only to be retired when the servicemen returned home needing jobs again. We listed incidences of obvious job discrimination, demonstrated clearly women's determination to work, and cited our own experience doing the very kinds of work our opponents were saying we were unfit to do. By the end of the day, council members who had started out in the morning brusque and sharp-edged in their manner toward the women speakers had warmed considerably to our point of view, and voted unanimously to pass the amendment, making California the first state to adopt such a plan. (The state of Washington set up an affirmative action program for women in the trades later in 1976.)

At the national level, similar challenges are being made. Currently, women's groups are suing and otherwise pressuring the Department of Labor to set goals for numbers of women to be employed in construction (3 to 7 percent over a three-year period), and for numbers of women to be taken in as apprentices in the trades (20 percent in the first year). A recent newspaper article stated that the Labor Department has "ordered builders who receive federal money to hire thousands of new female construction workers by next year." There is an accelerating quota over the next three years that these companies must fill with "female hard hats"—or risk losing their contracts. These quotas are tiny (3.1 percent, accelerating to 6.9 percent), but they represent a beginning.

Despite these efforts, there are still deplorably few women working in the trade unions, and the number of minority women is minuscule. Nationally, the 750,000-member carpenters union boasts a few handfuls of journeywomen at best; officials say records are not kept. Of the 7,356 apprentice carpenters in California, 98 (or 1.3 percent) are said to be women—and perhaps fewer, since some of us who are no longer working in the union are still on the lists. (Figures were supplied by the Division of Apprenticeship Standards, in San Francisco, for November 30, 1977.) Of the 34,089 apprentices in all 83 apprenticeable trades in California, 743 (or 2.1 percent) are women. Of the women apprentices, 22 percent are members of minority groups. At the journeyman level (the council balked at the term "journeywoman") only a handful of California's 90,000 dues-paying carpenters are women—"a few, scattered about," as one union official put it recently when I called to check.

One reason the amendment to the California Plan hasn't had more impact is that the goals were established, not on the basis of population parity (as they were in the case of minority groups), but on the basis of "availability and interest." But it is difficult for many women to be interested when they learn what life in the union is like for a woman. And groups that have as their purpose the outreach and qualification of women for apprenticeships and blue-collar work quite frankly don't urge untrained women to seek carpentry apprenticeships, knowing the resistance they will encounter. One such outreach group, the Women in Apprenticeship Program, Inc., located in San Francisco and funded by the Department of Labor, processes each year applications from about one thousand women, most wanting to work as carpenters. But many of these women, upon learning that becoming an apprentice carpenter means "continuous job-seeking, twelve months out of the year, selling her skills, knowing where to find employers and doing a p.r. job on them, will tend to pick another trade," according to Susie Suafai, executive director of the Women in Apprenticeship Program. "And we don't lead them down the primrose path."

In most of the state's apprenticeship programs, the union takes responsibility for keeping the apprentice working, as long as there is work available. And part of the program, in the electrical or plumbing trades, for instance, consists in making sure that the apprentice is placed in varied job situations, so that he or she will learn different facets of the trade. But the program for apprentice carpenters leaves it totally up to the apprentice to hustle jobs for the four-year apprentice period. This makes it much tougher for a woman, who is even less desirable to an employer than the minority male. Faced with the employer's skepticism about a woman's ability to do the work, along with his hostility at seeing a woman daring to seek work on his job site at all, the woman is hard put to convince him that she is a better person to hire than the numerous male apprentices competing with her for the same job. I attribute my getting union work completely to the fact that I was trained before I tried to enter as an apprentice. Hard as it was to convince contractors that this indeed was the case, once they suspected it might be true they hired me because they knew they were getting a good deal—essentially, a third-year apprentice at a beginner's wage. When the first job ends, the hustle starts over again, with each succeeding employer no more inclined to believe a woman can work and not "make trouble" on the crew, despite her growing record of union work. The night classes offered once or twice a week are not enough to enable an untrained apprentice to work on a job; the apprentice must find employers who are willing to train her on the job.

The situation is made more difficult by the fact that carpentry is increasingly specialized: to learn foundation work, framing and finish work, concrete work, and remodeling, one must usually switch employers, since many contractors do only one aspect of the work. Each time a woman seeks a job in a new line of carpentry work, the barriers are the same. The contractor shakes his head: "Well, you may have done finish work, but framing is something else again"; "Well, you may have framed houses, but concrete work is *hard.*"

It is not surprising that more women drop out of the carpentry apprenticeship program than any other. Most of the women who stay in the program work on federally funded jobs, where the contractor will lose his money if he does not comply with federal laws prohibiting discrimination on the basis of sex. The worst that can happen under the California Plan is that the state can mandate that the program accept a certain number of women, or threaten to strike the program from the state-approved training system. There are no teeth in California's affirmative action plan to put pressure on the individual contractor to hire women, or on the carpenters union to take more responsibility for seeing that women are hired. The less attractive the industry can make itself to women, the fewer women will apply, and the fewer will be hired. The system protects itself.

To the pounding sound of the pile
Driver's machine
We (Bang!)
Would like to apply
For a job (Bang!) as apprentice
Carpenters.
Are you (Bang!) hiring?
Crowds form around us
Jokes, jabs, disbelief
Questions . . .
"So you want to become
Carpenters?"
"No, we already are carpenters.
We are just looking for work."
Quickly I pull out the photos of our octagonal barn.
"We built this."
The atmosphere changes,
Respect passes across the eyes.
They look again.
Then come the questions—
Technical, testing,
They need proof.
"What kind of ties you use in those forms? Look like
flat ties to me." —"They're anaheims."
"You cut all those rafters yourselves?"
"Yeah, mostly by hand, each rafter was a compound angle."
"You sure your boyfriends didn't help?"
"It was built completely by women."
"Well," he says, "you have to belong to the union."
"Yes, we know," says one of us. "We have our letter
of subscription. We need a signature. We want to get in.
We need you to hire us."

"Well, I can't hire you till I have a job. You should
come back in thirty days, two weeks, next month. . . .
"Anyhow, the language is bad around here. Will you
flirt with the men?"
"We already have our crew."
"Why don't you put your name down here on this list."
"It's fifty people long, but I'll call you when you get to
the top."
Thanks, mister, we'll do that.

Days and days of men.
Laughing men
Suspicious men
Men with surprised faces
And lots of jokes.

We three women move
With substance across
Dirt-filled lots in
Brand-new territory.
Sure of ourselves, with
Years of experience, to make
Us tough—but tired
Of this long and funny joke.

—PRISCILLA DANZIG

The first job I had in the union was the best. There were just three or four of us on the crew, and sometimes just myself and the boss—a friendly, hardworking, twenty-one-year-old. On a small crew it's possible to learn a great deal; even an apprentice is in on most aspects of the work. And the fewer the people on the job, the sooner one is known for oneself and the less flack a woman takes. I worked for a man who, with his mother, shared the contractor's license his father had left when he died (a common way women become contractors), and we were building custom houses in Hayward, houses built with more thought and care than the tract houses that were being tossed up in the surrounding housing developments. But I had joined the union at a time when the construction industry, like the rest of the nation, was having a hard time. The union had just been on strike; unemployment was high, and rising. After four months, my boss turned to remodeling—a plentiful source of work even when new building comes to a standstill; the jobs were smaller, and he had no need of me. About a year later, he called me and asked if I wanted to come back and work for him, building an apartment building; but then I had another job, and by the time that was over, I was ready to leave the union for good. If I had stayed with him, I might have stayed in the union longer, for I was learning a great deal and being treated well. But by that time I had experienced too much of what union life was like; I had no heart to do it any longer.

A large part of my dislike of union work centered on the constant hassling I was subject to as the only woman on a job. A couple of times I was able to find work with Priscilla, but two of us still made an easy target for whatever garbage anyone wished to unload. By then, Connie had left union work to return to free-lancing, and the other seven union women we knew were looking for work in other parts of the Bay Area. At first the constant haranguing infuriated me. After a while it bored me. Toward the end, I was learning to return a good retort with humor, so that the offender got the point, but not with malice, and I got some fun from the exchange. Thinking up the rejoinders wasn't too difficult; comparing notes, we figured there were only about twelve variations on a theme, and I had all day to mull over my replies. Some of my worst tormentors eventually became my best friends, in a limited, union-style sense of the word.

But the fact remained: the union was a ripe breeding ground for divisiveness, whether it was between the sexes, between racial and ethnic groups, between the different trades, or between the different ranks within the trades—supervisor, foreman, journeyman, apprentice. One young man I had the misfortune to work with had a name for everyone: the boss was "the marine," and there were "Chink," "Blackie," "Okie," and so on. Only his pal, like himself a relative of one of the company bosses and a Wasp, was "brother." When workers are divided, they won't be able to make trouble for management and to make life better for themselves; no one cared that these attitudes were prevalent; no one ever tried to deal with the rampant hostilities or the pettiness.

In our case, it became clear that there was a special edge to the hostility. Not that women were hated more than blacks, say; we were just hated differently. It took me a while to realize that was what the reaction was: hatred. To the white males, at least, we represented a real threat—psychological and economic. They had been living with the belief that only men are capable of lifting, carrying, nailing, and figuring things out in the realm of building. Men are mighty and know about things of the world; a sense of this prowess seemed to be largely what separated them in their minds from women.

Before we found work, men were always asking us: Can you carry heavy beams? Can you put Sheetrock up on a ceiling? They forgot that a man does not carry a 14′, 4×12 beam by himself, or singlehandedly nail a piece of Sheetrock onto a ceiling. He gets help from someone else. When the work is really heavy, he gets a machine to do it for him. Once we were in the union, it became clear that everything a carpenter is required to do, a woman can do. Some things took us longer; some things we figured out alternative ways to do, because of lesser height or strength. But we needed no allowances that the union did not make all the time for a normal crew of carpenters, which includes older men, men with hurt backs, and short men. Some things we did better: I scrambled up a scaffold a lot quicker than a bulky partner; I was smarter than some; and, often, I cared more and tried harder. Things evened out between man and woman, just as they always had between man and man.

So here we were, doing "a man's work," and a good job of it. What to make of it? To many men, it meant either: men are not so great as they think they are, because everyone knows women are weak and incompetent; or: women are not weak and incompetent. Either was a threatening possibility to someone who derives much of his self-esteem from being the tough, strong-muscled "hard hat," who speaks of white-collar males as "pansies" and "namby-pambies." We were invading the fortress of his self-esteem.

In even the friendly overtures, there was an underlying sexual come-on, which I wearied of always having to turn aside. Despite the fact we were working alongside the men, doing equal work, they tried to relate to us as sexual objects. Getting no response, they looked upon us as freaks; we were not women as they were accustomed to knowing them. The idea that a woman might not want to be dependent on a man to support her, and that she might choose hard physical work as her means to independence, was foreign to them. We were creatures from another culture, small weirdos in tool belts and hard hats. "Aren't you married? *Why* aren't you married?" The question came again and again. I took to saying that I had twelve children, twelve hungry mouths at home to feed.

As if there were not enough cultural reasons for the men's hostility toward us, economics supplied some more. In fact, we were taking jobs that would have gone to men. We were watched by everyone who could possibly take the time to watch us, to see if we could do the work. Were we making the same pay for doing less? A valid question, and so was the reaction—a real anxiety in men's minds about the possibility of being flooded by a whole new work force they had never dreamed existed. Our presence meant more competition for jobs already scarce, in an economic system which thrives on competition between workers, which lowers the wage when there is a more abundant labor supply, and in which one's gain means another's loss.

Given the cultural and economic pressures, it was remarkable that women were treated as well as we often were. Often our best supporters were, ironically, our closest competition, those who, like ourselves, were "last hired, first fired"—the members of minority groups, who knew what we were up against because they were fighting it themselves. Many times a black or Chicano worker would take time to welcome me to a new job, or otherwise make it clear that he felt an allegiance toward me in my struggle to enter the white man's world. And many of the white males were helpful and supportive, despite the hostile undercurrent, particularly after I had worked on a job for a while. Respect would grow on my side, too, as I gained some understanding of the similarities in our positions, saw us all as small people in a system that was making maximum use of us to grind out the product, and sending the profits to someone else.

Part of what that system promotes is monotony and lack of challenge in the work situation; it seemed things were planned so that everyone worked at the lowest possible level of mental ability. The contractor made money that way. One person was paid to think on a job—the foreman; the rest of us were paid to *do*. Individual initiative was discouraged. Put a man to work and keep him at the same job until he performs it mechanically: that's profitable. For some workers, dope or alcohol, indulged in on the sly, helped counter the boredom; for others, gossip; for many, laughter and a wonderful way of making the best of things.

On one job the hoist operator (whom I had noticed made howling noises as he stood in his small cage and rode up and down the increasing height of the apartment building) became a friend when he found me one day making loud quacking noises as I was in my fourth month of putting in closet shelves, using a staple gun and wearing ear muffs to dull the sound. Working alone, as I did most of the eight months on that job, with no one about, and wearing the muffs against the rebound in the concrete cubicles, I was living out a philosophy something on the lines of: "I quack, therefore I am." In the hoist operator I found a kindred soul; after that, he and I ate lunch together sometimes in my truck in the company parking lot.

What impressed me was the large percentage of men who worked steadily and well, who did the best possible job and took such pride in it, no matter what the work they did. And I feel a lasting warmth for the many among them who found a friendly or a humorous word for me.

Journal Entry, July 1974

Now, on the job, the daily six o'clock waking, sleepy ride on a curving road to Hayward, fog lifting to sunshine by early morning, hot days, of hard trying. I have many now of the old bad feelings. I miss Tom's sensitivity and Bill's bawdy songs, graffiti on Mrs. B's walls, and, always, the stories we told. I am with little boys, in a world not mine: I am alien. Back again the old fears—that I am a poor worker, that I will be fired, and lately, that I will fall right off the roof down the forty feet to the ground where cement, nail-pricked boards, a gully, wait to catch me. Shame, that I cannot step lightly about on the tops of the unfinished walls, as the others do, on the 2×4 plates, on the edges of the joists. And the building grows ever higher. . . .

Yet the others sympathize while they tease, encourage. A boy, and his brother, and a friend, a mellow crew, to be sure. Barry tells of his first days up top, hugging the joists, and Jon says he crawled, as I did, on hands and knees. No word is ever said about my slowness, and today Jon told of his own first day on a roof, lugging 2×8's, taking six hours to lay a board, putting plywood in wrong. I miss the closeness of working with Tom and Bill, special friends, in Mendocino, or with Priscilla, Jean, Connie, drinking beer and dancing to the jukebox after we'd put up the shelves in a local women's bar, or the whole crew of us up on the roof playing about with sweet-smelling cedar shakes among the pointed pines in Albion. . . . Spoiled, for sure, I am. The job is fine; if he keeps me, I learn foundations through framing. It is not the fragmented, prefab, assembly-line production that Priscilla was doing, nailing 2×12's hour after hour, day after day, till her hands were raw and blisters running; not screwing on hardware in unit after unit of a condominium, as Connie did, for a boss who yelled at her in front of twenty watching men, crudely, unnecessarily. . . .

There are only a few people around, on my job, just we who work, a neighbor, some kids, and a chance woman worker who looks my way, stops on her rounds, asks, "How did you get this job? I just deliver eggs, but this winter, when work's harder to get, I'm going to work for United Parcel; joining the Teamsters." And another: "Been a mailwoman for five years. . . . Good to see you here. We're getting on."

Today, after driving through rain on a queasy stomach, reminiscent of morning feelings before on all those lonely jobs I went to for so many years, what seems so long ago, feeling bad, the Chinese neighbor caught us three as we stood waiting for the drizzle to turn torrent so we could leave for home, a holiday.

"You!" he said. "Lady carpenter! You skilled, more skilled than the men, you shine among them. The whole neighborhood watches you. Skilled! You are the queen, lording among them."

Tonight, after our rain-made holiday, P's and mine, spent in sleep and in reading aloud, I write about these things because Iva said, "It is more important, the real than the fiction," and because all of the women understood when I read and spoke the difference between what I do and what I want to do, and still taught me to value who I am, just now.

It is clear that a great deal of what bothered me about union work is not peculiar to the carpenters union, or even consciously promoted by union members. Women who have worked as carpenters for nonunion contractors, or for the railroad, have all the same complaints. Some of the greatest problems—the dehumanization of workers, the divisiveness and competition, the fragmentation and specialization of the work, the tedium and exploitation, sexism and racism—are problems I see as inherent in an economy in which profit-making is the prime mover. To think that we could put ourselves outside the faults of our society simply by working with women, outside the union, or in smaller companies, would be to fool ourselves badly. Nor is the problem simply one of working with men. Men gave me my start in carpentry; with interest and gentleness, strangers took me on in Mendocino, gave me the beginning chances I needed, and trained me well. There are real and important reasons I choose to work with women, but that is another story. The treatment I have complained about at the hands of union carpenters was never a part of my work with these other men, who became friends of mine.

Some of the problems with union work probably come about because of size. Men together in large numbers very likely acted more obnoxious than they would have had I had a chance to know them, and they me, on an individual basis. More important, large companies are more subject to specialization, hierarchy, and assembly-line methods—with all their unfortunate ramifications—than smaller companies, where work divisions may be less rigid, responsibility more shared. What did bother me specifically about the union was the power of the union bureaucracy. It made me extremely uncomfortable to be a dues-paying member of an organization that has and uses so much power to keep things in this country the way they are: to elect corrupt political figures, to shut out members of minority groups, to plan and build ugly structures that benefit the middle classes and ignore the poor. It felt wrong to be so powerless and yet to contribute my energies and money to a bureaucracy that had so great a hand in perpetuating so much I despise.

At its best, the union uses its power to ensure good wages and safe working conditions for thousands of workers. The wage is one of the highest in the trades: as of March 1978 a journeyman earns $11.70 an hour, or $15.20 with benefits—health and welfare, medical, pension, and vacation pay. A starting apprentice earned half of that, with the wage increasing as he or she acquires hours of work and apprenticeship schooling. After more than five years of earning my living as a carpenter, I still average only about $6.00 an hour, with no benefits. When a union carpenter is laid off a job, he or she can go on unemployment. It requires a powerful organization to stand up to the efforts of those with money to get the most work out of people for the least return, heedless of the costs in human welfare. Without the unions, we could expect to be working longer hours, for far less money, and performing tasks even more detrimental to our bodies and souls than lifting heavy wood and feeling bored. The horror stories of what life was like in this country before the unions were well documented; and there are plenty of examples still of appalling conditions in areas of work which are not unionized, or which lack unions strong enough to obtain what their members deserve and need.

"Our aim will be to promote and protect the interest of our membership, to elevate the moral, intellectual and social conditions of all working men and women to assist each other in sickness and distress," reads the preamble to the Bay Counties by-laws of the United Brotherhood of Carpenters and Joiners of America; ". . . to encourage apprenticeship and a higher standard of skill, to cultivate a feeling of friendship, and to assist each other to secure employment . . . to aid and assist all organizations to uphold the dignity of labor, resist oppression by honorable means. . . ." The goals are much the way I would write them, but in practice, the protection is for a few only, and limited at best.

In our own company, those of us with more experience are already training newcomers to the trade, we hope in a spirit very different from that which we encountered when we entered the apprenticeship program as union members. We make decisions about wages, work partners, and jobs collectively and share responsibilities and skills among ourselves. We struggle to find a balance between the methods of efficiency we learned in union work and our own human needs. With profits earned on some projects, we hope to be able to take on jobs that will benefit people other than ourselves—to train minority women, and perhaps build structures that could be useful to segments of the community. Some of us dream, however distantly, of one day forming a union of our own—a union of tradeswomen.

Meanwhile, I go on putting up bathroom tiles, laying drainpipe, installing french doors, building decks, and taping Sheetrock. I practice leading jobs and training other women with a minimum amount of hair-pulling, and study the ever-elusive art of estimating jobs. For each job I load enormous amounts of tools into my Volkswagen bug. My next step toward the big time: to afford a pickup truck.

FURNITURE MAKER/DESIGNER

I design and build custom furniture. My creations are usually one of a kind, lovingly crafted of hardwood and characterized by secret compartments, wooden nuts, bolts, hinges, and latches. I use high-quality construction with well-made joints that are often artful as well as functional. Some of the things I make are desks, tables, chairs, bookcases, cabinets, boxes, doors, beds, chopsticks, and store fixtures. I also do some antique reproductions. I am self-employed and have been supporting myself as a furniture maker for three years.

People often ask how I became involved in woodworking. There were a lot of reasons, but I guess the main one was that I have always liked making things. I love the satisfaction of creating things that are completely tailored to my needs and are an expression of my self. I have never been satisfied with the mass-produced products available in stores. Even as a child, I didn't like "store-bought" piggy banks and made a small wooden box for my treasures instead.

Another contributing factor is that I've always been involved in art and especially sculpture. When in college, I made several large pieces including a 20′-tall figure out of cardboard boxes. When that piece all but disintegrated after only one week of exposure to the weather, I decided to learn to work with a more durable material. By then I had had a small amount of experience with metal, plastic, and clay, but wood, with all its beauty, warmth, and workability, was the obvious choice. I never really considered anything else.

In spite of my love for the material, however, I had a minimal amount of experience with wood and woodworking tools. I knew how to use a saber saw, disk sander, drill, hammer, and handsaw. I had a lot to learn and soon discovered that it was hard to find someone to teach me everything I wanted to know. I knew that I could learn from books, but I needed to get the basics from a person first. I tried adult classes and friends, but no one was really willing to teach an "inexperienced female" (i.e., girl). I talked to several established woodworkers about the possibilities of an apprenticeship. One was nice and supportive but not doing exactly what I wanted to learn. The others were rude, told me to give up and get married, or simply refused to consider me. I didn't get discouraged, though; just more determined. Besides, after visiting several shops I realized that an apprentice at my level of experience would mostly do a lot of sanding. What I really wanted was a comprehensive teaching program that would give me a solid background in design and construction.

Finally, after several frustrating experiences, I heard about the carpentry and furniture-making classes available at the Peralta Community Colleges in the San Francisco Bay area. I moved there specifically to enroll in the two-year program for furniture and cabinetmaking at Laney College in Oakland. By this time I had narrowed my interest to furniture and sculpture and decided to put off building houses for a while. The program at Laney includes instruction in design and drafting and lectures on construction principles, as well as actual shop work. It is an organized curriculum with assigned projects, readings, exams, a well-equipped shop, and conscientious instructors. It was everything I wanted and really scary!

I was relaxed in classrooms at that point in my life, but this was no ordinary classroom. I was uncomfortable around the machinery and often resisted the urge to hide behind a desk. I felt intimidated by the teacher and by some of the men in the class and was so tense that I was exhausted at the end of every day. Luckily, there were several advanced women students who were supportive—and my own motivation was high. It was worth all of the fear, and after the first year I knew enough to begin to design and build my first piece of furniture. Unfortunately, I was still uncomfortable with the machinery.

My solution to that problem was to buy a table saw. I set it up in my garage and spent hours alone with it. I let myself go as slowly as I needed. That made all

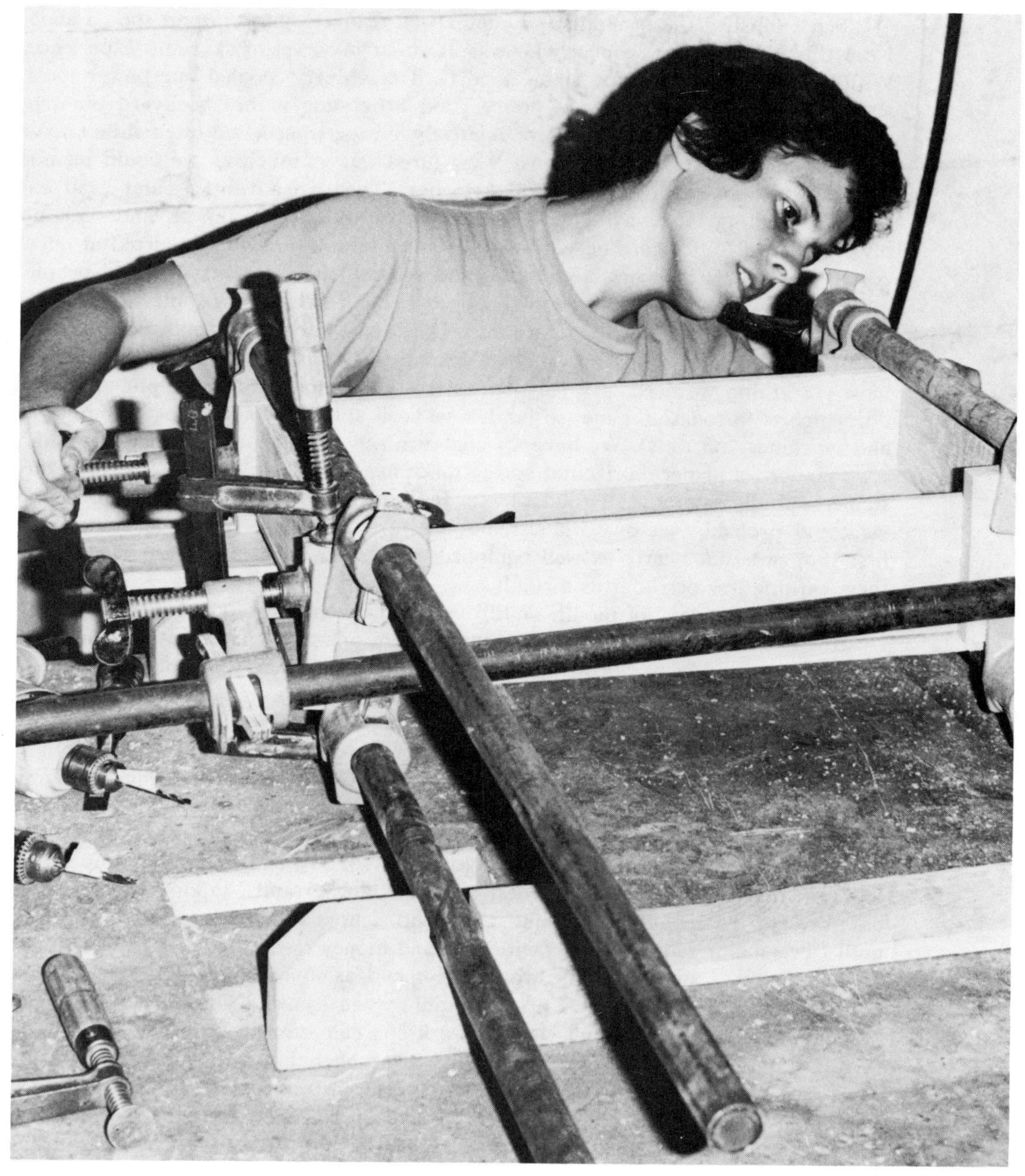

(Photo by Carol Newhouse)

of the difference for me; I developed confidence and a healthy respect for that saw that soon carried to all woodworking machinery. (If you decide to try running a table saw in your garage, first check into how much power you have. I had to turn off everything in my apartment including the refrigerator, in order to handle the initial surge of the motor.)

If the watchword for my first year of the program was "fear," for the second year it would have to be "confidence." I had proven to myself, my teacher, and my classmates that I was capable; I could finally relax. I experimented with different woods and joints, learned all kinds of tricks, developed some skill with hand tools, and began learning from books.

When I finished the program at Laney I was ready to jump in and start a shop. I did this with another woman who was at about my level of skill and experience. With a little effort we found some good used machinery, pooled our power tools, and managed to put together a pretty good little shop in her backyard building (600 square feet). Because we were relatively inexperienced, we often didn't have the answers to all of our questions. We figured out as much as we could on our own and then got help from friends and teachers. There are advantages and disadvantages to learning that way. I developed my own style and solutions and built up my confidence. But I got bogged down when too many problems occurred at once. Also I missed learning new techniques and secrets from more experienced people.

I have now moved on to a shop that I share with two other people whose skills are more extensive and varied than mine. They are a father-son woodworking team. I rent space in their shop and we share machinery, tools, and clamps. Together we have everything we need. My partners supply a lot more of the equipment than I do, which is fortunate for me (so far I have been able to avoid having to buy tools and machinery on time). We have a radial-arm saw, table saw, band saw, disk and edge sanders, a planer, horizontal boring machine, jointer, drill press, grinder, lathe, shaper, spindle sander, and router table. It is obviously a well-equipped shop and we could probably get by with less, but we take advantage of the machines. My first shop was not nearly as well equipped and I had to do a lot more handwork then, earning less per hour as a result.

I get work by word of mouth or through other woodworkers, friends, or family. I used to advertise, but perhaps because of where or how I placed my ads that was not worthwhile. My workload is fairly steady; I don't usually go for more than a few days without something to do. Sometimes, when there are gaps, I do work that I would rather avoid (such as building plywood cabinets, which is more financially lucrative than furniture making, but I get more satisfaction from the latter).

My hours are irregular. I find that I can work comfortably and efficiently only about six hours a day, but my five to six days a week are not always Monday through Friday. I have not totally come to terms with my guilt about not working a forty-hour week. But since I am self-employed, a lot of my hours are not shop hours (e.g., making bids, bookkeeping, designing, running errands, talking to clients). I haven't kept track of this time, but it adds up. I used to keep records on each job until I became more preoccupied with time and money rather than the joys of woodworking. Now I simply do the work and spend as much time as I need (within reason) to do a good job. Yet I do work more when I have a time limit or financial pressure and less when I don't want to work and can afford a little time off.

Money or the lack of it is an almost constant problem, but the challenge it presents can be enjoyable. The interesting part for me is that, instead of that same old paycheck every month, I know that I have the power to make it bigger or smaller. The worrisome part is that usually it's smaller. Furniture makers do not make a killing. Other means to supplement my income, such as teaching woodworking part-time, are a possibility.

There is much to learn about the business end of this work. In the beginning it was hard to make bids and harder to ask a fair price. I tried several bid-making techniques. First there was the "tedious" method: I tried to write down each operation and the time it would take to perform, totaled the hours, and multiplied by whatever my hourly rate was. Then I added that to the cost of materials and overhead, and "bingo"—I had a bid that had taken me several hours to complete.

Then there was the "reality" method: I guessed at what my client was willing to pay.

And now I use a loose combination of those methods called "a little guesswork": first I take about a minute to guess at how long the whole job will take in terms of weeks (e.g., 3½ weeks); I multiply that by my weekly rate (which currently works out to about $10 an hour including overhead), add the cost of materials and then

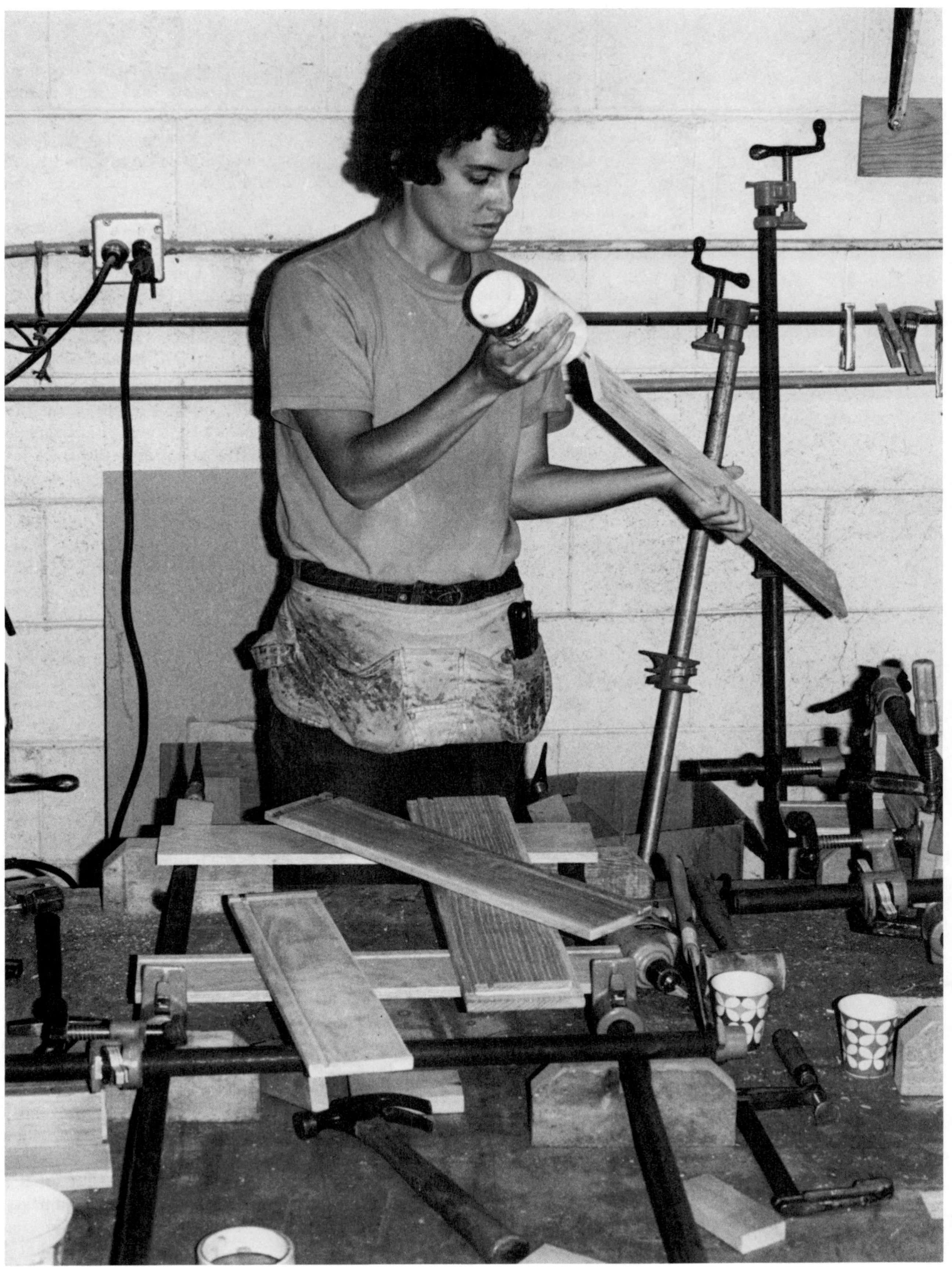

(Photo by Carol Newhouse)

combine those ingredients with the "reality" method. Usually adding the "reality" method causes me to adjust my bid downward a little; it rarely works in the other direction. Using the "guesswork" method, I've been more successful at getting out accurate and acceptable bids. I almost always give a firm bid before I do a job instead of working by the hour. I also try to make a sketch so that my clients and I know and agree on what I am going to build. This seems most agreeable to both of us.

When I make bids, I figure about $2 an hour for my overhead. That includes my shop rent, knife and blade sharpening costs, gas, electricity, phone, transportation, and repairs on tools and machinery. My space requirement is about 800 square feet if I work alone and a little less if I am sharing a large shop. The building needs to be safe, secure, dry, heated (for gluing, stability of the wood, and my health), light, and not too isolated from other craftspeople. It must also have a large, accessible door for easy loading of materials, machinery, and large pieces of completed furniture. Adequate power is essential; I need at least 110 and 220 volts at single-phase power with 100 amp service (more if I'm sharing the space). Hot water is important (for cleaning up excess glue) but not essential. In addition, I need a truck or easy access to one. My little station wagon has been satisfactory for about two thirds of my hauling needs, but I borrow a truck about two or three times a month. Obviously all of these things get expensive and need to be considered when I'm making bids.

Presenting bids to my customers has never been easy for me, although I am getting better at it. This is one of the subjects that I wish had been included in my woodworking education! The presentation is hardest when I am really interested in the job and know that my price will seem high to the client. I don't like having my clients assume that I am avaricious (the only time I will give an outrageously high bid is when the job is so unappealing that I need something to keep me interested). Basically I try to keep my prices reasonable. That doesn't mean that they are comparable to the prices of manufactured furniture, but then, neither is my work. I prefer to deal with only the customers who understand that. Potential clients often have a personal reaction to what they consider a high price, and I used to take their rejection of my bids personally. Now I give my clients a breakdown of my costs and labor so they can see that I am trying to be fair; I have learned to view their rejection of my price as their unwillingness or inability to pay what it costs. I don't like being talked down on a bid. I suspect that that happens to me more than most because people expect a woman to work for less money. It may also be that as a woman I have been socialized to have a low opinion of my value in the business world. Potential clients can usually sense that.

Some of my clients have been respectful, fair, and even encouraging. The customers with whom I have the most successful relationships are those who have come to me after having seen some of my work. They already know and like what I can do and usually have an idea of what I need to charge. Happily, such customers are more numerous the longer I am in this business.

When I first approached the wholesale hardware and lumber establishments I was afraid that they would never believe that I was in business for myself and therefore entitled to buy materials at wholesale prices. I was sure that my gender, size (small—5′2″), and the fact that I look younger than I am (twenty-five) would work against me. Actually, that hasn't been the case most of the time. I am treated pretty fairly in most of the wholesale places; the retail businesses are usually the ones with which I have trouble. In the wholesale places the salespeople assume that I must know what I'm doing or I wouldn't be there; in the retail stores they just figure that I am a "dumb broad" clumsily trying to build something. Those people are amazed to discover that I am often more knowledgeable than they. I am frequently greeted with curiosity and sometimes with disapproval. I am almost always aware that I am a novelty. Sometimes that helps, sometimes it hinders, and sometimes it doesn't make any difference. I used to be extra careful about asking for material with the proper terminology because I wanted to avoid the "dumb broad" label. I am more relaxed now and don't always worry about proving myself. Happily, I've found that the more certain I am of my abilities and worthiness as a furniture maker, the more likely I am to be treated respectfully. Still, on the days when I have to be out in the business world I am more tense than I am on shop days.

Tension and my health in general are of concern to me. I concentrate so hard when I am working that I am exhausted after six hours. Part of that tension, I'm sure, is related to the critic in my head who is hanging around waiting for me to make a mistake. And part of that tension is just the normal creative tension that builds as I build and is released when I'm finished. Healthwise, woodworking is not the greatest profession. Sawdust is a suspected carcinogen. I have allergic reactions to several types of wood. In spite of the fact that I wear a respirator, I have had to turn down some jobs because I get asthma from breathing the dust from redwood and Honduras mahogany. I do try to take safety precautions; I use a face shield, respirator, and ear protectors pretty regularly, and when I am too tired or upset to work I leave the shop. I learned to have good safety habits with machinery as a student, and I've continued to practice them in my shop.

In spite of the health hazards and financial worries, I like my work. It's exciting and challenging and a great outlet for creative energy. Having the power to design and build things is exhilarating for me. In addition, I use my body and have a reason to buy beautiful tools and wood. I am doing something different from most people and I love working for myself; bosses never did appeal to me. I like that I can sleep late when I want to. I enjoy adding the loving touches that make my work unique, such as little wooden latches and levers, and seeing the finished product after weeks or months of labor. Of course, there are days that are frustrating and days that are dull. A week of finish sanding may be bland, but it does offer relief from the pressures of always having to think. Besides, the results of a week's worth of sanding are incredible.

There are times when I think that woodworking would be great as a hobby, but for me it would have to be a full-time hobby. I am doing what I want and it is pretty much what I had always imagined I would do. In terms of my future as a furniture maker, there is a possibility that I will burn out. As I look at my co-workers, I wonder if they can do this work all of their lives. Several have burned out (and more eventually will) because custom furniture making involves so much work for so little money. As wonderful as the satisfaction is, it doesn't pay the bills. At times I have considered sacrificing some satisfaction for a little more money. I frequently reach a point where I have to decide whether to work fast (synonymous with sloppy for me) and therefore earn a higher hourly wage, or to work slowly and carefully and satisfy myself. Sometimes I do a little of each, and I almost always regret the work I've done in haste.

I know that I need more financial security than I currently have. Now that I have acquired my skills, I am beginning to focus on the problem of how to combine my financial needs with my desires for the future. My fantasy is to teach woodworking to women; I like to see us view ourselves as powerful beings. Teaching, in addition to bringing a steadier income, would also satisfy my need for more exposure to people. I would like to acquire other skills, among them carpentry, and learn to utilize different materials. I am currently learning to work with stained glass and want to do more experimenting with metal, plastic, and fabric. Furniture making cannot and does not satisfy all of my needs, such as my need to be outside more. It is entirely possible that in five or ten years I will give up woodworking as a vocation (although I would always have to have a home wood shop) and take on something completely new. For now, I am a furniture maker and pretty happy that way.

SKYLARK CONSTRUCTION CO.: BREAKTHROUGH FOR WOMEN

"Can you really do the work??"

"You mean you can hit a nail with a hammer?"

Although the first response when customers find that we are a women's crew is surprise and disbelief, our experiences working for our various customers usually turn out positive! Skylark Construction is the name I chose when I decided to become a licensed, bonded contractor and work for myself. Although I'd learned about carpentry and house building as I was growing up (my grandfather was a master carpenter and my father built several houses), I spent ten years as a computer programmer before deciding to take the risk and start doing the work I wanted to do. The chance to design and build a very special structure (see "Making Magic: Yurt Building") was a jumping-off point for me—a discovery that I could do the work I loved and make my living doing it. Applying for my contractor's license and forming Skylark was another step leading in the same direction. Under this name, I work with one other woman, La Rosa, doing all kinds of carpentry, from rough framing to finish work. We do remodeling, repairs, some design, new structures, additions, and even consulting for do-it-yourselfers (we give tips on making up plans, reducing costs, using tools and materials, and so on). With pretty minimal capital, a truck, and our tools, we're building a business that supports us—and that we enjoy.

Requirements for becoming a licensed contractor vary from state to state. Some states have a system of paper work and applications and fees that you must wade through; others require tests of your actual knowledge of construction plus lists of references and jobs you've done. Although I can share what I had to go through in terms of my own home state (Oregon), you'll have to look into requirements in yours if you're interested in becoming a contractor. But it can be done, if you're serious and persevere. Talk to other contractors, investigate the procedures in your area for licensing and bonding, keep records (both financial and photographic) of work you're doing, and think about how you want to set up and handle your business.

(Photo by La Rosa Ponderosa)

(Photo by La Rosa Ponderosa)

(Photo by La Rosa Ponderosa)

Having a contractor's license can open several doors to you in the world of carpentry and construction. You will be able to submit bids and work under contracts that bind both yourself and your customer; you'll be able to apply to finance building a house; you can list yourself as a minority contractor and apply for certain government jobs. You'll also probably find, as we have, that much of your work will continue to be remodeling, repairs—jobs small *and* large.

A contractor's building license to do business in the state of Oregon is gotten simply by purchasing it. No written or oral technical tests are required. The license is not transferable to another state. I purchased my license in 1977 and had to pay about $300 in fees, insurance, and bonding. Four different organizations provide the necessary paper work for the license: the Oregon State Builders Board in Salem; any independent insurance agent; the Department of Commerce (also in Salem); and the Federal Tax Office in Washington, D. C. The Oregon State Builders Board sells the actual license for about $30. However, before a license is issued, proof of liability insurance and bonding has to be shown.

To have insurance and bonding issued, the name of your business has to be registered with the Department of Commerce. This department checks to see if the name is already in use in the counties in which you plan to do business. Duplication of business names within counties is not allowed, so I was prepared to change my proposed name if necessary. With suggestions like Sky Blue and Sky High Construction being offered by friends, it's lucky that Skylark Construction was unique within the entire state of Oregon! The fee to register the name of your business is about $4.00, and an additional fee of $2.50 is paid for each county in which you plan to do business.

A suggestion for any woman signing for her "doing business" name is to consider what name she cares to show as owner. I have two names: one by which I am known within the women's community, and another on my social security card. Without thinking, I gave my social security card name to the Department of Commerce, and that is the name that now follows me everywhere for paper signing such as checks, verification of being in business, and so on. None of my customers or banks recognize the name, so I'm constantly having to explain. This is a nuisance you'll want to avoid!

The local insurance agent provides the insurance and acts as a bonding agency for you. The amount of insurance you need and the bond required is set by the state, with bonding varying slightly according to how much gross income you make or plan to make per year. Insurance is a set amount for all, regardless of income. To become bonded, you have to set up a bond account of at least $1,000, which is like having a bank account you can't touch. Even if you have this capital, it isn't always easy to find someone who will bond you—being a woman and starting out in the field are two points against you. Fortunately for me, a bright and sympathetic woman, Pat Jaffer, worked in my local insurance agency. She helped me to get both bonding and insurance. Even though I'd had insurance before (house and automobile), and in my previous profession had responsibility for large sums of money, she had to try several bonding agencies before one would accept me. I know that several male applicants with far less previous responsibilities had no difficulty at all in getting bonded. So if you're turned down, keep trying!

One further number on a piece of paper is necessary to complete getting your license. The Federal Revenue Department requires every tax-paying business to have a federal tax ID number (Employer Identification Number). This can be applied for simply by writing to the government and giving the DBN (doing business name) to them. In a few weeks, they send an enormous long number to you on a card. Keep it handy—you'll need it in the future.

What does it mean, legally, to have your license? First, in the state of Oregon, to be legally practicing the business of home building, you have to have the license (with associated liability insurance and bonding). The liability insurance and bonding are primarily for the customer's protection. For example, if we accidentally back our pickup truck through the customer's bay window, we have insurance to pay for the accident. This insurance does not cover us personally—that's another whole set of insurance, which is our option to have—but it does protect the customer's property. The bonding acts as another form of assurance for the customer. Some contractors have a way of walking out on the job, leaving the customer holding the bills. The bond is a way of covering these debts, at least up to a certain point.

Another reason why I decided to take a license was so that I could place Skylark Construction on a minority contractors list with the state, county, and city agencies. Governmental agencies are required by federal law to open bidding on construction jobs to minority contractors. A few extra points are given for minority contractors in their bidding. To date, we haven't bid on a state job because of the work locale. The city, however, has a low-income housing improvement loan system, and we've taken part in that.

Besides the legal implications or possibilities inherent, having a contractor's license might help you in buying tools, materials, and so on. Although in our area it doesn't seem that contractors are given discounts on tools or materials, in other areas it's customary. You may even live in an area with large building supply and tool stores that cater extensively to contractors. Here, lumber store or parts dealers will give extra consideration to our deliveries because they know we are working on time and keeping to a pretty close schedule. Our customers don't seem to care that we are licensed; they care more about a good job done at a reasonable cost and on time. In other areas (or even in this one, with different people perhaps), having a license might be an important factor in whether or not you get a particular job. Many people like the security of hiring someone who is licensed, and want to have the work done under a contract that specifies exactly what it will cost.

Here's a hint for any woman starting her own business: try to avoid "hiring" employees. The amount of taxes, insurances, and folderol required to maintain an employee makes it hard to function financially. This free-enterprise system is not set up to favor the small business! Seek other ways: you can hire on contract, form a collective, be a partnership, and so on. Since I work with one other woman, we share all the work responsibilities. Technically she is a "carpenter's helper," funded by CETA (Comprehensive Education and Training Act). Under this program, private sectors can file for grants that will help to pay an apprentice who learns on the job. This is something you should explore if your business provides enough work for other women, either regularly or periodically.

Responses from customers who don't know that we are an all-woman crew have been interesting. As the door is opened, we're often greeted with: "*Oh!* I thought your husband was coming!" or "Why—you're ladies!" But in spite of this, all of our customers have been supportive. Some are concerned about our falling off ladders, lifting heavy weights, walking on roofs, working in the cold. We are, too. Some try to help us until they see us wander by with a green 12′-long 4×6 on our shoulders—then they usually stand back and let us get to work. Sometimes we keep a green 12′-long 4×6 around the site even if it's not being used in the job, just to keep the area clear. . . .

Most of our customers have been women, many of whom responded to an article written about us in the local paper. Frequently, a woman customer will point out that she's called us because she felt that we would listen to her ideas about the job and that we would be sensitive to her needs. That's true on both counts. Many women want ideas about remodeling kitchens and baths, and we're interested in exchanging ideas, working on a personal level to design and build what is needed. We find, too, that we learn through this process, with many of the women we work for providing useful ideas and suggestions. A good percentage of our customers are retired. We appreciate their business because they usually have a great deal of practical information from years past, which they share with us. Also, their needs are usually very basic. We offer discounts to senior citizens and to low-income residents.

The novelty of an all-woman carpentry company attracts some customers, but overall, people seem to feel that women will do a better job for them. One of the biggest impressions we make on people is that we appear at the time we said we would to make the bid or to do the job. If an appointment can't be kept, we call and let the customer know we aren't coming and set up another time. Apparently these courtesies are largely lacking in the building industry, and people appreciate them.

Estimating or bidding is a hard part of the job. Building a new structure is somewhat easier in that a cost-per-square-foot formula can be applied—at least to give you a rough idea you can work with. This formula varies according to the economy of the country: the price of materials, of labor, and so forth. There are books published annually that break down building costs quite clearly and can be used as a reference (see "Resources"). Talking with other contractors and builders can help you here, too. Usually there's a more or less standard cost-per-square-foot figure floating around that will give you something to compare your own estimate with. For instance, right now in my area the cost for a new residential structure is about $40 per square foot, and for an uninsulated shed, about $8 per square foot.

Remodeling is much harder to bid on because often what appears to be a simple job turns into a bundle of knots and complications. For example, a customer may ask to have a small portion of her roof replaced. But once that section of the roof is off, it's evident that the rafters have rotted and need replacement also. And while we're at it, we might as well replace the gutters and soffit boards too. . . .

Variances in the bid may come from a change in the nature of the job, substitution of materials when the original planned-for materials are not available, and changes in the retail price of materials. It's not all that unusual for the customer to ask for a change in design as she sees the work progress. These changes may then alter the material needs. Because we work under contract, all changes must be negotiated and written up.

We begin a job by writing up an estimate for specific work and submitting it to our prospective customer. If it's accepted, we both (customer and builder) sign it, and it becomes a binding legal contract on both parts. Estimating work hours is very hard; even doubling what we originally think sometimes doesn't make it. But other contractors more experienced than us have assured us that they have the same problem: it is hard to know in advance how many hours of labor a job will entail. If you're remodeling, the task is even more difficult! Added to all the above difficulties is the changing weather. It rained for two weeks in the middle of the summer this year (unheard-of where we live). Of course, we were putting a roof on just as the clouds were building up. But that's the nature of the business—or the business of nature. . . .

All things considered, we're both enjoying our work, the structures we're getting to build (see the yurt in "Making Magic . . ."), and the customers we get to work with. And for those who still worry "Can you do the work?" La Rosa assures them: "We can do it better!"

part II

WOMEN'S WORK

Cabin by Rose Ann Carroll and Jeanne Tetrault. (Photo by Carol Newhouse)

Dome by Priscilla Danzig, Tirza Latimer, and Terri Miller, all now part of Seven Sisters Construction Co. (Photo by Carol Newhouse)

Octagonal barn by women's collective (including Jean Malley, Priscilla Danzig, Rose Ann Carroll, and CeCe Wells). (Photo by Lynda Koolish)

Room addition (entire right side of house, shingled) by Harriet Bye. (Photo by Lynda Koolish)

Plant shelves by Priscilla Danzig.
(Photo by Carol Newhouse)

Cabinets by Priscilla Danzig (now part of Seven Sisters). (Photo by Carol Newhouse)

Room addition by Jean Malley and CeCe Wells. (Photo by Carol Newhouse)

Stereo cabinet by Jan Zaitlin. (Photo by Lynda Koolish)

Chair by Jan Zaitlin. (Photo by Lynda Koolish)

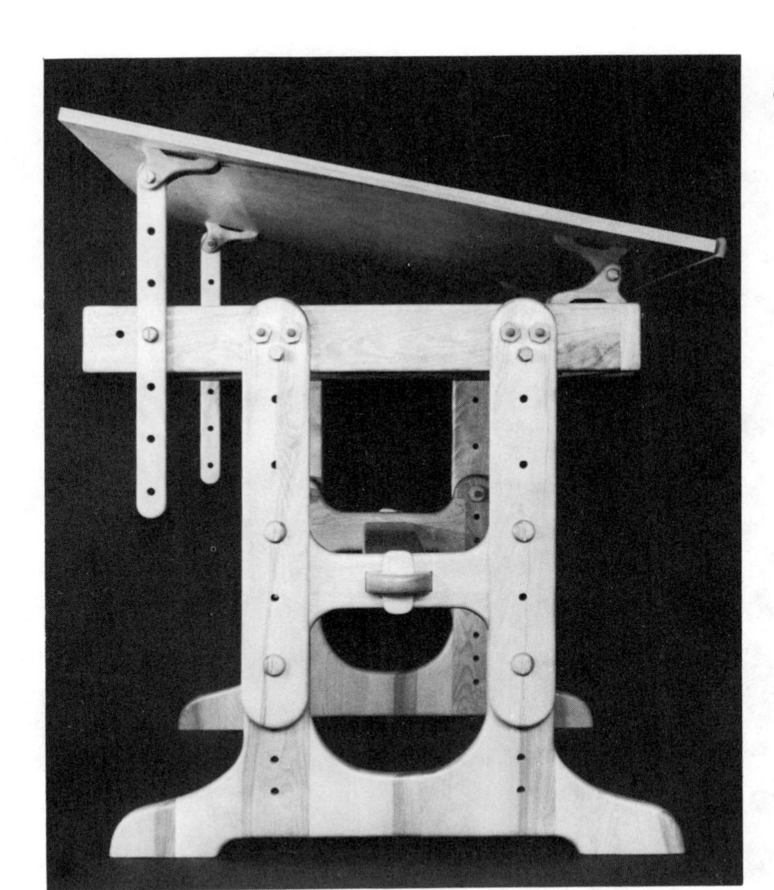

Side view of drafting table by Jan Zaitlin.
(Photo by Lynda Koolish)

Drafting table by Jan Zaitlin. (Photo by Lynda Koolish)

French doors with leaded windows by Sally Smith. (Photo by Carol Newhouse)

Room addition by Sally Smith; door by Shannon and Sally Smith. (Photo by Carol Newhouse)

Door by Shannon. (Photo by Dian)

Interior by Sally Smith. (Photo by Carol Newhouse)

House by Lynn Dorgan. (Photo by Janet Cole)

Room addition (upper part of house) by Sally Smith. (Photo by Carol Newhouse)

Small house by Hurriet Bye. (Photo by Carol Newhouse)

Small house by Cabbage Lane Women. (Photo by La Rosa Ponderosa)

Guest house by Lynn Dorgan. (Photo by Janet Cole)

House by Lynn Dorgan. (Photo by Janet Cole)

House by Lynn Dorgan. (Photo by Janet Cole)

(Photo by Janet Cole)

The Language of Carpentry: Terms and Definitions

As the language of carpentry is largely an unfamiliar one to many people, an attempt has been made throughout this book to explain and identify terms, products, and procedures as they have come up. The following section of definitions is offered as both a supplement and a reference. Some things are deliberately omitted: to redefine tools here, for example, would have been unnecessarily redundant.

With new products and new processes constantly being developed, the language of carpentry continually changes. If you hear of something you can't find defined in this book or another, ask around—it may turn out to be an entirely new material that's being mentioned. Regionalism, too, will produce a carpentry jargon that might differ from what you've read or heard before; don't be afraid to ask for explanations or definitions if you're feeling in the dark.

Remember, finally, that certain types delight in mystifying and confusing beginners, so don't be intimidated by the "mysteries" of carpentry. Exposure and experience will make you more and more comfortable not just with the tools, the materials, and the process, but with the language itself.

Terms and Definitions:

Acoustical tiles Square ceiling tiles designed to absorb sound and thus reduce noise transmission.

Adhesive Any substance that will cause materials to bond together when surface contact is made.

Anchor bolts J-shaped bolts that are embedded in concrete and act as an attachment between the foundation and the structural members of a building (also called *sill bolts*).

Apron A horizontal piece applied under the sill or stool of a window to hide the joint and "finish" the trim. Usually used on interior but occasionally applied to exterior trim.

Asbestos An indestructible fibrous mineral used in a multitude of construction materials (insulation, roofing shingles, joint compound, ceiling tiles, etc.). Asbestos fibers are known to be carcinogenic and materials containing them should be avoided whenever possible.

Aspenite 4′×8′ panels of bonded wood chips, usually used for interior wall covering.

Asphalt A petroleum residue insoluble in water and therefore widely used as a waterproofing material or moisture inhibitor in roofing materials, flooring underlayment, and others.

Base shoe A small molding piece used along a baseboard to cover the joint between the baseboard and the finished floor.

Batt or batten A thin strip of (usually) wood used to cover and protect a joint.

Batter board A simple, temporary setup of horizontal boards and vertical supports used in laying out the corners of a building.

Beam A heavy horizontal structural piece running between posts or from wall to wall and usually supporting a load such as roof or floor.

Bearing wall/bearing partition Any wall or partition that supports a vertical load plus its own weight.

Berm To heap up soil along a wall or in a certain area.

Bevel An angle or angled surface other than 90°; also, to cut a surface to any inclination/angle other than 90°.

Bevel siding An exterior siding applied horizontally. The wedge-like shape of the milled boards sheds water shingle-style.

Bird's-mouth The portion of a rafter that is notched out to fit over the top plate of the wall.

Blind-nail To drive a nail in such a way that the head will be covered by the next (adjoining) board. This system of nailing is used on tongue-and-groove flooring, for certain siding, and so on.

Board Any piece of lumber less than 2″ thick.

Board-and-batten A style of siding or paneling consisting of boards applied vertically with battens covering the joints between the boards.

Board foot A unit of measurement for lumber (1 board foot equals a piece 12″ long, 12″ wide, and 1″ thick).

Bracing A piece that is usually applied diagonally to stiffen or hold something in place.

Bridging Short pieces of 2×3 or 2×4 nailed in an X pattern between two joists to keep the joists from twisting (also helps to stiffen the floor).

Butt hinge A hinge consisting of two separate plates and a removable pin that holds them together. Commonly used on a door (one plate of each hinge is attached to the doorjamb; the second plate is attached to the edge of the door).

Butt joint The joint formed when one square surface abuts into another.

Casement A hinged window sash that swings open and closed.

Casing The frame for a door or window.

Caulking Any of a number of substances (usually plastic- or rubber-based) used to close and seal cracks and to exclude water and air.

Celotex A trade name for a building board made of compressed bagasse (dry pulp of sugar cane) and used for insulation and soundproofing.

Cement A fine, powdery material that is a complex mix of specially treated lime, alumina, silica sand, gypsum, etc. When mixed with water, it sets up (hardens). Cement is used as a base for other products, such as mortar and concrete.

Chamfer The corner of a board cut to a 45° angle. If two chamfered boards meet in a butt joint, a V joint occurs.

Collar beam A horizontal piece connecting a pair of rafters, forming a strong triangle that ties the roof together (helping to prevent it from spreading out). Also called a tie beam or tie collar.

Common rafter A rafter that runs from the top plate to the ridge piece (or to the matching rafter).

Compound cut A cut including both a miter and a bevel; usually done with a power saw rather than a handsaw.

Concrete A mixture of cement, water, sand, and gravel poured wet into forms or otherwise contained. When it hardens, it forms a strong, durable material.

Corner bead A metal reinforcing piece placed over corners when a room is being finished with gypsum wallboard. This term also may refer to any molding used to protect corners of a room.

Cornice A set of exterior trim pieces that enclose the meeting of roof and wall along the eave of a building.

Counterbore/countersink The upper part of a hole enlarged (counterbored) so that the head of a screw or bolt can be sunk below the surface of the material (countersunk).

Counterflashing A second layer of flashing applied over roofing or shingles when a first layer of flashing has already been installed between the roof sheathing and the roofing or shingles.

Course A row of shingles, roll roofing, etc.

Coving A curved face piece of wood molding used in right-angle corners to hide the joint.

Crawl space Space left under a house or structure between the ground level and the underpinnings of the building.

Crosscut To cut across the grain of the wood.

"d" A unit of measure for nails. *d* is spoken as "penny"; it was originally a term referring to the number of nails of a given size in a pound but now refers to the length of the nail.

Dado A rectangular groove cut into a piece of lumber; may be used as a verb (meaning to cut this groove) as well.

Dead load The weight of the structure itself (usually expressed in pounds per square foot).

Diagonal bracing *See* "let-in bracing."

Dimension lumber Material from 2″ to (and including) 4″ thick, 2″ or more wide, and any length.

Drip cap A piece of molding or flashing used to channel water away from parts of the exterior of a building. Usually applied over doors and windows.

Drop siding A general term for exterior siding whose joints are formed by specially milling the edges of the boards so that they overlap and fit in certain ways. V-rustic, shiplap, and tongue-and-groove are some examples of drop siding.

Dry wall A general term for wall finishes or coverings which are applied without first mixing with water (as opposed to plaster, which is applied wet). Also used as a specific term for gypsum wallboard (Sheetrock).

Eave The overhang along the lower edges of a roof.

Edge grain A term for grain in wood that forms a 45° angle with the surface of the board.

End grain The open grain exposed at the end of a cut-off piece of lumber.

End-nail To drive a nail through the face of one board into the end of another board.

Expansion joint A groove put in concrete to allow for expansion; a strip of fibrous material may be placed in this joint.

Face-nail To drive a nail perpendicular to the surface of the wood and/or to attach two boards or pieces with their width dimensions (faces) nailed together.

Factory edge The straight, milled edge of a piece of plywood, board, etc.

Fascia A board used flat across a joint or over the butt ends of other boards to conceal the joint or ends.

Felt paper Building or felt paper is a semiporous, strong, heavy-base material saturated with hot asphalt and rolled smooth. It comes in 36″-wide rolls of various weights. It is used with shakes and under sidewall shingles, used with hot asphalt for built-up roofing, and so on.

Fiberboard A panel (usually 4′×8′) of wood or cane fibers that are bonded together. Used for interior walls, cabinets, and others.

Fireblocking Horizontal 2×4's placed between studs in walls more than 8′ high. These are required by code in multistory buildings on all walls that carry floor loads of stories above. They prevent strong updrafts of air that might spread fire through the building.

Firmer chisel A chisel with a very thin blade.

Flashing Metal or asphalt-saturated paper strips placed strategically to help channel water away from certain areas (for example, where an adjoining roof line meets an existing exterior wall).

Flush A term for placing two pieces or surfaces even with one another so that neither projects beyond the other.

Footing The concrete base of a foundation wall, pier, etc.

Forms Temporary wood constructions that hold wet concrete until it sets up into a permanent shape.

Form stakes A metal stake with regular perforations that allow you to drive nails through the stake; used in construction of concrete forms.

Form tie wire Thin-gauge wire used to suspend rebar within forms.

Foundation The part of a building that supports it and connects it to the ground.

Furring Preparing a wall, floor, or other surface by adding thin strips of wood or metal which bring the wall, etc., out to a uniform level surface. Furring strips are the pieces used to do this.

Gable roof A roof that rises from two sides to a peak, forming a triangle at either end of the building.

Gain A notch or mortise cut to receive a hinge, the end of another piece of lumber, etc.

Gambrel roof A roof that rises from two sides to a peak but has two slopes per side (the old-fashioned barn roof).

Girder A structural piece similar to but usually heavier than a beam.

Glazing Installing glass into a wooden window sash or door.

Grade (lumber) A classification of lumber according to its strength and utility.

Grade (site) The level of the ground at the building site. May also refer to the level after construction is completed, or to a fixed point established during construction. May also refer to the mechanical act of smoothing and leveling a site or lot.

Gutter A metal, wood, or plastic trough attached along the eaves of a building to catch and channel away rain or melting snow.

Gypsum wallboard (Sheetrock) A widely used interior wall covering consisting of sheets of the mineral gypsum (combined with other materials) sandwiched between layers of heavy paper.

Hardboard Panels of densely packed wood fiber used in cabinetmaking, for interior wall covering, and so on.

Hardwood The wood from deciduous (leaf-shedding) trees. This wood is not necessarily *hard.*

Header A horizontal structural piece placed over door and window openings; it is designed to carry the loads that the normally placed studs would carry in these areas. Also called a *lintel.*

Header joist The joist that runs at either end of a house perpendicular to the regular joists. The other joists butt into the two header joists.

Heartwood The inner wood of a tree, which is usually more durable and of a higher grade than the outer sapwood.

Hip roof A roof that rises equally from all four sides to the common peak.

Insulation A material used to reduce heat, sound, or electricity transmission from one substance or surface or area to another.

Jalousie A shutter, window, or blind with adjustable horizontal slats for regulating incoming air and light.

Jamb The individual pieces of a doorframe or a window frame (usually two vertical or side pieces and one horizontal or top piece).

Joint compound A material used to fill and cover the joints between adjacent pieces of gypsum wallboard.

Joist A horizontal framing member that supports a floor and/or ceiling above.

Joist hanger A metal attachment used to secure a joist to a beam, girder, etc.

Kalwall A fiberglass product that comes in rolls about 4′ wide and 20′ long; it is used in solar heating devices.

Kerf The slot or groove cut into a piece of lumber by a sawblade.

Key A slot or groove in concrete made to accept a girder end, to lock sections of a slab together, and so forth.

Kiln-dried Term applied to wood that is dried with artificial heat and/or circulated air. It has a low moisture content and consequently will have little further shrinkage.

Kraft paper Two layers of paper with asphalt and reinforcing fiber sandwiched between. It is used as a vapor barrier and a draft stop.

Laminated A laminated beam, panel, etc., made of several layers glued, bonded or otherwise fastened together permanently.

Lath Thin strips of wood used as a base for plaster.

Ledger A piece attached (horizontally) to studs or wall surfaces, etc., which acts as a support for joists, shelving, and so on.

Let-in bracing Diagonal bracing set into notches so that the stud wall is left with a smooth surface. This bracing should run at a 45° angle from the top to bottom plate where possible.

Lineal foot A unit of length only, equaling one foot.

Lintel *See* "header."

Live load The variable weight placed in or on a building (such as snow on the roof, furniture inside). Expressed in pounds per square foot (estimated).

Lumber A general term for timber that has been sawed into boards, planks, structural pieces, etc.

Matched lumber Lumber milled with edges and possibly ends that interlock (such as tongue-and-groove).

Miter To cut on an angle.

Molding A piece of wood, usually quite narrow and light, used to cover joints, dress up a room, and so on. Moldings vary from simple curved face pieces to complex ornamental shapes.

Mortar A mixture of cement and sand (or cement, hydrated lime, and sand) used when laying brick, stone, or concrete blocks.

Mortise A rectangular cut made into a piece of wood; it receives the projecting tenon to form a strong, clean joint.

Nominal size The ordinary spoken dimension of a piece of lumber (e.g., a 2×4) that is not its actual size.

Nonbearing wall/non-bearing partition A wall or interior partition that supports only its own weight.

Nosing The part of a stair tread that projects over the riser of the step below it.

o.c. "On center," the common way to express a spacing of framing members. The measurement is from the center of one piece to the center of the next: e.g., 16″ o.c., or 24″ o.c. Nailing patterns may also be expressed this way.

Particleboard A panel of wood or cane pieces or particles bonded together under heat and pressure. It has various uses, both interior and exterior.

Plane To finish or smooth with a plane; to remove any rough spots or areas.

Plank A piece of lumber cut thicker than a board (a plank is usually 2″ material in widths of 6″ or more).

Plate The top plate in a stud wall: the horizontal, continuous 2×4 (doubled) along the top of the wall; the bottom or sole plate: the corresponding piece (not doubled) at the bottom of the wall.

Plumb Exactly vertical or perpendicular.

Plywood A built-up panel of wood whose layers or plies are laminated together.

Post anchor A metal attachment embedded in concrete, providing a fastening point for wood posts.

Post cap A metal attachment fastened to the top of a post, providing a secure attachment for the girder or beam resting on the post.

Psf Pounds per square foot.

Purlin A horizontal framing piece that supports the rafters of a roof.

Putty A soft, flexible compound used to seal holes in wood.

Rabbet A groove or cut along the edge of a piece of wood that allows another piece to fit into it, creating a rabbeted joint.

Rake The edges of the roof running parallel to the slope of the roof.

Rebar Steel reinforcing bar used in concrete walls, footings, and so on.

Resin Clear or translucent viscous substances (either of plant origin or synthetic) used as adhesives, varnishes, components of plastic, etc.

Rip To saw *with* the grain of a piece of wood; also, to cut a panel of wood or other material lengthwise.

Riser The vertical part of a stair between two consecutive treads.

Roof jack Device placed or set into a roof when a pipe (stovepipe, vent, or such) is to be fitted through the roof. It helps to hold the pipe rigid and allows for a waterproof joint around the pipe.

Roof nosing Metal flashing folded lengthwise to form a right angle corner; used along the eaves and rakes of the building under certain types of roofing.

Rosin A type of resin derived from pine tree sap and used, for example, to increase friction. Rosin-sized paper is used under oak flooring.

Rough sill Part of the framing of a window opening; the horizontal 2×4 or 2×6 that sits under the finished sill.

Sash The frame (usually wood or aluminum) of a window into which the glass panes are set.

Screed To level and smooth freshly poured concrete; also, the tool used for this work.

Shakes Hand-split pieces of wood used as roofing or siding material.

Sheathing Boards or panels attached to the exterior of a building (either to wall studs or rafters). The sheathing is usually then covered with siding (walls) or roofing (roof).

Shed Any small structure (free-standing or attached to a larger structure) used for storage or as a simple shelter.

Shed roof A shed roof has a single slope; resembles one half of a gable roof.

Sheetrock *See* "gypsum wallboard."

Shim A thin and usually tapered piece of wood, metal, or even paper that is driven or placed between two surfaces to fill space, help level of adjust one or both surfaces, and so on. Also, to place these pieces, level the surfaces, etc.

Shingle A thin piece of wood, asphalt-impregnated material, etc., used in a pattern of overlapping rows to roof a structure or to cover sidewalls. Also, to apply the materials.

Shiplap Boards cut with rabbeted (*see* "rabbet") edges that overlap to form a tight seal between adjacent pieces.

Siding The finishing or outer covering of the walls of the house.

Sill A horizontal piece that holds the upright portion of a frame (especially the base of a window). The *mudsill,* also called the sill, is the piece that rests on top of the foundation wall.

Softwood The wood from everbearing trees (conifers).

Sole plate *See* "plate."

Span The distance a piece covers or makes between supports.

Square Cut or placed at right or 90° angles. Also a unit of measurement for roofing materials (one square = 100 square feet). Also, the tool used to mark right angles (combination square or try square).

Stickers Thin strips of wood used to separate layers of lumber being air-dried. Stickers are sometimes used in construction of yurts, providing an interlacing frame that is incredibly light but strong.

Stool An interior trim piece that caps the windowsill.

Straightedge Any piece of straight lumber, metal, etc., used to guide a saw cut, mark a line, check layout, and so on.

Stress-graded lumber Lumber used in large buildings to make long spans, graded according to its strength.

Stud A vertical piece of framing lumber used in a wall.

Subfloor A first layer of flooring put over the floor joists and later covered with a finished floor.

Surfaced lumber Lumber that has been planed smooth on one or more sides.

Taper A gradual reduction of thickness from one end of a piece to the other, as in a tapered shingle.

Tenon The end of a piece of wood cut into a projection that fits a corresponding cavity (mortise) in another piece of work to form a joint.

Toenail To drive a nail at an angle through one piece of wood into another.

Top plate *See* "plate."

Transom A small hinged window installed above another window or a door; also, the horizontal piece this window is hinged to.

Tread The horizontal part of a stair or step.

Trimmer A 2×4 that is run vertically to support a header over a window or door.

Underpinnings The mudsill, pier blocks, girders, and posts used under the first floor of the house.

Valley The angled portion of a roof where two slopes meet.

Veneer A thin layer of wood or other material glued or bonded to a sublayer or set of sublayers.

Viga The Spanish term for a beam or girder.

Wainscoting A wall covering that goes over the lower half or two thirds of the wall. The lower paneling is separated from the upper part of the wall by a piece of horizontal trim called the *chair rail.*

Waler or whaler A horizontal piece used on a form wall to help keep the wall straight.

Water table Part of the exterior finish of a wood-frame house with a concrete foundation. It covers the joint between the exterior sheathing and the foundation, acting to make water drip free of the building rather than running down the face of the concrete or being pulled back under the mudsill.

Weatherstripping Thin strips of metal, felt, plastic, etc., installed around windows and doors to prevent drafts.

Yard lumber Any common-sized lumber used for regular light framing.

Building Code Tables

How to Use These Tables:

The following tables will allow you to choose the exact size of material and spacing for roof rafters, floor joists, ceiling joists, and so on. Although these tables look rather formidable at first, they are actually quite simple to use and will help you make plans for almost any building situation you encounter.

For rafters, first take Table 25-A-1 and find the "F_b" value for the species of lumber you'll be using. For example, if you plan to use Douglas fir for the roof rafters and are building in a snow area, the F_b value is found under "Roof Members" in the "Snow Loading" column across from the grade of Douglas fir you'll be working with. The F_b value for Construction grade in this instance is 1380. With this number, go to the appropriate table for your roof slope and the type of ceiling and roof covering you plan. For instance, you might be planning a high-slope roof (over 3 in 12) with a heavy roof covering; the information you need is in Table No. 25-T-R-11. Here you can check the possible spans for lumber of various sizes spaced at 12", 16", or 24" o.c., using the "1300" column that corresponds to the Construction-grade Douglas fir you'll be using. Depending upon the size of the structure you're building, you can choose an appropriate rafter size and spacing. Directly under the allowable span for each situation is a number that equals the "required modulus of elasticity," *E*. Check back to Table No. 25-A-1 (where you found your F_b value) to make sure that the *E* listed under the span you choose doesn't exceed the *E* listed there. If it does, you should go to a larger rafter or a closer spacing. In almost every instance, there will be no discrepancy,—but it's a good idea to double-check.

In choosing floor joists and ceiling joists, use the *E* value rather than the F_b; be sure to double-check your choice with the F_b value placed directly under each span. The procedure is the same: find the *E* value for the species and grade of lumber you'll be using, and then refer to the appropriate table for exact information. (Remember that ceiling joist spans may be broken by supporting—bearing—walls and that floor joists may rest on intermediate girders.)

Try to choose size and spacing of lumber that allow you to use material that is readily available and is manageable. Really large timbers can make enormous spans but may have to be special-ordered from the lumber store and actually end up costing you more than smaller pieces that need closer spacing or intermediate support. With the information in the tables and a phone call to check out local lumber prices, you can juggle your needs to find the most reasonable and economic solution.

TABLE NO. 25-A-1—WORKING STRESSES FOR JOISTS AND RAFTERS—VISUAL GRADING

These "F_b" values are for use where repetitive members are spaced not more than 24 inches. For wider spacing, the "F_b" values should be reduced 13 percent.

Values are for surfaced dry or surfaced green lumber, except for Southern Pine as indicated. Values apply at 19 percent maximum moisture content in use.

SPECIES AND GRADE[6]	SIZE	ALLOWABLE UNIT STRESS IN BENDING "F_b"			MODULUS OF ELASTICITY "E" 1×10^6 psi
		FLOOR OR CEILING MEMBERS	ROOF MEMBERS		
			SNOW LOADING	NO SNOW LOADING	
BALSAM FIR[1]					
EASTERN WHITE PINE					
EASTERN WHITE PINE (NORTH)					
ENGELMANN SPRUCE—ALPINE FIR (ENGELMANN SPRUCE—LODGEPOLE PINE)					
RED PINE					
Construction	2x4	800	920	1000	1.0(.9)
Standard	2x4	450	520	560	1.0(.9)
Utility	2x4	200	230	250	1.0(.9)
Studs	2x4	600	690	750	1.0(.9)
No. 1 & Appearance	2x6 and wider	1150	1320	1440	1.2(1.2)
No. 2	2x6 and wider	950	1090	1190	1.1(1.1)
No. 3	2x6 and wider	550	630	690	1.0(.9)
CALIFORNIA REDWOOD					
Construction	2x4	950	1090	1190	.9
Standard	2x4	550	630	690	.9
Utility	2x4	250	290	310	.9
Studs	2x4	700	800	880	.9
No. 1	2x6 and wider	1700	1960	2120	1.4
No. 1, Open grain	2x6 and wider	1350	1550	1690	1.1
No. 2	2x6 and wider	1400	1610	1750	1.3
No. 2, Open grain	2x6 and wider	1100	1260	1370	1.0
No. 3	2x6 and wider	800	920	1000	1.1
No. 3, Open grain	2x6 and wider	650	750	810	.9
ASPEN (BIGTOOTH—QUAKING)					
NORTHERN ASPEN[2]					
EASTERN WOODS					
Construction	2x4	750	860	940	.9(1.1)
Standard	2x4	425	490	530	.9(1.1)
Utility	2x4	200	230	250	.9(1.1)
Stud	2x4	575	660	720	.9(1.1)
No. 1 & Appearance	2x6 and wider	1100	1260	1380	1.1(1.4)
No. 2	2x6 and wider	900	1035	1125	1.0(1.2)
No. 3	2x6 and wider	525	600	660	.9(1.1)
COAST SITKA SPRUCE					
Construction	2x4	875	1010	1090	1.3
Standard	2x4	500	580	630	1.3
Utility	2x4	225	260	280	1.3
Stud	2x4	675	780	840	1.3
No. 1 & Appearance	2x6 and wider	1250	1440	1560	1.7
No. 2	2x6 and wider	1050	1210	1310	1.5
No. 3	2x6 and wider	600	690	750	1.3

(continued)

[1]Values for modulus of elasticity shown in ( ) apply to Balsam Fir.
[2]Values for modulus of elasticity shown in ( ) apply to Coast Species.
[3]Values for modulus of elasticity shown in ( ) apply to Northern Aspen.
[4]Values for modulus of elasticity shown in ( ) apply to Mountain Hemlock-Hem-Fir.
[5]Values for modulus of elasticity shown in ( ) apply to White Woods (Mixed Species).
[6]Utility grades of all species may be used only under conditions specifically approved by the Building Official.

SPECIES AND GRADE[6]	SIZE	ALLOWABLE UNIT STRESS IN BENDING "F_b"			MODULUS OF ELASTICITY "E" $1x10^6$ psi
		FLOOR OR CEILING MEMBERS	ROOF MEMBERS		
			SNOW LOADING	NO SNOW LOADING	
DOUGLAS FIR—LARCH					
DOUGLAS FIR—LARCH (NORTH)					
Construction		1200	1380	1500	1.5
Standard	2x4	675	780	840	1.5
Utility		325	370	410	1.5
Studs		925	1060	1160	1.5
No. 1 & Appearance		1750	2010	2190	1.8
Dense No. 2	2x6	1700	1960	2120	1.7
No. 2	and	1450	1670	1810	1.7
No. 3	wider	850	980	1060	1.5
DOUGLAS FIR SOUTH					
Construction		1150	1320	1440	1.1
Standard	2x4	650	750	810	1.1
Utility		300	340	380	1.1
Studs		875	1010	1090	1.1
No. 1 & Appearance	2x6	1650	1900	2060	1.4
No. 2	and	1350	1550	1690	1.3
No. 3	wider	800	920	1000	1.1
EASTERN HEMLOCK—TAMARACK					
EASTERN HEMLOCK (NORTH)					
Construction		1050	1210	1310	1.0
Standard	2x4	575	660	720	1.0
Utility		275	320	340	1.0
Studs		800	920	1000	1.0
No. 1 & Appearance	2x6	1500	1720	1880	1.3
No. 2	and	1200	1380	1500	1.1
No. 3	wider	725	830	910	1.0
COAST SPECIES[3]					
EASTERN SPRUCE					
Construction		875	1010	1090	1.1(1.2)
Standard	2x4	500	580	620	1.1(1.2)
Utility		225	260	280	1.1(1.2)
Studs		675	780	840	1.1(1.2)
No. 1 & Appearance	2x6	1250	1440	1560	1.4(1.5)
No. 2	and	1000	1150	1250	1.2(1.4)
No. 3	wider	600	690	750	1.1(1.2)
HEM—FIR					
HEM—FIR (NORTH)					
MOUNTAIN HEMLOCK—HEM—FIR[4]					
Construction		975	1120	1220	1.2(1.0)
Standard	2x4	550	630	690	1.2(1.0)
Utility		250	290	310	1.2(1.0)
Studs		725	830	910	1.2(1.0)
No. 1 & Appearance	2x6	1400	1610	1750	1.5(1.3)
No. 2	and	1150	1320	1440	1.4(1.1)
No. 3	wider	675	780	840	1.2(1.0)

(continued)

[1]Values for modulus of elasticity shown in ( ) apply to Balsam Fir.
[2]Values for modulus of elasticity shown in ( ) apply to Coast Species.
[3]Values for modulus of elasticity shown in ( ) apply to Northern Aspen.
[4]Values for modulus of elasticity shown in ( ) apply to Mountain Hemlock-Hem-Fir.
[5]Values for modulus of elasticity shown in ( ) apply to White Woods (Mixed Species).
[6]Utility grades of all species may be used only under conditions specifically approved by the Building Official.

SPECIES AND GRADE[6]	SIZE	ALLOWABLE UNIT STRESS IN BENDING "F_b"			MODULUS OF ELASTICITY "E" 1x10[6] psi
		FLOOR OR CEILING MEMBERS	ROOF MEMBERS		
			SNOW LOADING	NO SNOW LOADING	
IDAHO WHITE PINE WESTERN WHITE PINE WHITE WOODS (MIXED SPECIES)[5]					
Construction		775	890	970	1.2(.9)
Standard	2x4	425	490	530	1.2(.9)
Utility		200	230	250	1.2(.9)
Studs		600	690	750	1.2(.9)
No. 1 & Appearance	2x6	1100	1260	1370	1.4(1.1)
No. 2	and	925	1060	1160	1.3(1.0)
No. 3	wider	550	630	690	1.2(.9)
LODGEPOLE PINE					
Construction		875	1010	1090	1.0
Standard	2x4	500	580	620	1.0
Utility		225	260	280	1.0
Studs		675	780	840	1.0
No. 1 & Appearance	2x6	1300	1500	1620	1.3
No. 2	and	1050	1210	1310	1.2
No. 3	wider	625	720	780	1.0
MOUNTAIN HEMLOCK					
Construction		1000	1150	1250	1.0
Standard	2x4	575	660	720	1.0
Utility		275	320	340	1.0
Studs		775	890	970	1.0
No. 1 & Appearance	2x6	1450	1670	1810	1.3
No. 2	and	1200	1380	1500	1.1
No. 3	wider	700	800	880	1.0
NORTHERN PINE					
Construction		950	1090	1190	1.1
Standard	2x4	525	600	660	1.1
Utility		250	290	310	1.1
Studs		725	830	910	1.1
No. 1 & Appearance	2x6	1400	1610	1750	1.4
No. 2	and	1100	1260	1380	1.3
No. 3	wider	650	750	810	1.1
NORTHERN SPECIES					
Construction		775	890	970	.9
Standard	2x4	425	490	530	.9
Utility		200	230	250	.9
Studs		600	690	750	.9
No. 1 & Appearance	2x6	1150	1320	1440	1.1
No. 2	and	925	1060	1160	1.0
No. 3	wider	550	630	690	.9
NORTHERN WHITE CEDAR					
Construction		675	780	840	.6
Standard	2x4	375	430	470	.6
Utility		175	200	220	.6
Studs		525	600	660	.6
No. 1 & Appearance	2x6	1000	1150	1250	.8
No. 2	and	825	950	1030	.7
No. 3	wider	475	550	590	.6

(continued)

[1]Values for modulus of elasticity shown in ( ) apply to Balsam Fir.
[2]Values for modulus of elasticity shown in ( ) apply to Coast Species.
[3]Values for modulus of elasticity shown in ( ) apply to Northern Aspen.
[4]Values for modulus of elasticity shown in ( ) apply to Mountain Hemlock-Hem-Fir.
[5]Values for modulus of elasticity shown in ( ) apply to White Woods (Mixed Species).
[6]Utility grades of all species may be used only under conditions specifically approved by the Building Official.

SPECIES AND GRADE[6]	SIZE	ALLOWABLE UNIT STRESS IN BENDING "F_b"			MODULUS OF ELASTICITY "E" $1x10^6$ psi
		FLOOR OR CEILING MEMBERS	ROOF MEMBERS		
			SNOW LOADING	NO SNOW LOADING	
PONDEROSA PINE—SUGAR PINE(PONDEROSA PINE—LODGEPOLE PINE)					
Construction		825	950	1030	1.0
Standard	2x4	450	520	560	1.0
Utility		225	260	280	1.0
Studs		625	720	780	1.0
No. 1 & Appearance	2x6	1200	1380	1500	1.2
No. 2	and	975	1120	1220	1.1
No. 3	wider	575	660	720	1.0
SITKA SPRUCE					
Construction		925	1060	1160	1.2
Standard	2x4	500	580	620	1.2
Utility		250	290	310	1.2
Studs		700	800	880	1.2
No. 1 & Appearance	2x6	1300	1500	1620	1.5
No. 2	and	1050	1210	1310	1.3
No. 3	wider	600	690	750	1.2
SOUTHERN PINE (Surfaced dry)					
Construction		1200	1380	1500	1.4
Standard	2x4	700	800	880	1.4
Utility		325	375	410	1.4
Studs		950	1090	1190	1.4
No. 1		1750	2010	2190	1.8
No. 1 Dense		2050	2380	2590	1.9
No. 2	2x6	1200	1380	1500	1.4
No. 2 MG	and	1450	1670	1810	1.6
No. 2 Dense	wider	1650	1900	2060	1.7
No. 3		825	950	1030	1.4
No. 3 Dense		975	1120	1220	1.5
SPRUCE—PINE—FIR					
Construction		850	980	1060	1.2
Standard	2x4	475	550	590	1.2
Utility		225	260	280	1.2
Stud		650	750	810	1.2
No. 1 & Appearance	2x6	1200	1380	1500	1.5
No. 2	and	1000	1150	1250	1.3
No. 3	wider	575	660	720	1.2
WESTERN CEDARS					
WESTERN CEDARS (NORTH)					
Construction		875	1010	1090	.9
Standard	2x4	500	580	620	.9
Utility		225	260	280	.9
Studs		675	780	840	.9
No. 1 & Appearance	2x6	1300	1500	1620	1.1
No. 2	and	1050	1210	1310	1.0
No. 3	wider	625	720	780	.9

[1]Values for modulus of elasticity shown in ( ) apply to Balsam Fir.
[2]Values for modulus of elasticity shown in ( ) apply to Coast Species.
[3]Values for modulus of elasticity shown in ( ) apply to Northern Aspen.
[4]Values for modulus of elasticity shown in ( ) apply to Mountain Hemlock-Hem-Fir.
[5]Values for modulus of elasticity shown in ( ) apply to White Woods (Mixed Species).
[6]Utility grades of all species may be used only under conditions specifically approved by the Building Official.

TABLE NO. 25-T-R-1—ALLOWABLE SPANS FOR LOW OR HIGH SLOPE RAFTERS
20 LBS. PER SQ. FT. LIVE LOAD
(Supporting Drywall Ceiling)

DESIGN CRITERIA: Strength—15 lbs. per sq. ft. dead load plus 20 lbs. per sq. ft. live load determines required fiber stress. Deflection—For 20 lbs. per sq. ft. live load. Limited to span in inches divided by 240. RAFTERS: Spans are measured along the horizontal projection and loads are considered as applied on the horizontal projection.

RAFTER SIZE (IN)	RAFTER SPACING (IN)	Allowable Extreme Fiber Stress in Bending, "F_b" (psi)														
		500	600	700	800	900	1000	1100	1200	1300	1400	1500	1600	1700	1800	1900
2x6	12.0	8-6 0.26	9-4 0.35	10-0 0.44	10-9 0.54	11-5 0.64	12-0 0.75	12-7 0.86	13-2 0.98	13-8 1.11	14-2 1.24	14-8 1.37	15-2 1.51	15-8 1.66	16-1 1.81	16-7 1.96
	16.0	7-4 0.23	8-1 0.30	8-8 0.38	9-4 0.46	9-10 0.55	10-5 0.65	10-11 0.75	11-5 0.85	11-10 0.97	12-4 1.07	12-9 1.19	13-2 1.31	13-7 1.44	13-11 1.56	14-4 1.70
	24.0	6-0 0.19	6-7 0.25	7-1 0.31	7-7 0.38	8-1 0.45	8-6 0.53	8-11 0.61	9-4 0.70	9-8 0.78	10-0 0.88	10-5 0.97	10-9 1.07	11-1 1.17	11-5 1.28	11-8 1.39
2x8	12.0	11-2 0.26	12-3 0.35	13-3 0.44	14-2 0.54	15-0 0.64	15-10 0.75	16-7 0.86	17-4 0.98	18-0 1.11	18-9 1.24	19-5 1.37	20-0 1.51	20-8 1.66	21-3 1.81	21-10 1.96
	16.0	9-8 0.23	10-7 0.30	11-6 0.38	12-3 0.46	13-0 0.55	13-8 0.65	14-4 0.75	15-0 0.85	15-7 0.96	16-3 1.07	16-9 1.19	17-4 1.31	17-10 1.44	18-5 1.56	18-11 1.70
	24.0	7-11 0.19	8-8 0.25	9-4 0.31	10-0 0.38	10-7 0.45	11-2 0.53	11-9 0.61	12-3 0.70	12-9 0.78	13-3 0.88	13-8 0.97	14-2 1.07	14-7 1.17	15-0 1.28	15-5 1.39
2x10	12.0	14-3 0.26	15-8 0.35	16-11 0.44	18-1 0.54	19-2 0.64	20-2 0.75	21-2 0.86	22-1 0.98	23-0 1.11	23-11 1.24	24-9 1.37	25-6 1.51	26-4 1.66	27-1 1.81	27-10 1.96
	16.0	12-4 0.23	13-6 0.30	14-8 0.38	15-8 0.46	16-7 0.55	17-6 0.65	18-4 0.75	19-2 0.85	19-11 0.96	20-8 1.07	21-5 1.19	22-1 1.31	22-10 1.44	23-5 1.56	24-1 1.70
	24.0	10-1 0.19	11-1 0.25	11-11 0.31	12-9 0.38	13-6 0.45	14-3 0.53	15-0 0.61	15-8 0.70	16-3 0.78	16-11 0.88	17-6 0.97	18-1 1.07	18-7 1.17	19-2 1.28	19-8 1.39
2x12	12.0	17-4 0.26	19-0 0.35	20-6 0.44	21-11 0.54	23-3 0.64	24-7 0.75	25-9 0.86	26-11 0.98	28-0 1.11	29-1 1.24	30-1 1.37	31-1 1.51	32-0 1.66	32-11 1.81	33-10 1.96
	16.0	15-0 0.23	16-6 0.30	17-9 0.38	19-0 0.46	20-2 0.55	21-3 0.65	22-4 0.75	23-3 0.85	24-3 0.97	25-2 1.07	26-0 1.19	26-11 1.31	27-9 1.44	28-6 1.56	29-4 1.70
	24.0	12-3 0.19	13-5 0.25	14-6 0.31	15-6 0.38	16-6 0.45	17-4 0.53	18-2 0.61	19-0 0.70	19-10 0.78	20-6 0.88	21-3 0.97	21-11 1.07	22-8 1.17	23-3 1.28	23-11 1.39

[1]The required modulus of elasticity, *E*, in 1,000,000 pounds per square inch is shown below each span.

[2]Use single or repetitive member bending stress values (F_b) and modulus of elasticity values (E), from Table No. 25-A-1.

[3]For more comprehensive tables covering a broader range of bending stress values (F_b) and modulus of elasticity values (E), other spacing of members and other conditions of loading, see the Uniform Building Code.

[4]The spans in these tables are intended for use in covered structures or where moisture content in use does not exceed 19 percent.

TABLE NO. 25-T-R-2—ALLOWABLE SPANS FOR LOW OR HIGH SLOPE RAFTERS 30 LBS. PER SQ. FT. LIVE LOAD (Supporting Drywall Ceiling)

DESIGN CRITERIA: Strength—15 lbs. per sq. ft. dead load plus 30 lbs. per sq. ft. live load determines required fiber stress. Deflection—For 30 lbs. per sq. ft. live load. Limited to span in inches divided by 240. RAFTERS: Spans are measured along the horizontal projection and loads are considered as applied on the horizontal projection.

RAFTER SIZE (IN)	RAFTER SPACING (IN)	Allowable Extreme Fiber Stress in Bending, "F_b" (psi)														
		500	600	700	800	900	1000	1100	1200	1300	1400	1500	1600	1700	1800	1900
2x6	12.0	7-6 0.27	8-2 0.36	8-10 0.45	9-6 0.55	10-0 0.66	10-7 0.77	11-1 0.89	11-7 1.01	12-1 1.14	12-6 1.28	13-0 1.41	13-5 1.56	13-10 1.71	14-2 1.86	14-7 2.02
2x6	16.0	6-6 0.24	7-1 0.31	7-8 0.39	8-2 0.48	8-8 0.57	9-2 0.67	9-7 0.77	10-0 0.88	10-5 0.99	10-10 1.10	11-3 1.22	11-7 1.35	11-11 1.48	12-4 1.61	12-8 1.75
2x6	24.0	5-4 0.19	5-10 0.25	6-3 0.32	6-8 0.39	7-1 0.46	7-6 0.54	7-10 0.63	8-2 0.72	8-6 0.81	8-10 0.90	9-2 1.00	9-6 1.10	9-9 1.21	10-0 1.31	10-4 1.43
2x8	12.0	9-10 0.27	10-10 0.36	11-8 0.45	12-6 0.55	13-3 0.66	13-11 0.77	14-8 0.89	15-3 1.01	15-11 1.14	16-6 1.28	17-1 1.41	17-8 1.56	18-2 1.71	18-9 1.86	19-3 2.02
2x8	16.0	8-7 0.24	9-4 0.31	10-1 0.39	10-10 0.48	11-6 0.57	12-1 0.67	12-8 0.77	13-3 0.88	13-9 0.99	14-4 1.10	14-10 1.22	15-3 1.35	15-9 1.48	16-3 1.61	16-8 1.75
2x8	24.0	7-0 0.19	7-8 0.25	8-3 0.32	8-10 0.39	9-4 0.46	9-10 0.54	10-4 0.63	10-10 0.72	11-3 0.81	11-8 0.90	12-1 1.00	12-6 1.10	12-10 1.21	13-3 1.31	13-7 1.43
2x10	12.0	12-7 0.27	13-9 0.36	14-11 0.45	15-11 0.55	16-11 0.66	17-10 0.77	18-8 0.89	19-6 1.01	20-4 1.14	21-1 1.28	21-10 1.41	22-6 1.56	23-3 1.71	23-11 1.86	24-6 2.02
2x10	16.0	10-11 0.24	11-11 0.31	12-11 0.39	13-9 0.48	14-8 0.57	15-5 0.67	16-2 0.77	16-11 0.88	17-7 0.99	18-3 1.10	18-11 1.22	19-6 1.35	20-1 1.48	20-8 1.61	21-3 1.75
2x10	24.0	8-11 0.19	9-9 0.25	10-6 0.32	11-3 0.39	11-11 0.46	12-7 0.54	13-2 0.63	13-9 0.72	14-4 0.81	14-11 0.90	15-5 1.00	15-11 1.10	16-5 1.21	16-11 1.31	17-4 1.43
2x12	12.0	15-4 0.27	16-9 0.36	18-1 0.45	19-4 0.55	20-6 0.66	21-8 0.77	22-8 0.89	23-9 1.01	24-8 1.14	25-7 1.28	26-6 1.41	27-5 1.56	28-3 1.71	29-1 1.86	29-10 2.02
2x12	16.0	13-3 0.24	14-6 0.31	15-8 0.39	16-9 0.48	17-9 0.57	18-9 0.67	19-8 0.77	20-6 0.88	21-5 0.99	22-2 1.10	23-0 1.22	23-9 1.35	24-5 1.48	25-2 1.61	25-10 1.75
2x12	24.0	10-10 0.19	11-10 0.25	12-10 0.32	13-8 0.39	14-6 0.46	15-4 0.54	16-1 0.63	16-9 0.72	17-5 0.81	18-1 0.90	18-9 1.00	19-4 1.10	20-0 1.21	20-6 1.31	21-1 1.43

See Table 25-T-R-1 Footnote

TABLE NO. 25-T-J-1—ALLOWABLE SPANS FOR FLOOR JOISTS 40 LBS. PER SQ. FT. LIVE LOAD

DESIGN CRITERIA: Deflection—For 40 lbs. per sq. ft. live load. Limited to span in inches divided by 360. Strength—Live load of 40 lbs. per sq. ft. plus dead load of 10 lbs. per sq. ft. determines the required fiber stress value.

JOIST SIZE (IN)	SPACING (IN)	Modulus of Elasticity, "E", in 1,000,000 psi													
		0.8	0.9	1.0	1.1	1.2	1.3	1.4	1.5	1.6	1.7	1.8	1.9	2.0	2.2
2x6	12.0	8-6 720	8-10 780	9-2 830	9-6 890	9-9 940	10-0 990	10-3 1040	10-6 1090	10-9 1140	10-11 1190	11-2 1230	11-4 1280	11-7 1320	11-11 1410
	16.0	7-9 790	8-0 860	8-4 920	8-7 980	8-10 1040	9-1 1090	9-4 1150	9-6 1200	9-9 1250	9-11 1310	10-2 1360	10-4 1410	10-6 1460	10-10 1550
	24.0	6-9 900	7-0 980	7-3 1050	7-6 1120	7-9 1190	7-11 1250	8-2 1310	8-4 1380	8-6 1440	8-8 1500	8-10 1550	9-0 1610	9-2 1670	9-6 1780
2x8	12.0	11-3 720	11-8 780	12-1 830	12-6 890	12-10 940	13-2 990	13-6 1040	13-10 1090	14-2 1140	14-5 1190	14-8 1230	15-0 1280	15-3 1320	15-9 1410
	16.0	10-2 790	10-7 850	11-0 920	11-4 980	11-8 1040	12-0 1090	12-3 1150	12-7 1200	12-10 1250	13-1 1310	13-4 1360	13-7 1410	13-10 1460	14-3 1550
	24.0	8-11 900	9-3 980	9-7 1050	9-11 1120	10-2 1190	10-6 1250	10-9 1310	11-0 1380	11-3 1440	11-5 1500	11-8 1550	11-11 1610	12-1 1670	12-6 1780
2x10	12.0	14-4 720	14-11 780	15-5 830	15-11 890	16-5 940	16-10 990	17-3 1040	17-8 1090	18-0 1140	18-5 1190	18-9 1230	19-1 1280	19-5 1320	20-1 1410
	16.0	13-0 790	13-6 850	14-0 920	14-6 980	14-11 1040	15-3 1090	15-8 1150	16-0 1200	16-5 1250	16-9 1310	17-0 1360	17-4 1410	17-8 1460	18-3 1550
	24.0	11-4 900	11-10 980	12-3 1050	12-8 1120	13-0 1190	13-4 1250	13-8 1310	14-0 1380	14-4 1440	14-7 1500	14-11 1550	15-2 1610	15-5 1670	15-11 1780
2x12	12.0	17-5 720	18-1 780	18-9 830	19-4 890	19-11 940	20-6 990	21-0 1040	21-6 1090	21-11 1140	22-5 1190	22-10 1230	23-3 1280	23-7 1320	24-5 1410
	16.0	15-10 790	16-5 860	17-0 920	17-7 980	18-1 1040	18-7 1090	19-1 1150	19-6 1200	19-11 1250	20-4 1310	20-9 1360	21-1 1410	21-6 1460	22-2 1550
	24.0	13-10 900	14-4 980	14-11 1050	15-4 1120	15-10 1190	16-3 1250	16-8 1310	17-0 1380	17-5 1440	17-9 1500	18-1 1550	18-5 1610	18-9 1670	19-4 1780

[1]The required extreme fiber stress in bending, F_b, in pounds per square inch is shown below each span.

[2]Use single or repetitive member bending stress values (F_b) and modulus of elasticity values (E), from Table No. 25-A-1.

[3]For more comprehensive tables covering a broader range of bending stress values (F_b) and modulus of elasticity values (E), other spacing of members and other conditions of loading, see the Uniform Building Code.

[4]The spans in these tables are intended for use in covered structures or where moisture content in use does not exceed 19 percent.

TABLE NO. 25-T-J-6—ALLOWABLE SPANS FOR CEILING JOISTS 10 LBS. PER SQ. FT. LIVE LOAD
(Drywall Ceiling)

DESIGN CRITERIA: Deflection—For 10 lbs. per sq. ft. live load. Limited to span in inches divided by 240. Strength—Live load of 10 lbs. per sq. ft. plus dead load of 5 lbs. per sq. ft. determines required fiber stress value.

JOIST SIZE (IN)	SPACING (IN)	Modulus of Elasticity, "E", in 1,000,000 psi													
		0.8	0.9	1.0	1.1	1.2	1.3	1.4	1.5	1.6	1.7	1.8	1.9	2.0	2.2
2x4	12.0	9-10 710	10-3 770	10-7 830	10-11 880	11-3 930	11-7 980	11-10 1030	12-2 1080	12-5 1130	12-8 1180	12-11 1220	13-2 1270	13-4 1310	13-9 1400
	16.0	8-11 780	9-4 850	9-8 910	9-11 970	10-3 1030	10-6 1080	10-9 1140	11-0 1190	11-3 1240	11-6 1290	11-9 1340	11-11 1390	12-2 1440	12-6 1540
	24.0	7-10 900	8-1 970	8-5 1040	8-8 1110	8-11 1170	9-2 1240	9-5 1300	9-8 1360	9-10 1420	10-0 1480	10-3 1540	10-5 1600	10-7 1650	10-11 1760
2x6	12.0	15-6 710	16-1 770	16-8 830	17-2 880	17-8 930	18-2 980	18-8 1030	19-1 1080	19-6 1130	19-11 1180	20-3 1220	20-8 1270	21-0 1310	21-8 1400
	16.0	14-1 780	14-7 850	15-2 910	15-7 970	16-1 1030	16-6 1080	16-11 1140	17-4 1190	17-8 1240	18-1 1290	18-5 1340	18-9 1390	19-1 1440	19-8 1540
	24.0	12-3 900	12-9 970	13-3 1040	13-8 1110	14-1 1170	14-5 1240	14-9 1300	15-2 1360	15-6 1420	15-9 1480	16-1 1540	16-4 1600	16-8 1650	17-2 1760
2x8	12.0	20-5 710	21-2 770	21-11 830	22-8 880	23-4 930	24-0 980	24-7 1030	25-2 1080	25-8 1130	26-2 1180	26-9 1220	27-2 1270	27-8 1310	28-7 1400
	16.0	18-6 780	19-3 850	19-11 910	20-7 970	21-2 1030	21-9 1080	22-4 1140	22-10 1190	23-4 1240	23-10 1290	24-3 1340	24-8 1390	25-2 1440	25-11 1540
	24.0	16-2 900	16-10 970	17-5 1040	18-0 1110	18-6 1170	19-0 1240	19-6 1300	19-11 1360	20-5 1420	20-10 1480	21-2 1540	21-7 1600	21-11 1650	22-8 1760
2x10	12.0	26-0 710	27-1 770	28-0 830	28-11 880	29-9 930	30-7 980	31-4 1030	32-1 1080	32-9 1130	33-5 1180	34-1 1220	34-8 1270	35-4 1310	36-5 1400
	16.0	23-8 780	24-7 850	25-5 910	26-3 970	27-1 1030	27-9 1080	28-6 1140	29-2 1190	29-9 1240	30-5 1290	31-0 1340	31-6 1390	32-1 1440	33-1 1540
	24.0	20-8 900	21-6 970	22-3 1040	22-11 1110	23-8 1170	24-3 1240	24-10 1300	25-5 1360	26-0 1420	26-6 1480	27-1 1540	27-6 1600	28-0 1650	28-11 1760

See Footnote Table 25-T-J-1

TABLE NO. 25-T-R-7—ALLOWABLE SPANS FOR LOW SLOPE RAFTERS SLOPE 3 IN 12 OR LESS 20 LBS. PER SQ. FT. LIVE LOAD
(Unfinished Ceiling)

DESIGN CRITERIA: Strength—10 lbs. per sq. ft. dead load plus 20 lbs. per sq. ft. live load determines required fiber stress. Deflection—For 20 lbs. per sq. ft. live load. Limited to span in inches divided by 240. RAFTERS: Spans are measured along the horizontal projection and loads are considered as applied on the horizontal projection.

RAFTER SIZE (IN)	SPACING (IN)	Allowable Extreme Fiber Stress in Bending, "F_b" (psi)														
		500	600	700	800	900	1000	1100	1200	1300	1400	1500	1600	1700	1800	1900
2x6	12.0	9-2 0.33	10-0 0.44	10-10 0.55	11-7 0.67	12-4 0.80	13-0 0.94	13-7 1.09	14-2 1.24	14-9 1.40	15-4 1.56	15-11 1.73	16-5 1.91	16-11 2.09	17-5 2.28	17-10 2.47
	16.0	7-11 0.29	8-8 0.38	9-5 0.48	10-0 0.58	10-8 0.70	11-3 0.82	11-9 0.94	12-4 1.07	12-10 1.21	13-3 1.35	13-9 1.50	14-2 1.65	14-8 1.81	15-1 1.97	15-6 2.14
	24.0	6-6 0.24	7-1 0.31	7-8 0.39	8-2 0.48	8-8 0.57	9-2 0.67	9-7 0.77	10-0 0.88	10-5 0.99	10-10 1.10	11-3 1.22	11-7 1.35	11-11 1.48	12-4 1.61	12-8 1.75
2x8	12.0	12-1 0.33	13-3 0.44	14-4 0.55	15-3 0.67	16-3 0.80	17-1 0.94	17-11 1.09	18-9 1.24	19-6 1.40	20-3 1.56	20-11 1.73	21-7 1.91	22-3 2.09	22-11 2.28	23-7 2.47
	16.0	10-6 0.29	11-6 0.38	12-5 0.48	13-3 0.58	14-0 0.70	14-10 0.82	15-6 0.94	16-3 1.07	16-10 1.21	17-6 1.35	18-2 1.50	18-9 1.65	19-4 1.81	19-10 1.97	20-5 2.14
	24.0	8-7 0.24	9-4 0.31	10-1 0.39	10-10 0.48	11-6 0.57	12-1 0.67	12-8 0.77	13-3 0.88	13-9 0.99	14-4 1.10	14-10 1.22	15-3 1.35	15-9 1.48	16-3 1.61	16-8 1.75
2x10	12.0	15-5 0.33	16-11 0.44	18-3 0.55	19-6 0.67	20-8 0.80	21-10 0.94	22-10 1.09	23-11 1.24	24-10 1.40	25-10 1.56	26-8 1.73	27-7 1.91	28-5 2.09	29-3 2.28	30-1 2.47
	16.0	13-4 0.29	14-8 0.38	15-10 0.48	16-11 0.58	17-11 0.70	18-11 0.82	19-10 0.94	20-8 1.07	21-6 1.21	22-4 1.35	23-2 1.50	23-11 1.65	24-7 1.81	25-4 1.97	26-0 2.14
	24.0	10-11 0.24	11-11 0.31	12-11 0.39	13-9 0.48	14-8 0.57	15-5 0.67	16-2 0.77	16-11 0.88	17-7 0.99	18-3 1.10	18-11 1.22	19-6 1.35	20-1 1.48	20-8 1.61	21-3 1.75
2x12	12.0	18-9 0.33	20-6 0.44	22-2 0.55	23-9 0.67	25-2 0.80	26-6 0.94	27-10 1.09	29-1 1.24	30-3 1.40	31-4 1.56	32-6 1.73	33-6 1.91	34-7 2.09	35-7 2.28	36-7 2.47
	16.0	16-3 0.29	17-9 0.38	19-3 0.48	20-6 0.58	21-9 0.70	23-0 0.82	24-1 0.94	25-2 1.07	26-2 1.21	27-2 1.35	28-2 1.50	29-1 1.65	29-11 1.81	30-10 1.97	31-8 2.14
	24.0	13-3 0.24	14-6 0.31	15-8 0.39	16-9 0.48	17-9 0.57	18-9 0.67	19-8 0.77	20-6 0.88	21-5 0.99	22-2 1.10	23-0 1.22	23-9 1.35	24-5 1.48	25-2 1.61	25-10 1.75

See Footnote Table 25-T-R-1

TABLE NO. 25-T-R-11—ALLOWABLE SPANS FOR HIGH SLOPE RAFTERS SLOPE OVER 3 IN 12
30 LBS. PER SQ. FT. LIVE LOAD
(Heavy Roof Covering)

DESIGN CRITERIA: Strength—15 lbs. per sq. ft. dead load plus 30 lbs. per sq. ft. live load determines required fiber stress. Deflection—For 30 lbs. per sq. ft. live load. Limited to span in inches divided by 180. RAFTERS: Spans are measured along the horizontal projection and loads are considered as applied on the horizontal projection.

RAFTER SIZE (IN)	SPACING (IN)	Allowable Extreme Fiber Stress in Bending, "F_b" (psi)														
		500	600	700	800	900	1000	1100	1200	1300	1400	1500	1600	1700	1800	1900
2x4	12.0	4-9 0.20	5-3 0.27	5-8 0.34	6-0 0.41	6-5 0.49	6-9 0.58	7-1 0.67	7-5 0.76	7-8 0.86	8-0 0.96	8-3 1.06	8-6 1.17	8-9 1.28	9-0 1.39	9-3 1.51
	16.0	4-1 0.18	4-6 0.23	4-11 0.29	5-3 0.36	5-6 0.43	5-10 0.50	6-1 0.58	6-5 0.66	6-8 0.74	6-11 0.83	7-2 0.92	7-5 1.01	7-7 11.1	7-10 1.21	8-0 1.31
	24.0	3-4 0.14	3-8 0.19	4-0 0.24	4-3 0.29	4-6 0.35	4-9 0.41	5-0 0.47	5-3 0.54	5-5 0.61	5-8 0.68	5-10 0.75	6-0 0.83	6-3 0.90	6-5 0.99	6-7 1.07
2x6	12.0	7-6 0.20	8-2 0.27	8-10 0.34	9-6 0.41	10-0 0.49	10-7 0.58	11-1 0.67	11-7 0.76	12-1 0.86	12-6 0.96	13-0 1.06	13-5 1.17	13-10 1.28	14-2 1.39	14-7 1.51
	16.0	6-6 0.18	7-1 0.23	7-8 0.29	8-2 0.36	8-8 0.43	9-2 0.50	9-7 0.58	10-0 0.66	10-5 0.74	10-10 0.83	11-3 0.92	11-7 1.01	11-11 1.11	12-4 1.21	12-8 1.31
	24.0	5-4 0.14	5-10 0.19	6-3 0.24	6-8 0.29	7-1 0.35	7-6 0.41	7-10 0.47	8-2 0.54	8-6 0.61	8-10 0.68	9-2 0.75	9-6 0.83	9-9 0.90	10-0 0.99	10-4 1.07
2x8	12.0	9-10 0.20	10-10 0.27	11-8 0.34	12-6 0.41	13-3 0.49	13-11 0.58	14-8 0.67	15-3 0.76	15-11 0.86	16-6 0.96	17-1 1.06	17-8 1.17	18-2 1.28	18-9 1.39	19-3 1.51
	16.0	8-7 0.18	9-4 0.23	10-1 0.29	10-10 0.36	11-6 0.43	12-1 0.50	12-8 0.58	13-3 0.66	13-9 0.74	14-4 0.83	14-10 0.92	15-3 1.01	15-9 1.11	16-3 1.21	16-8 1.31
	24.0	7-0 0.14	7-8 0.19	8-3 0.24	8-10 0.29	9-4 0.35	9-10 0.41	10-4 0.47	10-10 0.54	11-3 0.61	11-8 0.68	12-1 0.75	12-6 0.83	12-10 0.90	13-3 0.99	13-7 1.07
2x10	12.0	12-7 0.20	13-9 0.27	14-11 0.34	15-11 0.41	16-11 0.49	17-10 0.58	18-8 0.67	19-6 0.76	20-4 0.86	21-1 0.96	21-10 1.06	22-6 1.17	23-3 1.28	23-11 1.39	24-6 1.51
	16.0	10-11 0.18	11-11 0.23	12-11 0.29	13-9 0.36	14-8 0.43	15-5 0.50	16-2 0.58	16-11 0.66	17-7 0.74	18-3 0.83	18-11 0.92	19-6 1.01	20-1 1.11	20-8 1.21	21-3 1.31
	24.0	8-11 0.14	9-9 0.19	10-6 0.24	11-3 0.29	11-11 0.35	12-7 0.41	13-2 0.47	13-9 0.54	14-4 0.61	14-11 0.68	15-5 0.75	15-11 0.83	16-5 0.90	16-11 0.99	17-4 1.07

See Footnote Table 25-T-R-1

TABLE NO. 25-T-R-13—ALLOWABLE SPANS FOR HIGH SLOPE RAFTERS SLOPE OVER 3 IN 12
20 LBS. PER SQ. FT. LIVE LOAD
(Light Roof Covering)

DESIGN CRITERIA: Strength—7 lbs. per sq. ft. dead load plus 20 lbs. per sq. ft. live load determines required fiber stress. Deflection—For 20 lbs. per sq. ft. live load. Limited to span in inches divided by 180. RAFTERS: Spans are measured along the horizontal projection and loads are considered as applied on the horizontal projection.

RAFTER SIZE (IN)	SPACING (IN)	Allowable Extreme Fiber Stress in Bending, "F_b" (psi)														
		500	600	700	800	900	1000	1100	1200	1300	1400	1500	1600	1700	1800	1900
2x4	12.0	6-2 0.29	6-9 0.38	7-3 0.49	7-9 0.59	8-3 0.71	8-8 0.83	9-1 0.96	9-6 1.09	9-11 1.23	10-3 1.37	10-8 1.52	11-0 1.68	11-4 1.84	11-8 2.00	12-0 2.17
2x4	16.0	5-4 0.25	5-10 0.33	6-4 0.42	6-9 0.51	7-2 0.61	7-6 0.72	7-11 0.83	8-3 0.94	8-7 1.06	8-11 1.19	9-3 1.32	9-6 1.45	9-10 1.59	10-1 1.73	10-5 1.88
2x4	24.0	4-4 0.21	4-9 0.27	5-2 0.34	5-6 0.42	5-10 0.50	6-2 0.59	6-5 0.68	6-9 0.77	7-0 0.87	7-3 0.97	7-6 1.08	7-9 1.19	8-0 1.30	8-3 1.41	8-6 1.53
2x6	12.0	9-8 0.29	10-7 0.38	11-5 0.49	12-3 0.59	13-0 0.71	13-8 0.83	14-4 0.96	15-0 1.09	15-7 1.23	16-2 1.37	16-9 1.52	17-3 1.68	17-10 1.84	18-4 2.00	18-10 2.17
2x6	16.0	8-4 0.25	9-2 0.33	9-11 0.42	10-7 0.51	11-3 0.61	11-10 0.72	12-5 0.83	13-0 0.94	13-6 1.06	14-0 1.19	14-6 1.32	15-0 1.45	15-5 1.59	15-11 1.73	16-4 1.88
2x6	24.0	6-10 0.21	7-6 0.27	8-1 0.34	8-8 0.42	9-2 0.50	9-8 0.59	10-2 0.68	10-7 0.77	11-0 0.87	11-5 0.97	11-10 1.08	12-3 1.19	12-7 1.30	13-0 1.41	13-4 1.53
2x8	12.0	12-9 0.29	13-11 0.38	15-1 0.49	16-1 0.59	17-1 0.71	18-0 0.83	18-11 0.96	19-9 1.09	20-6 1.23	21-4 1.37	22-1 1.52	22-9 1.68	23-6 1.84	24-2 2.00	24-10 2.17
2x8	16.0	11-0 0.25	12-1 0.33	13-1 0.42	13-11 0.51	14-10 0.61	15-7 0.72	16-4 0.83	17-1 0.94	17-9 1.06	18-5 1.19	19-1 1.32	19-9 1.45	20-4 1.59	20-11 1.73	21-6 1.88
2x8	24.0	9-0 0.21	9-10 0.27	10-8 0.34	11-5 0.42	12-1 0.50	12-9 0.59	13-4 0.68	13-11 0.77	14-6 0.87	15-1 0.97	15-7 1.08	16-1 1.19	16-7 1.30	17-1 1.41	17-7 1.53
2x10	12.0	16-3 0.29	17-10 0.38	19-3 0.49	20-7 0.59	21-10 0.71	23-0 0.83	24-1 0.96	25-2 1.09	26-2 1.23	27-2 1.37	28-2 1.52	29-1 1.68	30-0 1.84	30-10 2.00	31-8 2.17
2x10	16.0	14-1 0.25	15-5 0.33	16-8 0.42	17-10 0.51	18-11 0.61	19-11 0.72	20-10 0.83	21-10 0.94	22-8 1.06	23-7 1.19	24-5 1.32	25-2 1.45	25-11 1.59	26-8 1.73	27-5 1.88
2x10	24.0	11-6 0.21	12-7 0.27	13-7 0.34	14-6 0.42	15-5 0.50	16-3 0.59	17-1 0.68	17-10 0.77	18-6 0.87	19-3 0.97	19-11 1.08	20-7 1.19	21-2 1.30	21-10 1.41	22-5 1.53

See Footnote Table 25-T-R-1

TABLE NO. 25-T-R-14—ALLOWABLE SPANS FOR HIGH SLOPE RAFTERS SLOPE OVER 3 IN 12
30 LBS. PER SQ. FT. LIVE LOAD
(Light Roof Covering)

DESIGN CRITERIA: Strength—7 lbs. per sq. ft. dead load plus 30 lbs. per sq. ft. live load determines required fiber stress. Deflection—For 30 lbs. per sq. ft. live load. Limited to span in inches divided by 180. RAFTERS: Spans are measured along the horizontal projection and loads are considered as applied on the horizontal projection.

RAFTER SIZE (IN)	SPACING (IN)	Allowable Extreme Fiber Stress in Bending, "F_b"(psi)														
		500	600	700	800	900	1000	1100	1200	1300	1400	1500	1600	1700	1800	1900
2x4	12.0	5-3 0.27	5-9 0.36	6-3 0.45	6-8 0.55	7-1 0.66	7-5 0.77	7-9 0.89	8-2 1.02	8-6 1.15	8-9 1.28	9-1 1.42	9-5 1.57	9-8 1.72	10-0 1.87	10-3 2.03
	16.0	4-7 0.24	5-0 0.31	5-5 0.39	5-9 0.48	6-1 0.57	6-5 0.67	6-9 0.77	7-1 0.88	7-4 0.99	7-7 1.11	7-11 1.23	8-2 1.36	8-5 1.49	8-8 1.62	8-10 1.76
	24.0	3-9 0.19	4-1 0.25	4-5 0.32	4-8 0.39	5-0 0.47	5-3 0.55	5-6 0.63	5-9 0.72	6-0 0.81	6-3 0.91	6-5 1.01	6-8 1.11	6-10 1.21	7-1 1.32	7-3 1.43
2x6	12.0	8-3 0.27	9-1 0.36	9-9 0.45	10-5 0.55	11-1 0.66	11-8 0.77	12-3 0.89	12-9 1.02	13-4 1.15	13-10 1.28	14-4 1.42	14-9 1.57	15-3 1.72	15-8 1.87	16-1 2.03
	16.0	7-2 0.24	7-10 0.31	8-5 0.39	9-1 0.48	9-7 0.57	10-1 0.67	10-7 0.77	11-1 0.88	11-6 0.99	12-0 1.11	12-5 1.23	12-9 1.36	13-2 1.49	13-7 1.62	13-11 1.76
	24.0	5-10 0.19	6-5 0.25	6-11 0.32	7-5 0.39	7-10 0.47	8-3 0.55	8-8 0.63	9-1 0.72	9-5 0.81	9-9 0.91	10-1 1.01	10-5 1.11	10-9 1.21	11-1 1.32	11-5 1.43
2x8	12.0	10-11 0.27	11-11 0.36	12-10 0.45	13-9 0.55	14-7 0.66	15-5 0.77	16-2 0.89	16-10 1.02	17-7 1.15	18-2 1.28	18-10 1.42	19-6 1.57	20-1 1.72	20-8 1.87	21-3 2.03
	16.0	9-5 0.24	10-4 0.31	11-2 0.39	11-11 0.48	12-8 0.57	13-4 0.67	14-0 0.77	14-7 0.88	15-2 0.99	15-9 1.11	16-4 1.23	16-10 1.36	17-4 1.49	17-11 1.62	18-4 1.76
	24.0	7-8 0.19	8-5 0.25	9-1 0.32	9-9 0.39	10-4 0.47	10-11 0.55	11-5 0.63	11-11 0.72	12-5 0.81	12-10 0.91	13-4 1.01	13-9 1.11	14-2 1.21	14-7 1.32	15-0 1.43
2x10	12.0	13-11 0.27	15-2 0.36	16-5 0.45	17-7 0.55	18-7 0.66	19-8 0.77	20-7 0.89	21-6 1.02	22-5 1.15	22-3 1.28	24-1 1.42	24-10 1.57	25-7 1.72	26-4 1.87	27-1 2.03
	16.0	12-0 0.26	13-2 0.34	14-3 0.43	15-2 0.53	16-2 0.63	17-0 0.74	17-10 0.85	18-7 0.97	19-5 1.09	20-1 1.22	20-10 1.35	21-6 1.49	22-2 1.63	22-10 1.78	23-5 1.93
	24.0	9-10 0.19	10-9 0.25	11-7 0.32	12-5 0.39	13-2 0.47	13-11 0.55	14-7 0.63	15-2 0.72	15-10 0.81	16-5 0.91	17-0 1.01	17-7 1.11	18-1 1.21	18-7 1.32	19-2 1.43

See Footnote Table 25-T-R-1

Resources

General Carpentry Texts

Against the Grain: A Carpentry Manual for Women
Dale McCormick
Iowa City Women's Press Inc., Iowa City, Iowa, 1977
258 pp., paperback; $6.00

Written by a professional woman carpenter for other women, *Against The Grain* is a valuable reference for the beginning or the experienced builder. Each section of this book is clearly written and illustrated, with attention to process as well as to accurate technical detail. The tone of the book is serious but lively and personal. What a pleasant surprise to read a carpentry text that acknowledges that you might have trouble lifting a partition wall into place, for instance, and tells you what to do about it. Sections vary in depth, but some of the best ones cover stair building, joinery, and working with concrete.

Modern Carpentry
Willis H. Wagner
The Goodheart-Willcox Co., Inc., South Holland, Ill., 1973
480 pp., hardcover; approx. $10

Few carpenters that I know don't own a copy of this book. It's comprehensive enough to be considered the encyclopedia of carpentry. You can come back to this book for every imaginable detail and rarely be disappointed. Organization follows the usual "from the ground up" order, with sections on footings and foundations, framing (floor, wall and ceiling, and roof), roofing materials, and so on. There is a section on cabinetmaking and one on post and beam construction. If you buy only a few carpentry books, this one should be on your high-priority list.

Fundamentals of Carpentry: Vol. 2, Practical Construction
Walter E. Durbahn and Elmer W. Sundberg
American Technical Society, Chicago, 1969
504 pp., hardcover; $11.95

This is one of a series of textbooks on elements of construction produced by the American Technical Society. Like *Modern Carpentry,* it is a good general carpentry reference covering everything from "preparing for the job" through interior finish work. Certain chapters offer more depth than *Modern Carpentry*—for example, there's a very thorough chapter on using the builder's level and transit level—while others are less detailed.

Practical House Carpentry
J. Douglas Wilson
McGraw-Hill Book Co. Inc., New York, 1957
360 pp., paperback; $2.95

A contractor friend recommended this book to me, and it turned out to be an inexpensive but reasonably thorough basic text. It covers the usual material, with certain details (how to construct a form for tapered piers) not found elsewhere.

House-building Books

Illustrated Housebuilding
Graham Blackburn
Overlook Press, Inc., Woodstock, N.Y., 1974
155 pp., hardcover; $10

This is a breezy step-by-step book that is good for a first-time builder because it is definitely not intimidating. Each step is illustrated and reduced to simple, clear terms. If anything, the fault of the book is that it tends to oversimplify, which might prove troublesome if this was your only reference. Friends who have used this book (with little prior carpentry experience) recommend it.

From the Ground Up
John N. Cole and Charles Wing
Atlantic Monthly Press Book, Little, Brown and Co., Boston, 1976
244 pp., paperback; $7.95

Written by two owner-builders who went on to share and teach what they learned, *From the Ground Up* fills some vital gaps in previously existing building books. There is a lot of emphasis upon choosing a building site and then designing a house that fits both it and you. The design section includes the theory, formulas, and tables you need to select the structural beams for your house as well as floor joists and rafters. This section alone is worth the price of the book, but the authors go on to discuss the basics of building, heating, and wiring as well. While it won't really serve as a guide to general carpentry, the book offers a lot of unique and important information.

How to Build a Wood-Frame House
L. O. Anderson
Dover Publications, Inc., New York, 1973
223 pp., paperback; $3.50

Parts of this book are so thorough and clear that you won't need another reference and can work directly from the text. Other sections, unfortunately, skim much too lightly over the subject; the chapter on ceiling and roof framing, for example, talks about different types of roof framing but never tells you how to lay out a rafter. This is a book you might want to add to your library and use in conjunction with others.

How to Design and Build Your Own House
Lupe DiDonno and Phyllis Sperling
Alfred A. Knopf, New York, 1978
365 pp., paperback; $9.95

Here is a case of the best saved for last: this new book, written by two women who are both architects, is perhaps the finest of these house-building books, incorporating both theory of design and instructions in actual building. Although the builder in their book is still "he," women are encouraged to take an active part in the designing and making of their homes. A substantial portion of this book is devoted, naturally enough, to design. There is a section on drawing up preliminary plans and then making working drawings from these that is particularly valuable and not found elsewhere. Details for framing in and installing plumbing and heating units and a section on electricity add to the value of the book. The basic premise of this book is that a house should be built to suit your needs and fancies, and that "a house, like a person, should have its own mystique. It engages you, grows on you, endears itself to you in its own way."

Resources With More Specific Information/Orientation

Low-Cost Energy-Efficient Shelter for the Owner and Builder
ed. Eugene Eccli
Rodale Press, Emmaus, Pa., 1976
408 pp., hardcover; $10

This book addresses itself to designing structures that are environmentally sound, making minimum demands upon limited resources while still providing comfortable, attractive shelters to live in. Definitely a resource book you can go back to again and again, *Low-Cost . . . Shelter* contains a wealth of specialized material on solar greenhouses, solar water heaters, weatherproofing windows, using insulation, and so forth. The language of the book is simple enough and the illustration clear enough to make the subject matter feel approachable. There is a good section on converting parts of your existing house to take advantage of natural solar heat. There is also a list of "design groups"—people who are actively working in areas of low-cost housing, solar energy, and so on, and who are willing to work with you to solve problems and suggest alternative systems. The extensive bibliography will be valuable if you want to learn more than is contained in this book.

30 Energy-Efficient Houses . . . You Can Build
Alex Wade; photographs by Neal Ewenstein
Rodale Press, Emmaus, Pa., 1977
316 pp., paperback; $8.95

This is something of a companion volume to the above book. It is a book of plans and designs, photographs and suggestions; all houses included are space-consciously designed, heavily insulated, energy-saving. The houses range from very small to quite large, and each is discussed in terms of its actual good features, any problems created, actual heating bills, ups and downs of living in the house, and so on. Very specific details on, for example, types of siding and flooring materials used are included, along with both interior and exterior photographs. If you are building a house for yourself or doing professional carpentry, this book is a rich source of both ideas and contacts for actual building materials, economic comparisons, and so on.

Wood Structural Design Data, Vol. I
National Lumber Manufacturer's Assoc., Washington, D.C., 1957
Timber Design and Construction Handbook
Prepared by Timber Engineering Co., an affiliate of National Lumber Manufacturer's Assoc., F. W. Dodge Corp., New York, 1956
hardcover (out of print)

These two books may be found in the reference section of most libraries. They contain technical data on the structural properties and use of wood, and will be valuable if you are designing and drawing up your own plans.

Architectural Graphic Standards
Charles G. Ramsey and Harold R. Sleeper
John Wiley and Sons, New York, 1970
695 pp., hardcover; $50

This book, put out by the American Institute of Architects, is the reference work supreme for both carpenters and architects. It shows exact details and specifications for virtually any aspect of building you'll encounter. Ranging from the simplest to the most complicated work, this book can show you precisely how to cap wainscoting or wall paneling, install a skylight of any type, flash a difficult chimney or set of vents, install a suspension system for acoustical tile ceilings, and so on. This book has no real text and won't lead you through the steps of "how to"; it consists of pages of minutely detailed drawings with precise measurements and relationships of parts. If you're working as a professional carpenter and doing various types of jobs, this is a book you should become acquainted with and will someday want to own. You'll probably refer to it and learn from it for the rest of your career.

The Uniform Building Code
Dwelling Construction Under the Uniform Building Code
Both published by the International Conference of Building Officials, Whittier, Calif., periodically updated and revised

The Uniform Building Code is a comprehensive set of standards covering "the fire, life and structural safety of all buildings and related structures." It is used by most communities as the basis for building restrictions and procedures, and if you are building with a permit, you'll probably have to follow the code pretty exactly. Most libraries have a copy or two in closed stacks; you can ask for it and use it in the library. If you have a question about some aspect of the work you're doing and don't have immediate access to the *Code,* you can usually phone the local building department and ask specific questions. The *Code* usually costs about $17 but is quite hard to locate. A technical bookstore might carry it or order it for you.

Dwelling Construction is really a greatly shortened version of the *Code,* including most of the details for one- and two-story dwellings built with stud wall or masonry construction. This little book will provide you with details on headers, bracing, applying gypsum wallboard, and so forth. It is intended for "the convenience of the home builder who is interested in building a house in conformance with the building codes." Unfortunately it is very inconveniently hard to get hold of! Ask at your local building department or technical bookstore. It should cost only a couple of dollars and is well worth having if you are doing much building or remodeling.

National Construction Estimator
282 pp., paperback; $7.50
National Repair and Remodeling Estimator
Structures Cost Manual
238 pp., paperback; $10
All published by Craftsman Book Co., Solana Beach, Calif.

These three books are revised each year and reflect up-to-date costs for new construction and remodeling projects, based on labor costs and materials costs all over the country. They give cost per square foot for each aspect of building, plus ideas on labor-hours for specific jobs, estimates of materials needed, and so forth. These books are published by and can be ordered from Craftsman Book Company (see "Sources for Books . . .") and range from $7.50 to $10 (ask for current prices). If you are doing carpentry professionally, one of these books will help immeasurably in estimating jobs and giving bids. Don't take them entirely literally; a contractor friend says she uses them, but always has to adjust the figures some.

FHA Pole House Construction
U. S. Department of Housing and Urban Development, Washington, D.C.
29 pp.

This is a booklet on using poles for the foundation and framing of houses. While it doesn't cover the subject in much detail, it does provide the basics on making connections, bracing, embedding poles, and so on. There are enough diagrams of framing systems for you to improvise and work from this booklet. Tables with embedment depths for poles of various diameters in different types of soil are particularly useful if you're building this type of structure and want a strong foundation.

Pole Building Construction
Doug Merrilees and Evelyn Loveday
Garden Way Publishing Co., Charlotte, Vt., 1974
102 pp., paperback; $5.95

This is another book on the same subject, this one providing a little more detail (how to put on siding, how to tie rafters together, how to insulate suspended floors). There are working plans for a small, shed-roof cottage included.

I Built Myself a House
Helen Garvey
Shire Press, Box 40426, San Francisco, Calif. 94110
128 pp., paperback; $2.50

This is a very personal account of the building of a small country cabin using post and beam construction. If you're hesitating over taking on a similar project, this book may be the incentive you need. It has an air of enthusiasm that is hard to ignore. While this book won't tell you everything you need to know about this type of construction, it gives a good overview of the process of building a small structure and an introduction to the tools and materials you'll need.

Building with Logs
B. Allan Mackie
P.O. Box 1205, Prince George, British Columbia, Canada V2L 4V3
76 pp., paperback; $10

A fine teacher dedicated to the craft and creativity of building with logs, B. Allan Mackie has done an incredible work in Canada, inspiring a resurgence of active interest in the building of carefully crafted, permanent, and aesthetically beautiful log houses. His book has become the standard text on the subject and a best seller in that country. It covers choosing a site, financing, and planning, as well as the exact details of building a log house. Selection and cutting, peeling and storage of logs, plus instructions for handling heavy logs, framing in window and door openings, rafters, and even partitions are all included. There are excellent drawings and photographs throughout. The particular tools you need for this work are pictured, discussed, and shown in use. In some instances (as for log scribers), there are exact specifications for making your own tool for the job. Even if you've never thought of building a log house—or live in an area where it would be relatively impractical—you might well enjoy the book. The log houses Mackie writes of are ecologically sane, strong, durable, and handsome, and he talks about them with both precision and joy.

The Canadian Log House
ed. Mary Mackie
P.O. Box 1205, Prince George, British Columbia, Canada V2L 4V3
approx. 100 pp., paperback; $5.00 each

This yearly publication comes out each spring as an ongoing companion to *Building with Logs.* It includes writing by Mary Mackie, by her husband, and by other log house builders and enthusiasts. There are interviews with first-time builders as well as with professionals. Floor plans of different houses, reviews of related books, discussions of technical aspects of log house building all enrich this magazine; but above all of these, the photographs (both color and black and white) of log houses old and new will inspire you! These houses, for all their rugged simplicity, are astonishingly beautiful.

The B. Allan Mackie School of Log Building
P.O. Box 1205, Prince George, British Columbia, Canada V2L 4V3

If you are inspired by the above publications, you'll be pleased to learn that the Mackies run a registered trade school "teaching twentieth-century log construction skills"! This school is located in Canada but students from other countries are welcomed—and women are accepted as equal, active members of each class. (It was refreshing to find Mackie's brochure talking about the student as "he/she"—an assumption rarely made in *any* school, let alone a trade school devoted to construction!) If you're interested in learning how to build with logs, either for your own house or as a career, write for the latest brochure. There is also the option of traveling instructors. If you can set up a course with enough interested students in your area and pay for an instructor, the school will arrange to have a qualified log house builder come to teach and work with you.

Log Home Guide for Builders and Buyers
ed. Doris L. Muir
Muir Publishing Co. Ltd., Gardenvale, Quebec, Canada H0A IB0
$5.00 a year subscription in Canada; $6.00 elsewhere

This magazine is also devoted to the log home, but places more of an emphasis on precut log homes. It is a good source for locating companies that produce these homes and comparing their products. It also lists log-building schools, reviews related books, and has articles on and by people who have built or are building log houses. The illustrations and photographs are very well done.

The Timber Framing Book
Stewart Elliot and Eugenie Wallas
Housesmiths Press, P.O. Box 157, Kittery Point, Maine 03905, 1977
169 pp., paperback; $9.95

Like the Mackies' log-building publications, this book will start you thinking about a different way of building. The traditional timber-framed house is undergoing a revival under the hands of the Housesmiths, a group of builders devoted to constructing these elegantly simple homes. According to the authors, a timber-framed house can cost from 20 to 30 percent less to build than a stud-framed building of equivalent size. This book introduces you to timber framing and provides clear, well-illustrated instructions in the basic framing and joinery you will need to know. Each step is discussed in detail, with an emphasis on quality work for a durable as well as a handsome finished house. The "Raisin' Day" section includes some instruction on handling and lifting heavy beams that could easily be used for other projects.

When I first got my copy, I spent most of that evening reading it through—quite an accolade to a building book! Even though I had no immediate intentions of building a timber-frame house, the book held my attention and excited me. Now I'm wishing for an opportunity to put it to work as the handbook it is.

How to Create Interiors for the Disabled
Jane Randolph Cary
Pantheon Books (a division of Random House), New York, 1978
127 pp., paperback; $5.95

Ideas and specific instructions for adapting a house to the needs and comfort of the disabled person living there make this book a unique and much-needed resource. There are simple instructions for building ramps, creating windows that open more easily, designing a kitchen for a wheelchaired person, making pulley-operated doors, and so forth.

How to Build Decks
A Sunset Book, Lane Magazine and Book Co., Menlo Park, Calif., 1974
80 pp., paperback; $1.95

This book contains complete, clear, concise directions for planning and building a deck. Structural details are carefully illustrated so that you can see how something works as well as read about it. Excellent use of photographs for ideas and options, as well as tables for choosing sizes and spans for materials.

How to Rehabilitate Abandoned Buildings
Donald R. Brann
Easi-Bild Pattern Co. Inc., Briarcliff Manor, N.Y., 1975
285 pp., paperback; $3.50

If you do a lot of remodeling or are "rehabilitating" an older house, this book has some valuable information for you. It includes information on repairing sagging floors, fixing window sash, removing or adding partition walls, and so on. Because the book is written for the home handyperson, its language is simple and direct; each process is fully described and illustrated. You may want to skip some of the "philosophizing," but the "how to" is worth consulting.

Concrete Work Simplified
Donald R. Brann

Easi-Bild Pattern Co. Inc., Briarcliff Manor, N.Y., 1974, rev. ed. 1976
98 pp., paperback; $2.50

This book provides most of what you might want to know about working with concrete, from building forms to pouring slabs. It includes instructions for laying concrete block walls, building concrete steps, repairing and waterproofing basement walls, and more.

Floors: Selection and Maintenance
Bernard Berkeley
American Library Assoc., 1968
hardcover

Although this is an older book, it is fairly comprehensive in its coverage, particularly for wood floors of various types. Each type of finished floor (wood, resilient coverings such as linoleum and vinyl, carpets) is discussed in terms of initial cost, maintenance, installation, service life, and good and bad features. There is a full section on cleaning compounds, use of polishing and buffing equipment, and general floor care. Another section explains how to deal with scratches and stains on floors. This book is quite thorough and explicit—a useful handbook.

Wood Floors for Dwellings
Agricultural Handbook #204, U. S. Dept. of Agriculture Forest Service (order from Superintendent of Documents, U. S. Government Printing Office, Washington, D.C. 20402)
$.35

An inexpensive source of information on installing and maintaining wood floors, this booklet covers subfloors, systems for putting wood floors over concrete slabs, grading and installation of strip flooring, and so on. There is a good section on sanding and staining floors.

The Wall Book
Stanley Schuler
M. Evans and Co., Inc., New York, 1974
172 pp., paperback; $8.95

This book is devoted to the interior walls of a house and has a lot of very specific information that more general carpentry texts neglect. For example, it tells in detail how to install ceramic tiles (from putting up backer boards through maintaining and repairing finished walls). Other chapters cover wallpaper, plaster, wood paneling, rigid plastics, luminous walls, and more. Well illustrated by Marilyn Grastorf.

Home and Workshop Guide to Sharpening
Harry Walton
Popular Science Publishing Co., Harper & Row, New York, 1967
160 pp., paperback; $2.95

This book explains quite carefully how to sharpen woodworking tools, power-saw blades and handsaws, drill bits of various types, knives, and even garden tools. With the proper equipment (available from such companies as Woodcraft) and this book as a guide, you'll be able to keep a good cutting edge on all of your tools and save on sharpening fees.

Practical Farm Buildings
James S. Boyd, Ph.D.
Interstate Printers and Publishers, Inc., Danville, Ill., 1973
265 pp., paperback; $8.95

Although this text is intended for "on-the-job advice on farm building for non-engineers," it provides an excellent resource for designing your wood-frame house or other structure. Sections include "Computing Loads on Structures" and "Designing Joists, Beams, and Columns."

Roof Framing
H. H. Siegele
Drake Publishers, Inc., New York, 1972
175 pp., paperback; $3.95

Siegele discusses the principles of roof framing and their practical application. His manner is relaxed, so that reading the book is more like listening to a series of lectures by an interesting teacher. All aspects of framing are covered, and roof types vary from simple sheds or gables through octagons, gambrel roofs, and heavy timber designs. The book was written with the apprentice in mind, so everything is carefully spelled out, explained, and illustrated.

Roofers Handbook
William Edgar Johnson
Craftsman Book Co., Solana Beach, Calif., 1976
189 pp., paperback; $8.25

Whether you work as a professional carpenter, plan to reroof your own house, or just need to know how to deal with a persistently leaky roof, this is a book you can use. Written by an experienced professional roofer, it covers application of the basic types of shingles and shakes. There are sections on dealing with valleys and ridges of all sorts, handling leaks, and even operating a roofing company. This book is unusually well illustrated (by Laura Knight and the author), with enough attention to detail that you could easily take the book up on the roof with you and refer to it as you went along.

How to Work with Tools and Wood
ed. Robert Campbell and N. H. Mager
Pocket Books (a division of Simon and Schuster, Inc.), New York, 1971
paperback; $1.95

As the title indicates, this book covers choosing and working with basic hand tools and hand-held power tools, with particularly good sections on chiseling, using planes, and sharpening your tools. There is also a fine section on joining and holding wood, with specifics on making joints and using glues properly. If you don't have a friend who can teach you about tools and joinery, this book is doubly welcome as a resource!

The Complete Handbook of Power Tools
George R. Drake
Reston Publishing House, Reston, Va., 1975
415 pp., paperback; $4.95

This book covers eleven power tools (including drill press, sander-grinder, shaper, and table saw) in detail, telling how to operate, install, and maintain each tool. Tables for "trouble-shooting" give you a working reference for problems you might encounter with these tools: how to diagnose them and deal with them. Illustrated well with both drawings and photographs.

Work Is Dangerous to Your Health
Jeanne Stellman, Ph.D., and Susan Daum, M.D.
Vintage Books, 1973
paperback; $2.95

This book is subtitled: "A Handbook of Health Hazards in the Workplace and What You Can Do About Them"; it should be required reading for workers in all fields. Of particular interest to anyone doing carpentry (for oneself or professionally) is the section listing dangerous substances and materials one is likely to be exposed to; this section is broken down by occupation, and the carpenter's list is quite sobering. The bulk of the book is interesting enough reading on its own: each system of the body is discussed, with attention to keeping the whole body healthy. With the recent publicity about health hazards in asbestos manufacturing plants (as an example), this book poses some serious questions—and suggestions—that are critical to all of us. There is a particularly good section on protecting your back (one of the parts of the body most abused by many carpenters) and another good one on the impact of stress, such as noise and vibration, on the job. Besides awakening your consciousness about health hazards you might be exposing yourself to, this book will provide a great deal of useful and thought-provoking information. None of us can really afford not to read it!

Producing Your Own Power: How to Make Nature's Energy Sources Work for You
ed. Carol H. Stoner
Rodale Press, Emmaus, Pa., 1974
322 pp., hardcover; $8.95
The Solar Greenhouse Book
ed. James C. McCullagh
Rodale Press, Emmaus, Pa., 1978
328 pp., hardcover; $10.95
Goodbye to the Flush Toilet
ed. Carol H. Stoner
Rodale Press, Emmaus, Pa., 1977
paperback; $6.95

Rodale Press is taking serious steps to provide an up-to-date library of books on alternative solutions to contemporary problems. If you are concerned about designing and living in an energy-efficient house, these three books will provide some vital reading and resources.

Producing Your Own Power is a fine introduction to alternative power sources, such as windmills and wind generating plants, methane and water power plants. It is carefully done and contains a great deal of usable material. It also features a short section on "Conservation of Energy in Existing Structures."

The Solar Greenhouse Book should appeal to gardener and carpenter alike, as it offers a compilation of information and ideas not found anywhere else. Basically, the sections of this book cover designing, constructing, and managing the solar greenhouse. Contributors to this book live in various parts of the country, so that the different types of greenhouses shown and discussed are quite varied. Diagrams and photographs add to the text, and the overall result is an excellent resource, both in theory and in practical application of ideas.

Goodbye to the Flush Toilet is the first major publication on workable alternatives to our present water-wasteful waste system. This is an exploration of "water-saving alternatives to cesspools, septic tanks, and sewers" and should be read and considered by anyone involved in house building or remodeling or design. The book is full of very detailed information, including a careful comparison of present commercially manufactured composting toilets and instructions for building and using composting privies. There is a full section on saving household water by using flow-restricting devices, pressurized tank toilets, and so on. Sources for equipment and composting toilets are also given. This book may be a little ahead of its time in that bureaucrats in the building department may not be ready to listen or talk alternatives—but being well informed may help you convince them that you want to build a truly *workable* alternative.

Sources for Books, Tools, and Other Items

Craftsman Book Co.
542 Stevens Ave., Solana Beach, Calif. 92075

A mail-order source of practical reference books for builders, including hard-to-find books such as the *National Construction Estimator.*

The Woodcraft Supply Corp.
313 Montvale Ave., Woburn, Mass. 01801

Source for beautifully made tools for the carpenter, woodworker, instrument maker or cabinetmaker. Also offers a wide range of books covering everything from canoe building to timber framing, from violin making to Japanese architecture. Carries hard-to-find tools (including Japanese hand tools). Their $1.50 catalog will set you to daydreaming (and ordering!).

Brookstone Company
125 Vose Farm Rd., Peterborough, N. H. 03458

Another source for hard-to-find tools. Stock includes specialized tools such as a roofer's knife for cutting and trimming roll roofing or asphalt shingles. Also sells specialty items such as old-fashioned cut nails.

The Yurt Foundation
Bucks Harbor, Maine 04618
William Coperthwaite, Director

If "Making Magic" inspired you to want to build a yurt, the Yurt Foundation can provide you with the basic plans. It presently offers three plans: a "Standard Yurt" (17′ eaves diameter); the "Concentric Yurt" (32′ eaves diameter—actually a yurt within a yurt; order the "Standard" plan too); and the "Little Yurt" (12′ eaves diameter). The plans cost

$5.00 each and will arrive as a pleasant surprise—they are drawn up with hand lettering and are a nice alternative to regular blueprints. The Yurt Foundation also puts out a yearly calendar dedicated to "some person or group that has been concerned for simplicity." Those I've seen were tastefully done (one featured color photographs of yurts; the other with line drawings of Shaker furniture) and inexpensive.

Simpson Co.
General offices: 1470 Doolittle Drive, P.O. Box 1568, San Leandro, Calif. 94577
Manufacturer of quality metal connectors (Strong-Tie is their brand name product) such as joist hangers, post caps, and post anchors.

Burke Concrete Accessories, Inc.
General offices: 2655 Campus Drive, P.O. Box 5818, San Mateo, Calif. 94402
This company manufactures and sells an impressive variety of accessories for concrete and form work. It has branch offices all over the country, so its products are widely available. A source for form ties, special adhesives and waterproofing materials for concrete, among other items.

Zomeworks Corp.
c/o Steve Baer, P.O. Box 712, Albuquerque, N. Mex. 87103
Designs and plans for solar houses, research on retrofitting houses for use of solar energy. Ask for brochure.

Nelson Screw Products, Inc.
Logansport, Ind. 46947
Manufacturer of Little Giant floor jack.

Superintendent of Documents
U. S. Government Printing Office, Washington, D. C. 20402
Source for U. S. Department of Agriculture booklets and pamphlets, and many other publications.

Index

Acoustical tiles, 354
Adhesives, 92–93, 354 (*see also* Glues); cement, 92; kinds, uses, 92–93; mastics, 92, 93; panel, 296; resins, 92, 358; safety and care in use of, 93
Adjustable dado blade, 58
Adobe house building, 137–43; bricks, 139, 140, 142; doors and windows, 140–41; lintels, 141; plan, 140; vigas (roof beams), 137, 139 141–42; walls, 140, 141, 142
Adz, 244
A-frame house building, 259–67
Against the Grain: A Carpentry Manual for Women (McCormick), xii, 104, 168, 180, 183, 234–35, 374
Air-dried lumber, 81
Allen wrench, 55
Aluminum, 318; foil insulation, 318; metal drip edge, 269, 274; sheet metal shingles, 269
Anchor bolts (masonry anchors, post anchors, sill bolts), 89, 177–78, 185, 354, 358
Angle finder (movable bevel), 22
Apprenticeship programs, women and, 323–24, 329, 330; CETA, 339
Apron, defined, 354
Architectural-grade caulking, 90, 92
Architectural Graphic Standards (Ramsey and Sleeper), 237, 239, 375–76
Asbestos fiber products, 354; as carcinogens and warnings on use of, 93–94, 286, 291–92, 354; defined, 354; gypsum board, 286–92; insulation, 93–94
ASHRAE Handbook of Fundamentals, 145
Aspenite, defined, 354
Asphalt-based products, 268–71; defined, 354; roll roofing and, 273–77; shingles (*see* Asphalt shingles)
Asphalt shingles, 237, 268–71; amount needed, 268; colors, 269; grades, 268; laying, 269–71; underlayment for, 269
Attics, 93–95, 235; gable roof framing and, 235; insulation, 93–95
Awl, scratch, 28
Awning-type window, 151; light efficiency and, 151

Backing board, 291
Backsaw, 31–32
Balloon framing, 208
Ball peen hammer, 38
Band saw, 61–62
Base (baseboards), 83, 298–99; rubber, 299
Base shoe, 299, 354
Batt (batten), defined, 354. *See also* Board-and-batten
Batter boards, 165, 167–68; defined, 354; ledgers, 167; right-angle building, 167; size, 167; straight building, 167
Bay Area Women's Caucus, United Brotherhood of Carpenters and Joiners of America, 322
Beadwall, 145
Beam poles (purlins), 20; defined, 358; hexagon building and, 306, 307–14
Beams, 19, 79 (*see also* Girders; Posts; specific kinds, uses); collars (ties), 231, 236; defined, 79, 355; vigas (adobe house), 137, 139, 141–42, 360
Bearing walls (bearing partitions), 107–8, 208, 355 (*see also* Partition walls; Walls); adding window to, 133–36; determining (identifying), 107; removing and replacing with girder, 107–8
Bench saw. *See* Table saw
Berm, defined, 355
Bevel, 355; movable (angle), 22; sliding T, 25–26
Bevel siding, 248, 251–53; defined, 355
Bids (bidding, estimating), 161, 332–34, 338, 339, 340
Biological toilets, 147; kinds, 147
Bird's-mouth, defined, 355
Bits, electric drills and, 54–55; auger, 54; electrician's, 55; hole saws, 54–55; masonry, 54; "nest of," 130; power bore, 54; spade (flat), 54; twist, 54
Blind nail, 355
Blocking, 209, 212–13; exterior walls and, 209, 212–13; fire, 108, 209, 212–13; nailing and, 106; partition walls and, 108
Block plane, 34
Blueprints (blueprint reading), 157, 158, 161–63; elevations, 163
Blue stain, lumber, 82
Board-and-batten siding, 253–54; defined, 355
Board foot, 79; defined, 355; lumber priced by, 79
Board-on-board siding, 253–54; artificial, 253–54; plywood, 254
Boards, 79, 355 (*see also* specific kinds, materials, uses); batter (*see* Batter boards); defects, 81; defined, 355; floor joists, 298, 304–6 (*see also* Joists); wood paneling, 294–97
Body protection tools, 47–49. *See also* Fire hazards; Health hazards; specific kinds, uses
Bolt cutters, 46–47
Bolts, 88, 355 (*see also* Screws); anchor, 89, 177–78, 185, 354, 358; carriage, 88; countersinking, 88, 313–14, 355; defined, 355; lock washers, 88; machine, 88; molly, 88, 89; sizes, 88; stove, 88; toggle, 88–89
Books (reading, resources), 223–24, 374–80
Bookshelves, 103
Boots, work, 48
Bowing (lumber defect), 81
Boxes, 106; bracing, 106; form, for pier foundations, 176–77 (*see also* Form boxes)
Brace and bit, 39–41
Bracing (braces), 106, 184, 202, 209, 289; cross, 202–3; defined, 106, 355; form walls and, 184; gate building and, 117; let-in (diagonal), 208, 358; posts, 202–3; stud walls and, 208–9, 211–12, 213
Breast drill, 40
Bricklayer's chisel, 33–34
Bridging, defined, 355
Brookstone Company, 379
Budgeting expenses, 218, 223
Building codes, 157–58, 160, 181, 205, 220–21, 361–72, 376; inspections and, 72, 73; tables, 361–73; *Uniform Building Code,* 160, 172, 376
Building inspectors and inspections, 72, 73. *See also* Building codes
Building paper. *See* Paper; specific kinds, materials, uses
Building trades, women in. *See* Employment
Building with Logs (Mackie), 376–77
Burke Concrete Accessories, Inc., 380
Burlap, use on walls of, 296
Butt gauge, 28–29

Butt hinge, 355
Butt joints, 104, 355
Butyl caulking, 92

Cabbage Lane Women, 351
Cabinet Making and Millwork (Feirer), 57
Cabins, 97–101, 172, 190–99; log cabin, 97–101, 376–77; five-sided star cabin, 190–99
California Apprenticeship Council, 322
California Plan, 322, 323, 324
Canadian Log House, 377
Carpenters Union (United Brotherhood of Carpenters and Joiners of America), women in, 320–39
Carpentry (building, cabinetmaking, woodworking), 102–11; basic principles and terms, 106–11; building codes and (*see* Building codes); joints and splices, 103–5; language (terms and definitions), 354–60; learning, 102–11; materials, 78–96; resources (books), 374ff.; simple projects, 112–20; tools (*see* Tools); women and employment in (*see* Employment)
Carriage bolt, 88
Carroll, Rose Ann, 342, 343
Casement windows, 150–51, 355; light efficiency and, 150
Casing, 355 (*see also* Trim or casing, window); defined, 355; head, 136
Cat's-paw, 43
Caulking, 90–92; architectural, 90, 92; defined, 355; glazing compound, 91; gun, 90; kinds, uses, 90, 92; sealants, 90, 92; silicone-based compounds, 90–91, 92; spackling compound, 91; "tub and tile" caulk, 91; weather stripping, 95; Wet Patch, 90
C-clamp, 39
Ceilings, 163 (*see also* Floors; Roofs); adding a partition wall and, 108; false, 314; five-sided star cabin, 190, 198; gypsum board, 287–92; hexagon house, 314–15; insulation, 145; joists (*see* Joists); roof sheathing and, 237–39
Cellulose insulation, 94–95; companies producing, 95
Celotex, defined, 355
Cement, 355; roofing, 270–71, 273, 274–75, 276, 277; working with (*see* Concrete, working with)
CETA (Comprehensive Education and Training Act), 339
Chain saws, 134–35
Chalk line, 28; string, 165, 168–69; use of, replacing, 28
Chamfer (chamfered boards), defined, 355
Checks and splits, lumber, 81
Chemical toilets, 147
Chisels, 24, 33–34; bricklayer's, 33–34; care of, 33; cold, 33; firmer, defined, 356; proper use of, 33; sizes, 33; wood, 33
Circular house, 125. *See also* Round house building
Circular saws, portable, 50–54; adjusting mechanism, 50–51; bench (table), 57–59; blades, 51–53; care and safety in use of, 51–52; kinds, choosing, 50–51; protractor gauge, 51; rip guide, 51, 53
Clapboard siding, 248, 251–53
Clinching technique, driving nails and, 87
Close-grained woods, 82
Coarse-grained woods, 82
Codes, building. *See* Building codes
Cold chisel, 34
Collar beams (tie beams), 231, 236, 237, 355
Collection systems, solar, 145–46, 189; direct gain, 145; indirect gain, 145–46
Combination square, 26–27, 359
Commoner, Barry, quoted, 149
Common rafter, 355
Compass (keyhole) saw, 31
Complete Handbook of Power Tools, The (Drake), 378–79
Composting toilets, 146–47, 379
Compound cut, 355
Concrete, 178–80; cement and (*see* Cement); defined, 355; fasteners for, 89–90; forms, defined, 356; pier foundations (*see* Concrete pier foundations); working with, 178–80 (*see also* Concrete, working with)
Concrete, working with, 178–80, 182, 186–89, 224–25; chute handling, 186; consolidating, 186; determining amount needed, 178, 186; floating, 188; mixes and mixing, 178–79, 186, 224–25; moving, 188; pouring, 186–89; screeding, 188; setting-up process, 178; transit mix, 178, 186, 224–25; troweling, 188
Concrete pier foundations, 169–89, 301–4; casting one's own, 175ff.; continuous, 180–89 (*see also* Continuous concrete foundations); precast blocks, 173–75; with footings, 174, 175; without footings, 174, 175; working with concrete and, 178–80 (*see also* Concrete, working with)
Concrete Work Simplified (Brann), 377–78
Conductance, 145
Construction trades, women in. *See* Employment
Continuous concrete foundations, 161, 165, 171, 180–89, 224–25; floor joists, 205; footings, 180–82, 183; form walls, 182–89; foundation walls, 180ff.; framing the house and, 203–5; hexagon house building and, 301–4; mudsill (sill), 180–81, 182, 203, 205; stepped-off, 182; trenches, 181
Contour gauge, 29
Contractors, licensed, women as, 325, 336–40
Coperthwaite, Bill, 278
Coping saw, 32
Corner beads, 287, 288, 289–90, 355
Corner irons, 89; L-strap, 89; T-strap, 89
Corner posts, hexagon building and, 306
Corners, 250–51; finishing, siding and, 250–51, 257
Cornice, defined, 355
Corrugated fasteners, 88
Counterflashing, 356
Countersinking (counterboring), 313–14; defined, 355
Course, defined, 356
Coving (cove), 83; defined, 356
Craftsman Book Co., 379
Craftsman Woodturner, The (Child), 66
Crawl space, 180, 200, 356
Creosoted telephone poles, use of, 15
Crook (lumber defect), 81
Crosscut, defined, 354
Crosscut saw, 29–31; care of, 30; "points," 29; use of, 29–31
Cross-grained wood, 82
Crowbars (wrecking bars), 43–44
Crown, lumber, 81
Cup hooks, 88
Cupping (lumber defect), 81
Cutters, bolt, 46–47
Cutting, scoring, and shaping tools, 29–36. *See also* specific kinds, tools, uses
Cutting glass. *See* Glass

"D" ("penny"), nails and, 83, 356
Dado (dadoes), 104, 356; defined, 104, 356; kinds, 104; sets, 58; window, 135–36
Danzig, Priscilla, 342, 343, 344; poem by, 324–25
Dead load, defined, 356
Decay, lumber, 82
Decking, roof, 237–39, 240; weatherproofing and, 272 (*see also* Sheathing)
Decks, 169ff., 377; lumber for, 172; pier foundations and, 169ff.
Definitions and terms, carpentry, 106–11, 354–60
Demolition tools, 43–44. *See also* specific kinds, tools, uses
Design with Climate (Olgyay), 145
Diagonal bracing. *See* Let-in (diagonal) bracing
Digger, two-headed, 46
Dimension lumber, 79, 356; defects, 81
Door bottom sweep, 95
Doors, 118–20, 126–33, 206, 228; adjusting plumb, 130; building, 118–20; doorsill, 132; dutch, 118–20; flashing, 249–50; framing, 22, 131–32, 206, 210, 215–16; hanging a replacement, 127–30; hardware, 126, 127, 130; hinges, 126, 127, 128; insulation (weather stripping), 93–96, 133, 145, 249–50; interior and exterior, 138; jambs, 126, 132, 357; kinds, 131; locksets, 130; measurements, 126, 128, 131, 132; new, hanging, 131; preframed and ready-made, 126–27, 132; removing, 127; replacing, 126–33; sticky, 109; threshold, 95, 132, 298; tools, 130–31; trim or casing, 132–33
Door stop, 133
Dorgan, Lynn, 350, 352, 353
Double-hung window, 110–11, 150; light efficiency and, 150; replacing sash weights and cords, 110–11
Dozuki saw, 32
Drawknife, 36
Drill press, 62–73; attachments, 63; disadvantages, 63; head, 62; mortising attachment, 63; radial-arm, 63; uses, 62–63
Drills, 39–41 (*see also* Drill press; Electric drill); brace and bit, 39–41; breast, 40; care of bits, 41; eggbeater, 39, 40–41
Drip cap trims and moldings, 83, 136, 356
Droplights, 47
Drop siding, 356
Dry rot, lumber, 82
Dry wall, 286, 356. *See also* specific kinds
Dry wall saw, 289
Dutch door building, 118–20; plan, 118; window, 119–20
Dwelling Construction Under the Uniform Building Code, 376

Earmuffs, hearing protection and use of, 47, 50
Earning a living. *See* Employment
Eave, defined, 356

Edge-grain lumber, 81, 356
Eggbeater drill, 39, 40–41
Electrical trades, women in, 323
Electric drill, portable, 54–55 (*see also* Drill press; Drills); bits for, 54–55; burrs (rotary cutters), 55; chucks, 54; drill housing, 54; reverse drive, 54; use of, 55; variable-speed motor, 54
Electrician's bit, 55
Electrician's long-nosed pliers, 41–42
Electric tools. *See* Power tools; specific kinds, uses
Employment (employment opportunities and problems), women in building trades and, 320–40; apprenticeships and, 323–24, 329, 330, 339; furniture maker/designer, 330–35; as licensed contractors, 325, 336–40; unions and, 320–29
End grain, defined, 356
End-nail, defined, 356
Energy-efficient houses, 144–56; books on, 375, 379; food production and, 149; graywater disposal and recycling systems, 147–49; lighting and, 149–56; solar colection systems, 144–46; water waste and flush-toilet alternatives, 146–49; windows and light efficiency and, 149–56
Expansion joint, 356

Face-nail, defined, 356
Factory edge, defined, 356
Fascia, defined, 356
Fasteners, 88–90. *See also* Bolts, Nails; Screws; specific kinds, materials, techniques, uses
Feathering, gypsum board and, 290–91
Felt paper, 356. *See also* Paper; specific uses
Fences. *See* Gate building
FHA Pole House Construction (booklet), 173, 204, 376
Fiberboard, 356; wall sheathing and, 250, 286
Fiberglass insulation, 22, 23, 93–94, 95, 145, 228–29, 305, 315–16, 357; applying, 95; health warnings, 93–94, 305
Fiberglass shingles, 237, 268, 271
Fine-grained wood, 82
Fire blocking (fire stops), 108, 209, 212–13, 356
Fire hazards (fire-resistant products, fire protection), 268, 273; asphalt and fiberglass shingles, 268; fire blocking (fire stops) and, 108, 209, 212–13, 356; insulation and, 94–95 (*see also* Insulation; specific kinds); wallboard, 286, 287, 291, 296; woodburning stoves and fireplaces and, 229, 314; wood shingles or shakes and, 273
Fireplaces. *See* Woodburning stoves and fireplaces
Firmer chisel, defined, 356
Five-sided star cabin building, 190–99
Fixed window, 151; light efficiency and, 151
Flashing, 90, 92, 249–50, 269–70, 356; applying, 92; asphalt shingles and, 269–71; counterflashing, 356; defined, 356; metal drip edge, 92, 269, 274; roof nosing and, 92; siding and, 249–50, 257; waterproofing and, 249–50, 269–71, 273, 274–75, 356
Flat-grained lumber, 81
Flat roof, 231
Flatting timbers, 15, 17
Floor jacks, 46. *See also* Jacks
Floors (floorings), 297–99 (*see also* Ceilings; Roofs; specific kinds, materials); baseboards, 298–99; books on, 378; diagonal, 298; finishing (finished), 297–99; framing and, 205–8; hexagon house, 304; insulation, 93–95, 243, 305 (*see also* specific kinds, materials); joists, 18, 19 (*see also* Joists); kinds, choosing, 297–99; nailing, 298–99, 305–6; plywood, 297, 305–6; subfloors, 93–95, 206–8, 243, 297, 359; thresholds, 298; tongue-and-groove, 206, 225, 226, 297, 298, 306, 358; wood, 297–99, 305–6 (*see also* specific kinds, styles)
Floors: Selection and Maintenance (Berkeley), 378
Flush, defined, 356
Flush doors, 131
Flush toilets. *See* Toilets
Food production, energy-efficient systems and, 149
Footings, pier foundations, and, 171, 174–76, 178, 179, 180–82, 186; defined, 356; depth of holes for, 175–76, 181; form walls and, 182, 183, 186 (*see also* Form walls); post anchors and, 177–78; size of, 176
Fore plane, 35
Form boxes, pier foundations and, 176–77, 179–80, 356; lumber for, 176; post anchors and, 177–78; removing, 180; securing in place, 176–77; size of, 176; square, 176; tapered, 176
Forms, defined, 356
Form stakes, defined, 356
Form tie wire, 257
Form walls, 182–89; components of, 183–86; studs or stiffbacks, 183; working with concrete and, 186–89
Foundations, building a house and, 161, 168–89; continuous concrete, 180–89 (*see also* Continuous concrete foundations); defined, 357; digging holes for, 16; five-sided cabin and, 190–91; framing and, 200ff.; hexagon house and, 301–4; piers and, 169ff. (*see also* Piers); posts, replacing, 109–10
Four-foot T square, 27–28; choosing, care of, 27–28
Framing (house), 200–16, 226; balloon, 208; doors (*see* Doors); floors, 205–8; gable ends, 216; house on continuous foundations, 203–5; house on embedded poles, 204; house on piers, 202–3; partition walls, 108, 216, 226; platform, 208; roofs, 231–40, 268–70 (*see also* Rafters); walls, 108, 208–16, 226; walls with windows, 214–15; window openings, 22 (*see also* Windows)
Framing square (steel square), 27–28; choosing, care of, 27–28
Fretsaw, 32
From the Ground Up (Cole and Wing), 374–75
Fundamentals of Carpentry: Vol. 2, Practical Construction (Durbahn and Sundberg), 374
Furniture maker/designer, 330–35
Furring (furring out), 108; defined, 357

Gable roof house, 231–40, 357; attic, 235; collar (tie) beams, 231, 236, 237; framing, 231–40; framing walls, 208, 209, 216; gable end, 231, 235, 237; louvers, 235; rafters, 231–37; ridge piece (board), 231, 232, 233, 234, 236–37; sheathing, 237–40; trusses, 231, 236, 237; vents, 235; weatherproofing, 268–77
Gambrel roof, 231, 237; weatherproofing, 276–77
Garden gate. *See* Gate building
Gate building, 114–18; dog-eared tops, 117–18; framing ("rails"), 114; hinges, 115–16, 117; latches, 116, 117; lumber and materials, 114, 115–16; siding, 114; tools, 116; wood-frame and wire, 114; wood-sided, 114
Gauges, 28–29; butt, 28–29; contour, 29; stair, 28
Girders, 357 (*see also* Beams; specific kinds, uses); build-up, 173, 202; defined, 79, 357; framing the house and, 200, 201, 202–3, 204, 205–6; lifting atop foundation piles, 17; measuring and cutting, 202; pier foundations and, 171, 172, 173, 174, 182, 200, 201, 204, 205–6; posts and, 108; replacing bearing walls with, 107–8; sizes, 182
Glass (glass cutting), windows and light efficiency and, 149–50, 152–53, 154–55; cutter for, 154; glazing and, 357 (*see also* Glazing windows); procedure, 154–55
Glasses (goggles), safety, 47–48, 50
Glazing compound, 91
Glazing windows, 155–56, 357; back-puttying, 155; glazier's points for, 155; removing old putty, 155
Glues, 92–93. *See also* Adhesives; specific kinds, uses
Goodbye to the Flush Toilet (Stoner, ed.), 379
Grade (lumber), defined, 80, 357
Grade (site), defined, 357
Graywater disposal and recycling, 147–49; available systems, 148–49; slope conditions and, 148; soil types and, 148; use rate and, 148; water table and, 148
Green (freshly cut) lumber, 81
Greenhouses, 75–76; solar, 149, 379
Grout bed, 203, 204
Gutter, defined, 357
Gutter and lap seal, 92
Gypsum sheathing, 291
Gypsum wallboard (Sheetrock), 94, 286–92, 357; cautions and safety in use of, 286, 291–92; fire-resistant, 286, 287, 291, 296; insulating, 292; kinds, sizes, colors, 286–87, 291; materials, applying, 287–92; nailing, 287, 288, 289, 290–91, 292; needs, 287; repairing, 288, 292; sanding, 291–92; taping and mudding, 290–91, 292

Hacksaw, 32–33
Half lap joint, 104–5
Half round trims and moldings, 83
Hammer drill, rotary, 54
Hammers, 37–39; ball peen, 38; care of, 38; choosing, 37; curved-claw, 37; mason's, 44; materials, construction of, 37; nailing and, 37–38, 85, 86; nail set and, 38; shingling hatchet, 39; sizes, 37; sledgehammer, 44; straight-claw (rip), 37; tack, 38; wallboard, 38; wooden mallet, 38

Hand-held power tools. *See* Power tools, hand-held
Hand-powered staple gun, 90
Handsaws, 29–33; backsaw, 31–32; buying, 31; care of, 31; compass (keyhole), 31; coping saw, 32; crosscut, 29–31; dozuki, 32; fretsaw, 32; hacksaw, 32–33; miter box, 32; proper use of, 29–31; ripsaws, 29–31
Hand tools, 24–47. *See also* specific kinds, uses
"Handyman special" tools, avoiding, 24
Hardboard, 82, 357
Hardwood, 79–83, 357 (*see also* Lumber; specific kinds, uses); defects, 81; floors, 297–99; grading, 80; insulation, 93; kinds and uses, 79; nails and nailing, 84, 85; plain-sawed, 81; plywood, 82; quarter-sawed, 81
Header/deflection system, adding new window to existing wall and, 134–35
Header joist, 205; defined, 357
Headers (lintels), framing walls and, 208, 210, 215, 357; spans for windows and doors, 210
Health hazards, 355 (*see also* Fire hazards; Safety; specific kinds, products); body protection tools and, 47–49 (*see also* specific kinds, uses); books on, 379
Hearing protection, earmuffs and, 47, 50
Heartwood, 82, 357
Heating systems, 144–46 (*see also* Energy-efficient houses); food production and, 149; insulation and (*see* Insulation); solar, 144–46 (*see also* Solar heating); woodburning stoves, 229, 266, 314
Hexagon house building, 300–19; beam poles, 306, 307–14; corner posts, 306; costs, 300–1; floors, 304–6; foundation, 301–4; hat, 307–14; insulation, 305, 315–16, 317–18; lumber, 300–1, 303, 304–6, 315; plywood flooring, 297, 305–6; roof, 307–18; spacers, 315–16; tools, 305, 313–14; walls, 318
Hinges, 115–16; attaching, 117; butt, 355; door replacement and, 126, 127, 128; gate building and, 115–16; ornamental, 116; spring-loaded, 115; strap hinge, 115–16
Hip roof, 357
Holes, foundation digging and, 16
Hole saws, 54–55
Hollow (lumber warp defect), 81
Home and Workshop Guide to Sharpening (Walton), 378
Hopper window, 151; light efficiency and, 151
House building (home building): books (resources), 223–24, 374–80; carpentry and (*see* Carpentry); women as licensed contractors in, 325, 336–40
Housesmiths Press, 377
How to Build a Wood-Frame House (Anderson), 375
How to Build Decks, 172, 175, 377
How to Create Interiors for the Disabled (Cary), 377
How to Design and Build Your Own House (DiDonno and Sperling), 375
How to Rehabilitate Abandoned Buildings (Brann), 377
How to Work with Tools and Wood, 104, 105, 168, 180, 183, 378
Hydrochloric acid, mortaring rock walls and, 293, 294

I Built Myself a House (Garvey), 223, 376
Illustrated Housebuilding (Blackburn), 223–24, 374
Inspections (inspectors), building, 157, 161, 163, 181, 220–21. *See also* Building codes
Insulated roof decking, 237–39, 240; applying, 240
Insulating wallboard, 291
Insulation (insulating), 21, 22, 23, 76, 90ff., 198 (*see also* Waterproofing; Weatherproofing; specific kinds, materials, products, structures, uses); applying, 95; asbestos use and dangers, 93–94 (*see also* Asbestos fiber products); defined, 357; energy-efficient house and, 144ff.; fiberglass, 22, 23, 228–29 (*see also* Fiberglass insulation); floors (*see under* Floors); hexagon house, 305, 315–16, 317–18; kinds, materials, 93ff.; rigid, 94; roofs (*see under* Roofs); "R values," 93, 145; vapor barrier and, 94, 95; walls (*see under* Walls); weather stripping and, 95–96 (*see also* Weather stripping)
Intersecting (multiple shed) roofs, 231

Jack plane, 34–35
Jacks, kinds and uses of, 46; adding a window and, 134; floor, 46; Little Giant, 46, 110, 134; moving a house and, 245; raising walls and, 226; replacing foundation posts and, 110
Jalousie, 357
Jambs, 126, 132; defined, 357; door, 126, 132, 357; window, 135–36, 151–52
Jigsaw, electric, 55–56; use of, 56
Joint compound (Sheetrock mud), 287–88, 357; applying, 290–92
Jointer, electric, 63–64; buying, features, 64; care of, 64; cutterhead, 64
Jointer plane, 35. *See also* Jointer, electric
Joint knife (wall scraper), 288, 291
Joints, 103–5; butt, 104; dado, 104; defined, 104; expansion, 356; half lap, 104–5; miter, 104, 105–6; mortise and tenon, 104, 105, 358, 359; and putting wood pieces together, 103–5
Joist hangers, 89, 357
Joists, floor and/or ceiling, 205–7, 298, 304–6, 357; bridging (blocking) and, 206–7, 357; building code tables and, 361ff.; defined, 357; end, 205; framing house and, 205–6; header, 205, 357; pier foundations and, 171, 172, 173, 174, 182, 205–6; subfloors and, 206–7; trimmer (stringer), 205

Kalwall, 357
Kerf, 29, 30, 357; defined, 357
Kern, Ken, 67, 69
Key, concrete, defined, 357
Keyhole (compass saw), 31
Kiln-dried lumber, 81, 357; storing, 83
Kitchen house, 121–25
Knives: drawknife, 36; joint (wall scrapper), 288, 291; mat (utility), 34
Knots, lumber, 81–82
Kraft paper, 357. *See also* Paper; specific uses

Labor Department, United States, 323
Ladders, roof framing and, 235, 239. *See also* Scaffolding
Lag screws, 87
Laminated, defined, 357
Laney College (Oakland, Calif.), 330–32
Lap cement (roofing cement, roofing tar), 270–71, 273, 274–75, 276, 277
Latches, 116; gate building and, 116, 117
Lathe, wood, 65–66; accessories, 66; basic operations, 65–55; buying, features, 66; calipers, 66; faceplate turning, 65–66; headstock (live center), 65; spindle turning, 65, 66; turning tools, 66
Latimer, Tirza, 342
Layout, building (simple), 165ff.; batter boards, 165, 167–68; lines, 165–69; materials for, 165; measuring and making, 165, 166–69; pier foundations, 169; square corner placement, 166–67; strings, 165, 168–69, 181; 3–4–5 or 6–8–10 checking squareness of corners method, 166
Leaks (*see also* Waterproofing; Weatherproofing): caulking and flashing, 90, 92 (*see also* Caulking; Flashing); insulation, 93–95 (*see also* Insulation); weather stripping, 95–96
Let-in (diagonal) bracing, 208, 358; defined, 358
Levels, 44–46; care of, 45–46; kinds, choosing, 45; line, 45; spirit, 44–45; torpedo, 45; two-foot, 45; use of, 45
Licensed contractors, women as, 325, 336–40
Lights (lighting), 149–56; droplight, 47; energy-efficient houses and windows and, 149–56; spotlight, 47
Lineal foot, defined, 358
Linear foot, lumber priced and sold by, 79–80
Line level, 45
Lintels. *See* Headers
Live load, defined, 358
Locksets, 130; door replacement and, 130–31; installing, 131
Lock washers, 88
Log cabin (log house) building, 97–101, 376–77
Log Home Guide for Builders and Buyers (Muir, ed.), 377
Louvers, gable, 235
Low-Cost Energy-Efficient Shelter for the Owner and Builder (Eccli, ed.), 375
Low-water-use toilets, 147, 379
Lumber (wood parts and products), 78–83, 106–11, 161, 358 (*see also* specific houses, kinds, materials, parts, products, structures, techniques, uses); abbreviations, 80–81; actual and nominal sizes, 81, 358; adhesives, 92–93; A-frame house, 261, 262, 264–65; bidding on and buying, 161, 334–35; building code tables, 361ff.; caulking and flashing, 90–92 (*see also* Caulking; Flashing); converting thickenss of material to board feet, 79; crook, 81; crown, 81; decay-resistant, 78; defects, 81–82; defined, 358; dimensions, 79, 81, 358; dutch doors, 116, 117; edge-grain (flat-grain), 81; fasteners, 83–90; floors (subfloors), 297–99; framing and (*see* Framing; specific houses, kinds, structures); gate building and, 115; gen-

eral terms for, 79, 81–82; grading, 80, 357; green, 81; gypsum board, 286–92; handling and storage of, 83; hardboard, 82; hardwood, 79; hexagon building and, 300–1, 303, 304, 315; horizontal siding and, 251–53; insulation, 93–95 (*see also* Insulation); joints and splices, 103–5; joists, 205 (*see also* Joists); kiln-dried, 81; kinds of wood and uses, 78ff.; knots, 81–82; layout and, 165; log cabin building and, 97–101; measuring wood in "open space" and, 106–7; new door construction and, 132; new versus used, 221–22; particleboard, 82–83; piers and, 171, 173, 176, 182–85; plain-sawed, 81; planning needs and work schedule and, 161, 220; plumb on level positions, 106; plywood, 82, 254–55 (*see also* Plywood); prices, 79–80; putting pieces together, 103ff.; quartersawed, 81; roof framing and, 233, 237–40; rough, 80, 81, 115; running wild, 107; seasoned, 81; sizes, 79, 81, 358; softwood, 78–79; splits and checks, 81; surfaced, 80, 81, 115, 359; trims and moldings, 83; vertical measurements, 106–7; working with, 106–11; yard, 360

McCormick, Dale, xii, 234–35, 274
Machine bolt, 88
Mackie (B. Allan) School of Log Building, 376–77
Malley, Jean, 343, 345
Mansard roof, 231
Masonry bits, 54
Masonry sealer, 293–94
Masonry shields and anchors, 89. *See also* Anchor bolts
Mason's hammer, 44
Mastics, 92, 93; safety warnings, 93
Matched lumber, defined, 358
Mat knife, 34
Measuring tapes, 24–25; care and use of, 25; "refill," 25
Measuring tools, 24–29. *See also* specific kinds, uses
Mending plate, 89
Metal drip edge, 92; aluminum, 269, 274
Miller, Terri, 342
Mineral wool (rock wool) insulation, 93–94; applying, 95; health warnings, 93–94
Miter, defined, 358
Miter box, 32, 106
Miter joint, 104, 105–6
Miter square, 26
Modern Carpentry (Wagner), 172, 235, 239, 249, 374
Modern Woodworking (Wagner), 57
Molding cutters, 58
Moldings and trims, 83; drip cap, 83, 136, 356; kinds, shapes, and types, 83; molding, defined, 358
Molly screws (molly bolts), 88, 89
Mortar, 358; defined, 358; for rock walls, 293–94
Mortise and tenon joints, 104, 105, 358, 359
Movable bevel (angle finder), 22
Mudsill. *See* Sill

Nailer, defined, 108
Nail nippers, 43
Nails (nailing), 37–38, 83–87, 115; aluminum, 84; amount to use, 86; apron, 257; blind nailing, 85, 355; blocking and, 106; box, 84, 85, 115; casing, 84, 85; cement-coated, 84; choosing, 83–84; clinching, 87; common, 84, 85, 115, 207; cut, 84; "d" ("penny") units, 83, 356; "dimpling," 289; duplex (double-headed), 84; end nailing, 85, 356; face nailing, 85, 356; finishing, 84, 165, 166; floors, 298–99, 305–6; framing and, 209, 210, 212–13, 216; galvanized and ungalvanized, 22, 83–85, 115; gate building and, 115; gauge, 83; gypsum boards and, 287, 288, 289, 290–91, 292; hammers for, 85; helically ringed, 84; joining pieces of wood and, 106, 108; learning to drive (techniques for driving), 37–38, 85–86; masonry, 85; patterns expressed as "o.c.," 86–87; pulling out difficult, 86; "regular" diamond points, 84; ring-shank, 84; roof framing, 236, 239, 240; roofing, 84; rosin-coated, 84; shingles and shakes and, 272; siding, 252, 253, 254, 255; sinkers, 85; sizes, 83, 84; spikes, 84; spiral-shank, 84; staggering, 86; stainless-steel, 84; subfloors, 207; toenailing, 85, 304, 360; types and uses, 84; walls in place and, 216; weight, 83; wire (wire brads), 83, 84; wood paneling, 295, 296
Nail set, 38
National Construction Estimator, 376
National Repair and Remodeling Estimation Structures Cost Manual, 376
Nelson Screw Products, Inc., 380
Noise reduction, 287, 288; acoustical tiles, 354; earmuffs, 47, 50; gypsum boards, 287, 288; mountings, 288
Nonbearing walls (partitions), 358 (*see also* Partition walls); defined, 358
Nosing, defined, 358

"O.c." ("on center"), 171, 215; defined, 358; nailing and, 86–87
Octagon-shaped houses and structures, 124, 343
Oil toilets, 146
Open-grained wood, 82
Outside corner pieces, 83
Owner-Built Home, The (Kern), 67, 69

Panels (paneling), wood, 294–97. *See also* Wood paneling
Paper, building (felt, kraft, tar), 356; asphalt-based, 249, 273–77, 317–18; flooring and, 298; rock walls and, 293–94; roofing, 269, 273–77, 317–18; sheathing and siding and, 249, 257, 269; weatherproofing and, 269, 317–18
Particleboard, 358; defined, 358; wall sheathing and, 82–83, 250, 296
Partition walls (interior walls, nonbearing walls), 108–9, 208ff., 225–26, 286–97, 358 (*see also* Bearing walls; Walls); adding, 108, 216; burlap, 296; defined, 358; fire-resistant, 286, 287, 291, 296; framing, 108, 216, 226; gypsum wallboard, 286–92; hexagon house, 218; insulation, 93–95 (*see also* Insulation); patches, 292; repair cracks, 292; rock, 292–94; resurfacing, 288–92; Sheetrock, 286–92; stud framing and (*see* Stud wall framing); wainscoting, 295–96; wood paneling, 294–97
Peralta Community Colleges (San Francisco, Calif.), 330
Perlite insulation, 94
Phillips screw and screwdriver, 41, 87
Piers (pier foundations), 169–89; building, 169ff.; casting one's own, 175–80; concrete (*see* Concrete pier foundations); crawl space, 200; embedded poles, 173, 200; framing house and, 200–3; lumber, 171, 173; planning and measuring, 171ff.; post anchors, 177–78; precast, 173–75; spacing, 171–72, 173; span, 171; weight load formula, 172–73
Pine decking, floors of, 297
Pipes, 185 (*see also* Plumbing); form walls and, 185; wood heat stove, 229
Pit privy, 146
Planes, hand (planing), 34–35; adjusting and proper use of, 35; block, 34; care of, 35; defined, 358; fore, 35; jack, 34–35; jointer, 35, 63–64; router, 35; smooth, 35; spokeshaves, 35
Plank, defined, 79, 358
Planning hand-built house, 217–24, 241–42; A-frame, 259–61; budgeting, 218, 223; drawing plans (design), 218, 219–20; resources (books, reading), 218, 223–24, 374–80; site selection, 218–19 (*see also* Site); tools and materials, 218, 221–23
Plant shelves, 344
Plastic foam insulation, 94. *See also* specific kinds
Plastic roof patch, 92
Plate, 358; bottom (sole), 258; defined, 358; horizontal, 358; top, 358
Platform framing, 208
Pliers, 41–42; electrician's long-nosed, 41–42; locking plier-wrench, 41; vise-grip, 41
Plumb, defined, 358
Plumb bob, 28; care of, 28
Plumbing (*see also* Pipes): trade, women in, 32; "tub and tile" caulk, 91
Plywood, 82, 254–55, 358; applying siding, 254–55; bracing walls and, 209; core, 82; exterior, 82; faces, 82; floors (subfloors), 82, 206–7, 297, 305–6; form boxes, 176; form walls, 182–83; glues, 82; grades, 82; hardwood, 82; interior, 82; interior/exterior, 82; kinds and uses, 82, 254; nailing and joining, 106; plyscore, 207, 215; roofing and, 198, 227, 228, 316–18; sheathing, 237, 239, 248, 249, 250; siding, 164, 248, 249, 250, 254–55, 297; softwood, 82; veneer paneling, 296–97
Pole Building Construction (Merrilees and Loveday), 376
Pole House Construction (FHA booklet), 173, 204, 376
Poles (pole construction), 191–98, 222, 226; (*see also* Posts; Rafters); beam (*see* Beam poles); embedded, house on, 173, 200, 204, 205; five-sided star cabin, 190, 191–98; hexagon house, 300, 301, 306, 307–15; and pier foundations, 173, 200, 204; tops, leveling, 16–17; trees for, 18–19, 243–44, 245, 300, 301 (*see also* Timbers; Trees)
Polystyrene insulation, 94
Polysulfide caulking, 92

Polyurethane insulation, 94
Ponderosa, La Rosa, 336, 340
Porches (*see also* Decks): pier foundations and, 169ff.
Post anchors, 89, 177–78, 358
Post cap, 89, 358
Posts, 109–10 (*see also* Beams; Poles); anchors, 177–78 (*see also* Anchor bolts); bracing, 202–3; checking, 109–10; crawl space and, 200; cutting, 200; foundations and, 109–10, 171, 172, 173, 176, 177–78, 200–3; gate building and, 116; replacing, 109–10; trees, raising, 18–19
Power-actuated tool, 89
Power tools, hand-held (portable), 49–66 (*see also* specific kinds, uses); books on, 378–79; care of, 49–50; choosing, 49–50; safety and, 50; use of hand tools versus, 222–23; woodworking, 57–66
Practical Farm Buildings (Boyd), 378
Practical House Carpentry (Wilson), 374
Producing Your Own Power: How to Make Nature's Energy Sources Work for You (Stoner, ed.), 379
Protective devices (body protection tools), 47–49. *See also* Fire hazards; Health hazards; specific kinds, uses
Psf, defined, 358
Purlins. *See* Beam poles
Putting wood pieces together, 102–5
Putty, 358; glazing and, 155–56; window, 91, 155
Putty sticks, 296

Quarter round trim and moldings, 83
Quartersawed lumber, 81

Rabbet, defined, 358
Radial-arm drill press, 63
Radial-arm saw, 59–61; care and use of, 59–60
Rafters (roof rafters), 20–21, 102, 163 (*see also* Poles; Roofs); beam poles (purlins), 20, 306, 307–14, 358; bird's-mouth, 355; building code tables and, 361ff.; collar (tie) beams and, 231, 236, 237; common, 355; framing and, 163, 191–98, 216, 231–40; gable, 231–37; hexagon house and, 306, 307–16; laying out (stepping off), 233–36; plywood, 316–18; raising, 235–36, 311; span, 231–32; star cabin, 191–98; trusses, 231, 236, 237; weatherproofing and, 268–77; yurt building and, 281–82
Rafter square, 27
Rain, roofing and, 268–77. *See also* Leaks; Waterproofing; Weatherproofing; specific materials, techniques
Rake, roof, 358
Rebar (reinforcing rods), 178, 184–85, 186, 191, 358
Reciprocating ("all-purpose") saw, 56–57, 134
Rectangular house building, 13–23; adobe, 140; layout lines, 166–69
Reinforcing angle, metal, 89
Resins, 92, 358
Resources (books, tools), 223–24, 374–80
Respirators, use of, 48–49, 291–92; fiberglass insulation and, 94, 305; gypsum board and, 291–92
Rigid insulation, 94
Rip, defined, 358
Ripping bar (rip bar), 43–44
Ripsaw, 29–31; care of, 30; "points," 29; proper use of, 29–31
Riser, defined, 359
Rock walls, 292–94; backing, 293; choosing rocks, size, 293; mortar for, 293–94
Roll roofing (membrane roof), 237, 268, 271, 273–77, 317–18; amount needed, 275–76; applying, 273, 276–77; care of, 277; kinds, 273–74; Selvage Edge, 274; tools for, 276; underlayment, 274
Roofers Handbook (Johnson), 378
Roof Framing (Siegele), 231, 234, 378
Roofing nails, 84
Roof jack, 359
Roof nosing, 92
Roofs (roofing), 19–21, 231–40, 268–77 (*see also* specific houses, kinds, materials, parts, structures, techniques); beams (*see* Beams); books on, 378; caulking and flashing, 90–92; cement (*see* Cement); collar ties, 231, 236, 237; decking, 237–39, 240; flat, 331; framing, 163, 191–98, 231–40; gable, 99, 231–40 (*see also* Gable roof house); gambrel, 231, 276–77, 357; hip, 357; insulation, 21, 90–95, 237–39, 240, 245, 268–77, 315–18; intersecting (multiple shed), 231; kinds, 231; mansard, 231; membrane (roll roofing), 237 (*see also* Roll roofing); nailing, 236, 239, 240, 269, 272; pitch, 231–32; plank and beam, 237; plywood, 198, 227, 228, 316–18; rafters (*see* Rafters); replacement, 268–77; ridge, 163; rise, 231–32, 233; run, 231–32; sheathing, 237–40; shed, 231, 359; shingles (*see* Shingles); slope, 231–32, 271; trusses, 231, 236, 237
Rosin, defined, 359
Rotary hammer drill, 54
Rough lumber, 80, 115; actual and nominal sizes, 81
Rough sill, 215, 359
Round house building, 67–77; greenhouse, 75–76; USDA plans for, 77; yurt building, 278–85
Router plane, 35
"R" values, insulation expressed as, 93, 145

Saber saw, 55–56, 159
Safety (*see also* Fire hazards; Health hazards; specific aspects, materials, problems): body protection tools, 47–49 (*see also* specific kinds); glasses and goggles, 47–48, 50; respirators (*see* Respirators)
Sanding: disk plate, 58; respirator for, 291–92
Sapwood, 82
Sashes, 151, 152, 359; replacing weights and cords, 110–11
Saw guard, 31
Sawhorses, building, 112
Saws, hand-held, 29–33 (*see also* specific kinds, uses); buying, 31; care of, 31; miter box for, 32; power (*see* Saws, power); proper use of, 29–31
Saws, power, 50–62, 159, 222–23 (*see also* specific kinds, uses); reciprocating, 56–57, 134; window adding and, 134–35; woodworking, 56–62
Scabs, lumber, 202
Scaffolding, 160 (*see also* Ladders); hexagon roof and, 311–12; roof framing and, 235, 239
Scratch awl, 28
Screed (screeding), 188; defined, 359
Screwdrivers, 41; kinds, choosing, 41; Phillips, 41; ratchet-type ("Yankee"), 41; screws and, 41, 87 (*see also* Screws); use of and cautions, 41
Screw eyes, 88
Screw hooks, 88; L-shaped, 88
Screws, 87; brass, 87; countersinking, 313–14, 355; drilling holes for, 41, 87; flathead, 87; kinds, choosing, 41; lag, 87; oval-head, 87; Phillips head, 87; roundhead, 87; screwdrivers for, 41, 87 (*see also* Screwdrivers); sheet metal, 87; sizes, 87; slotted head, 87; wood, 87; zinc-coated (rust-resistant), 87
Sealants, 90 (*see also* Caulking); Mono, 90; Polyal One, 90; silicone rubber, 90, 92
Sealer, masonry, 293–94
Seasoned lumber, 81
Selvage Edge roofing, 274
Septic tanks, 147, 158, 265, 379; permits, 158; septic/leach field combination, 147
Seven Sisters Construction Co., Inc., 321, 342, 344
"Shake" nails, galvanized, 22
Shakes, wood, 271–73, 359 (*see also* Wood shingles); grades, 272; hand-split, resawn, 271; laying, 272–73; straight-split ("barn"), 271; taper-split, 271
Shannon, 349
Sheathing, 248–58, 359; defined, 359; fiberboard, 250; flashing and, 249–50; insulated, 94; insulated decking, 237–40; kinds, 248; paper, 249; particleboard, 82–83, 250, 296; plywood, 237, 239, 248, 249, 250; putting on, 239; roll roofing, 274; roof, 237–40; shingles, 237, 248, 249 (*see also* specific kinds); spaced, 237; walls, 248–58, 286–99 (*see also* Siding); weatherproofing, 268–77
Shed, defined, 359
Shed roof (single-pitch roof), 231, 359
Sheetrock, 164, 286–92, 359 (*see also* Gypsum wallboard); "fire-resistant," 286, 287, 291, 296; "tile backer," 164; W/R, 291
Shelter, (Shelter Publications), 308
Shields, masonry, 89. *See also* Anchor bolts
Shims, 107, 359; defined, 359
Shingles, 255–58, 359; applying, tools for, 257, 272; asbestos, 264; asphalt, 237, 268–71; dimensions, 255; double-coursing, 255–58; fiberglass, 237, 268, 271; kinds, grades, 255; quantities and costs, 255–56; roofing, 237 (*see also* specific kinds); sidewalls, 256–57; "squares," 257; wall designs, 258; wood (*see* Wood shingles); wood (wall) siding, 237, 248, 249, 255–58, 271–73, 297
Shingling hatchet (shingler's ax), 39, 257, 272
Shiplap: defined, 359; siding, 251, 254
Siding, 228, 248–58, 359; back-priming, 252; corners, 250–51, 257; drop, 356; flashing and, 249–50, 257; horizontal, 248, 249, 251–53; nailing, 252, 253, 254, 255; paper wall covering and, 249, 257; plywood, 164, 248, 249, 250, 254–55, 297; shingles, 237, 248, 249, 255–58, 268ff. (*see also* specific kinds); single versus double walls and, 248ff.; vertical, 248,

249, 253–54; walls, 248–58, 286–99; wood (*see* Wood siding)
Silicone rubber sealant, 90, 92
Sill (mudsill), 180–81, 182, 203–4, 215, 359; floor joists and, 205; rough, 215, 359
Sill bolts, 185
Sill horn, 135
Sills, window, 135–36, 151–52
Sill sealer, 203, 204
Simpson Co., 380
Site (building site), 181, 241; choosing, 15–16, 218–19; foundation work and, 181–82; grade, defined, 357
Skids, foundations and, 180
Sky, 16
Skylark Construction Co., 336–40
Skylids, 145
Slab floors, as solar collectors, 189
Slabwood, log cabin building and use of, 97
Sledgehammer (sledge), 44
Sliding T bevel (angle bevel), 25–26
Sliding window, 151; light efficiency and, 151
Smith, Sally, 348, 349, 350
Smooth plane, 35
Socket set, 42
Softwood, 78–83, 359; defects, 81; edge-grain, 81; floors, 297–99; grading, 80; kinds, uses, 79; nails and nailing, 85; plywood, 82
Solar Greenhouse Book, The (McCullagh, ed.), 379
Solar heating, 144–46; books on, 379, 380; collectors, 145–46, 189; food production and, 149; greenhouses, 149, 379; heat storage capacity and, 146; passive and active systems, 144–45, 294; rock walls and, 294; slab floors and, 189; space-heating systems, 145–46; thermal mass and, 146; water and, 144, 146; window placement and, 150
Sole plate. *See under* Plate
Space-heating solar systems, 145–46
Spackling compound, 91; wall repairs and, 292
Span, defined, 360
Spikes, 84
Spirit level, 44–45; use of, 45
Splices, 103–4
Splits and checks, lumber, 81
Spokeshaves, 35
Spotlights, 47
Spreaders, form walls and use of, 184
Square, defined, 359. *See also* Combination square; Steel square; Try square
Staircase fixtures, 27
Stair gauges, 27
Stair tread, 360; nosing of, defined, 358
Staple gun, 42–43; common gun tacker, 42; electric, 42; hammer tacker, 42; kinds, choosing, 42–43; staples, 43
Star cabin building: five-sided, 190–99; six-sided, 300–19
Steel square (framing square), 27, 232–34, 237; aluminum or steel, 27; body, 27; choosing, care of, 27; tongue, 27
Steel stud driver, 39, 89
Steps, 228; nosing, 358
Stickers, 359
Sticky doors, 109
Sticky windows, 111
Stool, window, defined, 359
Stop material, 83
Stove, woodburning. *See* Woodburning stoves and fireplaces
Stove bolts, 88
Straight-grained wood, 82
Strap hinge, 115–16
Strike-off board, concrete working and, 188
String: chalk line, 28, 165; layout, 165, 168–69
Stud finder, 107
Studs, 208, 209, 210, 215, 359; form walls and, 183; full-length, 215; gypsum board and, 288–89; locating (behind wall), 107; measuring and marking, 210; o.c. fillers (cripples), 215; placement of, 208, 210–11; tilted, remedying, 213; and wall framing (*see* Stud wall framing); yurt building and, 280–84
Stud wall framing, 208–9, 210–16; bracing, 208–9, 210–16; bracing, 208–9, 211–12, 213; fire blocking, 209, 212; gable ends, 216; lifting into place, 213; lumber, 208; measuring and marking, 210–11; partitions, 216, 288–89; plates, 208, 210, 215–16; procedure, 210–16; squaring and securing walls, 216; studs, 208, 209, 210–11 (*see also* Studs); walls with door openings, 215–16; walls with windows, 214–15
Suafai, Susie, 323
Subfloors (*see also under* Floors): defined, 359
Superintendent of Documents, U. S. Government Printing Office, 380
Surfaced (planed smooth) lumber, 80, 115, 359; actual and nominal sizes, 81, 359
Surforms, 35–36

Table saw (circular, bench saw), 57–59; arbor, 57–58; blades, 58; buying, 59; dado sets, 58; extension bars and tables, 58, 59; molding cutters, 58; safety guards, 58
Tack hammer, 38
Tape, gypsum board, 287, 290
Tape measures, 24–25; care and use of, 25; kinds, choosing, 25; "refill," 25
Taper, defined, 359
Taping and mudding, gypsum board and, 290–91, 292
Tar paper. *See* Paper; specific uses
Techfoam (foil-covered urethane foam) insulation, 21
Telephone poles, creosoted, use of, 15
Tenon and mortise joints, 104, 105, 358, 359
Tenoning jig, 58
Termite shields, 203, 204
Terms and definitions, carpentry and woodworking (alphabetical listing), 354–60
Tetrault, Jeanne, 342
T hinge, 115
30 Energy-Efficient Houses . . . You Can Build (Wade), 375
Threshold: and door-sweep combination, 95; interior, 132, 298
Tie beams. *See* Collar beams
Ties, for form walls, 184
Tie strap, 89
Timber Framing Book, The (Elliot and Wallas), 105, 377
Timbers, 79 (*see also* Lumber; Trees); flatting, 15, 17
Tin snips, 36
Toenail (toenailing), 85, 304, 360; defined, 360
Toggle bolts, 88–89
Toilets, 146–47; biological, 147; books on, 379; chemical, 147; composting, 146–47, 379; dry, 146–47; flush, 146–47, 379; graywater disposal and recycling, 147–49; low-water-use, 147, 379; reducing water waste and, 146–47
Tongue-and-groove wood, 206, 225, 226, 248, 251, 297, 298, 306, 358
Toolboxes, wooden, handmade, 46
Tools, 24–66, 159–60, 221, 222–23, 243 (*see also* specific kinds, materials, structures, uses); basic, 159–60, 222; books on and sources for, 378–80; borrowing versus buying, 222; buying, 24; care of, 24; power versus hand, 222; sharpening guide to, 378
Torpedo level, 45
Transom, 151, 360; light efficiency and, 151
Tread, stair, 358, 360; nosing of, 358
Trees (*see also* Timbers): felling, 243–44; hexagon house and, 300, 301; and poles, 243–44, 245, 300, 301; posts, raising, 18–19
Trenches, foundation work and, 181
Trimmer, defined, 360
Trim or casing, window, 136, 153; boards, 153; definition, 355; head casing, 136; head trim, 136
Trims and moldings, 83; drip cap, 83, 136, 356; kinds, shapes, and types, 83; molding, defined, 358
Trowels, 188; and troweling concrete, 178, 188
Trusses, 231, 236, 237
Try square, 26, 359
T square, four-foot, 27–28; choosing and care of, 27–28
"Tub and tile" caulk, 91
Two-foot level, 45
Two-headed digger, 46
Type X wallboard, 291

Underpinnings, defined, 360
Uniform Building Code, 160, 172, 376
Union work and wages, women and, 320–29
Urea formaldehyde insulation, 94
USDA (United States Department of Agriculture), 380; and round house building plans, 77; and wood floors, 378
U. S. Gypsum Co., 291; address, 291
Utility knife, 34

Valley, roof, defined, 360
Veneer, 360; plywood, 296–97
Vents: attic spaces, 235; form walls, 185
Vermiculite insulation, 94, 95
Vigas, 360; adobe house, 137, 139, 141–42; defined, 360
Vinyl products (plastics), 286, 287 (*see also* specific kinds, materials, products, uses); safety warnings, 286; wallboards, 286, 287
V-rustic siding, 251, 254

Wainscoting, 295–96, 360
Walers, form walls and, 184, 360
Wallboard, insulating, 94
Wallboard hammer, 38

Wall Book, The (Schuler), 288, 378
Walls, 107–9, 208–16 (*see also* Partition walls); adding partition to, 108–9; bearing, determining, and removing, 107–8 (*see also* Bearing walls); bracing, 208–9, 211–12, 213; form, 182–86; foundation, 180–86, 203–4, 208–16; framing, continuous foundation, 203–4, 208–16; gable ends, framing, 216; hexagon house, 318; insulating, 93–95, 145, 318 (*see also* Insulation); joining, 211, 215, 216, 226; locating studs behind, 107; log cabin, 99, 100; nonbearing, 358 (*see also* Partition walls); plywood, 254–55; raising and moving, 211, 216, 226, 244, 246; rectangular house, 21ff.; rough-framing, 208–16; sheathing and siding, 248–58, 286–99 (*see also* Sheathing; Siding); squaring and securing, 216; star cabin framing, 191; stud wall framing, 208–16; windows added to, 133–36; without windows, framing, 214–15; with windows, framing, 215–16; wood shingles, 255–58; yurt building, 280–81, 282–83, 284
Wane (lumber defect), 82
Warping (lumber defect), 81
Waste disposal: flush toilets and alternatives, 146–47 (*see also* Toilets); graywater recycling and, 147–49
Water, energy-efficient houses and, 144, 146–47; flush toilets and alternatives, 146–47 (*see also* Toilets); graywater disposal and recycling, 147–49; solar systems and, 144, 146
Waterproofing, 268–77 (*see also* Weatherproofing; Weather stripping); caulking and flashing, 90–92, 249–50, 269–71, 273
Water table, defined, 360. *See also under* Graywater disposal and recycling
Weatherproofing, 93–95, 268–77 (*see also* Insulation; Waterproofing; Weather stripping); asphalt shingles and, 268–71; caulking and flashing, 90–92; fiberglass shingles, 268; roll roofing, 273–77; windows, 136, 152; wood shingles and shakes, 271–73
Weather stripping, 95–96, 360; aluminum "retainer," 95–96; door bottom sweep, 95; matted felt strips, 96; spring metal, 95, 96; threshold and door-sweep combination, 95
Weed, Susun, 173
Wells, CeCe, 343, 345
Wet Patch, use of, 91
Wheelbarrows, concrete work and use of, 178
Windows, 110–11, 133–36, 149–56, 228; A-frame building, 265, 266; anatomy of, 151–52; caulking and flashing, 90–92, 249–50; clerestory, 152, 294; cutting glass and, 149–50, 152–53, 154–56; dadoes, 135–36; double-hung, replacing sash weights and cords, 110–11; drip cap, 152; drip groove, 151; dutch door, 118, 119–20; frame, 151; framing, 22, 23, 135, 210, 214–15; installing, 135–36; insulation (weatherproofing); 93–96, 136, 145, 152; jambs, 135–36, 151–52; light and energy efficiency and, 149–56, 294; muntins and lights, 152; preframed, 163; rot, checking, 155; sashes (*see* Sashes); sills, 135–36, 151–52; sizes, shapes, and types, 150–51; sticky, 111; top (head), 151–52; trim or casing, 136, 153; used, restoring, 155; yurt building and, 283
Wing divider, 29
Wire gate, building, 114–18
Wire mesh, for rock walls, 293–94
Women in Apprenticeshop Program, Inc., 323
Women United for Apprenticeship, 322
Wonder bar, 43
Woodburning stoves and fireplaces, 229, 266, 314
Wood chisels, 33; care and use of, 33; sizes, 33
Woodcraft Supply Corp., 379
Wooden mallet, 38
Wood Floors for Dwellings, 378
Wood lathe. *See* Lathe, wood
Wood paneling, 294–97; diagonal, 295; doors, 131; horizontal, 295; kinds, choosing, 295; lath strips, 296; nails, 295, 296; plywood veneer, 296–97; putting on, 295; vertical, 295; wainscoting, 295–96
Wood shingles, 237, 248, 249, 255–58, 271–73, 297; edge-grain (lumber), 81; exposure and, 272; grades, 272; laying, 272–73; and shakes, weatherproofing and, 271–73; tools for applying, 257, 272
Wood siding, 248–58, 268; back-priming, 252; bevel, 248; board-and-batten, 253–54; board on board, 253–54; clapboard, 248, 251–53; drop (rustic), 249, 251; horizontal, 248, 249, 251–53; kinds of wood, 251–52; nailing, 252, 253, 254, 255; plywood, 248, 249, 250, 254–55; shingles (*see* Wood shingles); tongue-and-groove, 248, 251; vertical, 248, 249, 253–54
Wood Structural Design Data, Vol. I, 375
Woodworking machinery, 57–66. *See also* specific kinds, uses
Work boots, 48
Working (employment), women in carpentry and woodworking. *See* Employment
Work Is Dangerous to Your Health (Stellman and Daum), 379
Wrecking tools, 43–44. *See also* specific kinds, uses
Wrenches, 41–42; adjustable-end, 41; Allen, 55; locking- plier-, 41; vise-grip, 41
W/R Sheetrock, 291

Yard lumber, defined, 360
Yurt building, 278–85; books on and resources, 379–80; cable for, 281; compression ring, 278, 279; double-concentric, 285; foundation system, 279–80; tension band (steel cable), 278, 279, 280–81; twelve-sided version, 278–85; walls, 280–81, 282–83
Yurt Foundation, The, 379–80